ADVANCES IN DIGITAL PATHOLOGY AND ARTIFICIAL INTELLIGENCE IN PATHOLOGY

ADVANCES IN DIGITAL PATHOLOGY AND ARTIFICIAL INTELLIGENCE IN PATHOLOGY

Edited by

ZAIBO LI
The Ohio State University, Columbus, OH, United States

ANIL V. PARWANI
The Ohio State University, Columbus, OH, United States

ELSEVIER

Elsevier
Radarweg 29, PO Box 211, 1000 AE Amsterdam, Netherlands
125 London Wall, London EC2Y 5AS, United Kingdom
50 Hampshire Street, 5th Floor, Cambridge, MA 02139, United States

ISBN: 978-0-443-40383-5

For information on all Elsevier publications visit our website at
https://www.elsevier.com/books-and-journals

Publisher: Megan Ball
Acquisitions Editor: Patricia Osborn
Editorial Project Manager: Priyamvada Jha
Production Project Manager: Vijayaraj Purushothaman
Cover Designer: Mark Rogers

Typeset by Integra Software Services

Dedication

To my wife Dakun Wang, for her unwavering support and constant encouragement; to my children Aidan and Brenna for their love and inspiration; to my parents for their lifetime guidance and support; and to my mentors, for their enduring guidance and collaboration.

Zaibo Li

I wish to acknowledge my wife, Namrata, for her steadfast support and encouragement; my children, Simran, Varun, and Sanam, for their inspiration and affection; my parents, who continue to guide me in life and have been pillars of strength; and my many mentors and colleagues, whose guidance and collaboration over the years have been invaluable.

Anil V. Parwani

Contents

6. Digital cytology

Swati Satturwar, Liron Pantanowitz, and Zaibo Li

7. Telepathology

Mauro Saieg and Rafael Parra-Medina

8. Artificial intelligence (imaging analysis) fundamentals

Abdul Rehman Akbar and Muhammad Khalid Khan Niazi

9. Applications of artificial intelligence in surgical pathology

Zaibo Li, Yueping Liu, William MacDonald, Shaoli Sun, Wei Chen, Ankush U. Patel, Anil V. Parwani, and Swati Satturwar

10. Artificial intelligence in cytology

David Kim, Swati Satturwar, Liron Pantanowitz, and Zaibo Li

11. Emerging advanced imaging technologies in digital pathology

Jeffrey L. Fine

12. Generative artificial intelligence technologies in pathology

Ehsan Ullah

13. Pixelomics: The pixelome as a foundation of precision pathology

Ehsan Ullah, Jingjing Hu, Hooman Rashidi, Liron Pantanowitz, Zaibo Li, and Anil V. Parwani

14. Large language models utilizing digital pathology workflows: Impact on pathology practice

Zhi Huang, Jialu Yao, and Tiancheng Bao

15. Usability of artificial intelligence tools in pathology practice

Andrey Bychkov and Yosep Chong

16. Regulations and ethical considerations in digital pathology and artificial intelligence

Rama R. Gullapalli and Anil V. Parwani

17. Conclusions

Zaibo Li and Anil V. Parwani

Contributors

Abdul Rehman Akbar Department of Pathology, School of Medicine, The Ohio State University Wexner Medical Center, Columbus, OH, United States

Tiancheng Bao Department of Pathology and Laboratory Medicine, Perelman School of Medicine at the University of Pennsylvania; Department of Bioengineering, University of Pennsylvania, Philadelphia, PA, United States

Marilyn Bui Department of Pathology, Moffitt Cancer Center, Tampa, FL, United States

Andrey Bychkov Department of Pathology, Kameda Medical Center, Kamogawa, Chiba, Japan

Wei Chen Department of Pathology, The Ohio State University, Columbus, OH, United States

Yosep Chong Department of Hospital Pathology, The Catholic University of Korea College of Medicine, Seoul, South Korea

Jeffrey L. Fine Pathology, University of Pittsburgh, Pittsburgh, PA, United States

Eric Glassy Department of Pathology, Affiliated Pathologists Medical Group, Rancho Dominguez, CA, United States

John Groth Endeavor Health Northshore, Chicago, IL, United States

Rama R. Gullapalli Department of Pathology, University of New Mexico Health Sciences, Albuquerque, NM, United States

Jingjing Hu University of California, San Diego, San Diego, CA, United States

Zhi Huang Department of Pathology and Laboratory Medicine, Perelman School of Medicine at the University of Pennsylvania, Philadelphia, PA, United States

David Kim Department of Pathology and Laboratory Medicine, Memorial Sloan Kettering Cancer Center, New York, NY

Zaibo Li Department of Pathology, The Ohio State University, Columbus, OH, United States

Yueping Liu Department of Pathology, Hebei Medical University Fourth Hospital, Shijiazhuang, China

Giovanni Mario Lujan The Ohio State University, Columbus, OH, United States

William MacDonald Department of Pathology, The Ohio State University, Columbus, OH, United States

Muhammad Khalid Khan Niazi Department of Pathology, The Ohio State University, Columbus, OH, United States

Liron Pantanowitz Department of Pathology, University of Pittsburgh Medical Center, Pittsburgh, PA, United States

Rafael Parra-Medina Instituto de Investigación, Fundación Universitaria de Ciencias de la Salud (FUCS), Bogotá, Colombia

Anil V. Parwani Department of Pathology, The Ohio State University, Columbus, OH, United States

Ankush U. Patel Department of Pathology, The Ohio State University, Columbus, OH, United States

Hooman Rashidi University of Pittsburgh, Pittsburgh, PA, United States

Ghulam Rasool Department of Machine Learning, Moffitt Cancer Center and Research Institute, Tampa, FL, United States

Mauro Saieg McGill University, Montreal, QC, Canada

Swati Satturwar Department of Pathology, The Ohio State University, Columbus, OH, United States

Rajendra Singh Department of Pathology and Laboratory Medicine, Perelman School of Medicine, University of Pennsylvania, Philadelphia, PA, United States

Shaoli Sun Department of Pathology, The Ohio State University, Columbus, OH, United States

Ehsan Ullah Health New Zealand -Counties District, Auckland, New Zealand; Department of Pathology, The Ohio State University, Columbus, OH, United States

Jialu Yao Department of Pathology and Laboratory Medicine, Perelman School of Medicine at the University of Pennsylvania, Philadelphia, PA, United States

Preface

Over the past decade, pathology has undergone a profound digital transformation. What began as the gradual adoption of whole slide imaging has rapidly evolved into a dynamic ecosystem that integrates high-resolution digital imaging, advanced computational analytics, artificial intelligence (AI), and increasingly, generative and multimodal models. These developments are not merely technological upgrades; they represent a paradigm shift in how pathologists visualize disease, extract diagnostic and prognostic information, collaborate across distances, and contribute to precision medicine.

Advances in Digital Pathology and Artificial Intelligence in Pathology was conceived to provide a comprehensive, practical, and forward-looking overview of this transformation. Our goal is to bridge foundational concepts with real-world implementation and emerging innovations, offering a resource that serves practicing pathologists, trainees, laboratory leaders, informaticians, data scientists, computational engineers, and industry partners alike.

This volume begins with foundational topics, including digital imaging technologies and the practical implementation of digital pathology workflows. It then explores clinical and nonclinical applications, digital cytology, and telepathology, highlighting both operational and global perspectives.

A major focus of the book is AI in pathology. Readers are introduced to the fundamentals of imaging analysis, followed by practical applications in surgical pathology and cytology. Emerging innovations such as advanced imaging technologies, generative AI, pixelomics, and large language and multimodal models are examined alongside critical discussions of usability, regulation, and ethical considerations.

The contributing authors are recognized leaders who bring diverse expertise spanning clinical practice, informatics, computational engineering, and regulatory science. Their collective insights reflect the multidisciplinary nature of digital pathology and AI.

Digital pathology and AI are no longer future concepts; they are reshaping everyday practice. We hope this book serves as a practical guide and an inspiration for continued innovation, rigorous validation, and responsible implementation as pathology enters a new, data-driven era.

Zaibo Li
Anil V. Parwani

Acknowledgments

This textbook has greatly benefited from the expertise and dedication of our contributors from across the United States, Canada, Asia, New Zealand, and other regions worldwide. We extend our sincere gratitude to each of them for generously their confidence in this project and for sharing their knowledge and experience.

The editors are also deeply appreciative of our publisher, Elsevier, for its continued support and encouragement throughout the development of this book. We offer special thanks to Priyamvada Jha, Content Manager at Elsevier, and Subash Balakrishnan, Copyright Coordinator at Elsevier, for their steadfast guidance and for skillfully and diligently overseeing the many complex details involved in the final stages of publication.

Zaibo Li
Anil V. Parwani

CHAPTER

1

Introduction

Zaibo Li and Anil V. Parwani

Department of Pathology, The Ohio State University, Columbus, OH, United States

Pathology sits at the center of modern medicine, providing definitive diagnoses that guide patient management, prognostication, and therapeutic decision-making. For more than a century, this discipline has relied on light microscopy and glass slides as its primary tools. While this paradigm has proven remarkably robust, it is increasingly challenged by rising case volumes, growing diagnostic complexity, subspecialization, workforce shortages, and escalating demands for accuracy, efficiency, and standardization. Against this backdrop, digital pathology (DP) and artificial intelligence (AI) have emerged as transformative technologies with the potential to fundamentally reshape pathology practice and its integration into clinical workflows.

DP refers to the acquisition, management, sharing, and interpretation of pathology information, most notably histologic and cytologic slides, in a digital environment. The digitization of glass slides into high-resolution whole-slide images (WSIs) has enabled pathologists to view, analyze, and annotate cases on computer workstations, untethered from the physical constraints of microscopes and slide trays. This transition mirrors earlier digital revolutions in radiology but comes with unique technical, regulatory, and cultural challenges stemming from the extreme image size, complexity, and diagnostic nuance of pathology specimens.

Over the past decade, advances in scanning hardware, image quality, data storage, network infrastructure, and viewing software have accelerated the adoption of DP worldwide. What initially began as a tool for education, research, and secondary consultation has progressively moved into routine clinical use, including primary diagnosis in pathology. Large academic medical centers, reference laboratories, and increasingly, community practices are implementing partial or fully digital workflows, driven by needs for remote sign-out, subspecialty expertise, quality assurance, and operational resilience—needs that were particularly highlighted during global disruptions such as the COVID-19 pandemic.

Concurrently, AI has rapidly evolved from experimental image analysis algorithms into clinically relevant tools capable of assisting pathologists with detection, classification, quantification, and prognostic assessment. Deep learning models trained on large collections of WSIs can now identify mitotic figures, grade tumors, quantify biomarkers, detect metastases, and uncover morphologic patterns that correlate with molecular alterations and clinical

Advances in Digital Pathology and Artificial Intelligence in Pathology
https://doi.org/10.1016/B978-0-443-40383-5.00002-9

outcomes. These capabilities position AI not as a replacement for the pathologist but as an augmentative technology that can enhance diagnostic consistency, reduce cognitive and workload burdens, and unlock new layers of information embedded within pathology images.

The convergence of DP and AI represents more than a technological upgrade; it signals a paradigm shift in how pathology is practiced, taught, and integrated into precision medicine. Digital workflows enable seamless image sharing across institutions and disciplines, facilitating multidisciplinary team discussions, telepathology, and global consultations. AI-driven image analysis extends the diagnostic reach of pathology into quantitative and predictive domains, supporting personalized treatment strategies and advancing the emerging field of pixelomics, in which imaging data are linked to genomic, transcriptomic, and clinical profiles.

Despite this promise, the path to widespread adoption is complex. Implementing digital pathology requires substantial investment, careful workflow redesign, robust validation, staff training, and adherence to regulatory and accreditation standards. Similarly, AI tools must be rigorously validated, transparently evaluated, and thoughtfully integrated into daily practice to ensure safety, usability, and clinical value. Ethical considerations, including data privacy, algorithmic bias, accountability, and the evolving role of the pathologist, must be addressed alongside technical innovation.

This book aims to provide a comprehensive, practical, and forward-looking overview of advances in DP and AI as they relate to clinical workflow. Beginning with the foundational technologies of digital imaging, it explores the implementation of digital pathology systems, their clinical and non-clinical applications, and the unique challenges of digital cytology and telepathology. Subsequent chapters delve into the fundamentals of AI, its applications across surgical pathology and cytology subspecialties, and emerging developments such as generative AI (GAI), large language models (LLMs), and pixelomics. The book also examines usability, regulatory frameworks, and ethical considerations, culminating in a synthesis of current knowledge and future directions.

By integrating technical, clinical, and operational perspectives, this book is intended for practicing pathologists, trainees, laboratory leaders, informaticians, and industry partners who seek to understand not only what DP and AI can do but also how they can be responsibly and effectively deployed to improve patient care. As pathology enters an increasingly digital and data-driven era, mastery of these tools will be essential to sustaining the discipline's central role in modern medicine.

AI disclosure

During the preparation of this work, the author(s) used ChatGPT to polish the writing. After using this tool/service, the author(s) reviewed and edited the content as needed and take(s) full responsibility for the content of the publication.

CHAPTER

2

Digital imaging technologies

Eric Glassy[a] *and John Groth*[b]

[a]Department of Pathology, Affiliated Pathologists Medical Group, Rancho Dominguez, CA, United States

[b]Endeavor Health Northshore, Chicago, IL, United States

Introduction: Transforming pathology for the digital age

For over 150 years, anatomic pathology has relied on the same fundamental tools: the light microscope and glass slides. While this analog foundation has served medicine well, it now limits pathology's potential in our interconnected, data-driven healthcare environment. We are experiencing a digital and artificial intelligence transformation that will revolutionize (Salto-Tellez et al., 2019) how we view tissue samples, collaborate, teach, and integrate computational tools into diagnostic practice.

Digital pathology encompasses the acquisition, management, sharing, and interpretation of pathology information in digital format. It represents far more than simple digitization (Abels et al., 2019; Evans et al., 2021b). Digital pathology offers unprecedented opportunities for remote consultation, quantitative analysis, artificial intelligence integration, and seamless collaboration across geographic boundaries. The COVID-19 pandemic accelerated digital workflow adoption as laboratories worldwide discovered they could maintain diagnostic excellence while enabling flexible, distributed practice models (Hanna et al., 2020).

This chapter examines the technologies driving this transformation and offers practical guidance for successful enterprise-level implementation. We explore the engineering principles behind whole slide imaging, analyze trade-offs between different technological approaches, and provide evidence-based recommendations for navigating the complex landscape of vendors, formats, and deployment strategies. Our goal is to equip pathology professionals with the knowledge needed to make informed decisions about digital transformation while avoiding common implementation pitfalls.

Advances in Digital Pathology and Artificial Intelligence in Pathology
https://doi.org/10.1016/B978-0-443-40383-5.00003-0

The strategic imperative for digital pathology

Addressing critical healthcare challenges

The case for digital pathology extends beyond technological capability to address fundamental challenges facing modern pathology practice. The global shortage of pathologists will worsen significantly over the next decade, creating an urgent need for tools that enhance productivity and enable resource sharing. Digital pathology directly addresses this challenge by allowing remote consultation, sharing subspecialty expertise, and enabling more efficient workflow management.

Quantifiable benefits

Digital pathology benefits include several key areas of improvement. Studies show significant efficiency gains in case turnaround times once digital workflows mature (Hanna et al., 2020). Digital systems provide significant error reduction by enabling standardized quality checks and automated artifact detection (Pantanowitz et al., 2016). These technologies also enhance collaboration by making real-time multiparty consultations and teaching sessions routine rather than exceptional. In addition, digital pathology provides future-proofing capabilities; digital formats enable artificial intelligence integration and advanced analytics that are impossible with traditional methods.

Market drivers and adoption landscape

Several converging factors accelerate digital pathology adoption. Healthcare consolidation creates multisite organizations requiring standardized workflows and expert resource sharing. Regulatory approval of digital systems for primary diagnosis began with the FDA's 2017 clearance of the Philips IntelliSite system, which eliminated the primary barrier to clinical adoption. Subsequent approvals of systems from Borowsky (2020) Bauer (2024), Iwuajoku (2025), and Wharton (2025), among others, have validated the technology's clinical readiness.

Cost trajectories favor adoption as scanner prices decrease while capabilities expand. Storage costs have decreased, making the massive file sizes associated with whole slide imaging more economically manageable. Cloud computing platforms provide scalable solutions that eliminate the need for significant upfront infrastructure investments, making advanced digital pathology accessible to smaller institutions.

Understanding whole slide imaging technology

The evolution from concept to clinical reality

The journey toward whole slide imaging digital pathology represents a gradual evolution punctuated by key technological milestones (Table 2.1). The first milestone can be traced back to the 1840s, when Alfred François Donné first used daguerreotype photography to capture microscopic images for pathology (Pantanowitz et al., 2016; Donné, 1844). This innovation laid the foundation for the principle that drives digital pathology today: the use of technology to reproducibly record, analyze, and share histologic images.

TABLE 2.1 Timeline to WSI.

Technological milestone	Year(s)
Daguerreotype photography	1840s
Television microscopy	1950–60s
Dynamic robotic microscopy	1980s
Whole-slide imaging	1990s—today

The 1950s and 1960s expanded this through the concept of telediagnosis—diagnosis at a distance—using television microscopy to review peripheral blood smears and urine sediments (Murphy and Bird, 1974; Weinstein et al., 2019). Further progress occurred with dynamic-robotic technology, when the term telepathology was coined in 1986 to define the practice of pathology by viewing an indirect image on a screen rather than examining a specimen directly through a microscope (Weinstein, 1986; Farahani and Pantanowitz, 2015; Weinstein et al., 1987). Although limited, these early systems demonstrated the potential to decouple diagnosis from physical location. They achieved early practical applications in remote consultations and educational settings, offering a glimpse of pathology beyond the physical slide.

The 1990s and 2000s marked a breakthrough with the development of virtual microscopy, which used software employing a client/server architecture to realistically emulate a high-power light microscope (Ferreira et al., 1997), and eventually whole slide imaging (WSI), involving scanning slides to produce digital images. The first patents were granted to Bacus Research Laboratory in 1997/98 (Pantanowitz et al., 2018). These systems could digitize entire glass slides at diagnostic resolution, creating navigable digital files that maintained the full spatial context of tissue sections. Although initially slow and costly, these early systems laid the technical groundwork for today's robust platforms.

The 2010s witnessed dramatic improvements in scanning speed, image quality, and software sophistication. By 2017, when the FDA approved the first WSI system for primary diagnosis, the technology had matured into a practical clinical tool with demonstrable benefits across multiple applications (Evans et al., 2018). In this process, a WSI system was defined as consisting of (1) an image acquisition subsystem that converts the content of a glass slide into a digital image file, and (2) a workstation environment for viewing the digital images (Food and Drug Administration Staff, 2016) (Fig. 2.1). Ongoing discussions between the Digital Pathology Association and the FDA are addressing the need to update this guidance, given the evolving technological landscape.

Core imaging components

A functional WSI implementation comprises several integrated systems working in concert (Table 2.2).

At its core lies sophisticated scanner hardware that functions as a robotic microscope, capturing the large amounts of image data necessary to digitally represent entire histologic glass slides. Table 2.3 lists performance benchmarks of a WSI system.

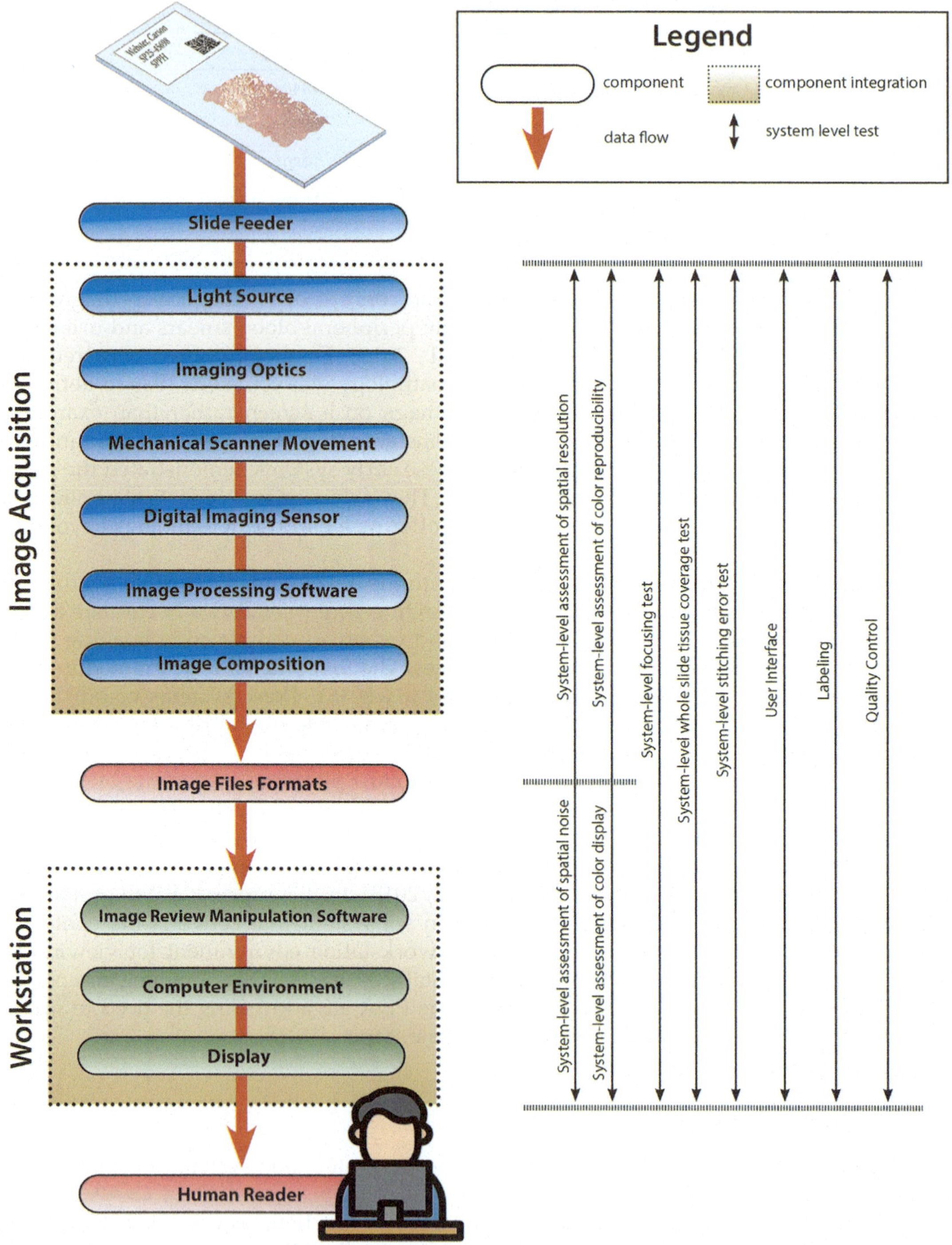

FIG. 2.1 Whole slide imaging system. *Modified from FDA; used with permission.*

TABLE 2.2 Whole slide imaging system components and performance specifications.

System category	Component	Function/Features	Typical specifications
Scanner hardware	Microscope optics	High-quality optical components with multiple magnifications	20×–40× objectives, NA 0.75–0.95, apochromatic correction
	Light source	Stable, uniform illumination system	LED arrays, 5000–6500 K color temperature, >50,000 h lifespan
	Digital Camera/ Sensor	High-resolution image capture	5–20 megapixel sensors, 12–16-bit depth, >60 fps
	Slide handling	Automated loading and positioning	1–1000 slide capacity, barcode reading, RFID tracking
	Stage control	Precision movement and positioning	Sub-micron accuracy, encoded feedback, vibration isolation
Software	Focusing system	Automated focus optimization	Dynamic or pre-scan mapping, real-time adjustment
	Image acquisition software	Scan orchestration and quality control	Multiresolution pyramids, real-time QC, metadata management
	Image viewer	Diagnostic interface for pathologists	Pan/zoom, annotations, multislide comparison, measurements
	Image management system (IMS)	Workflow and data lifecycle control	DICOM/HL7 integration, user management, audit trails
Display systems	Primary Monitor	High-resolution diagnostic display	4K–8K resolution, medical-grade calibration, >350 cd/m^2
	Second Monitor (optional)	Administrative and reference display	Standard resolution, color accuracy
Computing Infrastructure	Processing Server	Image handling and AI inference	Multicore CPU, 32–128 GB RAM, GPU acceleration
	Storage system	High-capacity, redundant storage	100TB–10PB capacity, RAID protection, tiered archiving
	Network infrastructure	High-bandwidth connectivity	10Gb Ethernet, managed switches, QoS*capabilities

** QoS: Quality of Service mechanisms or technologies control network traffic and ensure the performance of critical applications when network capacity is limited. It enables organizations to adjust their overall network traffic by prioritizing specific high-performance applications. See section on Quality of Service and Traffic Prioritization, page 37.*

Scanner hardware architecture

Modern digital pathology scanners integrate precision mechanics, advanced optics, and specialized electronics to automate the imaging process. The mechanical subsystem features slide-handling mechanisms that load specimens from racks or cassettes. This avoides damage to the slides while maintaining efficient throughput. The scanning stage provides precise X–Y–Z positioning, essential for proper alignment and focus throughout the scanning process. Sub-micron accuracy is often required.

TABLE 2.3 Performance benchmarks.

Parameter	Benchmark
Scanning speed	30–120 s per slide (20 × equivalent)
Image quality	<5% diagnostic discordance compared to glass slides.
System uptime	>95% availability with proper maintenance.
Storage efficiency	50–200 MB per slide (compressed)

Optical design considerations

The optical system is the scanner's most critical component; it directly determines image quality and diagnostic utility. High-quality microscope objectives, with 20× or 40× magnification, are selected based on their numerical aperture, which determines their resolving power. This creates a fundamental trade-off: higher numerical aperture objectives (0.75–0.95) provide superior resolution but create focusing challenges due to their shallow depth of field.

Illumination systems have undergone significant evolution, with LED sources replacing halogen lamps. LEDs offer superior longevity, thermal stability, and consistent spectral output, which are crucial for maintaining color fidelity across various tissue types and staining protocols. Modern systems often incorporate multiple LED wavelengths to optimize illumination for specific applications.

Digital capture systems

At the optical path's terminus sits a high-resolution digital camera, either a Charge-Coupled Device (CCD) or Complementary Metal-Oxide-Semiconductor (CMOS) sensor, which converts the analog light signal into digital data. Sensor characteristics, particularly pixel size and sensitivity, determine the final image quality and digital resolution. Area-scan sensors capture wide fields of view, while line-scan sensors enable faster, continuous scanning with reduced geometric distortion, as discussed below.

Scanning methodologies

Two primary strategies exist for acquiring image data across slides: area scanning (also known as tiling) and line scanning. Each represents distinct engineering approaches with specific advantages (Fig. 2.2).

Area scanning (tiling)

This method captures a series of individual, overlapping rectangular fields of view across the slide, much like overlapping photographs. The scanner moves the slide or the objective in a step-wise fashion, stopping to focus and capture each field of view. These tiles are later computationally stitched together to form the final whole slide image. Area scanning leverages familiar imaging techniques and handles tissue thickness variations within each tile. However, the stop-start motion can be slower and may introduce subtle artifacts at tile boundaries (Pantanowitz et al., 2015).

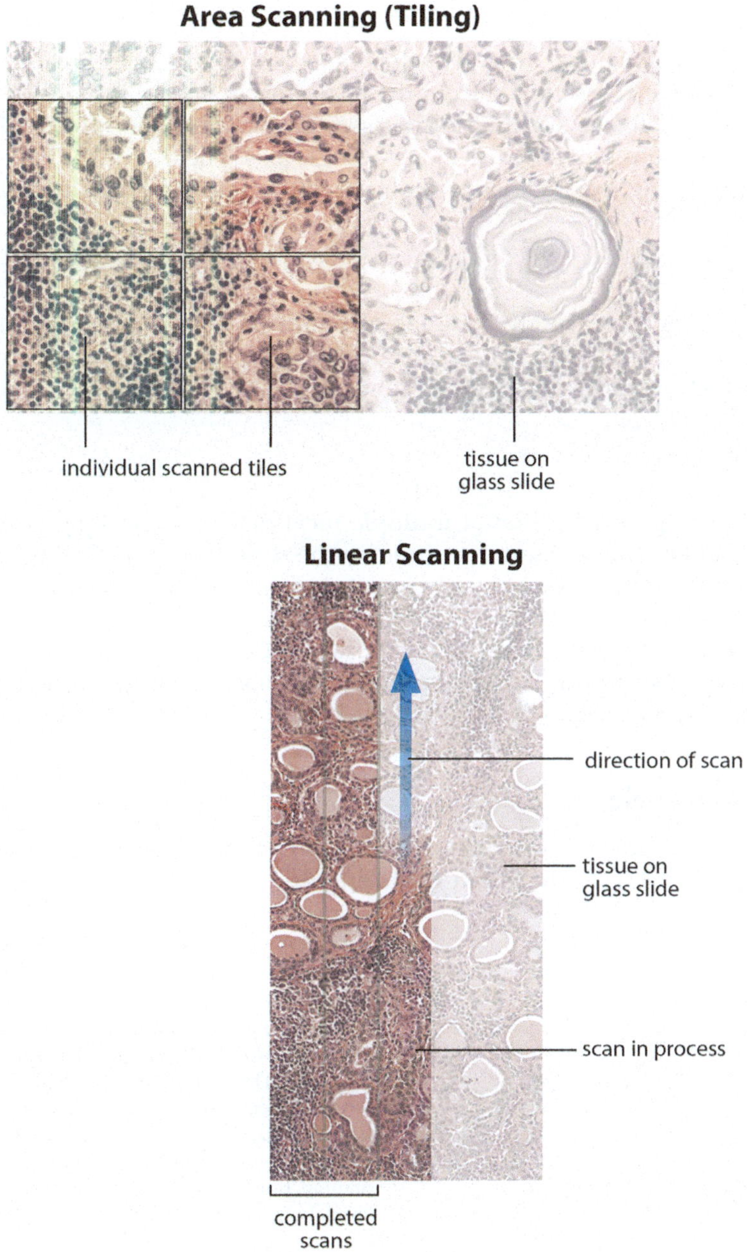

FIG. 2.2 Scanning techniques in whole slide imaging. Whole slide scanners use two main strategies to digitize tissue sections. In area scanning (tiling), the scanner acquires many small, high-resolution image tiles that are stitched together to form the complete virtual slide. This approach allows very precise focus at each tile but can introduce stitching artifacts and may be slower for large specimens. In linear scanning, the scanner captures the image in a continuous sweep across the slide, much like a flatbed scanner. Each method has trade-offs: tiling offers robust focus control but may produce visible seams, while linear scanning provides smoother navigation, is generally faster, and avoids stitching errors, but it is more sensitive to vibrations and can be less flexible in adjusting focus over uneven tissue.

Line scanning

Line scanning employs linear array detectors that continuously capture narrow strips of the image. The digital slide is assembled from these successive linear scans. This approach offers faster acquisition and fewer stitching artifacts due to continuous motion. The mechanical design is simpler because movement is along a single axis. The tradeoff is that line scanning is less effective at handling focus variations compared to some tiling implementations.

Hybrid approaches

Advanced systems increasingly combine elements of both methods, optimizing performance for specific applications. The choice often requires balancing throughput, image quality, and system flexibility.

Z-stacking

For slides requiring examination at multiple focal planes, Z-stacking refers to capturing images at different levels at the same position on the slide. These stacked images can then be navigated during viewing, allowing pathologists to adjust focus virtually, similar to turning the fine focus knob on a microscope. While Z-stacking offers significant advantages for applications like cytology specimens, which often contain three-dimensional cell clusters, each additional Z-level multiplies the scanning time and storage requirements (Fig. 2.3) (Hanna and Pantanowitz, 2022; Donnelly et al., 2021).

Extended depth of field

Extended Depth of Field offers an alternative approach to addressing multiple focal planes (Papazoglou et al., 2023). Rather than capturing and storing multiple Z-levels, Extended

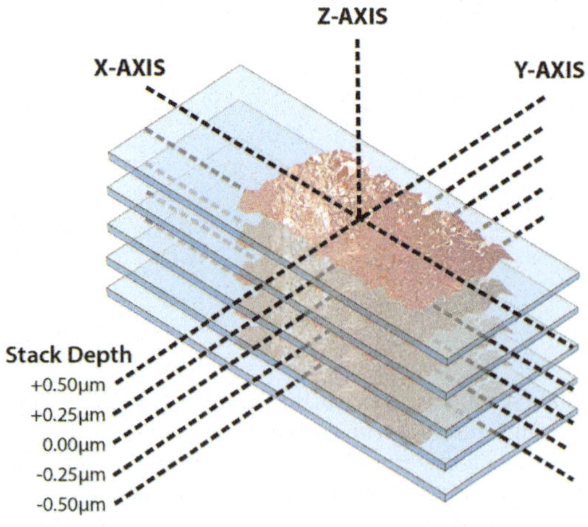

FIG. 2.3 Z-Stacking for Three-Dimensional Imaging. Z-stacking captures a series of images at incremental focal depths, creating a stack that spans above and below the tissue plane, typically in 0.25 μm increments, ranging from −0.50 to +0.50 μm relative to the primary focal plane (0.00 μm). This technique is particularly valuable for cytology and hematopathology, where diagnostically important structures may reside at different depths. Z-stacks allow users to scroll through the focal planes, simulating manual fine-focusing on a microscope. While powerful, Z-stacking increases scan time and storage requirements, and its utility must be weighed against workflow efficiency.

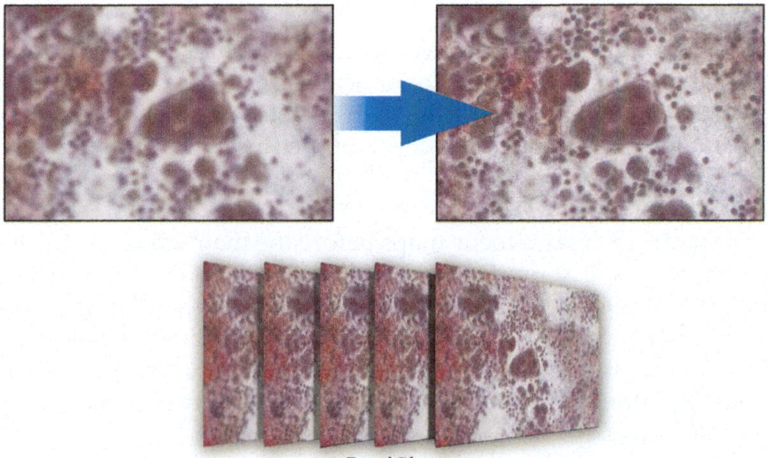

FIG. 2.4 Extended depth of field imaging. Extended depth of field (EDOF) imaging combines multiple focal planes into a single sharp composite, producing a uniformly in-focus image across the slide. Unlike Z-stacking, which preserves individual focal planes, EDOF is designed for efficient review of thick or uneven specimens without the need to scroll through layers. This approach is particularly helpful in cytology or microbiology, where three-dimensional cell clusters or organisms often exceed the depth of a single focal plane. The trade-off is the loss of depth information—making EDOF best suited for routine review rather than detailed 3D exploration.

Depth of Field algorithms analyze a Z-stack during scanning and computationally combine the in-focus portions from each level into a single composite image that appears entirely in focus (Fig. 2.4). This provides the advantage of enhanced focus across varying tissue depths without the storage and bandwidth penalties of full Z-stacking. The downside is that additional time is required to scan multiple planes. Cytology smears and bone marrow samples, which may have cells piled on top of each other, benefit from this technique. Detection of microorganisms is also improved. For example, this extended depth of field stacking technique has proven superior in identifying acid-fast organisms (Nurzynska et al., 2023).

The focusing challenge

Strategies to ensure diagnostic quality

Maintaining precise focus across the entire slide represents one of the most technically demanding aspects of whole slide imaging. Tissue sections are rarely perfectly flat, and variations in tissue thickness or mounting media create significant topographical challenges. Focus failures result in blurry images. WSI scanners employ sophisticated mechanisms to address this challenge through two fundamental strategies: dynamic focusing and pre-focus mapping (Montalto et al., 2021).

Dynamic focusing

Real-time focusing continuously assesses and adjusts focus as the scanner moves across the glass slide. Sensors measure focus quality at or near the capture point, adjusting the objective lens position accordingly. While offering the highest potential accuracy, this approach significantly impacts scanning speed due to the constant measurement and adjustment cycles.

Prefocus mapping

This alternative strategy creates focus maps before the main image acquisition begins. The scanner focuses on the tissue at predefined grid points across the slide surface, determining optimal Z-axis positions. During high-speed scanning, the system interpolates focus settings between the previously mapped points rather than continuously adjusting the focus. This dramatically increases scanning speed but depends on the accuracy of the initial focus map and interpolation algorithms. Unexpected tissue variations between map points may result in areas that are out of focus.

Hybrid solutions

Modern scanners increasingly employ hybrid approaches, creating initial focus maps while incorporating periodic dynamic checks. This balances speed with accuracy without significantly compromising either parameter. Regardless of methodology, efficient focusing relies heavily on accurate tissue detection algorithms that analyze low-resolution preview scans to identify tissue-containing regions, preventing wasted time on empty glass areas. After identifying tissue regions, the scanner selects a grid of points and performs rapid Z-stack acquisition at each point, capturing images at multiple focal planes. Image analysis algorithms then evaluate these stacks to determine the Z-position that yields the sharpest focus. These optimal Z-positions are then compiled into a three-dimensional map representing the slide's surface topography. This guides the objective lens during the subsequent high-resolution scan.

Critical performance parameters

Resolution: The foundation of diagnostic image quality

Resolution stands as the most fundamental parameter in digital pathology, directly influencing diagnostic confidence and accuracy. Optical resolution is the scanner's ability to distinguish closely spaced objects. The objective lens's numerical aperture and illumination wavelength govern it. A 20× objective with a 0.75 numerical aperture theoretically resolves structures 0.45 micrometers apart, which is sufficient for most routine diagnostic applications (Zarella et al., 2019).

Digital resolution, expressed as microns per pixel, represents the physical tissue area captured by each pixel. Resolution is not the same as magnification, which refers to the size of the resolved object (Sellaro et al., 2013). For diagnostic applications, digital resolution ranges from 0.5 μm/pixel (equivalent to 20× magnification) to 0.25 μm/pixel (equivalent to 40× magnification). The relationship between optical and digital resolution follows the *Nyquist sampling theorem,* which states that to faithfully capture all optical detail, the pixel size must be at most half the optical resolution limit. If this condition is not met, aliasing can occur, where false frequencies are introduced into the sampled data, jeopardizing the reconstruction process. Oversampling increases file size without adding meaningful information (Maciejewski et al., 2009).

Different pathology applications have distinct resolution requirements. For routine surgical pathology with H&E-stained specimens, scanning at 20× equivalent resolution suffices for primary diagnosis. This resolution adequately reveals nuclear details, mitotic figures, and architectural patterns critical for most diagnostic determinations. Cytology and hematopathology specimens, such as blood smears and bone marrow aspirates, benefit from higher resolutions to better evaluate nuclear chromatin patterns and cellular inclusions. For these applications, 40× or greater resolution is recommended (Borowsky et al., 2020).

The choice of scanning resolution has major implications for workflow efficiency. File size scales with the square of the resolution, which means that doubling the resolution quadruples the file size for the same tissue area. An H&E-stained surgical pathology slide scanned at 20× resolution produces an uncompressed file of approximately 1–2 GB, while the same slide at 40× generates a 4–8 GB file. Resolution thus directly impacts scanning time, storage requirements, network bandwidth needs, and viewing performance.

Scanning speed

Several factors influence whole slide scanner speed, including sensor sensitivity, chosen scanning parameters, focusing method, and specimen characteristics. Modern high-throughput scanners can digitize a standard slide at 20× magnification in 30–60 s under ideal conditions. Actual scanning times may be longer due to continuous autofocusing and other quality control processes (Genis et al., 2021).

When evaluating scan speeds for digital pathology implementation, laboratories must be cautious about relying solely on vendor-reported scanning times. These specifications reflect performance under idealized conditions. Scanning a standardized 1.5 cm × 1.5 cm tissue sample under ideal conditions may differ significantly from real-world practice. Specimens range from tiny biopsy fragments measuring just millimeters to large surgical sections that may span the entire slide. This variability in specimen size creates significant fluctuations in scanning times that vendor specifications fail to capture. A small biopsy might scan in under a minute, while a large colorectal tissue section would take significantly longer.

Besides size, there are other factors that influence scan time and the final file size. Variations in scanners and their settings may lead to differences in scanning time and "dropouts," where parts of the tissue are missed during scanning. A scanner's "first-time-right" rate, which measures the percentage of slides successfully scanned without needing a rescan, is an important metric for evaluating throughput (Mutter et al., 2022).

This disconnect between reported performance and real-world conditions can lead to serious miscalculations when determining the number of scanners required for a laboratory. Planning based solely on vendor specifications may result in insufficient scanning capacity. This leads to bottlenecks that compromise turnaround time. Laboratories must, therefore, conduct their own performance assessments using their actual specimen mix to make informed decisions about the number of scanners needed and how they may best be implemented (Fraggetta et al., 2021).

Batch scanning

Scanner throughput depends not only on raw scanning speed but also on its ability to process batches of slides and integrate into the laboratory workflow. Batch scanning functionality enables unattended operation. Entry-level systems handle 1–10 slides per batch, mid-range systems process 20–100 slides, and high-throughput systems can manage 200–400 or even up to 1000 slides in a single batch. The loading mechanism varies from manual loading, which requires operator intervention between each slide, to cassette-based systems with removable slide holders that can be preloaded and exchanged quickly. Continuous loading systems allow adding new slides while scanning is ongoing. Another factor to keep in mind is not just slide capacity for a single scanner but also simultaneous scanning using multiple smaller-capacity scanners. This leads to faster total slide turnaround times (Soliman et al., 2025).

File size and data management

The sheer scale of data captured by WSI is a fundamental characteristic that shapes every aspect of digital pathology implementation. A single tissue section, even moderately sized, scanned at diagnostic resolution produces a very large image file, easily reaching several gigapixels. For example, scanning a 20 × 15 mm area at 0.25 μm/pixel generates an image approximately 80,000 × 60,000 pixels, or 4.8 gigapixels. Uncompressed, such a file would be prohibitively large, at about 15 GB.

Managing this volume of data requires sophisticated enterprise strategies. WSI files are almost universally stored in a tiled, pyramid structure that addresses the impracticality of loading and displaying the entire gigapixel image at once. The full-resolution image forms the base layer of a multilayer pyramid. Successively lower-resolution versions of the image are generated and stored as upper layers. These down-sampled images simulate lower magnifications. Each layer is further subdivided into small, manageable square tiles (Fig. 2.5) (Krupinski et al., 2012).

When a pathologist views the virtual slide using imaging software, the system only loads and displays the tiles corresponding to the currently visible area on the screen at the appropriate resolution layer based on the zoom level. This dynamic loading of only the needed data is essential for smooth navigation and prevents overwhelming the computer's memory and processing power. The tiled pyramidal structure is a cornerstone of WSI file formats, enabling efficient streaming over computer networks.

Even with tiling, file sizes are huge. This makes image compression crucial for transmission and storage. Both lossless (reversible) compression, which allows perfect reconstruction of the original data, and lossy (irreversible) compression, which achieves greater size reduction by discarding some pixel information, are used. Lossless methods provide moderate reduction, making files 3–5 times smaller, while lossy methods enable much greater compression. Research shows that compression ratios up to 32:1, as created using JPEG2000, do not significantly impact diagnostic accuracy for routine applications (Papazoglou et al., 2023). However, other studies with different datasets and tissue types limit the generalizability of these findings (Fischer et al., 2025; Zanjani, 2019; Chen et al., 2020; Johnson et al., 2011; López, 2008; Doyle, 2010; Pantanowitz et al., 2017; Platiša, 2017; Kalinski et al., 2011; Jahn et al., 2020;

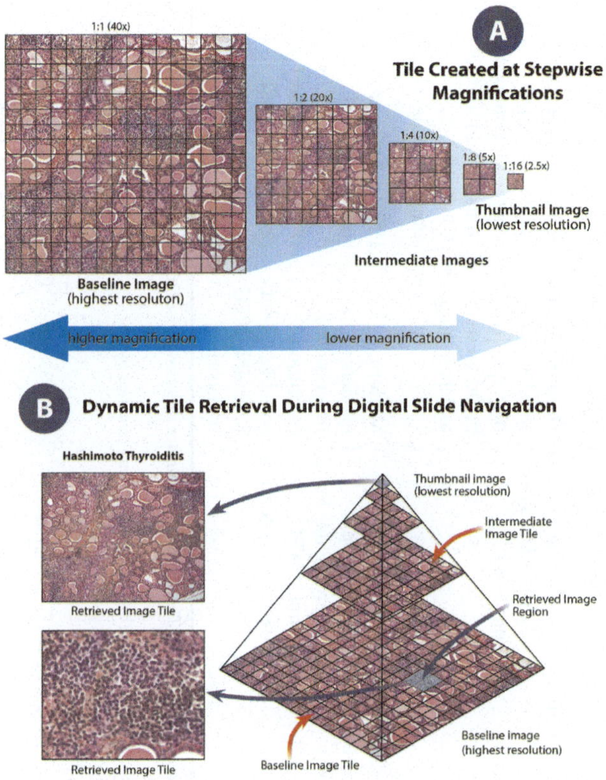

FIG. 2.5 Multiresolution pyramid structure: the foundation of whole slide imaging (WSI) file formats. (A) During scanning, each digital slide is stored as a pyramid of images at different resolutions. The highest-resolution baseline image (e.g., 40×) is progressively downsampled into intermediate layers (20×, 10×, 2.5×, etc.), culminating in a low-resolution thumbnail. This structure allows efficient storage and flexible viewing. (B) When navigating a digital slide, the viewer software dynamically retrieves only the tiles needed at the chosen magnification. At low power, only small thumbnail tiles are displayed, enabling rapid navigation across the slide. As the user zooms in, progressively higher-resolution tiles are retrieved and displayed in the region of interest. This pyramid system optimizes performance by balancing speed and image fidelity. This architecture underpins nearly all WSI file formats and is central to digital slide viewing.

Helin et al., 2018; Fischer et al., 2022). While JPEG and JPEG2000 are common in digital pathology, advances like JPEG XL and other modern transfer syntaxes represent significant improvements (Fischer et al., 2025). These improvements are reflected in the addition of lossless and general JPEG XL support through DICOM Supplement 232, which offers better color image compression, multiframe encoding, and flexible options (Anon., 2023). In addition, jpegli has been introduced as an advanced JPEG coding library that maintains high backward compatibility while also providing enhanced capabilities and a 35% improvement in compression ratio at high-quality settings (Szabadka et al., 2024). Consequently, pathologists need to stay informed about the evolving image compression landscape. Reversible compression can always be used, as it, by definition, does not impact image quality, while irreversible compression can be employed to reduce transmission time or storage space if the resulting quality is sufficient for reliable clinical performance (American College of Radiology, AAPM, SIIM, n.d.). Establishing appropriate compression parameters that balance file size with the preservation of diagnostic information remains a critical decision. Table 2.4 lists the common compression formats in clinical use and their various trade-offs.

Modern digital pathology systems have advanced beyond basic whole-image compression to adopt sophisticated methods that meet the specific needs of pathology workflows. These improved techniques, like progressive encoding and compression methods, acknowledge

TABLE 2.4 Current common clinical compression strategies and clinical applications.

Compression type	Technology	Compression ratio	Image quality*	File size (typical slide)	Clinical application	Pros	Cons
Uncompressed	Raw TIFF	1:1	Perfect	15–25 GB	Research, litigation	Perfect fidelity	Impractical size
Lossless	LZW, ZIP	3–5:1	Perfect	3–8 GB	Archive, reference	No information loss	Limited compression
Low-Loss JPEG2000	Wavelet	10–15:1	Excellent	1–2.5 GB	Primary diagnosis	Good balance	Some artifacts
Standard JPEG2000	Wavelet	20–32:1	Very Good	500MB–1.2GB	Routine diagnosis	Practical size	Minimal artifacts
High Compression	JPEG2000/HEVC	50–100:1	Good	150–500MB	Screening, education	Very small files	Visible artifacts
High-Quality JPEG (jpegli encoder)	DCT** (baseline/ progressive JPEG)	10–20:1	Very Good	1.0–2.0 GB	Routine diagnosis in legacy JPEG-based pipelines	Drop-in encoder; smaller files at high quality; broad viewer/ toolchain support	Still a JPEG (artifacts appear at higher compression ratios); less efficient than JPEG2000 at very high compression levels
JPEG XL (lossless or lossy)	VarDCT*** +modular (scalable); supports lossless JPEG recompression.	Lossless: ~2–3:1; Lossy: ~20–50:1	Perfect (lossless) to Excellent	Lossless: ~5–10 GB; Lossy: ~0.4–1.0 GB	Archive & distribution; emerging use for clinical WSI (requires validation)	Modern efficiency; progressive/ resolution-scalable; >8-bit; reversible JPEG transcoding	Uneven browser support; limited deployment in current WSI stacks; clinical use needs validation

* *Image Quality:* ***Perfect, Excellent****: No detectable diagnostic impact in validation studies.* ***Very Good****: <2% diagnostic discordance vs. uncompressed.* ***Good****: 2-5% discordance, suitable for specific applications*

*** DCT: Discrete Cosine Transform; a way to convert picture blocks into components so JPEG can store the important parts and compress the rest.*

**** VarDCT: Variable-Block-Size Discrete Cosine Transform; dynamically varies the size of the DCT blocks.*

that not all areas of a digital slide are equally important for diagnosis, enabling smarter data management without sacrificing clinical usefulness.

Progressive encoding is one such advancement. Images are transmitted in layers of increasing detail. This technique enables pathologists to begin examining slides almost immediately, with the initial low-resolution overview appearing quickly, while higher-resolution data continues loading in the background. For telepathology applications or remote consultations, where bandwidth is limited, this approach dramatically improves workflow efficiency by eliminating traditional wait times associated with large file transfers.

Region-based compression takes a more targeted approach by applying different compression parameters across the slide based on tissue characteristics and diagnostic importance. Areas containing actual tissue receive conservative compression settings that preserve fine morphological details essential for diagnosis. Meanwhile, background regions devoid of tissue—which can comprise substantial portions of many slides—undergo aggressive compression or may be represented simply as uniform color values. This selective approach can reduce overall file sizes by 40%–60% while maintaining full diagnostic quality in relevant areas.

Content-aware compression extends this concept further through intelligent analysis of image features. Emerging machine learning approaches automatically identify regions based on their diagnostic significance, potentially creating compression profiles tailored to specific tissue types and pathological processes (Aeffner et al., 2019). Regions containing critical diagnostic elements, such as nuclear detail, mitotic figures, or architectural patterns, receive minimal compression to preserve these essential features. Conversely, areas of uniform stroma or adipose tissue, which require less detailed examination, can tolerate more aggressive compression without clinical impact.[1]

File formats and standards

Proprietary formats and vendor lock-in

The proliferation of proprietary file formats is a major challenge in digital pathology, just as it was for radiology in the 1980s and 1990s. Each scanner manufacturer has historically developed optimized formats for their specific hardware and software ecosystems. These formats potentially offer faster loading or specialized features when used with a vendor's specific viewer, but this comes at the cost of reduced interoperability. A file created by one vendor's scanner may not be viewable by another vendor's software without conversion, which can be slow and require dedicated tools. Vendor lock-in remains a significant challenge for laboratories that use equipment from multiple vendors or plan to switch to a new system.

Among prevalent proprietary formats, Aperio SVS (.svs) is used by Leica Biosystems and Aperio scanners. It is a TIFF file (or, more specifically, a BigTIFF file if the file size exceeds 4 GB) with JPEG or JPEG2000 compression. It contains a pyramid of image tiles and some

[1] *Image Quality: Perfect, Excellent: No detectable diagnostic impact in validation studies. Very Good: <2% diagnostic discordance vs. uncompressed. Good: 2–5% discordance, suitable for specific applications.

metadata tags. SVS files are widely supported by software. Many third-party viewers can open SVS files because Aperio was an early leader in digital pathology. The NanoZoomer Hamamatsu scanners use NDPI files, which are a TIFF-based format with custom header information, often utilizing JPEG2000 or JPEG compression. NDPI files are well-supported by open libraries, although Z-stacks require special handling. Philips iSyntax (.isyntax) is used by Philips IntelliSite scanners. This is a proprietary format not based on TIFF. It is highly optimized for streaming in Philips' software. Unfortunately, no open-source libraries can currently read iSyntax. Users must convert iSyntax to another format for use elsewhere. 3DHISTECH (Sysmex) scanners use MRXS (.mrxs). This is a more complex format, consisting of a main file and a folder containing tile image data. It can use JPEG compression. The format is proprietary and partially encrypted, which hinders broad community support. Ventana BIF (.bif) is used by Roche/Ventana's DP200 and DP600 scanners. BIF stands for "Big Image Format". It's another proprietary container. Table 2.5 summarizes the various file formats.

Interoperability solutions

Vendor-Neutral Archives (VNAs) have emerged as an architectural approach to managing heterogeneous imaging data, including digital pathology. Rather than converting all images to a single format, VNAs store images in their native formats while providing a standardized interface for accessing and managing them. This approach allows organizations to maintain existing workflows while gaining many of the benefits of standardization (Pantanowitz et al., 2018). Interoperability solutions like OpenSlide and Bio-Formats have improved format compatibility by providing vendor-neutral access to major image formats (Linkert et al., 2010). Some formats, however, continue to be problematic.

Middleware solutions serve a similar function by providing translation layers between different systems and formats. These solutions can convert images between formats on demand, present standardized interfaces to viewing and analysis applications, and manage the complexity of working with diverse image sources. Some middleware platforms target digital pathology specifically, while others address enterprise imaging more broadly with pathology-specific modules or extensions.

These approaches provide practical paths to achieve interoperability in environments that have mixed systems and formats (Dash et al., 2021). They can be particularly valuable during transition periods when organizations are moving toward standards-based implementations but need to continue supporting legacy systems and data.

From a pathologist's perspective, these technical details relate to whether you can open a slide using your preferred viewer. In a multivendor environment, it may be necessary to perform conversions or ensure that the chosen software supports all required file formats. The lack of standardization has historically hindered digital workflow. A lab might have one scanner that produces SVS files and another that produces NDPI files. Trying to use a single viewer for both could be complicated if that viewer doesn't support one of the formats. Patients can also be affected if they seek a second opinion and the consultant does not have the appropriate viewing software. This lack of interoperability has been a major reason the pathology community promotes a standard format. This leads us to DICOM.

TABLE 2.5 Comparison of digital pathology file formats.

Format	Vendor/ Origin	Container type	Compression	Open-Source Support[a]	Strengths	Limitations
SVS (.svs)	Leica Biosystems/ Aperio	TIFF (BigTIFF)	JPEG, JPEG2000	OpenSlide, Bio-Formats	Widely used; early standard; good third-party support	File size can exceed 4GB; nonstandard TIFF metadata
NDPI (.ndpi)	Hamamatsu NanoZoomer	TIFF	JPEG, JPEG2000	OpenSlide, Bio-Formats	Well supported for 2D; open libraries can read; simple TIFF base	Limited support for Z-stacks (only one plane via OpenSlide); header quirks
MRXS (.mrxs)	3DHISTECH/ Sysmex	XML + folder	JPEG, JPEG-XR	Partial (some OpenSlide/ Bio-Formats)	Supports large multiresolution pyramids	Some versions are obfuscated/ encrypted; poor support for fluorescence and 16-bit images
iSyntax (.isyntax)	Philips IntelliSite	Proprietary binary	Custom (optimized)	(no OpenSlide/Bio-Formats support)	Highly optimized streaming in Philips' software	Fully proprietary; closed format; conversions to other formats needed
BIF (.bif)	Roche/ Ventana	Proprietary	JPEG	Partial (recent OpenSlide/ Bio-Formats)	Gaining third-party support	Older formats used TIFF or DICOM; BIF is less documented
VSI (.vsi)	Olympus	Multipage TIFF	JPEG	Bio-Formats	Compatible with OME-TIFF tools	Legacy; limited usage in pathology scanners
SVSLIDE (.svslide)	Sakura VisionTek	TIFF	JPEG	Limited	Lightweight; for small scanners	Not widely used or supported
CZI (.czi)	Zeiss	Proprietary XML	JPEG, JPEG-XR	Bio-Formats	Common in microscopy workflows: rich metadata	JPEG-XR decoding requires Windows codecs; not WSI-specific
SCN (.scn)	Leica (nonAperio)	TIFF-derived	JPEG	OpenSlide	Supported in OpenSlide; common in some Leica systems	Metadata inconsistencies
OME-TIFF	Open Microscopy Environment	TIFF + XML	Flexible (JPEG2000, LZW)	Broad support for Bio-Formats, QuPath, ImageJ, and others	Open, well-documented, supports multichannel time series and Z-stacks	Not all scanners export to this format; it is not natively supported by most commercial WSI vendors
DICOM WSI	DICOM WG26 standard	DICOM	JPEG2000, JPEG-LS, HEVC	Bio-Formats, DICOMizers[†], proprietary	Open standard; regulatory support (FDA); supports metadata-rich annotations	Complex structure (tiling, metadata hierarchy); vendor's native export is limited but increasing

[a] *OpenSlide is designed to manage large, pyramidal images used in digital pathology. Bio-Formats is a more versatile tool for reading and writing numerous digital image formats.*
[†] *DICOMizer refers to software or a process that converts medical images and related information into the DICOM format.*

DICOM as a unifying standard

The medical imaging community's experience with DICOM (Digital Imaging and Communications in Medicine) provides a roadmap for pathology standardization. DICOM, the international standard for medical imaging communication, was born out of the challenges of the early 1980s, when images from CT and MRI scanners were difficult to decode or share outside proprietary systems. In 1983, the American College of Radiology (ACR) and the National Electrical Manufacturers Association (NEMA) established a joint standards committee to address this need. Their first effort, ACR-NEMA 300 (1985), enabled point-to-point image transfer through a dedicated 16-bit interface, followed by a second version in 1988 that gained wider vendor adoption. Demonstrations of version 2.0 in 1990 at Georgetown University and RSNA showcased real interoperability in action. In 1993, the standard evolved to support Ethernet and TCP/IP networks; it was renamed the Digital Imaging and Communications in Medicine (DICOM) Standard and was formally published as NEMA PS3, laying the foundation for today's global medical imaging ecosystem (National Electrical Manufacturers Association (NEMA), n.d.). The subsequent creation of a framework for understanding teleradiology by the Society for Computer Applications in Radiology (SCAR), the predecessor to the Society for Imaging Informatics in Medicine (SIIM), further helped to solidify this version of DICOM as an essential standard for medical imaging (SCAR Technology Forum, 1994). DICOM's successful deployment in radiology, which enables vendor-neutral image storage, transmission, and integration, provides compelling reasons for its adoption in digital pathology (Herrmann et al., 2018; Digital Imaging and Communications in Medicine (DICOM), 2010).

One major benefit of DICOM is its management of metadata. Each file includes standardized fields for patient ID, case accession, staining method, objective power, and other relevant data. This ensures that a DICOM slide carries all necessary contextual information. Any DICOM-compatible system can access these fields without needing a proprietary database to interpret the file's identity. In pathology, key metadata such as specimen labels, stains, and barcodes can be stored and utilized. DICOM can also embed images of slide labels, annotations, and macro images linked to the case (Clunie et al., 2016).

Using DICOM makes pathology images compatible within healthcare ecosystems used by radiology, cardiology, and other subspecialties. If a healthcare system can store DICOM for radiology, it can also store pathology DICOM slides alongside cardiology and other scans. Combined viewers can display all these different types of images. This supports integrated diagnostics and helps with archiving, since many hospitals have long-term DICOM archives. In digital pathology, interoperability depends not only on exchanging images but also on ensuring that the accompanying metadata—the clinical, specimen, and workflow information that gives images their meaning—is consistently structured and transmitted. While DICOM defines the standard formats for images and metadata, the Integrating the Healthcare Enterprise (IHE) initiative builds on this foundation by specifying how that metadata should move across systems to support real-world use cases. Through its Pathology and Laboratory Medicine (PaLM) domain, IHE develops integration profiles, such as the Digital Pathology Image Acquisition (DPIA) profile, which define the required metadata elements and how they must be communicated at each stage of the workflow. In this way, IHE translates the technical capabilities of DICOM into practical, cross-vendor interoperability. This ensures that images and their context can be reliably used in patient care (Dash et al., 2021). Although adoption is still emerging, the first DICOM Connectathon for digital pathology in 2025 marked a critical milestone, demonstrating that multiple vendors could successfully exchange images and associated information through IHE-defined workflows. This achievement underscored how DICOM, guided by IHE profiles such as DPIA, provides the structure necessary to move digital pathology from vision to clinical reality.

International support for DICOM in pathology is gaining momentum. The European BigPicture project, a large pathology image repository initiative, has chosen DICOM as its standard (Bigpicture consortium, 2022) as have the Advanced Research Projects Agency for Health (ARPA-H) Imaging Data Exchange (INDEX) program (Digital Pathology Association, n.d.) and the Food and Drug Administration (FDA) (U.S. Food and Drug Administration, n.d.). Vendors are starting to add DICOM options, with some scanners now capable of outputting DICOM images directly. Software platforms, including viewing applications, have added DICOM WSI support, and conversion tools exist to transform legacy formats to DICOM.

The promise of DICOM is that a pathologist or institution can use different scanners and software without worrying about format incompatibilities; everything can speak DICOM. In practice, we are in a transition period where many existing archives remain in proprietary formats, and vendors still have proprietary features. Following the same trajectory that DICOM enabled in radiology, there is broad agreement that an open format is needed for long-term data integrity and cross-system use.

Software ecosystems: The digital microscope experience

Image management systems (IMS)/picture archiving and communication systems (PACS)

There are ongoing challenges with nomenclature in digital pathology, and terms are often used inconsistently across vendors, regulators, and health systems. In this section, we will do our best to clarify the situation. As described earlier, a whole slide imaging (WSI) system is currently defined by the FDA under the Technical Performance Assessment (TPA) guidance as consisting of (1) an image acquisition subsystem that converts the content of a glass slide into a digital image file, and (2) a workstation environment for viewing the digital images (Gorman et al., 2021). While the first part represents the scanner, the second consists of the image review and manipulation software, the computing environment, and the display. This image review software component is variably referred to in the digital pathology community as an Image Management System (IMS), a PACS, or, in FDA terminology, Digital Pathology Image Viewing and Management Software (product code QKQ) (U.S. Food and Drug Administration, n.d.). A Picture Archiving and Communication System (PACS), since its inception in the 1980s (McGeary, 2009), has been broadly described as a system that displays, archives, communicates, and manages digital medical images and related information, which may or may not include an acquisition device (Oosterwijk et al., n.d.). For our purposes in digital pathology deployment today, PACS is likely the most appropriate term, as it is already well established in radiology and other imaging domains and widely recognized within healthcare systems. By contrast, a Vendor Neutral Archive (VNA) is commonly referred to as a medical IT device that provides scalable image information and lifecycle management so that images and related information can be queried, stored, and retrieved. Unfortunately, there is no official definition of VNA functionality, and offerings vary considerably between vendors. The word "neutral" is also misleading, as most VNAs remain tightly linked to a PACS vendor. To add further complexity, some companies now provide viewers independent of their archiving solution, blurring functional boundaries and creating additional integration challenges. Nevertheless, the acronym VNA has become fixed in both the medical IT user and vendor communities, despite this ambiguity (Philips, n.d.). Importantly, in digital pathology, there is a growing blurring of roles between the PACS/IMS and the Laboratory Information System (LIS): the LIS remains the system of record for accessioning, tracking, and reporting, while the PACS/IMS increasingly serves as the workflow hub for image-centric diagnostics. This overlap underscores the need for clear definitions and seamless integration to avoid fragmented workflows. Table 2.6 describes definitions and functions of image management systems.

Digital pathology IMS/PACS functions as the central hub for organizing, tracking, and accessing large slide collections. These platforms must efficiently support terabyte-scale archives while offering user-friendly interfaces that integrate seamlessly with laboratory workflows.

Effective image management mirrors the logical structure of pathology practice, by organizing images by patient, case, and specimen. Metadata management captures the essential contextual details which includes demographics, clinical history, specimen information, and staining protocols. The software enables advanced search, filtering, and rapid retrieval from millions of stored images.

Modern systems employ variable on-premise, hybrid, or cloud-based database architectures tailored to the demands of digital pathology. Relational databases manage structured metadata and relationships, while high-performance storage solutions accommodate the large whole-slide image files. This approach combines robust querying capabilities with the speed necessary for large-scale image processing.

Workflow management features coordinate how cases progress through the diagnostic pipeline. These include generating worklists that assign cases to appropriate pathologists, monitoring case statuses across stages, and distributing tasks according to subspecialty expertise or workload. Advanced systems extend this functionality with rules-based automation, such as prioritizing urgent cases, routing subspecialty specimens to designated experts, or implementing tailored workflows for particular case types.

Access and accountability are maintained through role-based user management. For example, pathologists may be granted full diagnostic privileges, residents may receive restricted educational access, and clinicians may have view-only rights to relevant cases. Complementary audit trails record all system activity, such as who accessed which images, when, and what actions were taken. This supports both quality assurance and regulatory compliance.

In digital pathology, the debate over whether the LIS or PACS/IMS should serve as the central workflow engine closely mirrors the earlier transition in radiology from RIS-driven to PACS-driven practice. The issue is not about which system is "better," but rather about which environment pathologists primarily interface with to perform their daily work and how

TABLE 2.6 Definitions and functions of image management systems.

Term/system	Definition/Regulatory context	Strengths/Core functions	Limitations/Ambiguities
IMS (Image management system)	Vendor term for digital pathology platforms that manage WSI; not formally standardized	Organizes, stores, displays, and manages WSI with image review tools	Terminology is inconsistent; it overlaps with PACS and is not universally recognized
PACS (Picture Archiving and Communication System)	Originated in radiology (1980s); the FDA classifies pathology as equivalent under Product Code QKQ: Digital Pathology Image Viewing and Management Software	Displays, archives, communicates, and manages digital images; widely recognized across imaging	May or may not include acquisition; terminology imported from radiology but is variably applied in pathology
VNA (Vendor neutral archive)	Medical IT device marketed as providing scalable, long-term image and lifecycle management	Archives images and related information so they can be queried, stored, and retrieved across systems	"Neutral" is misleading—often linked to a PACS vendor; no official definition; functionality varies
Independent viewers	Some companies provide standalone viewers, separate from archiving solutions	Offers flexibility in visualization; it can be lightweight and faster to deploy	Adds integration complexity; fragments workflows if not tightly linked to PACS/IMS or LIS
LIS (Laboratory information system)	Legal system of record for pathology: accessioning, tracking, reporting	Ensures regulatory compliance (e.g., CLIA, CAP); anchors diagnostic records in the patient chart	Increasingly dependent on PACS/IMS for image-centric workflows, overlap requires seamless integration

effectively the two systems integrate. In radiology, the RIS remained the record of orders and reports, yet PACS became the tool radiologists lived in because it provided fast access to images, case worklists, and interpretation tools. The LIS in pathology continues to serve as the system of record, handling accessioning, specimen tracking, chain of custody, and final reporting while ensuring regulatory compliance (e.g., CLIA and the CAP) and anchoring the diagnostic record in the patient's chart. The PACS/IMS, by contrast, is optimized for image-centric tasks, offering rapid retrieval, navigation, annotation, collaboration, AI overlays, and orchestration of case worklists. This is what PACS did in radiology. However, for pathologists, this transition may prove more challenging, given the complexity of ordering and managing ancillary testing such as immunohistochemistry, molecular assays, and other specialized studies, which remain tightly embedded in LIS workflows. The future lies in tight LIS–PACS/IMS interoperability, supported by standards such as DICOM WSI, IHE PaLM profiles (e.g., DPIA), and HL7 FHIR to enable seamless exchange of images, metadata, and results. AI further magnifies this need: algorithms will operate within the PACS/IMS environment for image analysis, but must return validated results to the LIS to become part of the legal diagnostic report.

Integrating image management with the broader digital pathology workflow can follow several models (Table 2.7). In an **LIS-driven approach**, whole slide images are tightly coupled with accessioning and reporting functions, preserving continuity with established

TABLE 2.7 Comparison of image management and laboratory information systems.

Function	PACS/IMS (Image Management System/PACS)	LIS (Laboratory Information System)
Primary purpose	Storage, retrieval, display, and management of whole slide images; orchestration of image-centric workflows	Accessioning, specimen tracking, order entry, and reporting (system of record for pathology)
Core data managed	Whole slide images, annotations, and image metadata	Patient demographics, specimen details, orders, test results, and final reports
Workflow focus	Case review, slide navigation, image manipulation, consultations, AI integration	Specimen processing, chain of custody, case assignment, diagnostic report generation
Standards used	DICOM WSI, IHE PaLM profiles (e.g., DPIA)	HL7 v2, HL7 FHIR, IHE PaLM messaging
Integration point	Requires LIS integration for case context and reporting	Requires PACS/IMS integration for image viewing and analysis
User interface	Pathologist-facing viewer with panning, zooming, annotation, comparison, and AI result overlays	Pathologist and lab staff dashboards for order entry, case status, and report creation
Regulatory classification	FDA Product Code QKQ: Digital Pathology Image Viewing and Management Software	FDA-regulated as part of clinical laboratory systems (CLIA compliance in the United States)
AI integration	Hosts image analysis algorithms; displays AI overlays, heatmaps, and quantitative results within the workflow	Incorporates AI-derived results into structured pathology reports; ensures traceability and an audit trail
Blurring of Roles	Increasingly handles workflow lists, case routing, and sometimes reporting support	Still the legal system of record, but increasingly dependent on PACS/IMS for image-based diagnostic work

laboratory processes. A PACS/**vendor-based approach** leverages proprietary scanner software or platforms to manage image storage, navigation, and case assembly. Alternatively, an **intermediate or middleware platform** can serve as an orchestration layer between the LIS and imaging hardware, often providing multivendor interoperability, advanced visualization, and integration with computational pathology tools. Each strategy offers distinct advantages and limitations. Institutional priorities, such as scalability, interoperability, and regulatory compliance, guide the choice of model (Clunie, 2004; Digital Imaging and Communications in Medicine (DICOM), 2010; Lujan et al., 2022).

Viewing software: Redefining the microscope experience

Viewing software functions as the primary interface for pathologists when working with digital slides, effectively acting as a digital microscope. The design and features of these applications significantly impact diagnostic efficiency, accuracy, and user satisfaction.

Navigation should be intuitive in order to effectively review digital slides. Fundamental functions like panning, zooming, and quick access to regions of interest should be highly responsive and smooth, similar to traditional microscope stage movements. Various input methods are available, including standard mouse and keyboard controls, specialized devices like 3D mice, and touch interfaces on compatible displays.

Viewing enhancements improve the basic display of digital slides by adding features that support interpretation. Adjustments to contrast and brightness allow pathologists to optimize visualization of specific features, which is especially useful for poorly stained specimens. Color management tools enable adjustments in color balance and saturation, helping to offset variations in staining protocols or scanner characteristics. Digital zoom beyond the native scanning resolution can assist in examining borderline features, but it is important to remember that no additional detail is gained beyond the original scan resolution.

Specialized viewing modes cater to specific diagnostic needs. Thumbnail overviews provide orientation and context while inspecting details at high magnification. They also indicate which areas have been reviewed at particular magnifications. Side-by-side comparison enables direct evaluation of multiple slides, such as comparing a current biopsy with previous specimens or examining various immunohistochemical stains of the same tissue. Synchronization features maintain the same position and magnification across multiple slides during navigation, which is especially useful when comparing serial sections or different stains from the same block.

Importance of macro images

Modern WSI scanners feature automated tissue detection systems that identify tissue areas and restrict scanning to diagnostically important regions, as noted previously. These systems aim to boost efficiency by skipping blank slide areas, thus reducing scan times and file sizes. Although effective in principle, this automation can raise quality control issues that laboratories need to address systematically.

Automated tissue detection can fail in several predictable situations. Small tissue fragments, especially from needle biopsies or tiny specimens, might fall below the detection threshold and be mistaken for background. Likewise, pale or lightly stained tissues, such as certain fatty samples or mucin-rich lesions, may lack enough contrast for accurate automated identification. The situation worsens if operators don't manually verify that the tissue detection boundaries include all diagnostically important material.

These detection failures cause a mismatch between the physical glass slide and its digital version, which can result in missed diagnoses or delays. The macro image serves as a vital safeguard against such errors. This low-magnification overview shows the entire slide surface, including all tissue fragments and empty spaces, providing both automated systems and human operators with a comprehensive reference for tissue identification and selection.

Although they are important, macro images are not always automatically shown on WSI platforms. This difference in how vendors implement the feature means that pathologists and lab staff need specific training on how to access and interpret these overview images as part of regular quality checks. When pathologists receive digital cases, they should carefully review the macro image to verify that all visible tissue has been captured at high resolution.

Alternative quality assurance methods, such as directly comparing glass slides with digital images, are still available but tend to be impractical for everyday use. This manual verification adds a significant workload for histotechnologists and lab staff, conflicting with the efficiency goals of digital pathology implementation. Conversely, systematic macro image review by pathologists offers a more sustainable quality control approach that integrates more smoothly into diagnostic workflows (Fraggetta et al., 2018).

Metadata and annotations

Digital annotations are among the most valuable parts of modern pathology workflows. These digital markings, measurements, and comments applied directly to WSI include a wide range of diagnostic tasks, such as identifying regions of interest, outlining tumor margins, performing quantitative cell counts, taking morphometric measurements, making diagnostic classifications, and adding detailed textual notes. Moving from glass slides to digital platforms has transformed annotations from simple physical ink marks into advanced data objects that can improve diagnostic accuracy and support collaborative consultation.

Measurement tools provide a quantitative assessment of microscopic features. Linear measurements determine distances between points, which are useful for evaluating margins or assessing structural dimensions. Area measurements gauge the size of regions, such as tumor foci or specific tissue compartments. Cell counting tools help quantify cell populations, either through manual marking or semi-automated detection. Layering systems enable the grouping of related annotations that can be displayed or hidden. Classification schemes allow the categorization of annotations by type, importance, or other criteria. Filtering features display only annotations that meet specific criteria, reducing visual clutter during review.

Collaboration features enhance annotations beyond just individual review. Sharing options allow colleagues or trainees to view annotations, supporting consultation and education. Export functions enable annotations to be included in reports or presentations. Version control tracks changes to annotations over time, which is especially useful in educational or quality assurance settings where multiple reviewers may annotate the same slide (Hanna and Pantanowitz, 2022).

Several standardization efforts have arisen to solve annotation interoperability issues. The DICOM standard has grown to include annotation features via Presentation States and Structured Reports that connect annotation data directly to image files. These methods offer solid metadata management and integration with current medical imaging systems (Clunie, 2004).

Adding to the challenges of interoperability, traditional annotation formats designed for simple geometric shapes and text labels prove inadequate for representing sophisticated analytical outputs, such as AI-generated probability heatmaps, confidence intervals, or multilayered segmentation results. These advanced annotation types require specialized data structures and visualization tools that many standardized formats cannot accommodate without significant extension or modification. In addition, as images move through different processing stages or viewing systems, annotations may accumulate from multiple sources, including pathologists, AI algorithms, and automated image analysis tools. Managing these layered annotations while maintaining provenance, version control, and spatial accuracy requires sophisticated data management approaches that extend beyond simple file format considerations (Nurzynska et al., 2023).

User experience

The success of digital pathology implementation relies not only on image quality and technical performance but also on the design of viewing interfaces that support efficient diagnostic workflows. While advanced technical capabilities provide the basis for WSI systems, the viewer interface acts as the bridge between technology and clinical practice.

Usability testing with practicing pathologists represents an essential component of viewer development and refinement that is often undervalued by technology developers. Real-world testing reveals unintuitive interface elements that may seem logical to software engineers but prove cumbersome in actual diagnostic scenarios. These sessions identify workflow bottlenecks that emerge only under the time pressures and cognitive demands of routine pathology practice.

The most effective testing methods involve observing pathologists performing real diagnostic tasks instead of merely showcasing software features. This approach highlights the gap between impressive technical capabilities and genuinely useful functionality. Features that appear innovative during controlled demonstrations may introduce unnecessary complexity or mental strain, potentially slowing down the diagnostic process.

Successful viewing applications strike a balance between advanced features and simple operation. The interface should reduce cognitive load during diagnostics, helping pathologists focus on morphological interpretation rather than software navigation. This requires restraint in feature implementation. The software should only display capabilities that are broadly useful and avoid clutter from specialized tools that are not needed for regular diagnoses.

Viewing workstations and infrastructure

The standard configuration of a pathologist's workstation consists of a dual-monitor setup, where one display manages the laboratory information system (LIS), patient data, and administrative dashboards, while the dedicated imaging monitor handles WSI visualization and related diagnostic imagery. This separation allows pathologists to maintain access to clinical context while focusing primarily on histologic assessment.

Display technologies and monitor requirements

The viewing workstation is the pathologist's primary interface with the digital pathology system. It influences both diagnostic efficiency and user satisfaction. Careful specification and setup of these workstations are essential for successful digital pathology implementation. All medical displays have physical properties that directly influence image appearance, including luminance (brightness), the response function (how pixel values map to brightness), uniformity (consistency of brightness across the screen), and pixel pitch (spacing of pixels) (Silosky et al., 2022). Key luminance metrics include maximum luminance (Lmax), minimum luminance, and the luminance ratio. Again, we look to radiology for these characteristics to provide a baseline, along with recommended performance targets, which are detailed in the ACR–AAPM–SIIM Technical Standard for Electronic Practice of Medical Imaging (American College of Radiology, AAPM, SIIM, n.d.) and AAPM Report 270 (Bevins et al., 2019). These specifications are summarized in Table 2.8. While some of these characteristics may not apply, a current recommendation for pathology that is within scope is being developed by the International Electrotechnical Commission (IEC) (International Electrotechnical Commission, 2024), using the backdrop of medical imaging as a whole. This provides us with a baseline to follow.

Display quality directly influences diagnostic confidence and efficiency in digital pathology. Medical-grade monitors offer several benefits for pathology viewing, such as consistent brightness, accurate colors, higher resolution, and quality control features. Consumer-grade monitors may be adequate for some purposes but lack the calibration capabilities and quality assurance features found in medical-grade displays (Clarke et al., 2020). There is a significant price difference between medical-grade and consumer-grade monitors, however, and some pathologists actually prefer the viewing experience provided by the more affordable option.

Resolution requirements vary based on the viewing application and user preferences. The minimum resolution for primary diagnosis is considered to be 3–4 megapixels, while higher resolutions of 5–8 megapixels can provide additional detail that may benefit certain applications. Ultra-high-resolution displays can reveal larger tissue areas at high magnification, reducing the need for panning and potentially increasing efficiency.

Color accuracy is important in pathology because subtle stain differences often have diagnostic importance. Displays need to provide consistent color reproduction through proper calibration to accurately display stained tissues. Regular calibration with specialized equipment helps sustain this accuracy over time by adjusting for gradual changes that occur in all display technologies as components age.

The hardware supporting digital pathology viewing must provide enough performance for smooth navigation and image processing. Processor requirements are modest for basic viewing but increase significantly when using advanced features like real-time image

TABLE 2.8 Current radiology display recommendations.

Characteristic	Suggested performance level for medical displays
Ambient illuminance	25–75 lx
Ambient luminance	<0.2 cd/m^2
Minimum luminance	0.8 cd/m^2
Maximum luminance	350 cd/m^2
Luminance ratio	350
Luminance uniformity	LUDM $<15\%$, no clinically significant nonuniformities
Pixel pitch	Approximately 0.200 mm at typical radiology viewing distances

Modified from the International Electrotechnical Commission (2024). IEC 62563-3: Medical electrical equipment—Medical image display systems—Part 3: Evaluation methods for color displays (Draft). Retrieved from the IEC web store or an associated national standards body on September 6, 2025.

analysis (Murphy and Bird, 1974). Modern multicore processors with four or more cores and speeds of 3+GHz ensure sufficient performance for routine digital pathology tasks.

Graphics processing capabilities are needed for viewing digital slides. A dedicated graphics processing unit (GPU) with at least 4 GB of video memory significantly improves viewing performance compared to graphics performed by the central processing unit (CPU). The GPU handles the demanding tasks of decompressing, rendering, and manipulating large whole slide images, enabling smooth panning and zooming even with large or complex slides.

Memory requirements are high due to the large size of whole slide images. Basic setups include 16 GB of RAM, but 32 GB or more is recommended for advanced tasks or viewing multiple slides. Storage configurations use solid-state drives for the operating system and applications, enabling quick startup and smooth performance.

Network interface capabilities must match the available network infrastructure to prevent bottlenecks. Gigabit Ethernet connections are the minimum standard, while 10 Gigabit interfaces are beneficial in environments with optimized network infrastructure (see discussion of *Local Area Network* below). Wireless connections may be enough for secondary viewing, but they usually lack the consistent performance required for primary diagnostic work.

Quality assurance and regulatory considerations

Rigorous quality assurance ensures that digital images consistently meet diagnostic standards. To reliably replace traditional optical microscopy, digital images must consistenly meet diagnostic quality benchmarks. Laboratories adopting WSI conduct comprehensive validation studies comparing diagnoses on digital slides with those on glass slides across a representative case range, often following professional guidelines such as those from the College of American Pathologists (Evans et al., 2021a) and others (Hsu et al., 2025). Ongoing quality control is also crucial, including routine checks of scanner focus and color calibration, monitoring for scanning and stitching artifacts, and maintaining proper equipment upkeep (Kohlberger et al., 2019; Senaras et al., 2018; Hosseini et al., 2020).

Important clinical quality parameters for whole slide imaging devices include:

- Focus accuracy: Automated detection of out-of-focus regions
- Color fidelity: Consistent color reproduction across scanners and displays
- Artifact identification: Recognition of stitching errors, illumination variations, and compression artifacts
- Completeness verification: Confirmation that entire tissue areas are captured

Color calibration in digital pathology ensures accurate consistent color reproduction across scanners, viewing conditions, and over time. Common methods include color reference slides with standardized color patches that set baseline color values, IT8 targets designed for microscopy, and integrated color calibration tools build into scanner systems. The International Color Consortium (ICC) profile method, widely used in professional photography and printing, has been adapted for digital pathology to manage and characterize color throughout the imaging process.

Despite these efforts, standardizing color in digital pathology remains challenging. The lack of universally accepted standards and the inherent variability in tissue staining processes contribute to this difficulty. Research shows that color differences can influence diagnostic interpretation, emphasizing the importance of comprehensive color management strategies that encompass the entire imaging workflow, from slide preparation to scanning and display (Krupinski et al., 2019).

Image artifacts are another critical aspect of quality control. Various artifacts can appear during the WSI process, including focus issues that result in blurry areas, stitching artifacts that create visible seams between tiles, illumination inconsistencies causing uneven brightness across the image, and compression artifacts, especially with aggressive lossy compression. These artifacts can hinder diagnosis by masking true pathological features or creating misleading visual patterns (Patel et al., 2022). Fig. 2.6 shows four of the more common scanning errors.

Quality control procedures can spot and handle potential issues. Automated quality assessment algorithms can check focus quality, identify stitching errors, and detect other common artifacts. These automated checks support manual quality control performed by trained personnel who review images before they are released for diagnostic interpretation. Establishing clear quality thresholds and procedures for managing suboptimal scans is an essential part of a robust digital pathology workflow. Several vendors provide automated, AI-assisted QC tools.

Regulatory framework and compliance

Implementing digital pathology requires navigating a complex landscape of regulatory requirements and compliance factors that vary significantly by geographic location, institutional type, and specific clinical applications. The regulatory environment continues to evolve as digital pathology advances from experimental technology to routine clinical practice.

Digital pathology systems used for clinical diagnosis are regulated as medical devices in both the United States and the European Union (García-Rojo et al., 2019). In the U.S., the FDA classifies these systems as Class II devices, requiring 510(k) clearance. A significant milestone was reached in 2017 with the approval of the Philips IntelliSite Pathology Solution, setting a regulatory precedent for primary diagnosis. Subsequent systems from Leica (Aperio AT2 DX and GT450 DX) in 2019 and 2024 respectively, Hamamatsu (NanoZoomer S360MD) in 2022, and Roche Digital Pathology Dx (VENTANA DP 200 and DP600) in 2024 have received FDA clearance. Other scanners, such as several from 3DHISTECH, have obtained CE-IVD mark approval in Europe.

The FDA's oversight is organized around the "pixel pathway," which covers the entire imaging process, from tissue preparation to scanner performance, image processing, and display alidation and viewing. Each step must be thoroughly validated to confirm diagnostic equivalence with traditional microscopy. Ongoing quality control measures must be documented and maintained.

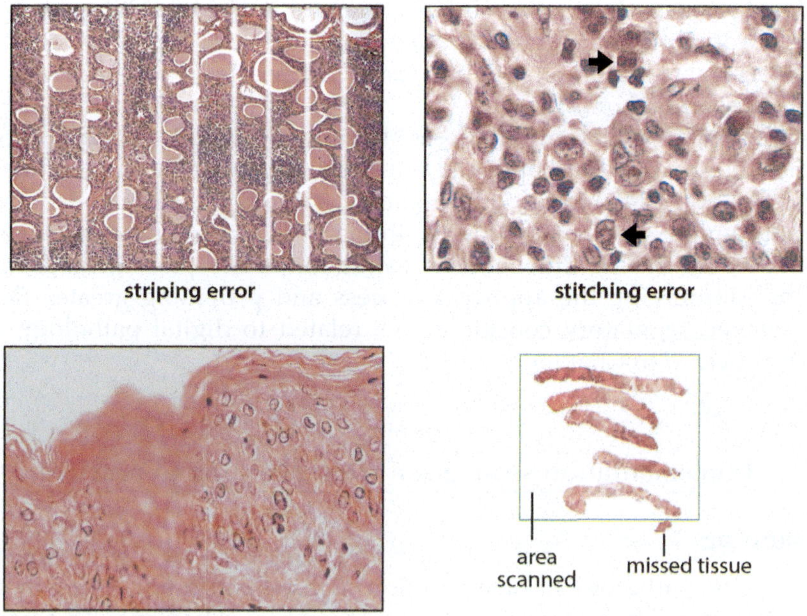

FIG. 2.6 Common scanning errors in whole slide imaging. Despite technical advances, digital slide scanning is prone to errors that can compromise diagnostic reliability. This compilation illustrates the four most common types of imaging errors that can occur during whole slide scanning, all of which require systematic quality control procedures to detect and manage. Striping errors appear as vertical bands caused by uneven illumination or sensor malfunction. Stitching errors arise when adjacent image tiles are misaligned, producing visual artifacts that obscure details. Focus errors result when the scanner fails to track tissue thickness accurately, leaving areas blurred. Finally, tissue finder errors represent perhaps the most critical failure, where automated tissue detection algorithms may miss diagnostically important material, creating a mismatch between the physical glass slide and its digital representation. Quality assurance processes are essential to detect and mitigate these issues before diagnostic use.

The pixel pathway requirements ensure that digital workflows do not introduce diagnostic risk and are equivalent to traditional glass slide review. The validation steps include:

- Preanalytical variables (fixation, staining, sectioning) must be controlled.
- Scanners must demonstrate resolution, color fidelity, and artifact control.
- Compression algorithms must be shown to preserve diagnostic information.
- Displays and viewers must be validated for consistent image quality and usability.
- Quality control procedures are mandated for ongoing performance assurance.

In the EU, digital pathology systems are regulated under the Medical Device Regulation (MDR) and require CE marking with Notified Body involvement. Classification as Class IIa or IIb devices depends on the intended use and associated risks, with strict documentation and clinical evidence requirements. In addition, the General Data Protection Regulation (GDPR) imposes significant obligations regarding patient data privacy, requiring a legal basis for data processing, support for individual rights, and data minimization. Cross-border data transfers and breach notifications are heavily regulated, impacting system design and operational deployment, especially in research and cloud-based environments (Decomplix, n.d.).

The current pixel pathway approach poses challenges when individual components reach the end of their lifespan, as has happened with several FDA-cleared monitors. The Digital Pathology Association's Regulatory and Standards Workgroup has collaborated with the FDA to promote a more modular approach, aiming to separate components within the pixel pathway. This would enable independent validation and replacement of hardware and software parts, simplifying the approval process and providing greater flexibility over time. More detailed regulatory considerations related to digital pathology and artificial intelligence are covered in Chapter 16.

Implementation strategies: From planning to practice

Strategic planning

Successful digital pathology implementation requires careful planning that aligns technical skills with organizational preparedness. This process begins with a comprehensive needs assessment and securing stakeholder buy-in on implementation goals and success metrics (Bruce et al., 2024). While many stakeholders are needed, a critical member is the information technology team. A concept embraced by health care systems is enterprise imaging, defined as a set of strategies, initiatives, and workflows implemented across a healthcare enterprise to consistently and optimally capture, index, manage, store, distribute, view, exchange, and analyze all clinical imaging and multimedia content to enhance the electronic health record (Roth et al., 2016). Enterprise imaging is rapidly emerging as a foundational paradigm for pathologists as digital pathology continues to advance. The concept was first articulated in the 2016 HIMSS-SIIM collaborative white paper, which outlined the essential elements of governance, infrastructure, workflows, enterprise viewers, and image exchange mechanisms necessary to support a truly enterprise-wide imaging ecosystem, of which pathology is a part (Roth et al., 2016). Fig. 2.7 More recently, the 2024 reflection on a decade of the HIMSS-SIIM Enterprise Imaging Community emphasized how whole slide pathology imaging has become a key domain within imaging informatics, requiring substantial investments in scanners, networking, cloud storage, advanced viewing applications, and inter-facility sharing to enable seamless integration (Roth et al., 2024a). Furthermore, the technical and standards-based challenges that must be addressed to scale image-based workflows across institutions and embed them into EHRs and clinical practice, especially DICOM adoption in pathology, were highlighted in a reflections and future directions assessment (Roth et al., 2024b). Taken together, these works underscore that enterprise imaging isn't just a radiology initiative; it is essential for pathology, laying the groundwork for interoperable, scalable, and multimodal image access that ultimately enhances diagnostic efficiency, collaboration, and long-term clinical value.

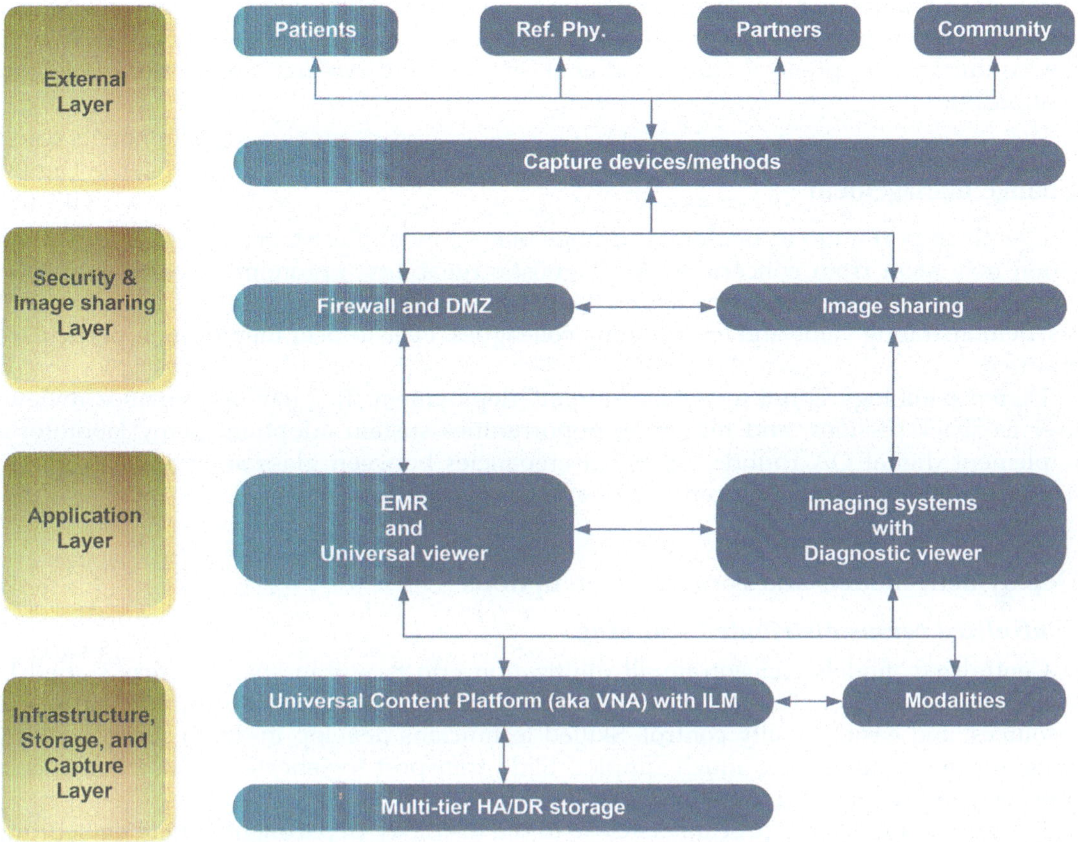

FIG. 2.7 Image enterprise imaging overview. enterprise imaging strategy endeavors in health. *Ref. Phy.* = *Referring Physician Courtesy of Eric Merchant, Endeavor Health.*

Storage infrastructure

Effective digital pathology storage infrastructure goes beyond primary storage systems to include multiple storage tiers with different performance and accessibility features. Active case storage needs high-performance, quickly accessible systems that support multiple users at once and enable fast image retrieval. Archive storage for completed cases can utilize more affordable, higher-capacity systems with longer access times but greater storage density. Backup and disaster recovery systems are part of the infrastructure that protect against data loss, but also increase storage needs. Many institutions use multiple backup tiers, such as local redundancy for immediate recovery and off-site replication for disaster scenarios. These safety measures can double or triple total storage needs but are essential for protecting irreplaceable diagnostic data. Staging areas for image processing, analysis, and quality control operations require additional high-performance storage that supports intensive data manipulation workflows. These temporary storage zones facilitate AI analysis, image enhancement, and batch processing operations without impacting primary diagnostic storage systems. The size and performance requirements of staging storage depend heavily on the institution's adoption of automated analysis tools and research applications (Williams et al., 2019).

Change management

Equally important to technology is change management. Resistance to digital implementation may stem from concerns over diagnostic reliability, ergonomics, or disruption of established routines. Champions within the pathology department can serve as early adopters, demonstrating value and encouraging colleagues. Peer-to-peer mentoring is particularly effective.

Digital pathology is not a "set-and-forget" deployment. Regular performance reviews, user feedback sessions, and retraining opportunities sustain adoption. Many laboratories implement digital QA rounds, where discrepancies between glass and digital reads are discussed openly, reinforcing confidence and continuous improvement.

Deployment models and configuration options

Centralized versus distributed scanning

Centralized models concentrate all digitization efforts within specific areas staffed by specialized personnel. This approach promotes standardized procedures, efficient use of resources, and easier quality control. Skilled technicians develop in-depth scanner knowledge, making maintenance much simpler. Slide transport logistics can cause bottlenecks, however, and require careful planning.

Distributed models place scanners close to the histology department, where embedding, sectioning, and staining occur. This reduces transportation needs and can shorten turnaround times by eliminating specimen movement delays. Integration with existing workflows might be easier, allowing technologists to start scanning during routine tasks. However, this complicates staffing and quality control across multiple sites.

Hybrid approaches combine elements of both strategies, recognizing that different case types may require different scanning methods. Urgent cases or specialized applications might utilize distributed systems for instant digitization, while routine specimens are processed through centralized high-capacity scanners.

Training and adoption

Transitioning from glass slides to digital workflows represents a fundamental shift in the daily practice of pathology. Beyond validating technical performance, the success of implementation depends on structured training and deliberate efforts to build confidence among pathologists and laboratory staff.

Digital systems demand new visual and interpretive skills. Pathologists must adapt to differences in navigation (scrolling, zooming) compared with a microscope, develop comfort with digital focusing tools, and learn to evaluate subtle features such as mitoses, nuclear detail, and microorganisms at high magnification. Early exposure to WSI during residency and fellowship accelerates adaptation, while continuing education modules, proficiency tests, and case-based workshops support practicing pathologists.

Histotechnologists, lab managers, and IT personnel must also be trained in scanner operation, image quality assessment, metadata capture, and troubleshooting. Training ensures that technical issues, such as poor focus, barcode mismatches, or incomplete image stitching, are recognized and corrected early.

Adoption is smoother when digital workflows are introduced in phases. Many institutions begin with educational archives or frozen sections before moving into primary diagnosis. Pilot programs allow users to gain experience, refine protocols, and provide feedback. This staged approach helps build institutional trust and addresses concerns about diagnostic accuracy or turnaround time.

Environmental and infrastructure requirements

The physical workspace environment is often overlooked when implementing digital pathology systems. Software features and hardware specs may capture all the focus, but the environmental conditions around digital workstations greatly influence both pathologists' performance and long-term system dependability. Achieving an ideal environment requires careful attention to lighting, layout, and ergonomic factors that promote sustained diagnostic work.

Lighting design and visual ergonomics

Ambient lighting conditions influence display visibility and user comfort during extended diagnostic sessions. Shifting from microscope-based pathology to monitor-based viewing under standard office lighting introduces new challenges for visual perception and eye strain management. Research shows that controlled ambient lighting levels between 20 and 40 lx offer optimal viewing conditions for medical displays, balancing visual comfort and contrast perception (Krupinski, 2014). This moderate lighting approach avoids extremes that can impair diagnostic accuracy. Extremely dark environments, while enhancing display contrast, can cause significant eye strain when users shift between bright screens and dark surroundings. Conversely, overly bright ambient lighting diminishes perceived contrast and can lead to reflections that hinder fine detail recognition. The ideal lighting environment supports both comfortable viewing and precise color perception. Display placement and antiglare measures are also key elements of lighting design. Workstations should be arranged to minimize direct reflections from windows, overhead lights, or other bright surfaces. Antiglare screen coatings can help reduce reflections but may slightly affect color accuracy, so they should be carefully evaluated in relation to diagnostic needs.

Spatial layout and workflow integration

Physical workspace design must support the entire pathology workflow rather than focusing only on digital viewing needs. Despite the digital shift, pathologists still need access to reference materials, consultation areas for colleague interactions, and integration points with any remaining nondigital processes. The workspace should enable smooth transitions between digital and traditional pathology tasks without causing workflow disruptions or physical obstacles.

Allocating space for collaborative activities is another important aspect often overlooked in workstation planning. Pathology diagnosis frequently involves consulting with colleagues, multidisciplinary team discussions, and teaching sessions that require flexible use of space. Digital workstations should be able to support multiple users when needed, while ensuring primary user comfort and efficiency.

Implementation approaches and environmental control

Institutional strategies for optimizing the physical environment vary widely depending on available resources, existing infrastructure, and organizational preferences. Some institutions invest in dedicated digital reading rooms with carefully controlled lighting, temperature, and acoustic conditions to enhance the digital pathology experience. These purpose-built spaces offer full environmental control but require significant capital investment and can limit workflow flexibility. Alternatively, some institutions retrofit existing pathologist offices or shared reading areas to support digital workstations while maintaining familiar work environments. This approach is often more cost-effective and can encourage user adoption by preserving established workflow patterns. However, retrofitting may necessitate compromises in environmental optimization and pose challenges in achieving consistent lighting and ergonomic conditions across multiple workstations (Stathonikos et al., 2019).

Scanner environmental requirements: Temperature and humidity control

While the workspace environment affects the comfort and performance of pathologists, scanner environmental conditions directly influence system reliability and image quality. WSI scanners are precise optical and mechanical instruments that require carefully controlled operating conditions to remain accurate and avoid premature component failure. Environmental changes that are hardly noticeable to people can significantly impair scanner performance and diagnostic image clarity.

Temperature stability is the most crucial environmental factor impacting scanner operation. Changes in temperature cause mechanical parts to expand and contract, which can disrupt the precise optical alignment needed for high-resolution images. Most manufacturers specify operating temperature ranges between 15°C and 30°C, but maintaining temperature stability is more important than the exact temperature. Rapid temperature swings exceeding ±2°C per hour can shift mechanical parts, which can affect focus accuracy and image registration across different areas on the slide.

Humidity control is equally important. Manufacturers recommend 30%–75% relative humidity in noncondensing conditions to prevent optical damage and static buildup.

Too much humidity encourages fungal growth on optical surfaces, causing image artifacts and costly cleaning or replacement. High humidity can also lead to condensation on internal optical parts, especially when there are significant temperature differences inside and outside the system.

On the other hand, very dry conditions generate static electricity, attracting dust particles to optical surfaces and electronic parts. This dust can lower image quality and interfere with automated slide handling and focus systems. Although static discharges are rare, they can damage sensitive electronics, leading to expensive repairs that proper humidity control could prevent.

The relationship between temperature and humidity adds further complexity to managing the environment. Rapid temperature changes can alter relative humidity, even if the actual moisture content stays the same, potentially pushing conditions outside acceptable ranges. HVAC systems supporting scanner areas need to coordinate both temperature and humidity controls, rather than managing each separately (Ardon et al., 2023).

Vibration control and structural considerations

Vibration isolation is one of the most technically challenging aspects of scanner environmental design. WSI scanners operate with micron-level precision, making them extremely sensitive to mechanical disturbances that might go unnoticed by human observers. The level of precision needed for high-resolution imaging means that vibrations measured in micrometers can cause focus drift, image blur, and registration errors.

Environmental vibration sources in healthcare facilities are numerous and varied. Building mechanical systems, such as HVAC equipment, elevators, and pumps, generate persistent low-frequency vibrations that can travel through building structures over long distances. Laboratory devices, like centrifuges, tissue processors, and automated analyzers, produce intermittent high-amplitude vibrations that can severely disrupt scanning operations when placed near WSI systems.

External vibration sources add further complexity to vibration management. Traffic, construction activities, and even heavy foot traffic can transmit vibrations through building foundations. Urban settings with subway systems, heavy truck traffic, or ongoing construction create particularly challenging vibration environments that may require advanced isolation techniques.

The frequency characteristics of different vibration sources determine their impact on scanner performance and help guide mitigation strategies. Vibrations below 10 Hz can cause large mechanical displacements, affecting slide positioning and focus stability. Vibrations between 10 and 100 Hz often cause image blur and registration issues. Higher-frequency vibrations above 100 Hz may have less effect on image quality but can accelerate mechanical wear and cause component fatigue (Philips, n.d.).

Mitigation strategies and installation approaches

Vibration mitigation strategies range from simple mechanical solutions to advanced active isolation systems, with the choice depending on both vibration severity and scanner sensitivity needs. Heavy, stable tables are the most basic method, using mass and rigidity to reduce vibration transfer. While effective in mild vibration environments, passive mass isolation has limitations in highly disturbed settings or for very sensitive scanning. Dedicated antivibration platforms feature specialized materials and engineering designs to deliver better isolation performance. These may include pneumatic isolators, spring-damper setups, or viscoelastic materials that absorb and dissipate vibrational energy. Active vibration control systems, though more costly, can offer exceptional protection by using sensors and actuators to counteract vibrations in real time.

Structural considerations are often as important as the isolation hardware in achieving proper vibration control. Ground-level locations experience less building-transmitted vibration than upper floors, where structural amplification can increase disturbances. Some facilities invest in structurally reinforced scanner areas with independent foundations to isolate the scanner environment from general building vibrations. The proximity of vibration sources also influences scanner placement planning. Effectiveness drops quickly as sources get closer, making spatial separation a primary, cost-effective mitigation strategy. When separation isn't possible, coordinated operation schedules can reduce conflicts between high-vibration activities and critical scanning periods.

Local area network considerations

Network infrastructure and performance optimization

The network infrastructure supporting digital pathology systems impacts system performance, user experience, and clinical workflow efficiency. Unlike traditional healthcare IT applications that mainly handle text-based data and small images, digital pathology generates large data volumes with distinct traffic patterns that challenge standard network designs. The smooth operation of WSI systems depends on carefully crafted network architectures that manage both the large file sizes and the interactive features of pathology viewing applications.

High-performance local area networks (LANs) form the foundation of an effective digital pathology infrastructure, connecting scanners, storage systems, viewing workstations, and archive components. Gigabit Ethernet (1 Gbps) is the minimum standard for digital pathology networks. However, this baseline quickly becomes insufficient for institutions with multiple concurrent scanners or high-volume operations. Ten Gigabit Ethernet connections are increasingly common for scanner uplinks and backbone infrastructure, offering the bandwidth needed for efficient large file transfers without causing network bottlenecks. These high-bandwidth needs highlight the challenge of moving multigigabyte image files efficiently through network systems. A single high-resolution WSI may take several minutes to transfer over a 1 Gbps connection, during which network capacity is mostly used by that one operation. Multiple simultaneous scanning tasks can quickly fill available bandwidth, causing queuing delays that reduce scanning efficiency and impact user experience at viewing workstations.

Traffic pattern analysis and network architecture

Digital pathology produces distinctly different traffic patterns that require specialized network architecture considerations. Scanner connections need sustained high upload bandwidth to transfer new images from acquisition systems to central storage locations. This traffic pattern differs significantly from most network applications because scanners generate consistent, high-volume data flows rather than the intermittent burst patterns of most healthcare applications.

Viewing workstations are characterized by frequent, smaller data requests as pathologists navigate slides and zoom between magnification levels. Each user interaction can trigger new image tile requests, creating variable bandwidth demands that can spike rapidly during active diagnostic sessions. The interactive nature of pathology viewing makes network latency as important as raw bandwidth, as delays in image loading directly affect diagnostic workflow efficiency.

Sustained high-volume uploads from scanners and variable interactive downloads to workstations require network architectures capable of managing both traffic types without mutual interference. Dedicated network segments or virtual LANs (VLANs) specifically assigned for digital pathology traffic offer isolation from other hospital network applications that might compete for bandwidth resources. This segmentation is especially valuable in shared hospital networks, where digital pathology must coexist with electronic health records, medical imaging, and other bandwidth-heavy clinical applications.

Quality of service and traffic prioritization

Quality of Service (QoS) mechanisms offer important network management features that ensure responsive performance during peak network usage. In digital pathology systems, QoS policies prioritize real-time viewing traffic over background tasks such as archiving, backing up, or nonurgent scanning. This prioritization ensures that pathologists who are diagnosing digital cases experience responsive system performance, even when network resources are contested by other components.

Effective QoS implementation involves careful analysis of different traffic types and their importance to clinical operations. Emergency consultation cases might be given the highest priority, followed by routine diagnostic viewing, with archival and administrative activities assigned lower priority levels. These policies must be applied consistently across all network elements to enable end-to-end traffic management that aligns with clinical priorities.

Infrastructure reliability and redundancy

Designing physical network infrastructure must focus on reliability and redundancy to avoid single points of failure. Redundant switches, multiple network paths, and fault-tolerant configurations offer the resilience needed for systems supporting patient care activities. Network component failures that might be minor inconveniences in other settings can halt diagnostic operations and delay patient care.

Comprehensive network monitoring systems are essential for providing visibility into performance metrics and proactively identifying potential issues before they affect clinical operations. These tools should monitor bandwidth utilization, latency, error rates, and the health of all network elements supporting digital pathology systems. Automated alerts ensure that technical staff are immediately notified of performance issues or component failures (Volynskaya et al., 2018).

Integration with laboratory information systems

Seamless integration of digital image files with Laboratory Information Systems (LIS) is needed for efficient clinical workflows (Alchami et al., 2023). This integration ensures that images are correctly linked to the right case, that relevant clinical data is accessible during image review, and that diagnostic findings from image analysis are incorporated into pathology reports. Integrating with the Laboratory Information System (LIS) is arguably the most important connection for digital pathology systems. It ensures proper pairing of digital slides with their case data, supports efficient workflow management, and enables comprehensive reporting that includes both diagnostic interpretations and relevant images. Several integration methods exist, ranging from simple to highly advanced. Unidirectional interfaces merely transmit case demographic data from the LIS to the digital pathology system, ensuring proper identification and organization of images. Bidirectional interfaces allow updating the LIS with information from the digital pathology system, such as confirming when slides have been successfully scanned and are ready for review. Full workflow integration incorporates digital pathology directly into the pathologist's LIS-based workflow, presenting cases for digital review at the right stage and capturing diagnostic data directly within the integrated environment. Standards-based integration methods utilize healthcare IT standards like HL7 (Health Level Seven) messages or FHIR (Fast Healthcare Interoperability Resources) APIs to exchange data between systems. These standards foster interoperability and reduce reliance on vendor-specific integration solutions. Sometimes, the absence of pathology-specific standards for certain workflow aspects requires custom integration development (Baidoshvili et al., 2018).

DICOM workflows leverage the standard's detailed patient and specimen identification to maintain links with LIS data. Barcode systems are key in this integration; slide barcodes scanned during digitization automatically link images to the correct case in the LIS. Some advanced systems also use these barcodes to retrieve relevant clinical data and display it alongside images during review. Integration also extends to reporting workflows. Digital annotations and measurements can be automatically added to pathology reports, reducing errors and enhancing efficiency. Images or links to images may be embedded directly into reports, enabling clinicians to view key findings without opening separate viewing systems (Brochhausen et al., 2020).

Wide area network and remote access solutions

Remote access features expand digital pathology beyond the main facility to support distributed workflows, external consultations, tumor boards at partner institutions, and remote work setups for pathologists. Wide Area Network (WAN) connections between facilities need to provide enough bandwidth for good viewing performance, although various optimization methods can lower bandwidth needs. Virtual Private Networks (VPNs) secure connections over public networks, enabling remote access while ensuring security and privacy. Traditional VPNs offer extensive security at the network level but may cause performance issues. Application-specific secure access solutions can improve digital pathology performance by being optimized for image data transmission. Edge caching and regional storage help enhance remote viewing by storing copies of frequently accessed images closer to users. These distributed caches automatically replicate data based on usage and policies. This reduces the need for large image transfers over limited WAN links during viewing. Progressive loading improves perceived performance by showing lower-resolution images initially while higher-resolution data loads in the background. Pathologists can then start to examine slides immediately without waiting for a full high-resolution transfer. When combined with smart prefetching of nearby regions based on navigation patterns, these techniques can deliver a responsive remote viewing experience even with limited bandwidth (Lauro et al., 2013). Some systems exchange data via HL7 messages between imaging systems and the LIS, while others use direct database connections.

Infrastructure support systems and maintenance

The reliable functioning of digital pathology systems relies on comprehensive infrastructure support that goes beyond just the main scanning and viewing parts. Power quality, physical security, and regular maintenance programs are essential support systems that directly influence scanner reliability, data accuracy, and long-term system performance. These infrastructure elements, while less obvious than the primary system components, often determine whether digital pathology implementations meet their intended clinical and operational goals.

Power quality and environmental systems

Power quality affects scanner reliability and is a common cause of equipment failure in digital pathology setups. WSI scanners contain sensitive optical and electronic parts that need clean, stable electrical power to work precisely. Power fluctuations, voltage spikes, and electrical noise can lead to focus issues, imaging artifacts, and early component failure.

Clean power delivery depends on multiple electrical quality factors beyond just voltage stability. Harmonic distortion from other lab equipment can disrupt scanner electronics, while transient voltage spikes from motor start-up or switching can harm delicate components. Many facilities provide dedicated electrical circuits for scanning systems, isolated from high-power lab devices that could cause electrical interference.

Uninterruptible power supplies (UPS) protect essential scanning systems. They offer both power conditioning and backup during outages. UPS units filter electrical noise, stabilize voltage fluctuations, and ensure continuous power during short outages that could otherwise disrupt long scanning processes. For high-volume scanning, UPS capacity should support not only the scanner but also related computing hardware, storage, and network systems needed for full operation.

Maintenance programs and service management

Regular maintenance is an important factor for achieving optimal scanner performance and durability. Systematic maintenance programs include routine cleaning of optical components, calibration procedures, and preventive replacement of wear parts such as lamps, filters, and mechanical components before they fail and disrupt operations. The precise nature of WSI systems makes preventive maintenance especially important, as component degradation can cause subtle image quality issues that may not be immediately visible but can affect diagnostic accuracy (Pantanowitz et al., 2018). A recent study from the Ohio State University Department of Pathology showed that scanner downtime was mostly due to hardware failures (Reagans et al., 2025). However, not all scanner downtime was hardware-related; other factors were related to software issues and slide contamination, where debris on the objective mounting media caused the creation of defective images. In contrast, system downtime was mostly caused by issues with image management.

Service contracts specify detailed maintenance schedules, response times for repairs, and the availability of loaner equipment during extended downtimes. These agreements should align with the operational needs of the institution. This ensures maintenance schedules fit peak scanning times and that repair response times minimize disruptions to clinical workflows. Contract terms should also address parts availability, software update procedures, and training requirements for new staff. Remote monitoring capabilities increasingly allow manufacturers to track system performance, potentially identifying developing issues before they lead to complete system failures. These systems can monitor usage patterns, component wear rates, and performance metrics that help plan preventive maintenance and long-term capacity needs. Proactive monitoring can significantly reduce unexpected downtime and improve maintenance resource allocation (Hosseini et al., 2020).

Specialized applications and use cases

Beyond routine histopathology

While brightfield histopathology is the primary use case, digital pathology increasingly supports specialized techniques requiring unique capabilities. One commonly deployed system is a fluorescence-capable scanner that acts as a digital fluorescence microscope, utilizing filter cubes or tunable light sources for specific excitation wavelengths. Applications include immunofluorescence (IF) and fluorescence in situ hybridization (FISH), where spectral features, not pigment-based contrast, are used for signal analysis (Webster, 2021). Unlike brightfield whole slide imaging (WSI), fluorescence applications often demand multichannel image stacks with specialized file formats like OME-TIFF and Bio-Formats-compatible types, alongside viewing software supporting three-dimensional data. Downstream analysis may involve signal measurement, spot detection, and spectral unmixing algorithms. Although less common in routine diagnostics, digital fluorescence scanning is gaining importance in nephropathology, dermatopathology, molecular oncology, and research environments. As whole slide imaging platforms advance, more vendors are incorporating fluorescence capabilities into hybrid or modular systems, allowing users to switch between brightfield and fluorescence modes within one device. This convergence supports integrated workflows for cases needing both routine H&E interpretation and molecular confirmation via fluorescent markers.

Other specialized digital pathology scanners have emerged to address specific needs outside routine histopathology (Salib et al., 2021). For cytology and hematology, dedicated scanners produce high-resolution images with clarity and accurate colors, as diagnosis often depends on nuclear chromatin texture, cytoplasmic granularity, and subtle staining differences. These systems support high numerical aperture objectives (e.g., 40× or 60× oil) and extended depth-of-field imaging for three-dimensional cell clusters and overlapping cells. Some vendors offer Z-stacking or extended focus algorithms tailored for cytologic specimens, enabling virtual navigation through different focal planes, as discussed earlier. These topics are also explored later in the book (see Chapters 6 and 7).

Frozen sections require quick turnaround and reliable performance, even with suboptimal slide preparation. Scanners used for intraoperative consultations focus on speed, ease of use, and seamless integration with remote viewing platforms. In these cases, image quality often takes a backseat to workflow speed, with newer systems balancing the need by starting with faster, lower-resolution scans and then performing targeted high-resolution rescans when necessary. Hybrid imaging systems combining robotic microscopes with whole slide scanning offer a flexible solution for time-sensitive and complex diagnostic applications such as frozen sections, touch imprints, and intraoperative consultations. Examples include the Leica LV1 (end of life), MicroScan, PreciPoint M8, Grundium Ocus, 3DHISTECH TrackView, and Motic EasyScan + TelePath. These systems are designed to provide interactive, rapid, and remote access to specimens, especially when a full WSI pipeline isn't feasible due to time, tissue quality, or resource constraints. In addition, compact personal scanners have become popular in decentralized or point-of-use settings. These desktop-sized, affordable devices are suited for education, quality assurance, and low-volume clinical use. They offer limited slide capacity and throughput but are compatible with standard file formats and cloud viewers, making them ideal for solo practitioners, smaller labs, or mobile applications. As digital pathology spreads into more diverse practice environments, these personal, task-specific scanners provide greater flexibility and accessibility without the infrastructure demands of enterprise systems.

Emerging technologies and future directions

Artificial intelligence integration

Modern scanners increasingly incorporate AI features for automated quality control, tissue detection, and initial analysis. These systems can identify scanning artifacts, adjust focus settings, and even provide preliminary screening results to prioritize urgent cases. Refer to other chapters in this book for detailed discussions on the many aspects of AI in digital pathology.

Cloud-based solutions

Cloud platforms provide scalable storage and processing options without significant investments in local infrastructure. These solutions are especially appealing for smaller institutions seeking advanced digital pathology features or multisite organizations that need standardized access. Cloud-based or hybrid infrastructure models offer potential benefits in scalability and cost control, but they also raise additional considerations related to data transfer, security, and vendor dependence (Webster et al., 2021).

Advanced imaging techniques

Emerging modalities include multispectral imaging for enhanced stain separation, polarized light imaging for specific tissue features, and three-dimensional imaging through techniques like photogrammetry or structured light scanning. These are discussed in other chapters.

Macroscopic (gross) specimen imaging and photomicroscopy

So far, this chapter has concentrated on whole slide imaging (WSI). However, before the advent of expensive scanners and computer-aided diagnoses, there were static images of microscopic slides and gross surgical pathology specimens. This is where digital pathology truly began.

Gross specimen photography

Macroscopic (gross) specimen imaging plays an increasingly important role in the digital transformation of pathology practice. Systematic digital documentation of gross specimens improves quality, facilitates consultation, and creates valuable resources for education and research. Modern gross photography systems range from simple handheld digital cameras to sophisticated integrated workstations specifically designed for pathology specimen documentation. Advanced systems feature standardized lighting, color calibration tools, scale markers, and specialized specimen stages to ensure consistent, high-quality images. Some systems feature multiple cameras for simultaneous capture from different angles or utilize specialized lighting options, such as polarized light or narrow-band illumination, to highlight specific features, such as visualizing crystals or vascular patterns (Hartman et al., 2017).

Software designed for managing gross photography simplifies image capture, annotation, organization, and integration with laboratory information systems. These applications allow for adding case information, measurements, and descriptive annotations directly to images. Integration with dictation systems enables pathologists to reference specific images during gross description dictation, establishing direct links between images and corresponding text descriptions in the pathology report. Automated image processing can be used to standardize appearance, fix color balance, or highlight specific features. Some systems include three-dimensional imaging capabilities through techniques like photogrammetry or structured light scanning to create 3D models of specimens that can be manipulated and offer additional spatial information beyond traditional photography (Turchini et al., 2018; Perez et al., 2023).

Artificial intelligence shows considerable promise in analyzing gross pathology images. A recent study demonstrated that deep learning-based analysis of gross features can significantly improve the sampling of ovarian epithelial tumors, especially during frozen section procedures (He et al., 2025). The AI model improved pathologists' ability to distinguish between borderline tumors and those with microinvasion or microinvasive carcinoma by identifying high-risk areas where microinvasion is most likely to occur. This capability allows pathologists to target their sampling more strategically during frozen section analysis, potentially enhancing diagnostic accuracy.

Effective integration of gross imaging into the laboratory workflow requires careful attention to process design and information flow. Optimal placement of imaging stations within the gross room minimizes disruption to specimen-handling procedures while providing convenient access for pathologists and pathology assistants. Barcode scanning or similar technologies helps link images to the correct case in the laboratory information system, reducing the risk of identification errors.

Standardized imaging protocols specify which specimens need photography, which views should be captured, and what additional documentation (such as scale markers or orientation indicators) should be included. These protocols ensure consistent documentation across personnel and shifts. Quality control procedures verify image quality, proper identification, and appropriate integration with case data before images are used for diagnostic purposes. The timing of imaging during the gross examination varies depending on institutional preferences and specimen types. Some laboratories capture images of intact specimens before dissection, while others document key dissection stages or specific findings during dissection. Many institutions use a combination approach, with protocols outlining the proper imaging sequence for different specimen types.

Digital gross imaging serves several important purposes in pathology practice. It provides objective documentation of the specimen's condition upon receipt, which can be especially valuable for medico-legal cases or when questions arise about specimen integrity. It also creates a permanent record of findings that might be altered or lost during subsequent processing, such as subtle mucosal lesions that may not be visible in microscopic sections. For complex resection specimens, gross images help connect macroscopic findings with microscopic sections by recording the location and orientation of sections taken for histologic analysis. This connection is especially important for cancer resections, where understanding the tumor's relationship to resection margins is crucial for staging and treatment planning. In consultation scenarios, gross images allow referring pathologists to share important macroscopic findings with consultants. For tumor boards and clinical conferences, gross images provide clinicians with valuable context for understanding the pathologic findings and their clinical implications.

Digital gross imaging also provides valuable resources for education and research. Teaching collections of well-documented gross specimens are essential for training pathology residents and medical students. Research uses include morphometric analysis of gross specimens and correlation studies between gross and microscopic findings.

Photomicroscopy

Modern microscope cameras range from entry-level consumer-grade digital cameras to high-performance cooled CCD or CMOS sensors optimized for microscopy. These cameras are mounted on the trinocular port of the microscope and integrated with image acquisition software. Most systems let users manually adjust exposure, white balance, gain, and other settings. This allows pathologists to fine-tune image capture for specific stains, lighting conditions, or specimen types.

Photomicroscopy is especially useful for targeting specific features of interest, such as mitotic figures, microorganisms, or immunostaining patterns. This method offers quick, on-demand imaging without the large file sizes and storage requirements of WSI. It is also commonly used to display unusual or classic findings in teaching materials, multidisciplinary tumor boards, journal submissions, or professional communication. In addition, it can be combined with artificial intelligence algorithms to clarify regions of interest, like lymphovascular invasion, perineural invasion, and tumor budding. PathChat from Modella AI (Lu et al., 2024) is one such example that can assist in primary diagnosis as well as teaching. Static images on pathology reports can include links to whole slide images, making for an enhanced, so-called smart report (Parwani et al., 2008).

Despite its usefulness, photomicroscopy has limitations. Field selection is inherently subjective, and reproducibility is restricted by manual focus and lighting adjustments. Without motorized stages or metadata, photomicrographs lack context and spatial reference to the entire tissue section. Nevertheless, many pathologists still rely on microscope photography as an efficient way to document findings, share images with colleagues, and build digital archives, especially in community or resource-limited settings. These same systems can also be used for live viewing of images, as discussed in the chapter on telepathology. As digital pathology develops, photomicroscopy remains a valuable complementary tool, especially in nonroutine situations, emerging markets, or personal learning environments. It continues to be relevant and demonstrates the adaptability of pathology practice and the importance of a high-quality microscope and a well-positioned camera.

Conclusion: Embracing digital transformation

Digital pathology is more than just a technological advancement; it allows for a fundamental rethink of pathology practices. It addresses major challenges and creates new opportunities to improve patient care (Pantanowitz et al., 2015). As this chapter demonstrates, successful adoption depends not only on understanding the technology itself but also on managing the complex interactions of organizational, regulatory, and workflow factors. Many of these topics are covered in detail elsewhere in this book.

The pathology community is at a pivotal point. Early adopters are seeing clear benefits in efficiency, collaboration, and diagnostic accuracy. The question is no longer whether digital pathology will become standard, but how fast institutions can adapt to stay competitive and effectively serve patients in a growing digital healthcare landscape. Successful implementation of digital pathology relies on understanding the complex interaction of imaging technologies, data management systems, software platforms, and infrastructure needs. Scanner technology keeps improving, with advances in speed, image quality, and automation. This makes digital pathology more practical for high-volume applications. Standardizing file formats, especially through DICOM adoption, improves interoperability and reduces vendor lock-in concerns that have historically slowed adoption.

Software systems are also evolving, with viewing platforms providing intuitive interfaces for pathologists and image management systems that integrate seamlessly with existing laboratory workflows. The quality of digital displays and viewing workstations continues to improve, giving pathologists tools that can match or exceed those of traditional microscopy.

To ensure successful adoption, implementation strategies must carefully consider organizational factors, workflow integration, and change management principles. The choice between centralized and distributed scanning, enterprise integration methods, and multisite deployment models significantly impacts both the immediate success and long-term viability of digital pathology initiatives.

Quality assurance and color management remain vital. Regulatory compliance adds complexity but also offers essential protections for patient safety and data security. Economic factors are also key, but the return on investment increasingly supports adoption as technology costs decline and benefits become clearer.

Looking ahead, digital pathology will facilitate innovative uses of artificial intelligence, advanced imaging, and cloud collaboration, further boosting its value. The integration of AI algorithms for automated screening, quantitative analysis, and decision support will improve pathologists' abilities in ways that traditional microscopy cannot achieve.

Ongoing standardization efforts, especially the adoption of DICOM for pathology, will encourage wider integration with healthcare IT systems and support new research, education, and clinical applications. Cloud-based solutions provide opportunities for smaller institutions to access advanced digital pathology tools without requiring significant local infrastructure investments.

The keys to successful implementation are shown in Table 2.9.

The foundation established by current digital pathology systems, including standardized imaging, interoperable formats, and validated workflows, will underpin the next wave of innovations in artificial intelligence, advanced analytics, and precision medicine.

TABLE 2.9 Strategies for successful implementation of a WSI system.

Parameter	Detail
Strategic planning	Clear goals, stakeholder alignment, and realistic timelines
Technical excellence	Appropriate technology selection and robust infrastructure
Change management	Comprehensive training and user support
Quality focus	Rigorous validation and ongoing monitoring
Economic sustainability	Sound business cases and financing strategies
Leadership commitment	Executive sponsorship and change management support
Technical infrastructure	Network capacity, storage capabilities, and IT support
Staff preparedness	Training requirements and change adaptation capacity
Workflow compatibility	Integration with existing laboratory processes
Financial planning	Total cost of ownership and return on investment modeling

Pathologists who master these basic technologies today will be best positioned to utilize tomorrow's computational breakthroughs for better patient outcomes. This transformation requires dedication, investment, and patience as organizations transition from traditional slide-based diagnosis to digital workflows. This transition may take years, and the shift is fraught with innumerable challenges, not the least of which is cost (Reagans et al., 2025). Yet, the growing evidence shows that institutions investing thoughtfully and systematically are achieving notable improvements in efficiency, quality, and capabilities that justify the effort and expense involved.

As digital pathology becomes more mainstream, the focus will move from basic implementation to optimization, advanced applications, and integration with new technologies. The path toward a fully digital pathology practice has already started, and success will come to those who embrace the change while maintaining the diagnostic excellence that defines our profession.

References

Abels, E., Pantanowitz, L., Aeffner, F., Zarella, M.D., van der Laak, J., Bui, M.M., Vemuri, V.N.P., Parwani, A.V., Gibbs, J., Agosto-Arroyo, E., Beck, A.H., Kozlowski, C., 2019. Computational pathology definitions, best practices, and recommendations for regulatory guidance: a white paper from the digital pathology association. J. Pathol. 0022-3417. 249 (3), 286–294. doi:10.1002/path.5331.

Aeffner, F., Zarella, M.D., Buchbinder, N., Bui, M.M., Goodman, M.R., Hartman, D.J., Lujan, G.M., Molani, M.A., Parwani, A.V., Lillard, K., Turner, O.C., Vemuri, V.N.P., Yuil-Valdes, A.G., Bowman, D., 2019. Introduction to digital image analysis in whole-slide imaging: a white paper from the digital pathology association. J. Pathol. Inform. 2153-3539. 10 (1), 9. doi:10.4103/jpi.jpi_82_18.

Alchami, F.S., Iqbal, Z., Björkhammer, C.N., Saeed, M.O., Ramakrishnan, R., Clelland, C., Ahmad, F., Charles, A., 2023. Whole slide imaging integration with lab information systems, a study of the requirements, processes and procedures enabling a reporting-based workflow. Pathol. Lab. Med. Int. 1179-2698. 15, 1–11. doi:10.2147/plmi.s388981.

American College of Radiology, AAPM, SIIM. n.d. ACR–AAPM–SIIM technical standard for electronic practice of medical imaging. Revised 2022. Available at: https://gravitas.acr.org/PPTS/GetDocumentView?docId=136. (Accessed 6 September 2025).

Anon. 2023. DICOM Standards Committee, 2023. Supplement 232: Non-Patient Instances Storage SOP Classes. Digital Imaging and Communications in Medicine (DICOM) Standard. National Electrical Manufacturers Association (NEMA). Rosslyn, VA, USA. Available from: https://www.dicomstandard.org/news-dir/progress/docs/sups/sup232.pdf.

Ardon, O., Labasin, M., Friedlander, M., Manzo, A., Corsale, L., Ntiamoah, P., Wright, J., Elenitoba-Johnson, K., Reuter, V.E., Hameed, M.R., Hanna, M.G., 2023. Quality management system in clinical digital pathology operations at a tertiary cancer center. Lab. Investig. 0023-6837. 103 (11), 100246. doi:10.1016/j.labinv.2023.100246.

Baidoshvili, A., Bucur, A., van Leeuwen, J., van der Laak, J., Kluin, P., van Diest, P.J., 2018. Evaluating the benefits of digital pathology implementation: time savings in laboratory logistics. Histopathology. 1365-2559. 73 (5), 784–794. Available from: http://onlinelibrary.wiley.com/journal/10.1111/(ISSN)1365-2559. doi:10.1111/his.13691.

Bauer, T.W., et al., 2024. A multicenter study to evaluate the analytical precision by pathologists using the Aperio GT 450 DX. J. Pathol. Inform. 15, p. 100401. doi:10.1016/j.jpi.2024.100401.

Bevins, N.B., Flynn, M.J., Silosky, M.S., Marsh, R.M., Walz-Flannigan, A.I., Badano, A., 2019. Display Quality Assurance (AAPM Task Group Report No. 270). American Association of Physicists in Medicine. doi:10.37206/183.

Bigpicture consortium. 2022 Report on unified open digital slide and annotation format specification (Deliverable D4.03). Task 4.5: technical implementation of open digital slide, annotation, and metadata format. In BIGPICTURE: central Repository for Digital Pathology.

Borowsky, A.D., et al., 2020. Digital Whole Slide Imaging Compared With Light Microscopy for Primary Diagnosis in Surgical Pathology. Archives of Pathology & Laboratory Medicine, 144 (10), 1245-1253. doi:10.5858/arpa.2019-0569-OA.

Borowsky, A.D., Glassy, E.F., Wallace, W.D., Kallichanda, N.S., Behling, C.A., Miller, D.V., Oswal, H.N., Feddersen, R.M., Bakhtar, O.R., Mendoza, A.E., Molden, D.P., Saffer, H.L., Wixom, C.R., Albro, J.E., Cessna, M. H., Hall, B.J., Lloyd, I.E., Bishop, J.W., Darrow, M.A., Gui, D., Jen, K.Y., Walby, J.A.S., Bauer, S.M., Cortez, D.A., Gandhi, P., Rodgers, M.M., Rodriguez, R.A., Martin, D.R., McConnell, T.G., Reynolds, S.J., Spigel, J.H., Stepenaskie, S.A., Viktorova, E., Magari, R., Wharton, K.A., Qiu, J., Bauer, T.W., 2020. Digital whole slide imaging compared with light microscopy for primary diagnosis in surgical pathology: a multicenter, double-blinded, randomized study of 2045 cases. Arch. Pathol. Lab. Med. 1543-2165. 144 (10), 1245–1253. Available from: https://meridian.allenpress.com/aplm/article/144/10/1245/427502/Digital-Whole-Slide-Imaging-Compared-With-Light. doi:10.5858/arpa.2019-0569-OA.

Brochhausen, M., Frankewitsch, T., Spyns, P., 2020. The digital pathology integrated platform and the potential benefits of semantic interoperability. Stud. Health Technol. Inform. 270, 1383–1384.

Bruce, C., Prassas, I., Mokhtar, M., Clarke, B., Youssef, E., Wang, C., Yousef, G.M., 2024. Transforming diagnostics: the implementation of digital pathology in clinical laboratories. Histopathology. 0309-0167. 85 (2), 207–214. doi:10.1111/his.15178.

Chen, Y., Janowczyk, A., Madabhushi, A., 2020. Quantitative assessment of the effects of compression on deep learning in digital pathology image analysis. JCO Clin. Cancer Inform. 2473-4276. 4 (4), 221–233. doi:10.1200/cci.19.00068.

Clarke, E.L., Munnings, C., Williams, B., Brettle, D., Treanor, D., 2020. Display evaluation for primary diagnosis using digital pathology. J. Med. Imaging. 2329-4310. 7 (2). Available from: http://medicalimaging.spiedigitallibrary.org/journal.aspx. doi:10.1117/1.JMI.7.2.027501.

Clunie, D.A., 2004. DICOM structured reporting: part 1. Overview and characteristics. Radiographics 24 (3), 859–868.

Clunie, D.A., Dennison, D.K., Cram, D., Persons, K.R., Bronkalla, M.D., Primo, H.R., 2016. Technical challenges of enterprise imaging: HIMSS-SIIM collaborative white paper. J. Digit. Imaging. 1618-727X. 29 (5), 583–614. doi:10.1007/s10278-016-9899-4.

Dash, R.C., Jones, N., Merrick, R., Haroske, G., Harrison, J., Sayers, C., Haarselhorst, N., Wintell, M., Herrmann, M. D., Macary, F., 2021. Integrating the health-care enterprise pathology and laboratory medicine guideline for digital pathology interoperability. J. Pathol. Inform. 2153-3539. 12 (1), 16. doi:10.4103/jpi.jpi_98_20.

Decomplix. n.d. CE marking for medical devices: the ultimate guide. https://decomplix.com/medical-device-ce-marking-guide/. (Accessed 16 August 2025).

Digital Imaging and Communications in Medicine (DICOM). 2010 Supplement 145: whole Slide Microscopic Image IOD and SOP Classes. DICOM Standards Committee.

Digital Pathology Association. n.d. Accelerating the Future of Digital Pathology: the ImagiNg Data EXchange (INDEX). Webinar Summary. Available at: https://digitalpathologyassociation.org/accelerating-the-future-of-digital-pathology-the-imaging-data-exchange-index. (Accessed 4 September 2025).

Donné, A.F., 1844. Cours de microscopie complémentaire des études médicales: Anatomie microscopique et physiologie des fluides de l'économie, J.-B. Baillière, Paris.

Donnelly, A.D., Mukherjee, M.S., Lyden, E.R., et al., 2021. Optimal z-axis scanning parameters for gynecologic cytology specimens. J. Pathol. Inform. 12, 5.

Doyle, S. 2010 Evaluation of effects of JPEG2000 compression on a computer-aided detection system for prostate cancer on digitized histopathology. In: 2010 IEEE International Symposium on Biomedical Imaging: From Nano to Macro.

Evans, A.J., Bauer, T.W., Bui, M.M., Cornish, T.C., Duncan, H., Glassy, E.F., Hipp, J., McGee, R.S., Murphy, D., Myers, C., O'Neill, D.G., Parwani, A.V., Rampy, B.A., Salama, M.E., Pantanowitz, L., 2018. US food and drug administration approval of whole slide imaging for primary diagnosis: a key milestone is reached and new questions are raised. Arch. Pathol. Lab. Med. 1543-2165. 142 (11), 1383–1387. Available from: http://www.archivesofpathology.org/doi/pdf/10.5858/arpa.2017-0496-CP. doi:10.5858/arpa.2017-0496-CP.

Evans, A.J., Brown, R.W., Bui, M.M., et al, 2021a. Validating whole slide imaging systems for diagnostic purposes in pathology: guideline from the College of American Pathologists Pathology and Laboratory Quality Center. Arch. Pathol. Lab. Med. 145 (3), 286–301.

Evans, A.J., Krupinski, E.A., Weinstein, R.S., Pantanowitz, L., 2021b. 2020 vision of digital pathology in action. J. Pathol. Inform. 12, 53.

Farahani, N., Pantanowitz, L., 2015. Overview of telepathology. Surg. Pathol. Clin. 1875-9181. 8 (2), 223–231. doi:10.1016/j.path.2015.02.018.

Ferreira, R., Moon, B., Humphries, J., Sussman, A., Saltz, J., Miller, R., Demarzo, A., 1997. The virtual microscope. J. Am. Med. Inform. Assoc. 1067-5027. 4, 449–453. Available from: http://jamia.oxfordjournals.org/content/22/e1.

Fischer, M., Neher, P., Schüffler, P., Ziegler, S., Xiao, S., Peretzke, R., Clunie, D., Ulrich, C., Baumgartner, M., Muckenhuber, A., Almeida, S.D., Gőtz, M., Kleesiek, J., Nolden, M., Braren, R., Maier-Hein, K., 2022. Deep learning on lossily compressed pathology images: adverse effects for imagenet pre-trained models. In: Medical Optical Imaging and Virtual Microscopy Image Analysis, 1, 73-83. doi:10.1007/978-3-031-16961-8_8.

Fischer, M., Neher, P., Schüffler, P., Ziegler, S., Xiao, S., Peretzke, R., Clunie, D., Ulrich, C., Baumgartner, M., Muckenhuber, A., Almeida, S.D., Gőtz, M., Kleesiek, J., Nolden, M., Braren, R., Maier-Hein, K., 2025. Unlocking the potential of digital pathology: novel baselines for compression. J. Pathol. Inform. 2153-3539. 17, 100421. doi:10.1016/j.jpi.2025.100421.

Food and Drug Administration 2016 Performance Assessment of Digital Pathology Whole Slide Imaging (WSI) Devices: guidance for Industry and Food and Drug Administration Staff. Available at the FDA media library.

Fraggetta, F., L'Imperio, V., Ameisen, D., Carvalho, R., Leh, S., Kiehl, T.-R., Serbanescu, M., Racoceanu, D., Mea, V. D., Polonia, A., Zerbe, N., Eloy, C., 2021. Best practice recommendations for the implementation of a digital pathology workflow in the anatomic pathology laboratory by the European Society of Digital and Integrative Pathology (ESDIP). Diagnostics. 2075-4418. 11 (11), 2167. doi:10.3390/diagnostics11112167.

Fraggetta, F., Yagi, Y., Garcia-Rojo, M., Evans, A.J., Tuthill, J.M., Baidoshvili, A., Hartman, D.J., Fukuoka, J., Pantanowitz, L., 2018. The importance of eSlide macro images for primary diagnosis with whole slide imaging. J. Pathol. Inform. 2153-3539. 9 (1), 46. doi:10.4103/jpi.jpi_70_18.

García-Rojo, M., De Mena, D., Muriel-Cueto, P., Atienza-Cuevas, L., Domínguez-Gómez, M., Bueno, G., 2019. New European Union regulations related to whole slide image scanners and image analysis software. J. Pathol. Inform. 2153-3539. 10 (1), 2. doi:10.4103/jpi.jpi_33_18.

Genis, M.Y., Remez, A.I., Untesco, M.I., Zhakota, D.A., 2021. Testing of actual scanner performance in a high-loaded UNIM laboratory environment. J. Pathol. Inform. 2153-3539. 12 (1), 39. doi:10.4103/jpi.jpi_4_21.

Gorman, B.L., Pantanowitz, L., Beckwith, B.A., 2021. Interoperability between anatomic pathology laboratory information systems and digital pathology systems: a systematic review. Arch. Pathol. Lab. Med. 145 (7), 840–850.

Hanna, M.G., Pantanowitz, L., 2022. Whole slide imaging in pathology: recommendations for practical adoption. Mod. Pathol. 35 (5), 585–595.

Hanna, M.G., Reuter, V.E., Ardon, O., Kim, D., Sirintrapun, S.J., Schüffler, P.J., Busam, K.J., Sauter, J.L., Brogi, E., Tan, L.K., Xu, B., Bale, T., Agaram, N.P., Tang, L.H., Ellenson, L.H., Philip, J., Corsale, L., Stamelos, E., Friedlander, M.A., Ntiamoah, P., Labasin, M., England, C., Klimstra, D.S., Hameed, M., 2020. Validation of a digital pathology system including remote review during the COVID-19 pandemic. Mod. Pathol. 1530-0285. 33 (11), 2115–2127. Available from: http://www.nature.com/modpathol/index.html. doi:10.1038/s41379-020-0601-5.

Hartman, D.J., Pantanowitz, L., McHugh, J.S., Piccoli, A.L., OLeary, M.J., Lauro, G.R., 2017. Enterprise implementation of digital pathology: feasibility, challenges, and opportunities. J. Digit. Imaging. 0897-1889. 30 (5), 555–560. doi:10.1007/s10278-017-9946-9.

He, D., Jin, L., Geng, H., Cao, L., 2025. Deep learning-based analysis of gross features for ovarian epithelial tumors classification: a tool to assist pathologists for frozen section sampling. Hum. Pathol. 0046-8177. 157, 105762. doi:10.1016/j.humpath.2025.105762.

Helin, H., Tolonen, T., Ylinen, O., Tolonen, P., Napankangas, J., Isola, J., 2018. Optimized JPEG 2000 compression for efficient storage of histopathological whole-slide images. J. Pathol. Inform. 2153-3539. 9 (1), 20. doi:10.4103/jpi.jpi_69_17.

Herrmann, M.D., Clunie, D.A., Fedorov, A., Doyle, S.W., Pieper, S., Klepeis, V., Le, L.P., Mutter, G.L., Milstone, D.S., Schultz, T.J., Kikinis, R., Kotecha, G.K., Hwang, D.H., Andriole, K.P., lafrate, A.J., Brink, J.A., Boland, G.W., Dreyer, K.J., Michalski, M., Golden, J.A., Louis, D.N., Lennerz, J.K., 2018. Implementing the DICOM standard for digital pathology. J. Pathol. Inform. 2153-3539. 9 (1), 37. doi:10.4103/jpi.jpi_42_18.

Hosseini, M.S., Brawley-Hayes, J.A.Z., Zhang, Y., Chan, L., Plataniotis, K., Damaskinos, S., 2020. Focus quality assessment of high-throughput whole Slide imaging in digital pathology. IEEE Trans. Med. Imaging. 1558-254X. 39 (1), 62–74. Available from: http://ieeexplore.ieee.org/xpl/RecentIssue.jsp?punumber=42. doi:10.1109/TMI.2019.2919722.

Hsu, Y.-H.R., Ahmed, I., Phlamon, J., Carment-Baker, C., Chan, J.Y.T., Prassas, I., Weiser, K., Zeidan, S., Clarke, B., Yousef, G.M., 2025. An adapted & improved validation protocol for digital pathology implementation. Semin. Diagn. Pathol. 0740-2570. 42(4), 150905. doi:10.1016/j.semdp.2025.150905.

International Electrotechnical Commission. (2024). IEC 62563-3: medical electrical equipment—Medical image display systems—Part 3: evaluation methods for colour displays (Draft). Retrieved from IEC web store or associated national standards body (Accessed 6 September 2025).

Iwuajoku, V., et al., 2025. An equivalency and efficiency study for one year digital pathology for clinical routine diagnostics in an accredited tertiary academic center. Virchows Arch, 487 (1), 3–12. doi:10.1007/s00428-025-04043-3.

Jahn, S.W., Plass, M., Moinfar, F., 2020. Digital pathology: advantages, limitations and emerging perspectives. J. Clin. Med. 2077-0383. 9 (11), 3697. doi:10.3390/jcm9113697.

Johnson, J.P., Krupinski, E.A., Yan, M., Roehrig, H., Graham, A.R., Weinstein, R.S., 2011. Using a visual discrimination model for the detection of compression artifacts in virtual pathology images. IEEE Trans. Med. Imaging. 0278-0062. 30 (2), 306–314. doi:10.1109/tmi.2010.2077308.

Kalinski, T., Zwönitzer, R., Grabellus, F., Sheu, S.-Y., Sel, S., Hofmann, H., Roessner, A., 2011. Lossless compression of JPEG2000 whole slide images is not required for diagnostic virtual microscopy. Am. J. Clin. Pathol. 1943-7722. 136 (6), 889–895. doi:10.1309/ajcpyi1z3tggaiep.

Kohlberger, T., Liu, Y., Moran, M., Chen, P.-.H.C., Brown, T., Hipp, J.D., Mermel, C.H., Stumpe, M.C., 2019. Whole-slide image focus quality: automatic assessment and impact on AI cancer detection. J. Pathol. Inform. 10, 39.

Krupinski, E.A., 2014. Human factors and human-computer considerations in imaging. Healthcare (Basel) 2 (1), 94–108.

Krupinski, E.A., Cheng, W.C., Nagy, P.G., et al., 2019. Impact of medical display calibration on observer performance. J. Digit. Imaging 32 (6), 972–979.

Krupinski, E.A., Johnson, J.P., Jaw, S., Graham, A.R., Weinstein, R.S., 2012. Compressing pathology whole-slide images using a human and model observer evaluation. J. Pathol. Inform. 2153-3539. 3 (1), 17. doi:10.4103/2153-3539.95129.

Lauro, G.R., Cable, W., Lesniak, A., Tseytlin, E., McHugh, J., Parwani, A., Pantanowitz, L., 2013. Digital pathology consultations—a new era in digital imaging, challenges and practical applications. J. Digit. Imaging. 0897-1889. 26 (4), 668–677. doi:10.1007/s10278-013-9572-0.

Linkert, M., Rueden, C.T., Allan, C., Burel, J.M., Moore, W., Patterson, A., Loranger, B., Moore, J., Neves, C., MacDonald, D., Tarkowska, A., Sticco, C., Hill, E., Rossner, M., Eliceiri, K.W., Swedlow, J.R., 2010. Metadata matters: access to image data in the real world. J. Cell Biol. 0021-9525. 189 (5), 777–782. Available from: http://jcb.rupress.org/content/189/5/777.full.pdf+html. doi:10.1083/jcb.201004104.

Lu, M.Y., Chen, B., Williamson, D.F.K., Chen, R.J., Zhao, M., Chow, A.K., Ikemura, K., Kim, A., Pouli, D., Patel, A., Soliman, A., Chen, C., Ding, T., Wang, J.J., Gerber, G., Liang, I., Le, L.P., Parwani, A.V., Weishaupt, L.L., Mahmood, F., 2024. A multimodal generative AI copilot for human pathology. Nature. 1476-4687. 634 (8033), 466–473. Available from: https://www.nature.com/nature/. doi:10.1038/s41586-024-07618-3.

Lujan, G., Ciarlo, M., Engelmann, B.E., 2022. Challenges in implementing a digital pathology workflow. J. Pathol. Inform. 13, 100089. doi:10.1016/j.jpi.2022.100089.

Maciejewski, M.W., Qui, H.Z., Rujan, I., Mobli, M., Hoch, J.C., 2009. Nonuniform sampling and spectral aliasing. J. Magn. Reson. 1096-0856. 199 (1), 88–93. doi:10.1016/j.jmr.2009.04.006.

McGeary, D., 2009. PACS - An overview. Biomed. Instrum. Technol. 0899-8205. 43 (2), 127–130. Available from: http://www.aami-bit.org/archive/0899-8205/43/2/pdf/i0899-8205-43-2-127.pdf. doi:10.2345/0899-8205-43.2.127.

Montalto, M., Casella, V., Verrill, C., et al, 2021. Digital pathology: a review of the technology and its integration into the workflow of a large academic center. J. Pathol. Inform. 12, 4.

Murphy, R.L., Bird, K.T., 1974. Telediagnosis: a new community health resource. Observations on the feasibility of telediagnosis based on 1000 patient transactions. Am. J. Public Health. 0090-0036. 64 (2), 113–119. doi:10.2105/AJPH.64.2.113.

Mutter, G.L., Milstone, D.S., Hwang, D.H., Siegmund, S., Bruce, A., 2022. Measuring digital pathology throughput and tissue dropouts. J. Pathol. Inform. 2153-3539. 13 (1), 8. Available from: www.jpathinformatics.org/browse.asp?sabs=n. doi:10.4103/jpi.jpi_5_21.

National Electrical Manufacturers Association (NEMA). n.d. History of DICOM®. DICOM Standard Available at: https://www.dicomstandard.org/history. (Accessed 4 September 2025).

Nurzynska, K., Li, D., Walts, A.E., Gertych, A., 2023. Multilayer outperforms single-layer slide scanning in AI-based classification of whole slide images with low-burden acid-fast mycobacteria (AFB). Comput. Methods Programs Biomed. 1872-7565. 234. Available from: www.elsevier.com/locate/cmpb. doi:10.1016/j.cmpb.2023.107518.

Oosterwijk, H., Devlin, S., Eid, M., Michaelides, V., Crawford, T.J., Gersten, J., Primo, H.R., Wirth, A.Editors, n.d. PACS Fundamentals, Fourth ed. Society for Imaging Informatics in Medicine.

Ópez C, L., 2008. Effects of image compression on automatic count of immunohistochemically stained nuclei in digital images. J. Am. Med. Inform. Assoc. 15, 794–798.

Pantanowitz, L., et al., 2016. Validating whole slide imaging for diagnostic purposes in pathology: guideline from the College of American Pathologists Pathology and Laboratory Quality Center. Arch. Pathol. Lab. Med. 140 (12), 1381–1392.

Pantanowitz, L., Farahani, N., Parwani, A., 2015. Whole slide imaging in pathology: advantages, limitations, and emerging perspectives. Pathol. Lab. Med. Int. 1179-2698. 23. doi:10.2147/PLMI.S59826.

Pantanowitz, L., Liu, C., Huang, Y., Guo, H., Rohde, G., 2017. Impact of altering various image parameters on human epidermal growth factor receptor 2 image analysis data quality. J. Pathol. Inform. 2153-3539. 8 (1). Available from: www.jpathinformatics.org/browse.asp?sabs=n. doi:10.4103/jpi.jpi_46_17.

Pantanowitz, L., Sharma, A., Carter, A.B., Kurc, T., Sussman, A., Saltz, J., 2018. Twenty years of digital pathology: an overview of the road travelled, what is on the horizon, and the emergence of vendor-neutral archives. J. Pathol. Inform. 2153-3539. 9 (1), 40. doi:10.4103/jpi.jpi_69_18.

Papazoglou, A.S., Karagiannidis, E., Liatsos, A., Bompoti, A., Moysidis, D.V., Arvanitidis, C., Tsolaki, F., Tsagkaropoulos, S., Theocharis, S., Tagarakis, G., Michaelson, J.S., Herrmann, M.D., 2023. Volumetric tissue imaging of surgical tissue specimens using micro-computed tomography: an emerging digital pathology modality for nondestructive, slide-free microscopy—clinical applications of digital pathology in 3 dimensions. Am. J. Clin. Pathol. 1943-7722. 159 (3), 242–254. Available from: http://ajcp.oxfordjournals.org/content/by/year. doi:10.1093/ajcp/aqac143.

Parwani, A.V., Mohanty, S.K., Becich, M.J., 2008. Pathology reporting in the 21st century: the impact of synoptic reports and digital imaging. Lab. Med. 0007-5027. 39 (10), 582–586. Available from: http://labmed.ascpjournals.org/content/39/10/582.full.pdf+html. doi:10.1309/LMAPF116QTKIDEBG.

Patel, A.U., Shaker, N., Erck, S., Kellough, D.A., Palermini, E., Li, Z., Lujan, G., Satturwar, S., Parwani, A.V., 2022. Types and frequency of whole slide imaging scan failures in a clinical high throughput digital pathology scanning laboratory. J. Pathol. Inform. 2153-3539. 13, 100112. doi:10.1016/j.jpi.2022.100112.

Perez, A.N., Sharif, K.F., Guelfi, E., Li, S., Miller, A., Prasad, K., Sinard, R.J., Lewis, J.S., Topf, M.C., 2023. Ex vivo 3D scanning and specimen mapping in anatomic pathology. J. Pathol. Inform. 2153-3539. 14. Available from: https://www.sciencedirect.com/journal/journal-of-pathology-informatics/about/aims-and-scope. doi:10.1016/j.jpi.2022.100186.

Philips. n.d. IntelliSite Pathology Solution Ultra Fast Scanner—Instructions for Use. Document 4522 207 41211; 2020-09-08. (Accessed 17 August 2025).

Platiša, L., 2017. Influence of study design on digital pathology image quality evaluation. J Med Imag (Bellingham, Wash) 4, 021108.

Reagans, R., Cevik, L., Kumar, H., Kellough, D., Limbach, A.L., Lujan, G., Parwani, A., Gokozan, H.N., 2025. Analysis of system and scanner downtime in a digital pathology–predominant institution: a 6-year experience. Am. J. Clin. Pathol. 0002-9173. 164 (4), 634–638. doi:10.1093/ajcp/aqaf094.

Roth, C.J., Lannum, L.M., Persons, K.R., 2016. A foundation for enterprise imaging: HIMSS-SIIM collaborative white paper. J. Digit. Imaging. 1618-727X. 29 (5), 530–538. doi:10.1007/s10278-016-9882-0.

Roth, C.J., Petersilge, C.A., Cram, D., Garriott, K., Lannum, L., Carey, C.K., Medina, N., Kwiatkoski, T., Whitfill, J.T., Towbin, A.J., 2024a. Celebrating 10 years of the HIMSS-SIIM enterprise imaging community and enterprise imaging informatics. J. Imaging Inform. Med. 2948-2933. 37 (6), 2722–2728. Available from: https://www.springer.com/journal/10278. doi:10.1007/s10278-024-01141-7.

Roth, C.J., Petersilge, C., Clunie, D., Towbin, A.J., Cram, D., Primo, R., Li, X., Berkowitz, S.J., Barnosky, V., Krupinski, E.A., 2024b. HIMSS-SIIM enterprise imaging community white papers: reflections and future directions. J. Imaging Inform. Med. 2948-2933. 37 (2), 429–443. Available from: https://www.springer.com/journal/10278. doi:10.1007/s10278-024-00992-4.

Salib, C., Bhardwaj, S., Hussein, S., Jamal, S.E., Petersen, B., Scigliano, E., Dembitzer, F.R., Teruya-Feldstein, J., 2021. Digital AI in hematology—Integration of the Scopio labs x100 scanner with newly implemented AI capabilities into routine clinical workflow. Blood. 0006-4971. 138 (Supplement 1), 4932. doi:10.1182/blood-2021-148821.

Salto-Tellez, M., Maxwell, P., Hamilton, P., 2019. Artificial intelligence—the third revolution in pathology. Histopathology. 1365-2559. 74 (3), 372–376. Available from: http://onlinelibrary.wiley.com/journal/10.1111/(ISSN)1365-2559. doi:10.1111/his.13760.

SCAR Technology Forum. 1994. Understanding Teleradiology Available at: https://ia801303.us.archive.org/23/items/UnderstandingTeleradiologySCAR1994/UnderstandingTeleradiology_SCAR_1994_ocr.pdf. (Accessed 4 September 2025).

Sellaro, T.L., Filkins, R., Hoffman, C., Fine, J.L., Ho, J., Parwani, A.V., Pantanowitz, L., Montalto, M., 2013. Relationship between magnification and resolution in digital pathology systems. J. Pathol. Inform. 2153-3539. 4 (1), 21. doi:10.4103/2153-3539.116866.

Senaras, C., Niazi, M.K.K., Lozanski, G., Gurcan, M.N., 2018. DeepFocus: detection of out-of-focus regions in whole slide digital images using deep learning. PLoS ONE 13, e0205387.

Silosky, M., Milman, R., Bevins, N.B., 2022. Features to consider when selecting displays for remote reading. J. Am. Coll. Radiol. 1558-349X. 19 (1), 73–75. Available from: http://www.elsevier.com/wps/find/journaldescription.cws_home/699814/description#description. doi:10.1016/j.jacr.2021.09.038.

Soliman, J., Weiser, K., Ahmed, I., Carment-Baker, C., Hockley, M., Prassas, I., Zeidan, S., Bruce, C., Clarke, B., Yousef, G.M., 2025. Selecting high-throughput scanners for clinical use: a multicenter institution experience. Am. J. Clin. Pathol. 0002-9173. 164 (4), 589–599. doi:10.1093/ajcp/aqaf083.

Stathonikos, N., Nguyen, T.Q., Spoto, C.P., Verdaasdonk, M.A.M., van Diest, P.J., 2019. Being fully digital: perspective of a Dutch academic pathology laboratory. Histopathology. 1365-2559. 75 (5), 621–635. Available from: http://onlinelibrary.wiley.com/journal/10.1111/(ISSN)1365-2559. doi:10.1111/his.13953.

Szabadka, Z., Bruse, M., Alakuijala, J. 2024. Introducing Jpegli: a new JPEG coding library. Google Open Source Blog. Available at: https://opensource.googleblog.com/2024/04/introducing-jpegli-new-jpeg-coding-library.html. (Accessed 4 September 2025).

Turchini, J., et al., 2018. Three‚ÄëDimensional pathology specimen modeling using photogrammetry. Am. J. Clin. Pathol. 142 (11), 1415–1423.

U.S. Food & Drug Administration. n.d. Product Classification: digital Pathology Image Viewing and Management Software (Product Code QKQ). Regulation Number 21 CFR 864.3700. Class II device. FDA. Accessed September 4, 2025.

U.S. Food and Drug Administration. (n.d.). Product classification: digital Pathology Image Viewing and Management Software (Product Code QKQ). Retrieved September 4, 2025, from: https://www.accessdata.fda.gov/scripts/cdrh/cfdocs/cfpcd/classification.cfm?id=5447.

Volynskaya, Z., Chow, H., Evans, A., Wolff, A., Lagmay-Traya, C., Asa, S.L., 2018. Integrated pathology informatics enables high-quality personalized and precision medicine digital pathology and beyond. Arch. Pathol. Lab. Med. 1543-2165. 142 (3), 369–382. Available from: http://www.archivesofpathology.org/doi/pdf/10.5858/arpa.2017-0139-OA. doi:10.5858/arpa.2017-0139-OA.

Webster, L.R., Degregorio, S.J., Conway, C., et al, 2021. Validation of whole-slide imaging for primary diagnosis of routine fluorescence in situ hybridization (FISH) for breast cancer. J. Pathol. Inform. 12, 38.

Weinstein, R., Holcomb, M., Krupinski, E., 2019. Invention and early history of telepathology (1985-2000). J. Pathol. Inform. 2153-3539. 10 (1). Available from: www.jpathinformatics.org/browse.asp?sabs=n. doi:10.4103/jpi.jpi_71_18.

Weinstein, R.S., 1986. Prospects for telepathology. Hum. Pathol. 0046-8177. 17 (5), 433–434. doi:10.1016/S0046-8177(86)80028-4.

Weinstein, R.S., Bloom, K.J., Rozek, L.S., 1987. Telepathology and the networking of pathology diagnostic services. Arch. Pathol. Lab. Med. 111 (7), 646–652.

Wharton, K.A., et al., 2025. Roche Digital Pathology Dx whole slide imaging system is comparable to traditional microscopy for primary diagnosis in surgical pathology. Am. J. Clin. Pathol. 164 (3), 367–384. doi:10.1093/ajcp/aqaf052.

Williams, B.J., Bottoms, D., Clark, D., Treanor, D., 2019. Future-proofing pathology part 2: building a business case for digital pathology. J. Clin. Pathol. 1472-4146. 72 (3), 198–205. Available from: http://jcp.bmj.com/. doi:10.1136/jclinpath-2017-204926.

Zanjani, F.G., 2019. Impact of jpeg 2000 compression on deep convolutional neural networks for metastatic cancer detection in histopathological images. J. Med. Imaging, 6, 027501.

Zarella, M.D., Bowman, D., Aeffner, F., Farahani, N., Xthona, A., Absar, S.F., Parwani, A., Bui, M., Hartman, D.J., 2019. A practical guide to whole slide imaging a white paper from the digital pathology association. Arch. Pathol. Lab. Med. 1543-2165. 143 (2), 222–234. Available from: http://www.archivesofpathology.org/doi/pdf/10.5858/arpa.2018-0343-RA. doi:10.5858/arpa.2018-0343-RA.

CHAPTER

3

Implementation of a digital pathology workflow

Zaibo Li and Anil V. Parwani

Department of Pathology, The Ohio State University, Columbus, OH, United States

Introduction

Over the past decade, pathology has entered a period of unprecedented technological transformation. The digitalization of histopathology, once considered an aspirational concept, has become a practical reality in many pathology laboratories worldwide (Stathonikos et al., 2013; Buck et al., 2014; Thorstenson et al., 2014; Stathonikos et al., 2019; Ardon et al., 2021; Babawale et al., 2021; Eloy et al., 2021; Temprana-Salvador et al., 2022; Ferreira et al., 2023; Clarke et al., 2025; Daniel et al., 2025; Iwuajoku et al., 2025; Montezuma et al., 2022; G.M. Lujan et al., 2021b; Retamero et al., 2020). Advances in whole-slide imaging (WSI) technology, coupled with robust image management platforms, have made it possible to digitize and review surgical pathology slides with a level of fidelity and convenience that rivals conventional microscopy. The global shift toward digital pathology has been further accelerated by the COVID-19 pandemic, which underscored the importance of remote diagnostic access, workflow flexibility, and secure data sharing in maintaining continuous pathology operations (Ardon et al., 2021; G.M. Lujan et al., 2021b).

At its core, digital pathology represents a paradigm shift in how pathologists interact with tissue specimens, data, and each other. Rather than viewing slides through optical microscopes, pathologists can now examine digital images rendered from high-resolution scans, often enriched by integrated computational tools for measurement, annotation, and artificial intelligence (AI)-assisted interpretation. Yet, beyond technological sophistication lies a complex process of organizational and workflow transformation. Implementing a digital pathology workflow is not merely a matter of purchasing scanners and software; it requires reengineering the entire diagnostic pathway, from specimen accessioning to image storage, reporting, and archiving.

Digital pathology workflows can be broadly classified into two models: partial digital implementation, in which only selected subspecialties (e.g., breast pathology, gastrointestinal pathology, or genitourinary pathology) or use cases are digitized (e.g., consultation, frozen

Advances in Digital Pathology and Artificial Intelligence in Pathology
https://doi.org/10.1016/B978-0-443-40383-5.00011-X

sections, or education); and full digital implementation, in which all routine diagnostic slides are scanned and reviewed digitally. Both models offer unique advantages and challenges. Partial implementation allows laboratories to pilot the technology, build confidence, and identify bottlenecks before large-scale deployment. Full digital adoption, while more ambitious, enables seamless integration with other digital health systems and unlocks efficiencies in workload distribution, telepathology, and computational analytics.

The motivations for adopting digital pathology are diverse. Many institutions pursue digital transformation to enhance diagnostic efficiency, reduce turnaround times, and facilitate remote sign-out capabilities. Others are driven by the desire to improve patient safety through standardized workflows and traceable digital records. Still others recognize digital pathology as a foundation for future innovations, particularly in AI-assisted diagnostics, computational image analysis, and precision oncology. Whatever the initial impetus, successful implementation requires alignment among multiple stakeholders: pathologists, laboratory managers, information technology teams, hospital administrators, and regulatory bodies.

Historically, pathology has lagged behind other medical disciplines in digital transformation, largely due to the technical and logistical complexities of digitizing glass slides at diagnostic resolution. However, several landmark institutional initiatives have demonstrated that digital pathology is both feasible and beneficial when carefully planned and executed (Stathonikos et al., 2013; Buck et al., 2014; Thorstenson et al., 2014; Stathonikos et al., 2019; Ardon et al., 2021; Babawale et al., 2021; Eloy et al., 2021; Temprana-Salvador et al., 2022; Ferreira et al., 2023; Clarke et al., 2025; Daniel et al., 2025; Iwuajoku et al., 2025; Montezuma et al., 2022; G.M. Lujan et al., 2021b; Retamero et al., 2020). Large academic institutions have reported measurable gains in workflow efficiency, cost savings through reduced slide handling and courier services, and improvements in multidisciplinary collaboration (Clarke et al., 2025; G.M. Lujan et al., 2021b; Ardon et al., 2024; G. Lujan et al., 2021a; Matias-Guiu et al., 2025). These experiences provide invaluable insights into the operational and strategic considerations that underpin successful digital adoption.

Among the key lessons learned from these early adopters is the critical importance of infrastructure readiness. The implementation of a digital pathology workflow demands a robust network capable of handling terabyte-scale data traffic, secure and redundant storage systems, and scalable image management software capable of seamless integration with existing laboratory information systems. Furthermore, validation of digital systems for clinical use is essential to ensure diagnostic equivalence between digital and optical modalities. This requires rigorous quality assurance protocols, standard operating procedures, and regulatory compliance with guidelines issued by professional bodies such as the College of American Pathologists (CAP), the Digital Pathology Association (DPA), and international agencies (Evans et al., 2022; Williams et al., 2020; Eloy et al., 2025; Zarella et al., 2019; Filippo Fraggetta et al., 2021b).

Equally important is the human factor. Digital transformation changes how pathologists work, communicate, and interact with their environment. Training programs must address not only the technical aspects of using digital systems but also the cognitive adaptation required for interpreting digital images. User acceptance, workflow redesign, and sustained leadership support are crucial determinants of success. Institutions that have achieved smooth digital transitions often emphasize early and continuous engagement of pathologists in planning, system selection, and validation.

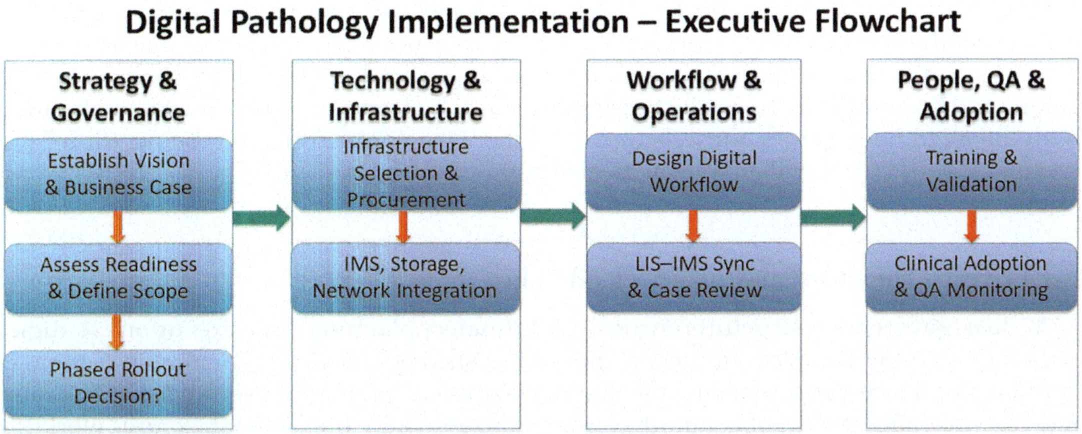

FIG. 3.1 Digital pathology implementation executive flowchart.

Finally, digital pathology implementation must be understood as a long-term investment rather than a one-time project. Beyond initial hardware and software acquisition, ongoing costs include system maintenance, data storage expansion, software updates, and cybersecurity. A comprehensive business case is essential to secure institutional funding and demonstrate the economic value of digital pathology, whether through direct operational savings or indirect benefits such as improved efficiency, reduced turnaround time, and enhanced academic and research capacity.

This chapter presents a comprehensive roadmap for implementing a digital workflow for clinical surgical pathology practice (Fig. 3.1). Drawing from institutional case studies and published evidence, it delineates the technical, operational, and human dimensions of digital adoption. Each section addresses a core domain of implementation, beginning with strategic planning and extending through technical infrastructure, workflow design, integration, validation, staffing, training and education, adoption, and continuous quality improvement. The goal is to provide a practical guide that can be adapted by institutions of varying sizes, complexities, and readiness, ensuring that the promise of digital pathology translates into sustainable improvements in diagnostic practice.

Strategic planning and scope of implementation

Establishing vision and rationale

The foundation of any digital pathology initiative is a clearly defined institutional vision. This vision must articulate why digital transformation is being pursued and what success will look like. Effective strategies arise from collaboration among clinical, technical, and administrative leaders.

Institutions adopt digital pathology for various reasons: improving operational efficiency, supporting remote workflows, enabling computational pathology and education, or advancing precision medicine initiatives (Jones-Hall, 2021; Baidoshvili et al., 2018). Whatever the drivers, leadership should explicitly link digital transformation to institutional priorities, such as quality improvement, workforce optimization, and telepathology expansion.

Strong and visible support from senior leadership is critical. Pathology chairs, administrators, and hospital executives must endorse the initiative and commit to sustained funding. Without alignment across these groups, digital transformation risks being viewed as a technical project rather than a strategic institutional advancement.

Governance structure and project leadership

A clear governance structure ensures coordinated planning and execution. A digital pathology steering team/committee, composed of stakeholders across anatomic pathology, LIS, IT, compliance, procurement, and clinical subspecialties, should oversee major decisions and resource allocation. Subcommittees may address validation, infrastructure, workflow design, and training.

Equally important is the appointment of a project champion, ideally a pathologist with both clinical credibility and technical interest. The project champion acts as a liaison between users and administrators, articulates the clinical needs of the department, and advocates for sustained investment.

Assessing institutional readiness

Implementing digital pathology requires a candid assessment of institutional readiness. A readiness assessment typically evaluates three interdependent domains: technological infrastructure, operational workflow, and cultural acceptance.

Technological readiness involves determining whether the existing IT environment can support the high bandwidth, data storage, and security demands of digital pathology. This includes assessing network capacity between laboratories and reading rooms, evaluating server capabilities, and confirming the institution's data redundancy and disaster recovery policies. Operational readiness focuses on workflow logistics—specimen volume, slide production capacity, staff expertise, and turnaround time targets. Laboratories must map the current state of slide processing, labeling, distribution, and archiving to identify how digital workflows can enhance or replace existing steps. Cultural readiness pertains to the willingness of staff to embrace digital systems. Even with excellent technology, implementation can stall if users are skeptical or reluctant to change. Conducting surveys or focus groups early in the process helps identify potential barriers, such as ergonomic concerns, resistance to new technology, or uncertainty about image quality.

The assessment should culminate in a document identifying gaps, risks, and required resources. This report often informs funding requests and the prioritization of deployment steps.

As institutions expand from pilots to enterprise-wide use, readiness evaluations must also consider scalability—forecasting slide volume growth, storage requirements, software upgrades, and alignment with enterprise imaging strategies. Standardized policies for scanning, storage, and access ensure consistent quality as digital pathology scales.

Defining the scope: Full versus partial digital implementation

Institutions must determine whether to pursue a full digital workflow or a partial (hybrid) implementation. Each approach has distinct implications for cost, complexity, and timeline.

Full digital implementation involves scanning all histopathology slides for primary diagnosis. This model offers the greatest long-term efficiency and interoperability benefits, as all data are natively digital. However, it requires substantial upfront investment in scanners, storage, and computing infrastructure, as well as validated workflows for every specimen type. Institutions that have adopted full digital sign-out often do so in phases, beginning with high-volume, low-complexity specimens (e.g., biopsies) before expanding to resections, cytology, and frozen sections (Clarke et al., 2025; Bruce et al., 2024).

Partial implementation refers to the selective digitization of specific use cases or subspecialties. Common examples include digitizing slides for remote consultation, multidisciplinary tumor boards, or teaching archives. This model allows laboratories to gain experience and demonstrate value before committing to a full-scale transition. Partial implementation is often favored in the early phases of digital transformation because it mitigates financial risk and provides flexibility to refine processes.

The decision between full and partial adoption depends on institutional priorities, funding availability, and technical readiness. Importantly, partial digitalization should not be viewed as a failure to commit but rather as a strategic step in a gradual evolution toward full digital practice.

Phased rollout models

- Digital pathology implementation is most successful when approached as a phased journey rather than a single event. Common rollout models include:
- **Pilot Phase:** A limited-scale pilot allows the institution to evaluate the technology under real-world conditions. The pilot should include a representative case mix and involve enthusiastic pathologists who can provide constructive feedback. Key metrics during this phase include image quality, scanning turnaround time, file transfer speed, and user satisfaction.
- **Validation Phase:** Once pilot performance is satisfactory, formal validation is required to demonstrate diagnostic equivalence between digital and glass-slide modalities. Institutions often follow professional guidelines to structure this validation, ensuring that both intraobserver and interobserver concordance are acceptable for clinical sign-out.
- **Incremental Expansion:** After validation, digital workflows can be extended to additional subspecialties or specimen types. Expansion should be accompanied by continuous training and user support to maintain momentum.
- **Full Integration:** The final phase involves complete integration with the LIS, standardized use across all subspecialties, and optimization of storage and retrieval systems. Continuous quality improvement processes are instituted to monitor performance and address issues such as image latency or data security.

Each phase should include predefined performance indicators to track progress and justify further investment. These indicators may include turnaround time, slide-handling errors, staff productivity, or user satisfaction.

Funding and resource planning

Implementing a digital pathology program requires substantial financial resources. Budget planning must be comprehensive, encompassing both direct and indirect costs (Ho et al., 2014).

Direct costs include scanner acquisition, workstation hardware, monitors compliant with diagnostic standards, image management software, and data storage infrastructure. Service contracts for maintenance, calibration, and software updates must also be factored in.

Indirect costs often exceed initial estimates and may include increased staff, training, workflow redesign, increased IT support, and eventual expansion of storage capacity. In addition, hidden costs, such as increased electrical power consumption or air conditioning needs for server rooms, should not be overlooked.

Securing funding can be achieved through multiple sources. In large academic institutions, funding may derive from departmental budgets, hospital capital funds, grants, or strategic innovation initiatives. In some cases, partnerships with industry vendors provide cost-sharing arrangements, particularly when institutions participate in technology development or validation studies. Demonstrating a **return on investment (ROI)** is key to securing sustained financial support. The ROI may be quantified through cost savings in courier services, slide storage, and labor efficiency, or through intangible benefits such as enhanced collaboration and research output (https://digitalpathologyassociation.org/roi-calculator, 2026).

A comprehensive business plan should outline the scope, timelines, expected outcomes, and sustainability, including future hardware refresh cycles and storage needs.

Procurement and vendor selection

Once funding is secured, institutions enter the procurement phase. Vendor selection should be guided by both technical performance and interoperability. Important considerations include scanning speed, image resolution, color fidelity, and system reliability. Equally critical is compatibility with the institution's LIS and adherence to open standards, such as Digital Imaging and Communications in Medicine (DICOM) for pathology.

Vendor evaluations typically involve technical demonstrations, pilot testing, and reference checks from other institutions. During this process, attention should be given to software usability, workflow automation, and vendor responsiveness to customization requests. Institutions should avoid locking themselves into proprietary ecosystems that limit future integration or scalability.

Negotiation with vendors should include service-level agreements that define uptime expectations, maintenance schedules, and escalation procedures for technical issues. Data ownership clauses must be carefully reviewed to ensure that the institution retains full control of digital assets.

Workflow mapping and process redesign

Implementing digital pathology fundamentally changes how slides move through the laboratory. Therefore a detailed mapping of existing workflows is essential before introducing digital steps. This mapping exercise identifies potential bottlenecks and informs decisions about scanner placement, slide labeling, and data routing (Ho et al., 2013).

Typically, the workflow begins with tissue processing and slide preparation, followed by slide scanning, image quality assurance, digital distribution, and reporting. Each of these steps must be clearly defined and standardized. For instance, laboratories may need to adopt barcoding systems to link physical slides to digital images accurately.

The scanning step introduces new variables—scanner throughput, file size, and scanning error rates—that must be monitored daily. Quality control personnel must be trained to verify that all slides are successfully digitized before they are filed or discarded. Similarly, the LIS must be configured to record the digital slide identifier, ensuring traceability between physical and virtual assets.

Digital workflows also necessitate the redesign of physical workspaces. Reading rooms may need larger monitors, ergonomic furniture, and high-speed network connections. Conversely, physical slide storage areas may be reduced over time, allowing reallocation of space to other laboratory functions.

Stakeholder engagement and communication

The success of digital pathology implementation depends heavily on human engagement. Pathologists, laboratory technicians, and IT professionals must all feel invested in the process. Effective communication strategies help build trust and mitigate resistance to change (Mikkelsen et al., 2022).

Institutions that succeed in digital transformation typically maintain transparent communication channels throughout the project. Regular updates, town hall meetings, and training sessions help to keep staff informed and involved. Early involvement of end-users in system testing and validation fosters a sense of ownership.

Pathologists may initially be skeptical about the diagnostic reliability of digital slides or the ergonomics of prolonged screen use. Demonstrating equivalence through validation studies, sharing success stories, and providing adequate technical support can alleviate these concerns. In addition, creating forums where users can share feedback and report issues encourages continuous improvement.

Building the business case and defining success metrics

Digital pathology represents a significant capital investment, including scanners, servers, software licenses, display workstations, storage solutions, network upgrades, and training programs. The business case must articulate both tangible and intangible benefits. Tangible benefits may include greater operational efficiency and reduced manual slide handling; lower courier and storage costs; increased capacity for remote consultation; and acceleration of research and AI development. Intangible benefits may include enhanced pathologist experience and ergonomic workflows; improved teaching and training resources; and strengthened institutional reputation and competitiveness.

Defining measurable success criteria ensures accountability and guides continuous improvement. Common metrics are summarized in Table 3.1.

TABLE 3.1 Common metrics used for measuring success in digital pathology implementation.

Success metric category	Description/indicators
Operational efficiency	Turnaround time reduction, optimized case tracking, and fewer lost or misplaced slides
Diagnostic accuracy	Concordance between digital and glass-slide diagnoses
Economic impact	Reduced courier costs, decreased glass slide storage needs, and labor efficiencies
User satisfaction	Survey results assessing pathologists' confidence, workload, ergonomics, and perceived image quality
Educational and research impact	Number of digitized teaching sets, use of images for AI development, research productivity, and publications

Technical infrastructure and system integration

Overview of digital pathology infrastructure

A digital pathology program relies on a coordinated ecosystem of whole-slide imaging (WSI) scanners, IMS, LIS, data storage platforms, network architecture, and diagnostic workstations. Each component must be properly configured and integrated to support efficient workflows, high diagnostic reliability, and seamless data exchange.

A well-designed digital infrastructure must meet several key requirements, including: (1) high-throughput image acquisition aligned with laboratory slide volumes; (2) secure and efficient image management with full traceability and interoperability; (3) scalable data storage and backup capable of handling terabytes to petabytes of images; (4) tight integration with LIS and institutional EHR systems; and (5) flexibility to support future growth, AI tools, and evolving clinical standards.

Achieving these objectives requires close collaboration between pathologists, IT professionals, and system vendors. It also demands proactive planning to anticipate future expansion—what may seem like ample capacity at launch can quickly become inadequate once full digitization begins.

Whole-slide imaging systems

WSI scanners are the entry point of digital pathology. They convert glass slides into high-resolution digital images by capturing sequential tiles and stitching them into a seamless whole-slide file. Modern scanners vary widely in capacity, throughput, and automation features, and their selection depends on case volume, specimen diversity, and desired turnaround time.

Scanner types and throughput

Scanners can be broadly classified as high-throughput, medium-throughput, and single-slide systems (Soliman et al., 2025; Patel et al., 2021). High-throughput scanners are capable of digitizing hundreds or thousands of slides per batch and are suited for large academic centers or centralized reference laboratories. These scanners typically operate autonomously, with barcode recognition and automated loading trays that minimize manual handling. Medium-throughput scanners are ideal for departmental or subspecialty use, supporting lower daily volumes but offering rapid turnaround for specific workflows such as frozen sections or consultations. Single-slide or on-demand scanners are often deployed in specific workflows such as frozen sections, consultations, and rapid on-site evaluation (ROSE) in cytology, research, or educational settings, where flexibility and immediate access outweigh batch efficiency.

Scanner speed depends on multiple variables, including magnification level (20× vs. 40×), file format, focus algorithms, and the presence of z-stacking for thicker sections. Institutions implementing full digital sign-out should ensure that scanner throughput matches daily slide production, ideally allowing complete digitization within 24 h of slide preparation to avoid diagnostic delays.

Some digital pathology systems have received United States FDA clearance or authorization for primary diagnosis (Table 3.2).

Whole slide imaging file compression, formats, and standardization

WSI files are extremely large, often ranging from 1 to 3 gigabytes per slide at 40× magnification. Efficient image compression is, therefore, critical to manage storage costs and optimize network performance. Compression can be lossless (preserving full

TABLE 3.2 Summary of digital pathology systems with United States FDA clearance/approval.

Scanner (system)	Manufacturer	FDA pathway/510(k) or De Novo	Year (FDA decision)	Short intended-use/notes
IntelliSite Pathology Solution (PIPS)	Philips	De Novo (PMA-like pathway for novel device)	2017	First WSI system authorized for primary diagnosis (automated slide creation, viewing, IMS)
Aperio AT2 DX System	Leica biosystems	510(k) K190332	2019	High-throughput scanner + viewing workstation; cleared for primary diagnosis on FFPE surgical pathology slides
Aperio GT 450 DX	Leica biosystems	510(k) K232202	2024	High-throughput scanner; supports native DICOM output in addition to SVS
NanoZoomer S360MD Slide Scanner System	Hamamatsu photonics	510(k) K213883	2022	Automated digital slide system intended as an aid for pathologists in FFPE surgical pathology (not intended for frozen sections or cytology)
Roche VENTANA DP 200	Roche (Ventana)	510(k) K232879	2024	Digital pathology Dx system (VENTANA DP 200) cleared to aid in the review and interpretation of digital images of FFPE slides
Roche VENTANA DP 600	Roche (Ventana)	510(k)—Additional clearance for high-volume scanner (DP 600)	2025	High-capacity (≈240-slide) scanner added to Roche's Digital Pathology Dx portfolio; cleared for primary diagnostic use

diagnostic fidelity) or lossy (reducing file size with minimal visual degradation). For primary diagnosis, lossless or minimally lossy formats are preferred to ensure diagnostic integrity. Institutions should validate that the chosen compression settings do not compromise image interpretability.

One of the early challenges of digital pathology has been the diversity of proprietary image formats (e.g., .svs, .ndpi, .mrxs, .iSyntax) associated with different scanner vendors. Each manufacturer typically developed its own file structure, compression algorithms, and metadata schemes (Table 3.3).

TABLE 3.3 Whole-Slide Imaging (WSI) file format comparison.

Format	Primary vendors	File extension (s)	Compression type	DICOM compatibility	Metadata support	Interoperability
SVS	Leica/ Aperio	.svs, .tif	JPEG, JPEG2000 (lossy/ lossless)	Partial (convertible)	Strong—stores pyramid levels, scan parameters, and labels	Moderate; widely supported by IMS and viewers
NDPI	Hamamatsu	.ndpi	JPEG	Limited (requires conversion)	Good—includes optical settings, magnification, z-levels	Moderate; well-supported by major viewers
MRXS	3DHISTECH	.mrxs	JPEG, JPEG2000, PNG	Limited (convertible)	Extensive—supports layers, annotations, macro images	Moderate; requires specific libraries
SCN	Leica (older systems)	.scn	JPEG, JPEG2000	Limited	Good—contains acquisition, camera data, pyramid levels	Low–moderate (vendor-specific)
BIF	Ventana/ Roche	.bif	JPEG	Limited	Strong—supports layers, metadata blocks	Low; fewer third-party tools
QPTIFF	Philips	.tif	JPEG	Limited; Philips also supports native iSyntax	Good—TIFF-based metadata	Moderate; TIFF-compatible but large
iSyntax	Philips (current)	.isyntax	JPEG-XR (lossy/ lossless)	Emerging support (via middleware)	Rich metadata with XML structure	Low–moderate (proprietary)
OME-TIFF	Open standard (Glencoe, OME)	.ome.tif	JPEG, JPEG2000	Convertible to DICOM	Excellent—open, extensible metadata schema	High; cross-platform
DICOM-WSI	Multivendor open standard	.dcm	JPEG, JPEG2000, JPEG-XR	**Native**	Excellent—uniform structured metadata	High—designed for full interoperability

To promote interoperability, the Digital Imaging and Communications in Medicine (DICOM) Working Group 26 extended the DICOM standard to encompass WSIs (Clunie et al., 2018). This standardization allows for consistent encoding of images, annotations, and metadata, supporting cross-platform compatibility and long-term accessibility (Clunie et al., 2018). Transitioning to DICOM requires infrastructure readiness and vendor collaboration, but institutions that adopt it benefit from uniform data handling, reduced vendor lock-in, and easier integration with radiology imaging archives. As radiology and pathology converge within enterprise imaging strategies, DICOM compliance becomes increasingly valuable for creating unified patient imaging records (Kuzmak and Dayhoff, 2000; Singh et al., 2011).

Image quality and calibration

Diagnostic fidelity in digital pathology hinges on accurate color reproduction, sharpness, and focus consistency. Regular calibration of scanners—using standardized color targets and focus slides—is essential to maintain image quality. Calibration routines should be incorporated into daily or weekly maintenance schedules, and quality control logs should be maintained for regulatory compliance (Inoue and Yagi, 2020; Clarke and Treanor, 2017; Bautista et al., 2014).

Pathologists are sensitive to color variations, especially in hematoxylin and eosin (H&E) staining. Consequently, scanner software should support color normalization algorithms to minimize variability across batches. Periodic review of scanned slides by designated quality assurance staff helps identify artifacts such as stitching errors, debris, or incomplete scans.

Image management systems (IMS)

Once slides are digitized, the management and accessibility of digital images become central to workflow success. The IMS serves as the central hub of the digital pathology environment. It is analogous to a picture archiving and communication system (PACS) in radiology but tailored to the specific needs of pathology. A robust IMS functions as the digital counterpart to the traditional slide archive, enabling secure, efficient, and scalable handling of WSIs. Effective image management supports diagnostic review, education, research, quality assurance, and archiving, ensuring that users across roles and departments can access the right data at the right time.

The IMS should manage large image files efficiently, support diverse file formats if possible, and provide seamless performance across a wide user base. The IMS must also be interoperable with the LIS, electronic medical record (EMR), and ancillary systems such as AI analytics or image annotation tools. Achieving this level of integration requires careful planning of database structure, network bandwidth, caching strategies, and access control.

An effective IMS provides rapid image access and smooth navigation, reliable linkage between images and LIS metadata, annotation and measurement tools for clinical and research use, audit trails and role-based permissions for regulatory compliance, and multi-user collaboration capabilities for tumor boards, consultations, and teaching. A wide range of IMS platforms is commercially available, among which several have obtained FDA clearance. Table 3.4 summarizes the key systems.

TABLE 3.4 Summary of commercially available IMS platforms.

Company	Country	Product	Closed/open	Regulatory status
Philips	The Netherlands	Philips IntelliSite Pathology Solution	Closed (scanner-dependent)	FDA-cleared
PathPresenter	United States	PathPresenter Platform	Open (Hamamatsu NanoZoomer S360MD)	FDA-cleared
Sectra	Sweden	Digital pathology solution	Open (Leica AT2 DX and Aperio GT450 DX)	FDA-cleared
PathAI	United States	AISight Dx	Open (Hamamatsu NanoZoomer S360MD, Leica Aperio GT 450 DX, and Roche VENTANA DP 200/DP 600).	FDA-cleared
Inspirata	United States	Dynamyx	Open	FDA-cleared
Paige	United States	FullFocus™ digital viewer	Open (Philips Ultra-Fast Scanner, Leica Aperio GT 450 DX, and Hamamatsu NanoZoomer S360MD)	FDA-cleared
Indica Labs	United States	HALO AP Dx	Open (Hamamatsu NanoZoomer S360MD and Leica Aperio GT 450 DX)	FDA-cleared
Proscia	United States	Concentriq AP-Dx	Open (Hamamatsu NanoZoomer S360MD)	FDA-cleared
Epredia	United States/Hungary	E1000 Dx Digital Pathology Solution	Open (E1000 Dx Scanner)	FDA-cleared
Leica Biosystems	United States/Germany	ImageScope/eSlide Manager	Open	NOT FDA-cleared
Hamamatsu	Japan	NDP Viewer/NDP.serve	Open	NOT FDA-cleared
Aiforia	Finland	Aiforia Clinical Suite	Open	NOT FDA-cleared
AIRA Matrix	India	AIRADHI	Open	NOT FDA-cleared
Crosscope (now part of Proscia)	India	Crosscope Dx	Open	NOT FDA-cleared
Qritive	Singapore	PANTHEON	Open	NOT FDA-cleared
Pramana	United States/India	Vista	Open	NOT FDA-cleared
INFINITT	South Korea/United States	INFINITT Digital Pathology Solution (IDPS)	Open	No FDA clearance

IMS performance and scalability

IMS performance directly affects user satisfaction. Pathologists expect images to load and pan smoothly without noticeable latency. Achieving this requires high-speed network connectivity between servers and workstations, as well as optimized image streaming protocols. Modern IMS platforms employ *tile-based streaming*, which loads only the visible portion of an image rather than the entire file, significantly reducing bandwidth requirements.

Scalability is another critical consideration. As the number of scanned slides grows, the IMS must handle millions of image files efficiently. System architecture should, therefore, be modular, allowing the addition of servers or cloud nodes as demand increases. Vendors should be evaluated on their ability to support horizontal scaling without service interruption.

Integration with LIS

The LIS remains the authoritative source of patient and specimen data in anatomic pathology. Seamless integration between the LIS and IMS ensures that each digital slide is accurately linked to its corresponding case, block, and stain. This linkage is typically achieved through barcode identifiers embedded during slide labeling. When a pathologist opens a case in the LIS, associated digital slides should appear automatically in the IMS viewer, eliminating manual searching (F. Fraggetta et al., 2021a).

Bidirectional communication between systems is essential. When a pathologist makes annotations or measurements in the IMS, these should be exportable to the LIS or report generation system. Likewise, case status updates (e.g., "in review," "signed out") should synchronize between platforms. True integration minimizes duplicate data entry and prevents mismatches between patient records and digital images.

Supporting research and AI integration

Beyond diagnostic utility, the storage and management of WSIs underpin research innovation, particularly in computational pathology and AI development. Structured data storage and metadata tagging facilitate cohort selection and large-scale image retrieval for algorithm training. Institutions can enable research access through sandbox environments that isolate research datasets from clinical operations to maintain compliance and data integrity (Lianas et al., 2025).

As AI algorithms evolve, the IMS may serve as the central platform for inference deployment—linking diagnostic workflows with automated image analysis modules that detect, quantify, or classify histologic features. Selecting an appropriate IMS that can accommodate such extensions ensures that future AI incorporation needs can be met without fundamental redesign or replacement.

Data storage, security, and retention

Storage architecture is one of the most technically demanding aspects of digital pathology implementation. A fully digital laboratory can generate tens to hundreds of terabytes of image data annually, depending on case volume and scan resolution. Therefore storage systems must balance capacity, speed, redundancy, and cost (Clarke et al., 2025; G.M. Luján et al., 2021b).

Storage tiers

Most institutions adopt a tiered storage model. **Primary storage (Tier 1)** includes high-performance disks used for active cases under review. These systems prioritize speed and reliability, often using redundant array configurations (RAID) for data protection. **Secondary storage (Tier 2)** consists of slower but more cost-effective systems for recently completed cases. **Archival storage (Tier 3)** is used for long-term, low-cost options such as magnetic tape, optical media, or cloud-based cold storage. These archives are optimized for durability rather than retrieval speed. Policies should define how long images remain in each tier before migration, depending on institutional retention requirements and case reaccess frequency.

On-premises versus cloud solutions

Institutions face a key decision between on-premises storage and cloud-based storage. **On-premises** storage solutions provide greater control and potentially lower latency but require significant capital investment and ongoing maintenance. Institutions often deploy redundant arrays of independent disks (RAID), storage area networks (SANs), or object-based storage systems with built-in data replication to prevent data loss. **Cloud storage** offers scalability and elasticity, allowing institutions to expand storage capacity without major infrastructure overhauls. Cloud providers also offer advanced redundancy and geographic replication, though institutions must ensure compliance with health information privacy regulations. Hybrid models—combining on-premises high-speed storage for active cases with cloud archiving for long-term data retention—are increasingly common and cost-effective.

Hybrid models are increasingly popular, combining on-premises systems for active data and cloud storage for archives or disaster recovery. When using cloud services, institutions must ensure compliance with healthcare data protection standards, such as HIPAA, GDPR, or local equivalents.

Data backup and disaster recovery

Robust backup policies are essential to protect against data loss. Backups should occur at regular intervals and be stored in geographically separate locations. Automated verification processes confirm backup integrity. Disaster recovery plans should define recovery time objectives (RTO) and recovery point objectives (RPO) appropriate to clinical needs. For mission-critical systems, failover mechanisms ensure continuity of access even if the primary server fails.

Data security and privacy

Patient data contained within digital pathology systems is subject to stringent privacy regulations, such as the Health Insurance Portability and Accountability Act (HIPAA) in the United States or the General Data Protection Regulation (GDPR) in Europe. Ensuring compliance requires a combination of technical, administrative, and physical controls.

Technical safeguards include encryption of stored images (data at rest) and encrypted transmission channels (data in motion). Administrative safeguards involve access policies, staff training, and periodic security audits. Physical security encompasses server room access restrictions, environmental controls, and redundant power and network systems.

Institutions must also define protocols for image anonymization and de-identification, particularly when sharing data for research or AI development. Metadata that might inadvertently identify patients—such as accession numbers, dates, or annotations—should be scrubbed before export. Implementing automated anonymization workflows can prevent data breaches while facilitating ethical secondary use of data.

Data retention and archiving

Regulatory requirements determine how long pathology materials must be kept. In the U.S., CAP mandates a minimum 10-year retention for glass slides and blocks (https://webservices.ncleg.gov/ViewDocSiteFile/33038, 2026) but no formal guideline yet exists for digital WSIs. As a result, institutions set their own policies: some delete images after 3–6 months, while others keep them much longer to support research and innovation.

With broader digital pathology adoption, managing the balance between image accessibility and storage costs has become essential, as the cost of storage increases linearly every year. Effective WSI Life Cycle Management (WSI-LCM) policies are, therefore, critical.

Key WSI-LCM considerations include why images are retained, how long they are needed, how quickly they must be accessed, and the cost of storing them. Common retention patterns include:

- **Clinical diagnostic cases:** Keep for up to 3 months; requires rapid, high-performance storage. After sign-out, these cases can be moved to the next level of storage (prior cases after sign-out).
- **Prior cases after sign-out:** Duration may vary by subspecialty (an average period of 2–3 years). Moderate access times are acceptable, with relatively slower and less expensive storage.
- **Educational cases:** These cases need to be kept indefinitely with fast access.
- **Legal/regulatory cases:** Often stored for up to 10 years, retrieved rarely, and suitable for slow, low-cost storage.

Network architecture and performance

The transition to digital pathology substantially increases network demands. WSIs, even when compressed, are significantly larger than typical clinical data files. Therefore network design must provide sufficient bandwidth, low latency, and high reliability.

Bandwidth requirements

Bandwidth requirements depend on case volume, image size, and concurrent users. As a general guideline, institutions should provide at least 1 Gbps connections between scanners, storage servers, and pathologist workstations, with 10 Gbps backbones preferred in high-volume environments. Network performance should be tested under peak loads to identify bottlenecks.

Latency and caching

Even minor latency can affect pathologists' productivity. To mitigate this, IMS systems often use local caching, temporarily storing recently accessed images on workstations or local servers. This allows near-instantaneous reloading of slides during viewing sessions. Network optimization techniques, such as content delivery networks (CDNs) or distributed caching, can further enhance performance across geographically dispersed campuses.

Network security and access control

Because digital pathology involves protected health information (PHI), network security must meet stringent standards. Access control mechanisms must ensure that only authorized personnel can view, annotate, or modify images. Role-based access control (RBAC) frameworks assign permissions according to user roles (e.g., pathologist, resident, technician, researcher). Audit trails log every access event, ensuring accountability and supporting regulatory compliance.

For remote access, secure virtual private network (VPN) connections or institutionally managed web portals are used. Multifactor authentication, encryption in transit (via HTTPS or TLS), and data integrity checks are mandatory safeguards.

Diagnostic workstations and displays

The quality of the viewing workstation directly influences the diagnostic experience. Pathologists transitioning from microscopes to monitors require high-resolution, color-calibrated displays that accurately reproduce histological detail.

Display specifications

A fully digital pathology transition also requires a diagnostic display/monitor in the pathologists' workstation. Digital pathology displays can be divided into three categories: medical grade (MG), professional grade (PG), and consumer-off-the-shelf (COTS). A pathology-oriented MG display recently received FDA approval (https://www.barco.com/en/product/mdpc-8127, 2026) and preliminary attempts to benchmark these instruments for primary diagnosis were published (Abel et al., 2020; Clarke et al., 2020). Other studies have compared the performance of other displays with MG displays (Cazzaniga et al., 2025).

Generally speaking, diagnostic monitors need at least 27-inch screens with 4 K resolution or higher, providing a sufficient field of view for efficient navigation. Color calibration should conform to medical imaging standards, and luminance should be stable across viewing sessions. Some institutions deploy dual/multiple-monitor setups, with one for image viewing and the second one for the LIS or report interface, to enhance multitasking efficiency.

Ergonomics

Digital sign-out shifts the physical posture of pathologists from microscope-based to screen-based work. Proper ergonomic setup, including adjustable monitor height, supportive chairs, and optimal lighting, reduces fatigue and prevents musculoskeletal strain (Stathonikos et al., 2019). Training in ergonomic practices should be part of onboarding for all digital users.

Peripheral devices

Annotation tablets, programmable keyboards, and high-precision mice can improve navigation speed and comfort. Some users prefer trackballs or touch interfaces for faster panning and zooming (Alcaraz-Mateos et al., 2022). Institutions should accommodate individual preferences to enhance productivity and satisfaction (Rogers et al., 2025).

IT support, maintenance, and lifecycle management

Digital pathology systems require ongoing IT support far beyond traditional histology operations. Dedicated IT personnel—ideally embedded within the pathology department—should oversee hardware maintenance, software updates, and user support (Clarke et al., 2025).

Routine maintenance includes:

- Monitoring scanner performance and updating firmware.
- Verifying data backups and storage integrity.
- Updating antivirus definitions and applying security patches.
- Reviewing system logs for errors or performance degradation.

Lifecycle management involves planning for hardware refresh cycles (typically every 3–5 years), ensuring compatibility between legacy and new systems, and budgeting for software upgrades. A long-term maintenance roadmap prevents sudden obsolescence and guarantees a consistent user experience.

Future considerations

Digital pathology infrastructure should not be designed solely for current needs but also for future scalability. Anticipated developments include integration with AI-driven diagnostic tools for image classification, tumor quantification, and predictive analytics; expansion of cloud-native systems that enable real-time collaboration across institutions; adoption of edge computing to process large images closer to acquisition sites, reducing latency; and implementation of data lakes that combine digital slides with molecular, genomic, and clinical data for precision medicine research. Institutions should design modular systems that can incorporate these innovations without requiring complete reengineering. Periodic technology reviews ensure that infrastructure remains aligned with emerging standards.

Digital workflow and operational integration

While infrastructure and technology establish the foundation for digital pathology, success ultimately depends on how effectively digital processes are integrated into day-to-day laboratory operations. Implementation requires reengineering traditional workflows to accommodate new steps such as slide scanning, digital assignment, and electronic sign-out. A well-designed workflow ensures that digitization enhances, rather than disrupts, the efficiency of diagnostic services (Filippo Fraggetta et al., 2021b; Ameisen et al., 2014; Fabián et al., 2025; Fraggetta et al., 2017; Hartman and Imaging, 2020).

Mapping the current workflow

The first step toward digital integration is to map the current glass-slide workflow, from specimen accessioning to report sign-out. This map identifies all stakeholders (histotechnologists, pathologists, transcriptionists, IT staff) and documents the flow of specimens, slides, and data. Typical steps include specimen accessioning, grossing, tissue processing, embedding, sectioning, staining, coverslipping, labeling/barcoding, and delivery to pathologists.

Digital conversion introduces new nodes into this flow—namely, slide scanning, image quality control, digital image upload, and viewing. Workflow mapping enables laboratory managers to visualize where scanning logically fits and where automation or redesign can improve throughput.

Designing a digital slide workflow

Slide preparation

The quality of a digital image begins with the quality of the glass slide. Slides must be clean, uniformly coverslipped, and correctly labeled/barcoded. Even minor imperfections, such as excess mounting medium, debris, or air bubbles, can distort scanning.

Barcode labeling is essential. Each slide should carry a 2D barcode encoding the accession number and slide ID, enabling automatic linkage to the case record. Laboratories should standardize label placement to ensure reliable recognition by scanner cameras.

Scanning logistics

Modern histology laboratories typically include IHC facilities as part of the routine workflow. Ideally, whole-slide scanners are positioned between the H&E and IHC staining areas so that slides can be scanned immediately after coverslipping. This strategic placement minimizes manual handling and transport, reduces the risk of slide loss or damage, and ensures close temporal alignment between the physical and digital slides. In high-throughput laboratories, assigning dedicated scanning personnel is recommended to ensure continuous operation, rapid troubleshooting, and consistent quality control (Ardon et al., 2021; Clarke et al., 2025; G.M. Lujan et al., 2021b).

Slides are usually loaded into racks or cassettes, verified via barcode, and automatically queued for scanning. Contemporary scanners incorporate automated tissue detection, focus mapping, and calibration algorithms; however, technician oversight remains essential. All scanned fields must be reviewed to confirm adequate focus and image fidelity before acceptance. Key quality control checkpoints include focus accuracy, scan completeness, absence of scanning or staining artifacts, and appropriate image color balance.

File naming and metadata

The IMS assigns a unique digital identifier to each image, linked to metadata such as case number, tissue type, stain, and block/slide ID. Consistency in naming conventions is critical for preventing mismatches. Ideally, metadata is imported directly from the LIS via HL7 interfaces to prevent manual entry errors.

Image ingestion and verification

Upon completion of scanning, digital files are automatically uploaded to the IMS. The system performs checksum verification to confirm file integrity, followed by thumbnail generation and indexing. In high-volume environments, this process is parallelized across multiple scanners and servers.

Some laboratories employ automated dashboards that display scanning progress, error rates, and throughput metrics in real time. These tools enable supervisors to identify bottlenecks early, such as scanner downtime or poor slide quality.

LIS-IMS synchronization

Integration between the LIS and the IMS is the backbone of digital workflow. Without reliable synchronization, pathologists cannot seamlessly access digital slides within their case worklists.

Case creation and slide association

When a case is accessioned in the LIS, a unique case ID is generated. During scanning, barcode data links each digital image to this case ID, allowing automatic association. Once ingestion is complete, the LIS receives a confirmation message from the IMS that slides are available for review.

In a fully integrated environment, the pathologist can open the case in the LIS and directly launch the digital viewer, which loads all corresponding slides in the correct order and with the correct stain identifiers (Eloy et al., 2021; Clarke et al., 2025; G.M. Luján et al., 2021b).

Real-time status updates

Two-way communication between LIS and IMS allows real-time updates of case status—for example, whether slides are "in scan," "available," or "awaiting QC." This transparency supports efficient workload management and prevents premature case assignment.

Case assignment and pathologist review

Case distribution

In a digital environment, cases can be distributed electronically rather than physically. This capability allows dynamic workload balancing among pathologists, regardless of physical location. For example, cases from a high-volume satellite site can be digitally reassigned to available subspecialists at the main campus.

Review interface

Pathologists review slides using a digital viewer integrated within the LIS or as a standalone application. Features typically include smooth pan-zoom navigation replicating microscope control, slide thumbnails and overview maps, annotation and measurement tools, side-by-side comparison of serial sections or stains, and layered viewing for immunohistochemical panels.

Performance is critical: latency must be minimal, and color representation must match the original slide.

Sign-out and reporting

After review, the pathologist signs out the case electronically within the LIS. Some systems allow annotations or image snapshots to be embedded in the pathology report for documentation and communication with clinicians.

Handling of special stains and frozen sections

Special stains and IHC slides may present unique challenges due to color complexity and variable contrast. Scanning parameters (exposure time, color calibration) should be optimized for each stain type.

Digitizing frozen sections allows remote intraoperative consultation (telepathology). Scanning systems must deliver rapid turnaround—ideally within five minutes from tissue cutting to image display. High-speed scanners or live telepathology cameras are employed for this purpose. Institutions implementing digital frozen section workflows must ensure redundant network connectivity, as delays can impact surgical decision-making. Validation should demonstrate diagnostic equivalence to optical review under time-critical conditions.

Workflow metrics and performance monitoring

Continuous monitoring of workflow performance ensures sustainability and highlights areas for optimization. Key performance indicators (KPIs) include scanning throughput (slides/hour per scanner), error rates (proportion of failed or rescanned slides), turnaround time (time from coverslip to image availability), storage utilization (growth rate and access frequency), and user activity metrics (number of cases reviewed digitally versus on glass). Dashboards and automated analytics support daily operational oversight. Regular review of metrics enables laboratories to balance staffing, adjust scanning schedules, and forecast storage requirements.

Hybrid workflows and transitional strategies

Few institutions transition instantly to full digital operation; most begin with hybrid workflows combining glass and digital review (Buck et al., 2014; Thorstenson et al., 2014; Stathonikos et al., 2019; Ardon et al., 2021; Babawale et al., 2021; Eloy et al., 2021; Temprana-Salvador et al., 2022; Clarke et al., 2025; Iwuajoku et al., 2025; G.M. Lujan et al., 2021b; Retamero et al., 2020; Bruce et al., 2024; Fabián et al., 2025; Flach et al., 2022).

Hybrid workflows are exemplified as the following: (1) pilot subspecialties (starting digital workflows with subspecialties that benefit most, such as dermatopathology or gastrointestinal pathology); (2) remote consultations (using digital slides for second opinions while maintaining glass archives); (3) deploying digital systems in academic or training contexts before diagnostic rollout.

During the hybrid phase, laboratories must manage cases that include both digital and glass slides. Clear labeling, LIS indicators, and staff training are required to avoid confusion. Standard operating procedures should specify when digital images are used for diagnosis and when glass review remains mandatory.

Quality control of scanning process

Routine QC checks

Each scanning batch should undergo systematic QC inspection. This includes verifying focus, color fidelity, completeness of tissue capture, and the absence of artifacts. Institutions often designate dedicated QC personnel who review thumbnail grids or random samples (Ardon et al., 2023; Chong et al., 2022; Ho et al., 2006).

Automated algorithms can assist by detecting focus failures, tissue misregistration, or color anomalies. Regular QC reports feed back into histology and scanning teams for continuous improvement.

Slide re-scanning protocol

Slides failing QC should be flagged automatically for rescanning. Reasons for rescanning, such as out-of-focus regions or labeling errors, must be logged for tracking. Persistent trends may indicate issues with tissue preparation or scanner calibration.

Record keeping and audit trails

Every image and QC event should be logged with a timestamp and operator ID. These records form part of the laboratory's quality management system, satisfying regulatory and accreditation requirements.

Workflow optimization

The digital transition offers a unique opportunity to apply lean management principles to pathology operations. Lean methods emphasize the elimination of waste, continuous improvement, and value creation for the end user. In the digital context, "waste" can refer to redundant manual steps, delays in image availability, or inefficiencies in data transfer.

Process metrics should be defined and monitored regularly—such as average scanning turnaround time, rescan rate, system uptime, and diagnostic reporting time.

Feedback loops between end-users and administrators ensure that workflow refinements are data-driven. Regular process review meetings encourage transparency and collective ownership of improvement initiatives.

Staffing and training

Staff roles and responsibilities

Digital pathology expands the traditional roles within the pathology laboratory, introducing new responsibilities for both technical and professional staff. Successful implementation requires a clear definition of roles, cross-functional training, and effective communication among teams (Clarke et al., 2025; G.M. Lujan et al., 2021b).

Pathologists remain the ultimate arbiters of diagnosis, but their interaction with cases evolves dramatically in a digital environment. They must learn to navigate WSIs, annotate findings, and manage digital workflows through IMS dashboards. Digital sign-out also demands familiarity with case tracking systems, electronic templates, and remote collaboration tools. Pathologists increasingly serve as advocates for workflow optimization, providing feedback on usability and performance.

Digital pathology technologists play a pivotal role in the digital pipeline. Their tasks include preparing slides for scanning, ensuring optimal staining quality, labeling slides with machine-readable barcodes, and verifying scanning completeness. They may also handle scanner operations, image quality checks, and rescanning if needed. In smaller laboratories, these duties may be assigned to histotechnologists; however, appropriate training and competency assessment are essential before they assume this role (Clarke et al., 2025).

IT professionals manage server configurations, network performance, and system integrations with LIS/IMS. They ensure that the infrastructure meets uptime, latency, and security requirements. Pathology informaticians serve as liaisons between laboratory operations and IT, translating clinical needs into technical solutions. Together, they handle maintenance, software upgrades, and interface monitoring.

Administrative personnel may take on new tasks related to user access management, system licensing, vendor coordination, and performance reporting. Workflow analysts and project managers are often essential for coordinating implementation timelines and cross-departmental communication.

Large institutions frequently establish a digital pathology committee/team to oversee the transition. This committee typically includes pathologists, digital pathology technologists, IT specialists, administrators, and quality officers. Its responsibilities encompass policy creation, resource allocation, milestone tracking, and post-implementation evaluation.

Training and competency development

The introduction of digital tools requires structured education to ensure confidence and competence across all staff levels. Training should not be viewed as a one-time event but as an ongoing process integrated into departmental culture (Williams et al., 2020; Browning et al., 2020).

Training programs for pathologists should include hands-on sessions covering WSI navigation, annotation tools, measurement utilities, and digital reporting interfaces. Pathologists must also be trained in recognizing image artifacts that may affect interpretation, such as focus errors or compression artifacts. Institutions that conduct side-by-side comparisons between glass and digital slides during the transition period help pathologists build diagnostic confidence. Peer mentoring, internal workshops, and formal courses provided by professional organizations further enhance skill development. Many institutions adopt a "digital champions" model—identifying early adopters who mentor their colleagues and serve as internal advocates for the technology.

Technologists require instruction on scanner operation, file management, and routine maintenance. Training should also emphasize quality control, such as checking for scanning completeness, labeling accuracy, and image clarity. Creating standard operating procedures (SOPs) for each task ensures consistency across staff and shifts.

IT teams must understand the clinical implications of downtime, data loss, and latency. Training focuses on system redundancy, data security protocols, and interoperability standards. Cross-training IT personnel with laboratory staff promotes a shared understanding of workflow priorities and fosters rapid issue resolution.

Competency assessments should be conducted periodically, with documentation maintained as part of institutional quality management programs. As technologies evolve, refresher courses and continuing education are critical to maintaining proficiency. Professional societies, such as the College of American Pathologists (CAP) and the Digital Pathology Association (DPA), often provide valuable educational resources, webinars, and workshops.

Validation and quality assurance in digital pathology

Validation and quality assurance (QA) are cornerstones of digital pathology implementation. Unlike traditional glass-slide microscopy, digital pathology introduces new variables—scanners, image formats, display devices, network performance, and software—that can influence diagnostic accuracy. Robust validation protocols and ongoing QA programs ensure that whole-slide images (WSIs) reliably replicate the optical review experience, support regulatory compliance, and maintain diagnostic confidence.

Validation

Validation encompasses technical performance (image fidelity, scanner calibration, and system reliability) and clinical performance (diagnostic concordance, reproducibility, and error tracking).

Technical validation begins with evaluating hardware and software capabilities, including resolution and focus accuracy (image sharpness matches optical microscopy across the entire tissue section), color fidelity (the staining hues, contrast, and intensity are accurately reproduced), tissue coverage (all tissue regions are captured), and scanning speed and throughput. Test slides representing typical tissue types, staining variations, and known artifacts are scanned and analyzed. Any deficiencies prompt calibration, software adjustments, or changes in scanning protocols. Multiple scans of the same slide under different conditions—time of day, operator, or scanner unit—should yield consistent images (Buck et al., 2014; Babawale et al., 2021; Williams et al., 2020; Ho et al., 2006; Hsu et al., 2025).

Clinical validation assesses whether pathologists can accurately render diagnoses using WSIs compared to glass slides. Key components include case selection, diagnostic concordance, and washout periods. Validation studies should encompass the full spectrum of cases expected in routine practice, including benign, malignant, inflammatory, and technically challenging specimens. Subspecialties may require separate validation due to unique diagnostic nuances (e.g., dermatopathology, hematopathology). Pathologists independently review cases in both digital and glass formats. Concordance rates are calculated to assess agreement. High concordance validates that WSIs faithfully replicate traditional microscopy. Any systematic discrepancies must be investigated and mitigated. A minimum washout period between the glass slide read and the digital read using WSIs should be set.

If digital pathology is used for telepathology or remote sign-out, validation must replicate anticipated network conditions and display configurations. This ensures that diagnostic accuracy is maintained across locations and devices.

Regulatory and professional guidelines shape the validation process. Institutions typically refer to standards set forth by bodies such as the College of American Pathologists (CAP) in the US (Table 3.5), the Royal College of Pathologists (RCPath) in the UK, or other regional authorities/accreditation agencies (Evans et al., 2022; Williams et al., 2020). These guidelines outline recommendations for scanner performance, image resolution, display quality, case selection, minimum discordance rate, training, QA, and workflow documentation. Compliance ensures legal and professional alignment, particularly when digital slides are used for primary diagnosis or remote sign-out.

Quality assurance (QA)

Quality assurance (QA) is fundamental to sustaining diagnostic integrity in digital pathology.

As digitization introduces new variables that do not exist in conventional glass-slide microscopy, laboratories must implement systematic processes to monitor and maintain image quality. Both technologists and pathologists play critical roles in daily or batch-level inspections of scanned slides, evaluating focus errors, blurring, color distortions, tissue

TABLE 3.5 Strong recommendations from the CAP/ASCP 2021 guideline for validating whole-slide imaging systems.

Strong recommendation	Description/rationale
Sample Size	Use at least 60 cases for one use case (e.g., H&E of fixed tissue), plus 20 additional cases for each other application (e.g., IHC) that was not part of the 60-case set
Diagnostic Concordance	Establish intraobserver concordance between digital and glass slides. If concordance is below 95%, investigate and address the causes
Washout Period	Include a washout period of at least two weeks between reviewing glass slides and digital slides

truncation, and missing sections. Barcode accuracy and metadata integrity must also be verified to ensure proper case assignment and seamless integration with laboratory information systems. Modern scanners and imaging platforms increasingly incorporate automated QC algorithms capable of detecting focus or registration deficiencies, enabling timely corrective actions before diagnostic review (Ardon et al., 2023; Chong et al., 2022; Ho et al., 2006; Kim et al., 2021).

Monitoring rescan rates is another essential operational metric. Elevated rescan frequencies may indicate upstream problems in tissue processing, coverslipping, or staining, or may reflect deficiencies in scanner calibration or operator technique. Likewise, display calibration is critical, as diagnostic interpretation relies heavily on the accurate rendering of color, brightness, and contrast. Regular calibration using photometric or colorimetric tools ensures that WSIs are viewed consistently across users and workstations, maintaining diagnostic reproducibility.

Digital pathology software (IMS, etc.) QA processes, such as functional validation, regression testing, bug tracking, and version control, ensure stability and reliability. Institutions should maintain a formal change-management policy to vet software updates and patches before they are deployed clinically, preventing unintended disruptions in workflow or diagnostic performance.

Data integrity is another cornerstone of digital pathology QA. Modern systems maintain comprehensive audit trails that document scanning times, operator identifiers, user access logs, annotations, modifications, case assignments, and sign-out timestamps. In the event of a diagnostic discrepancy or technical failure, audit data enables efficient root-cause analysis and targeted remediation.

Despite rigorous QA protocols, occasional errors in scanning, data handling, or interpretation are inevitable. A robust error-management framework helps institutions address these issues systematically. This includes the timely identification of errors through automated alerts, QC checkpoints, or user reporting; classification of error types (technical, operational, interpretive); implementation of corrective actions such as rescanning, retraining, or workflow modification; and preventive strategies guided by trend analysis.

Ongoing performance monitoring using quantitative metrics provides an objective measure of digital pathology system performance. Key indicators include diagnostic concordance between digital and glass slides, scanner uptime, image rejection and rescan rates, digital turnaround time, and user satisfaction from pathologists. Tracking these metrics longitudinally enables laboratories to verify that digital pathology platforms continue to meet clinical, operational, and regulatory standards.

The above-discussed QA items are summarized in a checklist below (Table 3.6).

Validation and QA establish the foundation for safe, reliable, and effective digital pathology practice. Technical validation confirms that scanners and software produce high-fidelity images, while clinical validation ensures diagnostic concordance with traditional microscopy. Continuous QA, performance monitoring, and error management safeguard diagnostic integrity over time. By embedding these practices into the digital workflow, institutions mitigate risk, maintain regulatory compliance, and reinforce trust among pathologists, clinicians, and patients.

TABLE 3.6 Quality assurance checklist.

QA category	Checklist items
1. Image quality	□ No focus errors or blurring □ Accurate color rendering □ No tissue truncation or missing sections □ Barcode readable and metadata correct
2. Automated QC tools	□ Automated focus/tissue detection functioning □ Scanner pre-checks completed □ Alerts reviewed and resolved before sign-out
3. Rescan management	□ Rescan rate monitored □ Causes of elevated rates investigated □ Corrective action documented
4. Display calibration	□ Monitors calibrated for color, contrast, brightness □ Calibration documented at scheduled intervals
5. IMS/viewer software QA	□ Functional testing after updates □ Version control maintained □ Change management review completed □ No unresolved critical bugs
6. Data integrity/ audit trails	□ Audit logs active and retained □ Operator ID and scan timestamps recorded □ Image access and annotations tracked
7. Error management	□ Errors detected and logged □ Errors classified (technical/operational/interpretive) □ Corrective and preventive actions implemented
8. Performance metrics	□ Diagnostic concordance reviewed regularly □ Scanner uptime meets target □ Image rejection and rescan rates within thresholds □ Digital turnaround time acceptable □ User satisfaction feedback collected
9. Routine maintenance	□ Scanner optics cleaned □ Dust/debris removed from slides and racks □ Stain excess minimized before scanning

Clinical adoption

Resistance to change is a common barrier in any digital transformation. Pathologists and laboratory staff, accustomed to decades of glass-slide practice, may express skepticism about digital accuracy or workflow efficiency. Effective change management strategies are, therefore, crucial.

Institutions should prioritize transparent communication and involve stakeholders early in the planning process. Demonstrating the benefits, such as faster turnaround, remote flexibility, and integration with AI, helps build enthusiasm. Pilot projects can serve as proof-of-concept to showcase tangible improvements in daily work.

Leadership plays a central role in cultivating a culture of innovation. Recognizing early adopters, sharing success stories, and celebrating milestones sustain morale and momentum. Providing avenues for feedback ensures that concerns are heard and addressed promptly. Adoption is facilitated by structured management strategies that emphasize early engagement of stakeholders, transparent communication of benefits and expectations, training and education to familiarize staff with the new system and its advantages, pilot projects to demonstrate workflow efficiency, and identification of "digital champions" to mentor colleagues and advocate for adoption.

Phased implementation

Most institutions adopt a phased approach to digital pathology deployment (Ardon et al., 2021; Clarke et al., 2025; G.M. Luján et al., 2021b).

- **Pilot Phase:** Implement a limited workflow, often within a single subspecialty or for consultation cases, to evaluate technology performance, staff training, and workflow impact.
- **Hybrid Phase:** Expand scanning to selected routine cases, creating a mix of glass and digital workflows. This stage identifies bottlenecks, validates systems, and refines SOPs.
- **Full Digital Phase**: Transition all diagnostic cases to a digital workflow, supported by established quality control measures, IT infrastructure, and trained staff.

Each phase is accompanied by performance monitoring, staff feedback, and iterative process improvement. Phased implementation reduces risk, promotes staff acceptance, and builds institutional confidence in the digital system.

Adoption metrics and performance monitoring

Institutional adoption can be assessed using multiple performance indicators, including the volume of cases reviewed digitally, diagnostic turnaround time, user satisfaction surveys, cost tracking, and quality metrics such as rescan rates, diagnostic concordance, and QC compliance. Regular review of these metrics ensures that adoption remains aligned with institutional goals and identifies areas for further optimization.

Sustainability and long-term strategy

Sustainable adoption requires integration of digital pathology into the institution's long-term operational and strategic planning. Key considerations include scalable infrastructure (planning for increased slide volume and storage demands), continuous quality improvement (monitoring KPIs and updating SOPs as technology evolves), ongoing training (maintaining staff competency and addressing turnover or new hires), vendor partnerships (ensuring support for equipment, software updates, and interoperability standards), and strategic alignment (aligning digital pathology initiatives with enterprise imaging, precision medicine, and research objectives). Institutions that prioritize sustainability create a resilient and future-ready pathology service capable of adapting to evolving clinical, technological, and regulatory landscapes.

In summary, the adoption of digital pathology is a multidimensional process that combines technology, finance, operational redesign, and human factors. Building a compelling business case, securing funding, and implementing structured governance support sustainable adoption. Phased implementation, stakeholder engagement, and performance monitoring facilitate smooth integration into routine practice. When approached strategically, digital pathology delivers operational efficiency, educational value, research opportunities, and positioning for future innovations, while ensuring patient safety and diagnostic excellence.

Lessons learned and key considerations

Experience from early adopters and more than two decades of published work demonstrate that digital pathology (DP) is a transformative yet inherently complex undertaking. Major findings from these early adopters are summarized in Table 3.7.

Across institutions, several recurring lessons have emerged. First, automation—especially barcode tracking, automated ingestion, and standardized metadata—significantly reduces friction and minimizes errors. Second, user acceptance hinges on performance; even brief image-loading delays erode confidence, making investment in high-performance networks and computing infrastructure essential. Third, hybrid phases must be explicitly defined, as unclear expectations about what is or is not signed out digitally create operational confusion. Fourth, QA cannot be treated as an afterthought; routine image QC, scanner checks, and performance monitoring are necessary to maintain diagnostic confidence and regulatory compliance. Finally, close communication between IT and pathology is indispensable. Institutions that treat DP as a joint clinical–technical initiative, rather than a laboratory project, are consistently more successful.

Digital pathology implementation represents a comprehensive modernization of diagnostic practice. It requires seamless integration of scanners, image management systems,

TABLE 3.7 Summary of digital pathology implementation studies.

Published studies	Key findings and experiences
Buck et al., 2014 (Community Hospital Validation) (Buck et al., 2014)	• Primary diagnosis using WSI is feasible in smaller community settings. • Robust validation protocols and user training are essential for pathologists' confidence. • Demonstrated early diagnostic concordance with glass slides.
Thorstenson et al., 2014 (Sweden, large-scale routine use) (Thorstenson et al., 2014)	• One of the first large-scale routine DP deployments. • Critical success factors included workflow standardization, high-throughput scanning, and strong institutional commitment. • Highlighted scanner uptime and IT scalability as ongoing challenges.
Stathonikos et al., 2019 (Dutch academic center, fully digital) (Stathonikos et al., 2019)	• Completed transition to 100% digital sign-out. • Key lessons: end-to-end workflow redesign, integration with LIS, and investment in continuous training. • Noted improved efficiency but cautioned that slide quality and scanning errors still require monitoring.
Ardon et al., 2021 (NYC tertiary cancer center, COVID-19) (Ardon et al., 2021)	• Rapid scaling of DP enabled remote diagnosis during COVID-19. • Identified the importance of flexible governance, emergency workflows, and the rapid deployment of home workstations. • Confirmed DP supports high-volume oncology workflows.

TABLE 3.7 Summary of digital pathology implementation studies—cont'd

Published studies	Key findings and experiences
Babawale et al., 2021 (All Wales experience) (Babawale et al., 2021)	• National-level validation demonstrated concordance across multiple sites and subspecialties. • Emphasized standardized validation protocols, cross-site training, and the need for robust QC checkpoints.
Eloy et al., 2021 (IPATIMUP implementation) (Eloy et al., 2021)	• Successful deployment required careful workflow mapping, barcode tracking, and standardized slide preparation. • Emphasized cultural change management and continuous pathologist feedback.
Temprana-Salvador et al., 2022 (Catalan Health Institute, 8 hospitals) (Temprana-Salvador et al., 2022)	• Large network transition underscored the value of centralized governance, unified standards, and phased rollout. • Early results showed workflow transparency improvements and stable turnaround times.
Ferreira et al., 2023 (Private Lab, CEDAP) (Ferreira et al., 2023)	• Private practice implementation requires cost-conscious decisions, scalable storage strategies, and strong vendor support. • DP enabled enhanced productivity and facilitated AI/automation readiness.
Clarke et al., 2025 (Large-scale DP implementation review) (Clarke et al., 2025)	• Summarized global lessons: organizational governance, change management, robust IT architecture, and quality assurance are the pillars of successful DP adoption. • Highlighted persistent challenges (storage, integration, scanning QC).
Daniel et al., 2025 (Remote Canadian regions) (Daniel et al., 2025)	• Remote-region deployment showed DP enables equitable access and subspecialty support. • Lessons included handling network limitations, scan logistics, and leveraging DP for regional collaboration.
Iwuajoku et al., 2025 (1-year equivalency and efficiency, academic center) (Iwuajoku et al., 2025)	• One-year of experience confirmed that DP maintains diagnostic accuracy and improves efficiency once workflows mature. • Identified early-stage bottlenecks (scanning, viewer performance) that diminish over time.
Montezuma et al., 2022 (Private practice challenges/ opportunities) (Montezuma et al., 2022)	• Private practices face unique challenges (cost, staffing, high case-mix variability). • Opportunities include faster turnaround, improved client service, and competitive differentiation.
Lujan et al., 2021b (Large academic institution during COVID-19) (G.M. Lujan et al., 2021b)	• COVID accelerated DP adoption, validating remote primary diagnosis and remote tumor boards. • Lessons included rapid IT mobilization, temporary regulatory flexibility, and increased acceptance among pathologists.
Retamero et al., 2020 (Multicenter hospital network) (Retamero et al., 2020)	• Demonstrated full routine digital diagnosis across multiple hospitals. Standardized staining and scanning protocols, as well as centralized storage, were critical. • DP improved workload distribution and consultation turnaround.

laboratory information systems, displays, and storage architecture. Data management and security strategies—including tiered storage, backup protocols, and disaster recovery—are foundational to reliable operation. Human factors are equally important: workflow redesign, staff training, and change-management strategies help build user confidence and efficiency, while identifying "digital champions" accelerates cultural acceptance. Rigorous validation and ongoing quality assurance ensure diagnostic equivalence and sustained reliability. Ultimately, phased implementation supported by sound financial planning, governance, and alignment with institutional strategy enables organizations to scale effectively and achieve long-term value.

The transition to digital workflows provides substantial clinical, operational, and academic advantages. Efficiency improves through reduced physical slide handling and faster case distribution. Diagnostic flexibility increases as remote access and telepathology support collaboration across sites and time zones. Digital archives expand educational resources and enable large-scale research, including AI development. Operational resilience is strengthened, allowing uninterrupted service during disruptions such as pandemics or natural disasters. Integrated QA processes and detailed audit trails further enhance quality and patient safety. Collectively, these benefits position pathology departments for innovation and long-term strategic growth (Clarke et al., 2025; G.M. Lujan et al., 2021b; Matias-Guiu et al., 2025; Baidoshvili et al., 2023).

Despite its advantages, DP adoption presents practical challenges. Substantial initial investment is required for scanners, storage, networking, and software. Cultural resistance may arise as staff transition from traditional microscopy to digital platforms, underscoring the need for training and engagement. Technical complexity—ensuring scanner calibration, image fidelity, and system interoperability—demands careful planning and ongoing support. Regulatory requirements for data privacy, retention, and validation must be met consistently. Addressing these challenges through structured planning, phased rollout, and continuous monitoring is essential to achieving a reliable and sustainable digital pathology practice (Clarke et al., 2025; G.M. Lujan et al., 2021b). Major key lessons learned from recently published studies are summarized in Table 3.8.

TABLE 3.8 Key lessons learned from recent studies.

Category	Key lessons reported across studies
Strategic planning & governance	• Strong leadership and clear institutional vision are essential. • Early stakeholder engagement (pathologists, IT, lab staff) is a critical success factor. • Phased or hybrid rollout approaches reduce disruption and allow learning.
Technology selection & infrastructure	• Scanner reliability, uptime, and throughput are major determinants of success. • Robust network bandwidth and high-availability storage are mandatory for large-scale adoption. • Viewer performance significantly affects pathologists' acceptance.
Workflow integration	• DP implementation must align with LIS, barcode systems, and specimen tracking for efficiency. • Automation (tracking, QC flags, load balancing) improves consistency and reduces errors.

TABLE 3.8 Key lessons learned from recent studies—cont'd

Category	Key lessons reported across studies
Operational challenges	• Slide quality variability, scanning failures, and rescan rates are persistent issues. • Staffing adjustments may be needed. • Remote-region implementations face unique logistical challenges.
Pathologist adoption & training	• Early skepticism is common but declines with experience. • Training and education accelerate acceptance. • DP benefits (flexibility, improved efficiency, AI integration, ergonomics, etc.) support long-term adoption.
Validation & regulatory compliance	• Validation for primary diagnosis is achievable across academic, community, and private settings. • Validation needs to be tailored to institutional workflows and subspecialty needs. • Regulatory compliance requires ongoing documentation and quality control.
Clinical performance & diagnostic impact	• Diagnostic concordance with microscopy is consistently high across studies. • Turnaround time is often maintained or improved once workflows stabilize. • Remote access, consultation, faster case sharing, and AI integration are major benefits.
Scalability & large-scale deployment	• Enterprise-scale systems require centralized governance, standardized protocols, and continuous QA. • Multihospital networks benefit from harmonized staining, scanning, and reporting processes.
Cost, storage, & sustainability	• Storage planning (tiered storage, image lifecycle management) is essential for long-term sustainability. • Upfront investment is high, but efficiency gains and reduced physical slide handling offset the costs over time.

Conclusions and future directions

The journey to a fully integrated digital pathology workflow is complex, requiring a careful balance of technology, people, processes, and policy. Institutions that approach implementation systematically—combining robust infrastructure, validation, quality assurance, training, and strategic planning—are well positioned to realize the full benefits of digital pathology.

Ultimately, the adoption of digital workflows represents a paradigm shift. By embracing digital pathology, healthcare institutions enhance diagnostic capability, improve patient outcomes, and open new horizons for research, education, and innovation. The lessons learned during early adoption will continue to inform best practices, guiding pathology departments worldwide toward a future that is more efficient, precise, and patient-centered.

Digital pathology is poised to become an integral component of precision medicine and modern laboratory practice. Future developments include:

- Artificial Intelligence and Computational Pathology: AI algorithms will assist with case triage, quantification of histologic features, predictive modeling, and biomarker discovery, potentially enhancing diagnostic accuracy and efficiency.
- Enterprise Imaging Integration: Seamless connection of pathology with radiology, genomics, and clinical imaging will support comprehensive patient profiles, enabling truly integrated diagnostics.
- Standardization and Interoperability: Widespread adoption of DICOM and other open standards will facilitate multivendor compatibility, long-term data accessibility, and regulatory compliance.
- Sustainability and Cost Optimization: As digital pathology matures, institutions will optimize resource allocation, reduce waste, and demonstrate ROI more clearly, supporting broader adoption.

AI disclosure

During the preparation of this work, the author(s) used ChatGPT to polish the writing. After using this tool/service, the author(s) reviewed and edited the content as needed and take(s) full responsibility for the content of the publication.

References

Abel, J.T., Ouillette, P., Williams, C.L., Blau, J., Cheng, J., Yao, K., Lee, W.Y., Cornish, T.C., Balis, U.G.J., McClintock, D.S., 2020. Display characteristics and their impact on digital pathology: a current review of pathologists' future "Microscope". J. Pathol. Inform. 21533539. 11 (1). Available from: https://www.sciencedirect.com/science/journal/21533539. doi:10.4103/jpi.jpi_38_20.

Alcaraz-Mateos, E., Hernández-Gómez, R., Calvente, E.R., Sánchez-Campoy, N., Martínez González-Moro, I., Caballero-Alemán, F., Poblet, E., 2022. Comparison of muscle activity while using different input devices in digital pathology. Revista Espanola de Patologia. 1988561X. 55 (1), 19–25. Available from: http://www.sciencedirect.com/science/journal/16998855. doi:10.1016/j.patol.2021.02.005.

Ameisen, D., Deroulers, C., Perrier, V., Bouhidel, F., Battistella, M., Legrès, L., Janin, A., Bertheau, P., Yunès, J.-B., 2014. Towards better digital pathology workflows: programming libraries for high-speed sharpness assessment of Whole Slide Images. Diagn. Pathol. 1746-1596. 9 (S1). doi:10.1186/1746-1596-9-s1-s3.

Ardon, O., Asa, S.L., Lloyd, M.C., Lujan, G., Parwani, A., Santa-Rosario, J.C., Van Meter, B., Samboy, J., Pirain, D., Blakely, S., Hanna, M.G., 2024. Understanding the financial aspects of digital pathology: a dynamic customizable return on investment calculator for informed decision-making. J. Pathol. Inform. 21533539. 15. Available from: https://www.sciencedirect.com/science/journal/21533539. doi:10.1016/j.jpi.2024.100376.

Ardon, O., Labasin, M., Friedlander, M., Manzo, A., Corsale, L., Ntiamoah, P., Wright, J., Elenitoba-Johnson, K., Reuter, V.E., Hameed, M.R., Hanna, M.G., 2023. Quality management system in clinical digital pathology operations at a Tertiary Cancer Center. Lab. Investig. 00236837. 103 (11), 100246. doi:10.1016/j.labinv.2023.100246.

Ardon, O., Reuter, V.E., Hameed, M., Corsale, L., Manzo, A., Sirintrapun, S.J., Ntiamoah, P., Stamelos, E., Schueffler, P.J., England, C., Klimstra, D.S., Hanna, M.G., 2021. Digital pathology operations at an NYC Tertiary cancer center during the first 4 months of COVID-19 pandemic response. Acad. Pathol. 23742895. 8. 23742895211010276. doi:10.1177/23742895211010276.

Babawale, M., Gunavardhan, A., Walker, J., Corfield, T., Huey, P., Savage, A., Bansal, A., Atkinson, M., Abdelsalam, H., Raweily, E., Christian, A., Evangelou, I., Thomas, D., Shannon, J., Youd, E., Brumwell, P., Harrison, J., Thompson, I., Rashid, M., Leopold, G., Finall, A., Roberts, S., Housa, D., Nedeva, P., Davies, A., Fletcher, D., Aslam, M., 2021. Verification and validation of digital pathology (whole slide imaging) for primary histopathological diagnosis: all wales experience. J. Pathol. Inform. 21533539. 12 (1), 4. doi:10.4103/jpi.jpi_55_20.

Baidoshvili, A., Bucur, A., van Leeuwen, J., van der Laak, J., Kluin, P., van Diest, P.J., 2018. Evaluating the benefits of digital pathology implementation: time savings in laboratory logistics. Histopathology. 13652559. 73 (5), 784–794. Available from: http://onlinelibrary.wiley.com/journal/10.1111/(ISSN)1365-2559. doi:10.1111/his.13691.

Baidoshvili, A., Khacheishvili, M., van der Laak, J.A.W.M., van Diest, P.J., 2023. A whole-slide imaging based workflow reduces the reading time of pathologists. Pathol. Int. 14401827. 73 (3), 127–134. Available from: http://onlinelibrary.wiley.com/journal/10.1111/(ISSN)1440-1827. doi:10.1111/pin.13309.

Bautista, P.A., Hashimoto, N., Yagi, Y., 2014. Color standardization in whole slide imaging using a color calibration slide. J. Pathol. Inform. 21533539. 5 (1), 4. doi:10.4103/2153-3539.126153.

Browning, L., Colling, R., Rittscher, J., Winter, L., McEntyre, N., Verrill, C., 2020. Implementation of digital pathology into diagnostic practice: perceptions and opinions of histopathology trainees and implications for training. J. Clin. Pathol. 0021-9746. 73 (4), 223–227. doi:10.1136/jclinpath-2019-206137.

Bruce, C., Prassas, I., Mokhtar, M., Clarke, B., Youssef, E., Wang, C., Yousef, G.M., 2024. Transforming diagnostics: the implementation of digital pathology in clinical laboratories. Histopathology. 0309-0167. 85 (2), 207–214. doi:10.1111/his.15178.

Buck, T.P., Dilorio, R., Havrilla, L., O'Neill, D.G., 2014. Validation of a whole slide imaging system for primary diagnosis in surgical pathology: a community hospital experience. J. Pathol. Inform. 21533539. 5 (1), 43. doi:10.4103/2153-3539.145731.

Cazzaniga, G., Mascadri, F., Marletta, S., Caputo, A., Guidi, G., Gambaro, G., Eccher, A., Tos, A.P.D., Pagni, F., L'Imperio, V., 2025. Benchmarking digital displays (monitors) for histological diagnoses: the nephropathology use case. J. Clin. Pathol. 0021-9746. 78 (11), 798–804. doi:10.1136/jcp-2024-209418.

Chong, Y., Bae, J.M., Kang, D.W., Kim, G., Han, H.S., 2022. Development of quality assurance program for digital pathology by the Korean Society of Pathologists. J. Pathol. Transl. Med. 23837845. 56 (6), 370–382. Available from: http://jpatholtm.org/journal/view.php?doi=10.4132/jptm.2022.09.30. doi:10.4132/jptm.2022.09.30.

Clarke, B., Carment-Baker, C., Bruce, C., Hanna, K., Yousef, G.M., 2025. Large scale implementation of DP for clinical diagnoses: experience, challenges, and lessons learned. Crit. Rev. Clin. Lab. Sci. 1040-8363. 1–15. doi:10.1080/10408363.2025.2549309.

Clarke, E.L., Munnings, C., Williams, B., Brettle, D., Treanor, D., 2020. Display evaluation for primary diagnosis using digital pathology. J. Med. Imaging. 23294310. 7 (2). Available from: http://medicalimaging.spiedigitallibrary.org/journal.aspx. doi:10.1117/1.JMI.7.2.027501.

Clarke, E.L., Treanor, D., 2017. Colour in digital pathology: a review. Histopathology. 13652559. 70 (2), 153–163. Available from: http://www3.interscience.wiley.com/journal/118511588/home. doi:10.1111/his.13079.

Clunie, D., Hosseinzadeh, D., Wintell, M., De Mena, D., Lajara, N., Garcia-Rojo, M., Bueno, G., Saligrama, K., Stearrett, A., Toomey, D., Abels, E., Apeldoorn, F.V., Langevin, S., Nichols, S., Schmid, J., Horchner, U., Beckwith, B., Parwani, A., Pantanowitz, L., 2018. Digital imaging and communications in medicine whole slide imaging connectathon at digital pathology association pathology visions 2017. J. Pathol. Inform. 21533539. 9 (1). Available from: www.jpathinformatics.org/browse.asp?sabs=n. doi:10.4103/jpi.jpi_1_18.

Daniel, M., Nowak, K., Vajpeyi, R., Clarke, B., Evans, A., Carment-Baker, C., Weiser, K., Martin, M., Girard, N., Fyfe, K., Zeidan, S., Bruce, C., Yousef, G.M., 2025. From microscopes to monitors: unique opportunities and challenges in digital pathology implementation in remote canadian regions. Diagnostics. 2075-4418. 15 (16), 1983. doi:10.3390/diagnostics15161983.

Eloy, C., Fraggetta, F., van Diest, P.J., Polónia, A., Curado, M., Temprana-Salvador, J., Zlobec, I., Purqueras, E., Weis, C.A., Matias-Guiu, X., Schirmacher, P., Ryška, A., 2025. Digital transformation of pathology—The European Society of Pathology expert opinion paper. Virchows Arch. 14322307. 487 (5), 971–981. Available from: https://www.springer.com/journal/428. doi:10.1007/s00428-025-04090-w.

Eloy, C., Vale, J., Curado, M., Polónia, A., Campelos, S., Caramelo, A., Sousa, R., Sobrinho-Simões, M., 2021. Digital pathology workflow implementation at IPATIMUP. Diagnostics. 2075-4418. 11 (11), 2111. doi:10.3390/diagnostics11112111.

Evans, A.J., Brown, R.W., Bui, M.M., Chlipala, E.A., Lacchetti, C., Milner, D.A., Pantanowitz, L., Parwani, A.V., Reid, K., Riben, M.W., Reuter, V.E., Stephens, L., Stewart, R.L., Thomas, N.E., 2022. Validating whole slide imaging systems for diagnostic purposes in pathology guideline update from the College of American Pathologists in Collaboration with the American Society for Clinical Pathology and the Association for Pathology Informatics. Arch. Pathol. Lab. Med. 15432165. 146 (4), 440–450. Available from: https://meridian.allenpress.com/aplm/article/146/4/440/464968/Validating-Whole-Slide-Imaging-Systems-for. doi:10.5858/arpa.2020-0723-CP.

Fabián, Švajdler, Jirásek, 2025. Integration of digital pathology workflow in the anatomic pathology laboratory. Ceskoslovenska Patologie 61 (1), 22–28.

Ferreira, I., Montenegro, C.S., Coelho, D., Pereira, M., da Mata, S., Carvalho, S., Araújo, A.C., Abrantes, C., Mário Ruivo, J., Garcia, H., Oliveira, R.C., 2023. Digital pathology implementation in a private laboratory: the CEDAP experience. J. Pathol. Inform. 21533539. 14, 100180. doi:10.1016/j.jpi.2022.100180.

Flach, R.N., Fransen, N.L., Sonnen, A.F.P., Nguyen, T.Q., Breimer, G.E., Veta, M., Stathonikos, N., van Dooijeweert, C., van Diest, P.J., 2022. Implementation of artificial intelligence in diagnostic practice as a next step after going digital: the UMC Utrecht perspective. Diagnostics. 20754418. 12 (5). Available from: https://www.mdpi.com/2075-4418/12/5/1042/pdf?version=1650523969. doi:10.3390/diagnostics12051042.

Fraggetta, F., Caputo, A., Guglielmino, R., Pellegrino, M.G., Runza, G., L'imperio, V., 2021a. A survival guide for the rapid transition to a fully digital workflow: the "caltagirone example. Diagnostics. 20754418. 11 (10). Available from: https://www.mdpi.com/2075-4418/11/10/1916/pdf. doi:10.3390/diagnostics11101916.

Fraggetta, F., Garozzo, S., Zannoni, G.F., Pantanowitz, L., Rossi, E.D., 2017. Routine digital pathology workflow: the Catania experience. J. Pathol. Inform. 21533539. 8 (1), 51. doi:10.4103/jpi.jpi_58_17.

Fraggetta, F., L'Imperio, V., Ameisen, D., Carvalho, R., Leh, S., Kiehl, T.-R., Serbanescu, M., Racoceanu, D., Mea, V. D., Polonia, A., Zerbe, N., Eloy, C., 2021b. Best practice recommendations for the implementation of a digital pathology workflow in the anatomic pathology laboratory by the European Society of Digital and Integrative Pathology (ESDIP). Diagnostics. 2075-4418. 11 (11), 2167. doi:10.3390/diagnostics11112167.

Hartman, D.J., Imaging, W.-S., 2020. Clinical workflows and primary diagnosis. Adv. Anat. Pathol. 15334031. 27 (4), 236–240. Available from: http://www.anatomicpathology.com. doi:10.1097/PAP.0000000000000272.

Ho, J., Ahlers, S.M., Stratman, C., Aridor, O., Pantanowitz, L., Fine, J.L., Kuzmishin, J.A., Montalto, M.C., Parwani, A. V., 2014. Can digital pathology result in cost savings? A financial projection for digital pathology implementation at a large integrated health care organization. J. Pathol. Inform. 21533539. 5 (1), 33. doi:10.4103/2153-3539.139714.

Ho, J., Aridor, O., Glinski, D.W., Saylor, C.D., Pelletier, J.P., Selby, D.M., Davis, S.W., Lancia, N., Gerlach, C.B., Newberry, J., Anthony, L., Pantanowitz, L., Parwani, A.V., 2013. Needs and workflow assessment prior to implementation of a digital pathology infrastructure for the US Air Force Medical Service. J. Pathol. Inform. 21533539. 4 (1), 32. doi:10.4103/2153-3539.122388.

Ho, J., Parwani, A.V., Jukic, D.M., Yagi, Y., Anthony, L., Gilbertson, J.R., 2006. Use of whole slide imaging in surgical pathology quality assurance: design and pilot validation studies. Hum. Pathol. 37 (3), 322–331. doi:10.1016/j.humpath.2005.11.005.

Hsu, Y.-H.R., Ahmed, I., Phlamon, J., Carment-Baker, C., Chan, J.Y.T., Prassas, I., Weiser, K., Zeidan, S., Clarke, B., Yousef, G.M., 2025. An adapted & improved validation protocol for digital pathology implementation. Semin. Diagn. Pathol. 07402570. 42 (4), 150905. doi:10.1016/j.semdp.2025.150905.

Inoue, Yagi, 2020. Color standardization and optimization in whole slide imaging. Clin. Diagn. Pathol. 4 (1). doi:10.15761/cdp.1000139.

Iwuajoku, V., Ekici, K., Haas, A., Khan, M.Z., Kazemi, A., Kasajima, A., Delbridge, C., Muckenhuber, A., Schmoeckel, E., Stögbauer, F., Bollwein, C., Schwamborn, K., Steiger, K., Mogler, C., Schüffler, P.J., 2025. An equivalency and efficiency study for one year digital pathology for clinical routine diagnostics in an accredited tertiary academic center. Virchows Arch. 14322307. 487 (1), 3–12. Available from: https://www.springer.com/journal/428. doi:10.1007/s00428-025-04043-3.

Jones-Hall, Y., 2021. Digital pathology in academia: implementation and impact. Lab Anim. 0093-7355. 50 (9), 229–231. doi:10.1038/s41684-021-00828-6.

Kim, Y.J., Roh, E.H., Park, S., 2021. A literature review of quality, costs, process-associated with digital pathology. J. Exerc. Rehabil. 22881778. 17 (1), 11–14. Available from: e-jer.org/. doi:10.12965/jer.2142018.009.

Kuzmak, P.M., Dayhoff, R.E., 2000. Use of Digital Imaging and Communications in Medicine (DICOM) in the integration of imaging into the electronic patient record at the Department of Veterans Affairs. J. Digit. Imaging. 08971889. 13 (2), 133–137. doi:10.1007/bf03167644.

Lianas, L., Rio, M.D., Pireddu, L., Aspegren, O., Giunchi, F., Fiorentino, M., Leo, S., Zelic, R., Vincent, P.H., Destefanis, N., Zugna, D., Richiardi, L., Pettersson, A., Akre, O., Frexia, F., 2025. An open-source platform for structured annotation and computational workflows in digital pathology research. Sci. Rep. 2045-2322. 15 (1). doi:10.1038/s41598-025-13546-7.

Lujan, G.M., Savage, J., Shana'ah, A., Yearsley, M., Thomas, D., Allenby, P., Otero, J., Limbach, A.L., Cui, X., Scarl, R. T., Hardy, T., Sheldon, J., Plaza, J.A., Whitaker, B., Frankel, W., Parwani, A.V., Li, Z., 2021b. Digital pathology initiatives and experience of a large academic institution during the coronavirus disease 2019 (COVID-19) pandemic. Arch. Pathol. Lab. Med. 15432165. 145 (9), 1051–1061. Available from: https://allen.silverchair-cdn.com. doi:10.5858/arpa.2020-0715-SA.

Lujan, G., Quigley, J.C., Hartman, D., Parwani, A., Roehmholdt, B., Van Meter, B., Ardon, O., Hanna, M.G., Kelly, D., Sowards, C., Montalto, M., Bui, M., Zarella, M.D., LaRosa, V., Slootweg, G., Retamero, J.A., Lloyd, M.C., Madory, J., Bowman, D., 2021a. Dissecting the business case for adoption and implementation of digital pathology: a white paper from the digital pathology association. J. Pathol. Inform. 21533539. 12 (1). Available from: https://www.sciencedirect.com/journal/journal-of-pathology-informatics/about/aims-and-scope. doi:10.4103/jpi.jpi_67_20.

Matias-Guiu, X., Temprana-Salvador, J., Lopez, P.G., Kammerer-Jacquet, S.F., Rioux-Leclercq, N., Clark, D., Schürch, C.M., Fend, F., Mattern, S., Snead, D., Fusco, N., Guerini-Rocco, E., Rojo, F., Brevet, M., Tellez, M.S., Tos, A.D., Di Maio, T., Ramírez-Peinado, S., Sheppard, E., Bannister, H., Gkiokas, A., Arpaia, M., Dhia, O.B., Martino, N., 2025. Implementing digital pathology: qualitative and financial insights from eight leading European laboratories. Virchows Arch. 14322307. 487 (4), 815–826. Available from: https://www.springer.com/journal/428. doi:10.1007/s00428-025-04064-y.

Mikkelsen, M.L.N., Frederiksen, M.H., Marcussen, N., Williams, B., Kidholm, K., 2022. Prior to implementation of digital pathology—assessment of expectations among staff by means of normalization process theory. Int. J. Environ. Res. Public Health. 16604601. 19 (12). Available from: https://www.mdpi.com/1660-4601/19/12/7253/pdf?version=1655185645. doi:10.3390/ijerph19127253.

Montezuma, D., Monteiro, A., Fraga, J., Ribeiro, L., Gonçalves, S., Tavares, A., Monteiro, J., Macedo-Pinto, I., 2022. Digital pathology implementation in private practice: specific challenges and opportunities. Diagnostics. 2075-4418. 12 (2), 529. doi:10.3390/diagnostics12020529.

Patel, A., Balis, U.G.J., Cheng, J., Li, Z., Lujan, G., McClintock, D.S., Pantanowitz, L., Parwani, A., 2021. Contemporary whole slide imaging devices and their applications within the modern pathology department: a selected hardware review. J. Pathol. Inform. 21533539. 12 (1), 50. doi:10.4103/jpi.jpi_66_21.

Retamero, J.A., Aneiros-Fernandez, J., Del Moral, R.G., 2020. Complete digital pathology for routine histopathology diagnosis in a multicenter hospital network. Arch. Pathol. Lab. Med. 15432165. 144 (2), 221–228. Available from: https://www.archivesofpathology.org/doi/pdf/10.5858/arpa.2018-0541-OA. doi:10.5858/arpa.2018-0541-OA.

Rogers, J., Vedaraju, Y., Hsu, J., Kinskey, J., Wesley Long, S., Christensen, P., 2025. A comparative usability assessment of computer input devices for navigating digital whole slide images. J. Pathol. Inform. 21533539. 18, 100449. doi:10.1016/j.jpi.2025.100449.

Singh, R., Chubb, L., Pantanowitz, L., Parwani, A., 2011. Standardization in digital pathology: supplement 145 of the DICOM standards. J. Pathol. Inform. 21533539. 2 (1), 23. doi:10.4103/2153-3539.80719.

Soliman, J., Weiser, K., Ahmed, I., Carment-Baker, C., Hockley, M., Prassas, I., Zeidan, S., Bruce, C., Clarke, B., Yousef, G.M., 2025. Selecting high-throughput scanners for clinical use: a multicenter institution experience. Am. J. Clin. Pathol. 0002-9173. 164 (4), 589–599. doi:10.1093/ajcp/aqaf083.

Stathonikos, N., Nguyen, T.Q., Spoto, C.P., Verdaasdonk, M.A.M., van Diest, P.J., 2019. Being fully digital: perspective of a Dutch academic pathology laboratory. Histopathology. 13652559. 75 (5), 621–635. Available from: http://onlinelibrary.wiley.com/journal/10.1111/(ISSN)1365-2559. doi:10.1111/his.13953.

Stathonikos, N., Veta, M., Huisman, A., van Diest, P.J., 2013. Going fully digital: perspective of a Dutch academic pathology lab. J. Pathol. Inform. 21533539. 4 (1), 15. doi:10.4103/2153-3539.114206.

Temprana-Salvador, J., López-García, P., Castellví Vives, J., de Haro, L., Ballesta, E., Abusleme, M.R., Arrufat, M., Marques, F., Casas, J.R., Gallego, C., Pons, L., Luis Mate, J., Fernández, P.L., López-Bonet, E., Bosch, R., Martínez, S., Ramón Y Cajal, S., Matias-Guiu, X., 2022. DigiPatICS: digital pathology transformation of the Catalan Health Institute Network of 8 Hospitals—planification, implementation, and preliminary results. Diagnostics. 2075-4418. 12 (4), 852. doi:10.3390/diagnostics12040852.

Thorstenson, S., Molin, J., Lundström, C., 2014. Implementation of large-scale routine diagnostics using whole slide imaging in Sweden: digital pathology experiences 2006-2013. J. Pathol. Inform. 21533539. 5 (1), 14. doi:10.4103/2153-3539.129452.

Williams, B.J., Williams, B.J., Treanor, D., Treanor, D., 2020. Practical guide to training and validation for primary diagnosis with digital pathology. J. Clin. Pathol. 14724146. 73 (7), 418–422. Available from: http://jcp.bmj.com/. doi:10.1136/jclinpath-2019-206319.

Zarella, M.D., Bowman, D., Aeffner, F., Farahani, N., Xthona, A., Absar, S.F., Parwani, A., Bui, M., Hartman, D.J., 2019. A practical guide to whole slide imaging a white paper from the digital pathology association. Arch. Pathol. Lab. Med. 15432165. 143 (2), 222–234. Available from: http://www.archivesofpathology.org/doi/pdf/10.5858/arpa.2018-0343-RA. doi:10.5858/arpa.2018-0343-RA.

CHAPTER

4

Clinical applications of digital pathology

Giovanni Mario Lujan, Anil V. Parwani, and Zaibo Li

Department of Pathology, The Ohio State University, Columbus, OH, United States

Introduction

Digital pathology (DP) is widely regarded as one of the most consequential shifts in diagnostic medicine since the mainstreaming of immunohistochemistry and molecular testing. For centuries, the light microscope defined both the identity and the daily cadence of pathology: slides were prepared in histology, handed off to the pathologist, and examined through the eyepiece, field by field, until a diagnosis could be stated confidently. The method is robust and familiar, with all its rituals—the click of objective lenses, the pencil sketch of a gland or a granuloma, the handwritten note in the margin—interwoven with the profession's culture.

The 21st century, however, has recomposed the diagnostic toolkit. With whole slide imaging (WSI), the glass slide can be rendered as a navigable digital object, enabling enterprise networks, teleconsultation (Pantanowitz et al., 2018), and algorithmic analysis. The microscope's field of view becomes a zoomable canvas, and the pathologist becomes not only a morphologist but also a curator of data, a convener of multidisciplinary conversations, and, increasingly, the "conductor" who coordinates inputs from computation, imaging, and molecular assays.

Before digitization, the analog strengths and weaknesses of pathology were both obvious. The microscope, elegant in its simplicity, requires no network, no login, and no license server. Yet slides are fragile; stains fade; labels detach; and archives expand relentlessly into basements and closets. Retrieving a case may involve ladders and cardboard rather than search bars and filters. Sharing material is slow and expensive, relying on couriers, customs clearance, cold chains for certain special stains, and the perpetual anxiety that an irreplaceable block might go missing.

Meanwhile, radiology completed its digital transformation decades earlier. Picture archiving and communication systems (PACS) digitized images, standardized formats such as DICOM enabled interoperability (Leng et al., 2024), and instantaneous sharing across health

Advances in Digital Pathology and Artificial Intelligence in Pathology
https://doi.org/10.1016/B978-0-443-40383-5.00004-2

systems became routine. Pathology lagged for understandable technical reasons: WSI files are massive, tissue is three-dimensional and variably stained, and image fidelity must support subtle diagnostic distinctions (Huisman et al., 2010). Yet this delay also conferred an advantage. When digitization finally arrived, pathology could leapfrog into a more integrated, enterprise-aligned ecosystem rather than retracing radiology's early missteps.

The early days of DP were, fundamentally, telepathology days. Static photomicrographs transmitted over dial-up connections and robotic microscopes controlled remotely formed the first generation of digital consultation (Weinstein et al., 1987). Scandinavian and Canadian programs deployed these systems for frozen sections at small or rural hospitals, allowing an on-call pathologist in a city to guide a technologist hundreds of kilometers away through a difficult margin assessment (Henriksen et al., 2020).

These early pilots proved two enduring principles. First, distance should not determine quality: a patient in a rural clinic deserves the same subspecialty insight as a patient treated near a university hospital. Second, pathologists adapt readily when the patient's benefit is clear, even when early technology is cumbersome or imperfect.

The inflection point came when scanners could capture entire slides at diagnostic resolution quickly enough for routine clinical work. What began as tools for education and tumor boards matured into clinical platforms capable of supporting primary diagnosis. The 2017 FDA clearance of the Philips IntelliSite Pathology Solution for primary diagnosis was not merely a regulatory milestone but a cultural one. It reassured health systems that DP was safe and reliable and motivated vendors to harden products for high-volume laboratories operating under clinical constraints (Evans et al., 2022).

Today, DP is not simply a digital substitute for the microscope. It functions as an integration layer (Hanna et al., 2022). Images link directly to laboratory information systems (LIS) and electronic medical records (EMR) (Lloyd et al., 2016) tumor boards navigate WSI alongside CT scans and molecular reports; residents learn from curated digital teaching sets; and researchers treat slides as analyzable data rather than static artifacts (Bueno et al., 2016).

Three broad shifts describe this transformation.

From locality to ubiquity

A slide is no longer "in the drawer" but exists within a secured, shared digital namespace. A consultant across town—or across the world—can review it within minutes rather than days.

From description to quantification

Counting mitoses, estimating tumor cellularity, or assigning ER and HER2 scores become measurable, reproducible, and auditable, supported by visual overlays and digital records.

From the pathologist alone to the diagnostic ensemble

The pathologist coordinates morphology with radiology, genomics, and computation, shaping a coherent diagnostic narrative for the clinical team.

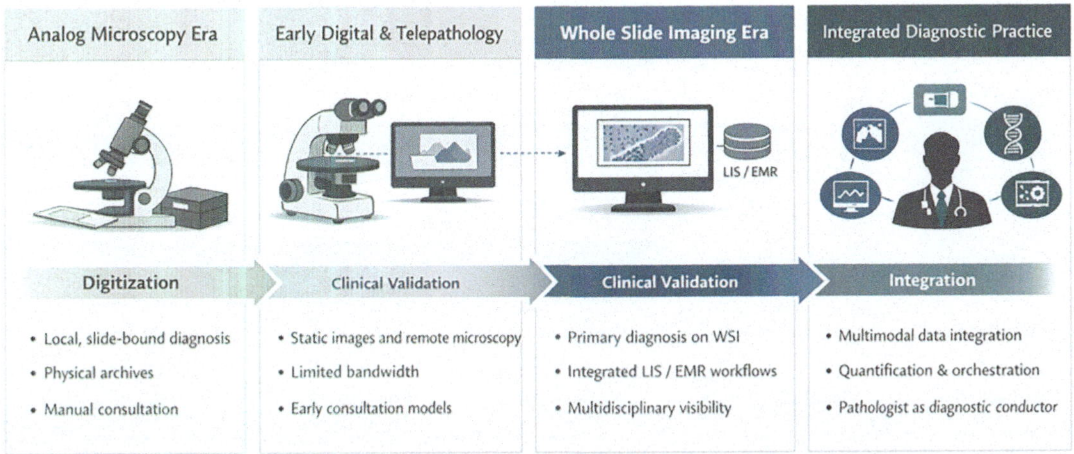

FIG. 4.1 Evolution of pathology practice from microscope-centered diagnosis to digitally integrated diagnostic workflows. Traditional analog microscopy relied on localized slide review and physical archives. Early digital pathology enabled static image sharing and remote consultation but remained limited in scope. Modern whole-slide imaging supports primary diagnosis, enterprise integration, and multidisciplinary collaboration. In integrated diagnostic practice, the pathologist coordinates morphology, radiology, molecular data, and computational tools into a unified diagnostic narrative.

These transitions from localized microscopy to digitally integrated diagnostic practices are summarized in Fig. 4.1.

In short, DP relocates pathology from a practice bound to the microscope to a discipline embedded within the broader data flows of precision medicine.

Primary diagnosis in surgical pathology

Validation of equivalence

The threshold question for any laboratory considering DP is blunt: can a diagnosis made on a screen be trusted as much as one made at the microscope? After more than a decade of systematic study, the answer is yes, with caveats that closely mirror those of analog practice.

The clinical digital pathology workflow supporting primary diagnosis is illustrated in Fig. 4.2.

Systematic reviews encompassing thousands of cases consistently report 95%–99% concordance between WSI and conventional microscopy (Goacher et al., 2017). The residual discrepancies are overwhelmingly minor and rarely alter clinical management. The multicenter randomized noninferiority trial conducted by Mukhopadhyay and colleagues was decisive (Mukhopadhyay et al., 2018). Spanning nearly 2000 cases across multiple subspecialties, the study demonstrated that WSI was noninferior to conventional microscopy for primary diagnosis. This trial, together with supporting validation studies, underpinned regulatory clearance and emboldened health systems to invest in digital workflows.

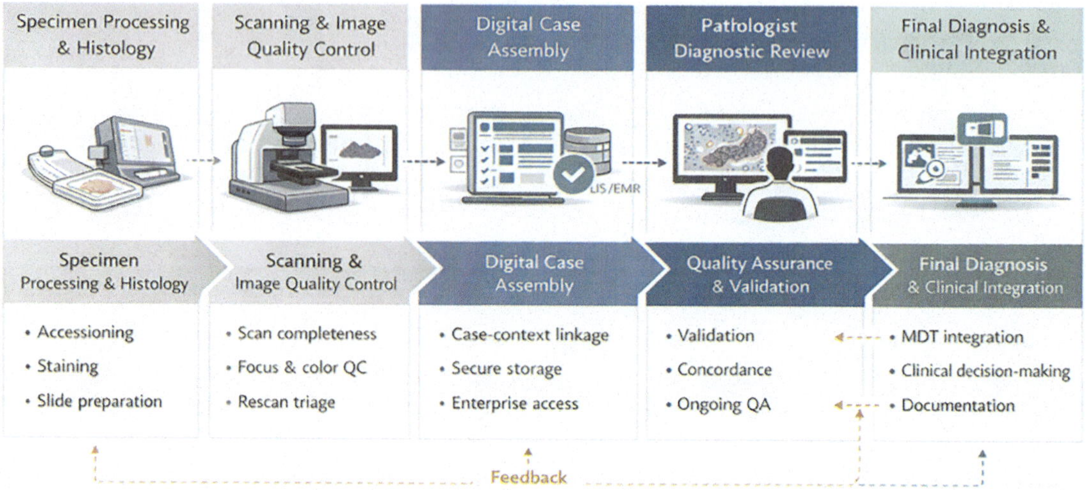

FIG. 4.2 End-to-end digital pathology workflow supporting primary diagnosis. Following specimen processing and histology, slides are digitized using whole slide imaging with defined image quality control and rescan triage. Digital slides are assembled with case metadata and integrated into laboratory and enterprise systems. Pathologists perform diagnostic reviews using calibrated workstations with annotation and navigation tools. Quality assurance and governance processes, including validation and concordance reviews, provide continuous feedback to upstream steps. Final diagnoses are delivered through structured reports and integrated into multidisciplinary team (MDT) clinical workflows.

Real-world validations extended these findings beyond carefully controlled trials. The All Wales digital pathology program analyzed over 7,400 slides across seven hospitals and reported a concordance rate of 97.1% (Mukhopadhyay et al., 2018). In Eastern Québec, telepathology supported frozen sections and primary diagnoses in remote hospitals, directly influencing surgical decision-making on the same day (Têtu et al., 2014). Dutch centers, including Utrecht and Radboud University Medical Center, documented the transition to fully digital sign-out, detailing not only diagnostic accuracy but also the workflow redesign, scan triage, rescan policies, monitor calibration, and training required to sustain it (Stathonikos et al., 2013).

Together, these experiences highlight a practical lesson: diagnostic equivalence is not achieved abstractly but locally—by validating one's own case mix, stains, scanners, and pathologists, and by allowing sufficient time for digital navigation to become intuitive.

Surgical subspecialty perspectives: What actually changes

The differential impact of digital pathology across subspecialties is summarized in Fig. 4.3.

Dermatopathology: Melanocytic nuance and collaborative vision

If any subspecialty invites skepticism about pixels, it is dermatopathology. Melanocytic lesions demand meticulous attention to nuclear detail, pagetoid scatter, nesting patterns, and the architecture of the epidermal junction. Early concerns centered on whether compression

	Case Volume	Imaging Needs	Molecular / Biomarker Integration	Tumor Boards / MDT	Collaboration Needs	AI Opportunity
Surgical Pathology	✓ Large & diverse case load	✓ Full slide imaging	✓ Mandatory component in oncology	✓ Core part of cancer MDTs	✓ Multidisciplinary review, remote sign-out	✓ TMA analysis, tumor grading
Hemato-pathology	✓ High case complexity	✓ Immuno-phenotyping integration	✓ NGS, flow cytometry	✓ Integral to hem/onc consultations	✓ Molecular corrolate dual reviewers	✓ Decision support tools risk assessment
Dermatopat-hology	✓ High case throughput	✓ High-resolution imaging	✓ Molecular subtyping (evolving)	✓ Occasional consultative role	✓ Remote practice, referral cases	✓ Triage, classification guidance
Cytopathology	✓ Rapid onsite evaluations	✓ Digital slide cytology	✓ Liquid biopsy, genetic testing	✓ Variable role in some cases	✓ Cytologist-pathologist review, education	✓ Triage, classification guidance
Molecular Pathology	✓ NGS & companion dx testing		✓ Central coordination in precision medicine	✓ Drives genomics based-tumor boards	✓ Interdisciplinary reporting, bioinformatics	✓ Variant classification prognosic models

FIG. 4.3 Differential impact of clinical digital pathology across pathology subspecialties. The clinical value of digital pathology varies according to case volume, imaging requirements, molecular and biomarker integration, multidisciplinary team (MDT) participation, collaboration needs, and opportunities for computational augmentation. Surgical pathology, hematopathology, dermatopathology, cytopathology, and molecular pathology each leverage digital workflows in distinct ways, reflecting differences in diagnostic complexity, integration demands, and clinical context.

artifacts or color drift might obscure nuance. In practice, concordance between WSI and glass is high, and DP adds tangible capabilities (Goacher et al., 2017).

These include side-by-side display of multiple biopsies, precise bookmarking of critical regions, and melanoma tumor boards in which dermatologists, oncologists, and pathologists view the same digital field simultaneously, rather than crowding around a multiheaded microscope. As one senior dermatopathologist remarked to a trainee, "You miss less context because you see more at once."

Genitourinary pathology: Gleason grading, standardization, and the helpful algorithm

Prostate cancer grading remains one of the most consequential and debated tasks in surgical pathology. Digital platforms support reproducible Gleason scoring and enable the integration of computational tools that pre-annotate suspicious glands or quantify pattern distributions. High-impact studies demonstrate that such tools can achieve expert-level performance and serve as consistency-enhancing adjuncts rather than replacements for human judgment (Nagpal et al., 2020; Bulten et al., 2020).

Consider a common vignette: a community hospital uploads a nine-core prostate biopsy set late in the day. A genitourinary specialist at a referral center reviews the WSI remotely, adds targeted annotations, and an algorithm highlights a subtle focus of higher-grade disease. The final diagnosis and management plan are aligned by the following morning without courier delays.

Breast pathology: Quantification for therapies that hinge on numbers

Therapeutic decisions in breast oncology depend on quantitative biomarkers such as ER, PR, HER2, and Ki-67. Digital image analysis provides consistent, auditable measurements with visual overlays that show precisely which cells were counted (Cornish, 2020). In multidisciplinary conferences, oncologists readily grasp a Ki-67 index when it is presented alongside distribution plots and representative tiles, and they can interrogate thresholds in real time. The pathologist can demonstrate how biologically plausible cutoffs were chosen and how staining variability was controlled (Sato et al., 2021).

Neuropathology: Rarity rewards connectivity

Neuropathology's rarity and complexity make it particularly well suited to digital workflows (Rosado et al., 2025). Validation studies confirm accuracy for primary diagnosis, but the value of DP is magnified by connectivity (Alassiri et al., 2020). A pediatric glioma discussed at an international conference can include the entire WSI rather than selected thumbnails. When a rapid second opinion is required, digital sharing compresses what was once a week of shipping into hours of review, allowing surgery to proceed with greater confidence.

Hematopathology: Layered interpretation and synchronized views

Bone marrow and lymph node interpretation is inherently integrative, combining morphology with immunohistochemistry, flow cytometry, and clinical context. WSI has been validated for marrow and lymphoid tissue, and digital viewers make correlative review intuitive. H&E, reticulin, and immunostains can be opened in synchronized panes, flow plots displayed alongside annotated regions of interest, and quantitative outputs incorporated directly into structured reports. The diagnostic narrative becomes layered rather than fragmented (Shean and Rets, 2025; Wang et al., 2022).

Practical lessons from early programs

Adoption of DP is as much sociological as it is technical. Early adopters consistently reported temporary efficiency losses that resolved with experience and ergonomic optimization. High-quality monitors, color calibration, responsive input devices, and dual-screen configurations proved critical. Histology workflows improved when scanning moved upstream and triage rules were explicit—for example, prioritizing biopsies and same-day cases or flagging blocks likely to require rescans.

Governance emerged as a cornerstone of trust. Written validation plans, representative case lists, clear definitions of major versus minor discordance, and routine QA reviews fostered confidence. Perhaps most importantly, programs that engaged pathologists early—through listening sessions, pilot projects, and subspecialty champions—encountered less resistance than those that announced digital transitions by administrative decree.

Case vignettes from the COVID-19 era

The COVID-19 pandemic served as an unplanned stress test for DP. At major cancer centers, stay-at-home orders forced immediate pivots to remote sign-out. Secure VPNs, home workstations with calibrated displays, and rapid validation protocols were implemented within days. Skeleton crews staffed histology laboratories, while pathologists finalized cases remotely. Turnaround times were maintained, and coverage expanded when staff were quarantined (G.M. Lujan et al., 2021b).

At academic centers, including large public university hospitals, DP workflows scaled rapidly. Biopsies were scanned the same day, frozen sections were digitized when second opinions were needed, and even consumer-grade monitors were validated using defined luminance and color criteria. The experience clarified a central lesson: resilience in the face of systemic disruption is itself a patient-safety feature.

Remote primary diagnosis (digital sign-out)

The digital infrastructure supporting remote primary diagnosis and operational resilience is illustrated in Fig. 4.4.

The extension of digital pathology from consultation to remote primary diagnosis—commonly referred to as remote sign-out—represents a natural but consequential evolution. In this model, the pathologist renders the final diagnosis without ever handling the glass slide, relying entirely on WSI viewed outside the physical laboratory.

The barriers to remote sign-out have historically been regulatory, cultural, and psychological rather than technical. Concerns centered on diagnostic accuracy, monitor quality, network reliability, and medico-legal exposure. Yet, experience has steadily eroded these objections. Validation studies demonstrate that, when appropriate quality controls are in place, remote diagnosis is equivalent to on-site digital diagnosis and to conventional microscopy (G.M. Lujan et al., 2021b; Hanna et al., 2020).

The COVID-19 pandemic accelerated adoption dramatically. Institutions that had planned gradual rollouts were forced to implement remote sign-out within weeks. Secure virtual

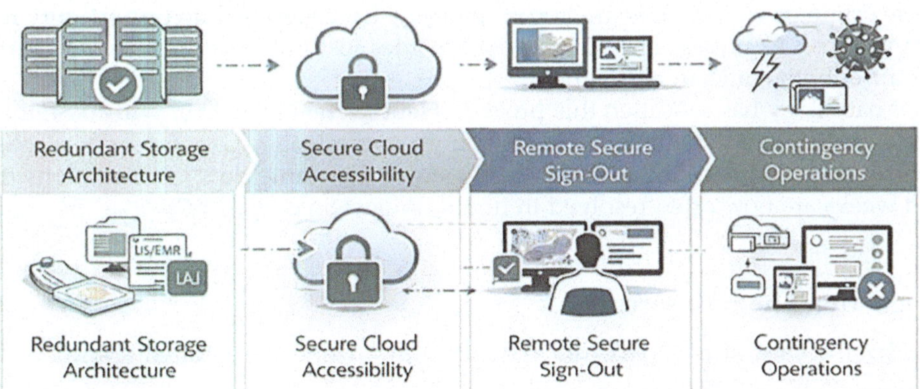

FIG. 4.4 Digital pathology infrastructure supporting remote primary diagnosis and operational resilience. Redundant storage architecture, secure cloud accessibility, and validated remote sign-out enable pathologists to perform primary diagnostic work outside the physical laboratory. These capabilities support the continuity of clinical operations during disruptions such as public health emergencies, natural disasters, or infrastructure outages, while maintaining diagnostic quality and governance.

private networks, defined workstation specifications, and rapid validation protocols enabled continuity of care despite physical distancing. At large academic centers, concordance remained high, and turnaround times were preserved, underscoring that remote digital sign-out is not an emergency workaround but a viable long-term model.

Despite these successes, the cost of entry remains significant. Whole slide scanners require substantial capital investment, storage infrastructure grows continuously, and dedicated informatics support is essential. For smaller community hospitals, return on investment may not be immediately apparent. While reductions in courier costs, physical slide storage, and turnaround time are tangible, benefits such as improved collaboration, workforce flexibility, and institutional resilience are harder to quantify. As a result, the adoption of remote sign-out often follows initial on-site digital deployment rather than preceding it (Giovanni Lujan et al., 2021a).

Sustainability considerations add another dimension. Data centers consume significant energy, and the carbon footprint of large-scale digital storage is increasingly scrutinized. Without deliberate strategies for energy-efficient computing and data lifecycle management, digital pathology risks increasing the environmental burden of healthcare. Institutions are beginning to weigh these factors alongside clinical and economic considerations.

Secondary consultation and expert review

From couriers to cloud-based collaboration

For much of the 20th century, the process of secondary consultation in pathology was defined by the physical movement of glass. A clinician suspecting a rare sarcoma or an unusual lymphoma would request a second opinion; the pathology department would pack the slides into a box, fill out forms, and send them via courier to a reference center. Days or weeks later, the slides would arrive; the expert would review them, dictate a report, and then the fragile originals would make the return journey. Each step was an opportunity for delay, cost, or even loss. Countless anecdotes exist of slides lost in transit or arriving shattered, to the despair of both clinician and patient.

Digital pathology has rewritten this process (Henriksen et al., 2020). With WSI, the consultation is instantaneous. A pathologist in Pittsburgh can upload a case and invite a colleague in Berlin, Shanghai, London, or São Paulo to review it that same day. Consultations that once spanned weeks are now often resolved in hours (Weinstein et al., 1987).

Case vignette: The rare sarcoma

Consider the case of a 34-year-old patient with an undifferentiated spindle cell tumor resected in a regional hospital. The local pathologist suspects synovial sarcoma but is uncertain. In the analog world, slides would be mailed to a reference sarcoma center; the turnaround might take ten days. In the digital world, WSI are uploaded to a secure consultation portal. Within 24 h, the expert responds: morphology and immunohistochemistry are consistent with synovial sarcoma, confirmed by a FISH study for the SYT-SSX translocation.

The patient's oncologist, now equipped with a definitive diagnosis, initiates targeted therapy without waiting weeks.

Such vignettes illustrate the tangible, patient-level impact of digital consultation. Time is not just convenience; it is an opportunity for earlier therapy, fewer clinic visits, and reduced anxiety for families.

Multidisciplinary team (MDT) integration

The role of pathology in MDTs

Cancer care is increasingly multidisciplinary (Hanna et al., 2022). Tumor boards bring together surgeons, oncologists, radiologists, and pathologists to decide on complex treatment plans. In the analog era, pathologists presented static images—a few photomicrographs selected to illustrate the case. This was inherently limiting; colleagues saw only what the pathologist chose to photograph.

With DP, tumor boards are interactive experiences. The pathologist navigates the WSI in real time, zooming into mitotic figures, highlighting margins, or showing biomarker staining patterns. Radiology images can be displayed side by side, creating a holistic view of the patient's disease (Leng et al., 2024).

Case vignette: The breast cancer tumor board

A 52-year-old woman with invasive ductal carcinoma is discussed at the tumor board. The radiologist shows the mammogram and MRI, emphasizing multifocality. The surgeon describes surgical options. The pathologist projects the WSI: an H&E at low magnification reveals tumor spread across lobules; zooming in demonstrates high-grade nuclei. An overlay of HER2 immunostain, quantified digitally, shows strong positivity in 40% of cells. In real time, the oncologist understands the molecular rationale for trastuzumab. The team leaves the meeting aligned on both surgical and systemic therapy plans.

This is not just convenience—it is data democratization. Nonpathologists see the evidence with their own eyes, fostering trust and collaboration (Hanna et al., 2022).

Evidence of impact

- In a prospective trial, the integration of DP into MDTs reduced preparation time by nearly 9% and discussion time by 21%, while improving standardization (Chang et al., 2025).
- In another large retrospective study of patients with colorectal liver metastasis, the MDT discussions were found useful in providing the patient with all available options in the community (Milana et al., 2022).
- During COVID-19, virtual boards maintained the quality of cancer care despite restrictions, with many clinicians preferring hybrid models even after the restrictions were lifted (G.M. Lujan et al., 2021b).

Archiving and data management

The data explosion

One of the less glamorous but most formidable challenges of digital pathology (DP) is data storage. A single WSI may be 1–5 GB; a large hospital producing 200,000 slides annually generates hundreds of terabytes per year, quickly approaching petabyte scale. Unlike radiology, where a CT scan may be hundreds of megabytes, pathology deals with larger magnitudes (Pantanowitz et al., 2018).

Strategies in practice

- Tiered storage: "Hot" storage for recent cases (fast-access servers), "warm" storage for 1–2-year-old cases (network-attached), and "cold" storage (tape, optical, or deep-freeze cloud) for long-term archives.
- Hybrid cloud models: Increasingly common, balancing scalability with cost. One European lab stores recent cases locally for speed while archiving older slides to cloud providers with redundancy across countries (Subramanian and Subramanian, 2022).
- Disaster recovery: Air-gapped backups and immutable storage are being tested, with blockchain being explored for audit trails (Subramanian and Subramanian, 2022).

Case vignette: The archive room transformed

At a major academic center, a basement once filled with rows of slide cabinets is now half-empty. Older cases have been digitized and stored in cold cloud storage. Residents who once descended into dusty stacks now retrieve cases with a keyword search. The time required to prepare teaching sets has dropped from weeks to hours. For researchers, the ability to query by biomarker or morphology has unlocked new projects that were previously impossible.

Regulatory landscape

Retention policies vary among different regions. For example, CLIA requires at least 10 years in the United States (CLIA., 1992) and it requires up to 30 years in Germany. Ensuring compliance with such rules demands robust metadata, cryptographic checksums, and auditable access logs (Evans et al., 2022). The FAIR principles (Findable, Accessible, Interoperable, Reusable) are increasingly applied to pathology archives, ensuring that data are not only stored but also usable for future care and discovery (Wilkinson et al., 2025).

Quality assurance (QA) and continuous improvement

Using DP for routine diagnostic QA processes is convenient and efficient in a paperless and glassless way. In addition, institutions run digital–glass concordance studies on representative cases. Errors are categorized as major or minor, with continuous review. Many labs report that once pathologists acclimate to digital workflows, concordance with glass improves because the digital environment facilitates side-by-side review and easy comparison with prior cases (Evans et al., 2022).

AI is emerging as a QA tool in conjunction with DP. Algorithms can detect staining variability, flag atypical patterns in daily workloads, or highlight discordances between digital biomarker scoring and historical averages. For example, at one institution, a pathologist at a large institution reported that an algorithm flagged an unusual cluster of HER2-negative breast cancers in one week; further review discovered a staining machine malfunction.

CAP requires documentation of validation, competency, and QC records. ISO 15189 emphasizes reproducibility and traceability. ESDIP has issued best-practice recommendations for Europe (Fraggetta et al., 2021). These frameworks ensure that digital workflows are not only validated at go-live but are also continuously monitored.

Lessons from global implementations

Representative global patterns of digital pathology implementation are summarized in Fig. 4.5.

Digital pathology is no longer confined to pilot projects or isolated academic laboratories. Over the past decade, it has been implemented at scale across continents, with each region adapting the technology to its own healthcare structures, regulatory environments, and clinical needs. Collectively, these experiences paint a picture of a technology that has reached clinical maturity.

Europe

The Netherlands: First to go fully digital

The Netherlands is frequently cited as a global pioneer in digital pathology adoption. Institutions such as Radboud University Medical Center and University Medical Center Utrecht have pursued an ambitious goal: the elimination of routine glass slide sign-out. Early efforts focused on high-volume subspecialties such as prostate and breast pathology, demonstrating equivalence with conventional microscopy and building confidence among pathologists.

Resistance was initially encountered. Senior pathologists expressed concern about screen fatigue, while information technology leaders worried about storage costs and system reliability. These challenges were addressed through phased rollouts, robust local validation, and continuous feedback. Within a few years, Radboud was operating as one of the first fully digital pathology departments worldwide. The Dutch experience illustrates that success depends as much on cultural change management as on technology (Stathonikos et al., 2013; Bulten et al., 2020; Baidoshvili et al., 2018).

Spain: Regional champions and oncology integration

In Spain, regional champions drove adoption. At Hospital Universitario Virgen de las Nieves in Granada, digital pathology was tightly coupled with breast oncology workflows.

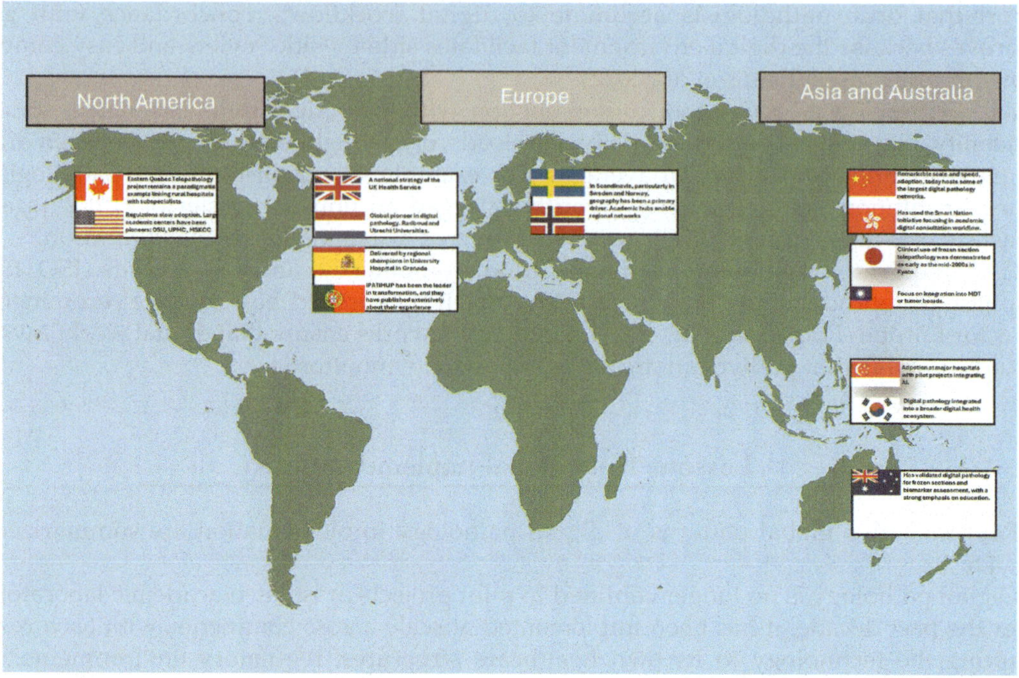

FIG. 4.5 Global implementations of clinical digital pathology across diverse health systems. Adoption patterns reflect regional differences in healthcare organization, regulatory frameworks, infrastructure maturity, and workforce models. Large-scale national and regional implementations in Europe, North America, and parts of Asia demonstrate the feasibility of enterprise digital pathology for primary diagnosis, consultation, and multidisciplinary care. In other regions, digital pathology is frequently deployed to support telepathology, subspecialty access, and diagnostic equity. Together, these global experiences highlight multiple implementation pathways rather than a single universal model.

The digitization of immunohistochemistry slides enabled reproducible HER2 testing and accelerated case sharing during tumor boards. Surgeons and oncologists reported increased trust in pathology interpretations when biomarker heatmaps and annotated slides were presented visually rather than described textually.

Here, adoption was driven less by pathologists seeking convenience and more by multidisciplinary colleagues demanding richer, data-supported diagnostic communication (Retamero et al., 2020a, 2020b).

United Kingdom: A national strategy

Few countries have pursued digital pathology with the same national ambition as the United Kingdom's National Health Service. Government-supported consortia, including PathLAKE, have aimed to embed digital pathology and artificial intelligence into cancer workflows (Coulter et al., 2022; Browning et al., 2021; Colling et al., 2019). The All Wales validation study became a landmark, demonstrating high concordance across seven hospitals and thousands of slides (Babawale et al., 2021).

Beyond validation, the Welsh experience proved that digital pathology could be deployed across a public health system with constrained budgets. In routine practice, digital workflows now support breast, lung, and prostate cancer tumor boards across multiple NHS trusts. Cultural change has followed success: consultants who initially resisted digital sign-out have become advocates, citing flexibility and improved inter-hospital collaboration.

Scandinavia: Geography as a catalyst

In Sweden and Norway, geography has been a primary driver. Dispersed populations and rural hospitals have made telepathology essential for equitable care. Regional networks centered around academic hubs have enabled frozen section interpretation and second opinions for remote hospitals. Surgeons have reported reduced operative delays and fewer re-operations, reinforcing the value of digital connectivity. In these systems, digital pathology was not a luxury but a necessity (Henriksen et al., 2020).

France, Germany, Italy, Portugal, and the Rest of Europe: Precision and policy

In the rest of Europe, there are several initiatives, many led by the European Society for Digital and Integrative Pathology (Eloy et al., 2021b). In France, digital pathology has been closely linked to cancer center networks and telepathology initiatives. Multicenter studies validated its use for frozen sections and biomarker testing, paving the way for broader deployment. In Germany, leading institutions such as Charité Berlin and Heidelberg University Hospital have integrated digital pathology into both clinical workflows and quality assurance programs, emphasizing national standards and harmonization across laboratories. In Portugal, IPATIMUP has been the leader in transformation, and they have done a great job publishing their experiences (Eloy et al., 2021a).

Asia

Japan

Japan has been an early adopter of digital pathology, with telepathology playing a central role in its development and clinical uptake. Early work in Japanese institutions demonstrated routine generation and use of both macroscopic and microscopic digital images for diagnostic purposes, with virtual slide (VS) technology initially introduced for education and conference use and progressively expanded into consultation and routine diagnosis (Tsuchihashi, 2011). Clinical use of virtual slide systems in intraoperative frozen section telepathology was demonstrated as early as the mid-2000s at regional centers in Kyoto, underscoring both feasibility and operational value for real-time remote diagnosis (Tsuchihashi et al., 2008). The establishment of organized groups such as the Japanese Research Society of Telepathology and Pathology Informatics fostered collaboration across major hospitals, and quantification of immunostains using VS images has gradually been adopted in diagnostic practice.

China: Scale and speed

China's adoption of digital pathology has been remarkable in scale. The UPMC–KingMed Diagnostics collaboration demonstrated the feasibility of large-scale, cross-border digital consultation, enabling thousands of cases to be reviewed with preserved diagnostic accuracy. These networks addressed a pressing need: the uneven distribution of subspecialty expertise across a vast population (Zhao et al., 2015).

Today, China hosts some of the largest telepathology networks globally. Provincial hospitals routinely upload cases to urban centers, and artificial intelligence pilots are layered onto these platforms to triage workloads. China's experience demonstrates what is possible when infrastructure investment and political will align (He et al., 2021).

Taiwan: Efficiency through MDT integration

Taiwan has focused on integrating digital pathology into multidisciplinary tumor boards. A prospective study demonstrated reductions in preparation time and discussion duration when digital tools were used, without compromising decision quality (Chang et al., 2025). The cultural impact was significant, as oncologists and surgeons valued direct access to pathology evidence in the same visual format as radiology.

Singapore and Hong Kong: Smart nation models

Singapore's Smart Nation initiative has supported the adoption of digital pathology at major hospitals, with an early emphasis on lung, breast, and prostate cancer workflows. Pilot projects integrating AI-assisted biomarker quantification have highlighted the importance of data governance and privacy (Teo et al., 2025). In Hong Kong, academic digital consultation workflows have been developed with particular attention to legal and ethical frameworks for cross-border data sharing (Cho, 2022).

South Korea: Infrastructure-driven innovation

South Korea has integrated digital pathology into a broader digital health ecosystem. Institutions such as Samsung Medical Center have achieved high levels of imaging informatics maturity and deployed advanced telepathology solutions, including high-speed network-enabled frozen section consultation. These efforts illustrate how infrastructure can enable digital pathology to leapfrog traditional constraints.

North America

United States: Innovation amid regulation

In the United States, adoption has been shaped by regulatory rigor. The FDA clearance of the Philips IntelliSite system marked a turning point, but laboratory-level validation remains demanding. Despite this, leading centers have pushed forward. Remote sign-out was validated at major cancer centers during the COVID-19 pandemic, achieving high concordance. Academic institutions demonstrated large-scale remote workflows and validated nontraditional display devices under defined criteria.

The US experience reflects innovation despite regulatory complexity, with academic centers leading adoption and community hospitals following cautiously.

Canada: Bridging distance

Canada's vast geography has made telepathology indispensable. The Eastern Québec Telepathology Project remains a paradigmatic example, linking rural hospitals with subspecialists and enabling real-time frozen section support (Têtu et al., 2014). Subsequent expansion across provinces has reinforced the role of digital pathology in delivering equitable care (Bruce et al., 2024).

Other Regions

Australia

Australian centers validated digital pathology for frozen sections and biomarker assessment, with a strong emphasis on education. Digital repositories have become integral to pathology training, particularly in rural settings where exposure to rare cases is limited.

South America

In Latin America, Brazil and Argentina have emerged as regional leaders. Institutions reported high concordance in dermatopathology and gastrointestinal pathology, and they emphasized digital platforms for remote education and consultation.

Middle East

Saudi Arabia integrated digital pathology into national cancer modernization initiatives, extending subspecialty expertise across hospitals and reducing disparities between urban and rural centers.

Africa

Adoption in Africa remains at an earlier stage, but pilot programs in Uganda, South Africa, and Nigeria demonstrate feasibility. Collaborations with European centers have enabled remote consultations for pediatric and rare tumors, highlighting the potential of digital pathology to mitigate workforce shortages in low-resource settings.

Challenges

Despite proven feasibility and safety, large-scale adoption of digital pathology remains uneven. The reasons are multifactorial, arising from the interaction of technology, culture, regulation, economics, and equity.

Technical challenges

Data scale remains the most visible hurdle. A single WSI may exceed several gigabytes, and annual volumes can quickly reach petabyte levels (Pantanowitz et al., 2018). While storage costs continue to decline, operational and environmental costs remain significant.

Scanning speed, reliability, and integration into histology workflows are equally critical. Bottlenecks arise when scanning is not carefully aligned with laboratory processes, particularly in high-volume settings (Hanna et al., 2022).

Interoperability poses another challenge. Although DICOM-WSI exists as a standard, implementation remains inconsistent across vendors, complicating multivendor environments and collaborative research efforts. Network infrastructure can further limit adoption, particularly in low-resource settings with unstable connectivity (Leng et al., 2024).

Cultural and workforce challenges

Pathology is deeply rooted in tradition. For many senior practitioners, the microscope is not merely a tool but a symbol of professional identity. Transitioning to digital workflows can feel disruptive or even threatening. Generational differences are apparent, with trainees often adapting more readily than established consultants.

Training and ergonomics are, therefore, critical. Navigating WSI differs from glass microscopy, and without structured training, frustration can impede adoption. Poorly designed workstations can introduce new ergonomic risks, underscoring the need for investment in high-quality displays, adjustable furniture, and appropriate lighting.

Regulatory and legal challenges

Regulatory frameworks remain fragmented. In the United States, laboratory validation requirements persist despite device-level approval (Evans et al., 2022). In Europe, CE-IVDR provides a pathway, but national interpretations vary (Fraggetta et al., 2021). In Asia, regulatory maturity differs widely between countries. Cross-border consultation raises unresolved medico-legal questions regarding licensure and liability, which can slow adoption in the absence of clear guidance.

Equity challenges

Perhaps the most profound challenge is equity. Digital pathology holds the promise to democratize access to expertise, yet adoption remains concentrated in high-income settings. Low- and middle-income countries face infrastructure, budgetary, and workforce constraints. Without deliberate global investment, digital pathology risks reinforcing existing disparities rather than alleviating them.

Future directions and conclusion

Digital pathology has reached a tipping point. Validation studies have settled questions of safety, national programs have demonstrated scalability, and global collaborations have illustrated its equity potential. The most consequential story, however, concerns how digital pathology will redefine the professional identity of the pathologist.

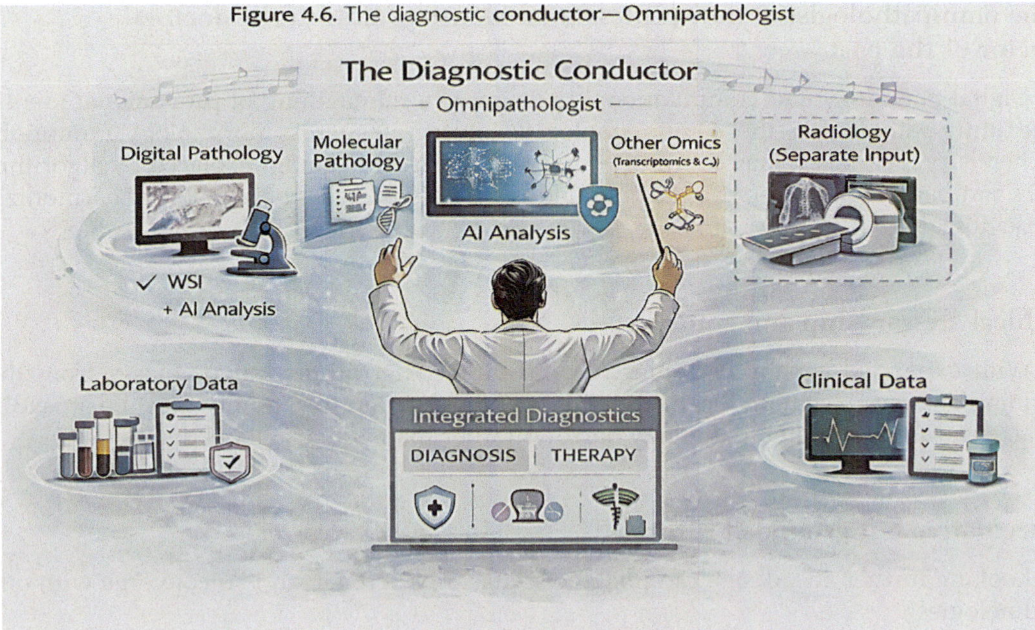

FIG. 4.6 The diagnostic conductor—omni-pathologist. In an integrated diagnostic environment, the pathologist functions as a central coordinator who synthesizes diverse, discipline-specific data streams into a coherent diagnostic narrative. Digital pathology and whole slide imaging, computational and artificial intelligence–assisted analyses, laboratory and molecular data, clinical information, and radiologic findings—generated and interpreted within their respective specialties—are integrated to support diagnosis, prognosis, and therapeutic decision-making. This model emphasizes orchestration and interpretation across domains rather than ownership of individual data sources.

From microscopist to diagnostic conductor (Fig. 4.6)

The metaphor of the diagnostic conductor captures the emerging reality. Modern diagnosis generates signals from morphology, immunohistochemistry, molecular testing, radiology, and computation. Without integration, these signals risk cacophony. Digital pathology amplifies the pathologist's role as an orchestrator, ensuring that diverse data streams coalesce into a coherent clinical narrative.

Toward unified diagnostic reporting: Omnipathology

The logical extension of this model is unified, interactive diagnostic reporting. Such reports integrate digitized slides with biomarker overlays, molecular annotations, radiologic correlations, and computational risk models. For clinicians, this enhances transparency and trust; for patients, it transforms abstract results into comprehensible visual narratives.

The omnipathologist, a new professional identity, akin to the doctors' doctor of the past

Digital pathology and computational tools invite a reimagining of professional identity. The future pathologist—the "Omnipathologist"—integrates across morphology, immunohistochemistry, genomics, imaging, and computation. Far from being displaced by algorithms, the Omnipathologist becomes indispensable as the only professional capable of harmonizing these domains.

Ethical stewardship and equity

With expanded capability comes responsibility. Algorithms inherit biases from their training data, necessitating local validation and diverse datasets. Equity must be an explicit goal, ensuring that digital pathology narrows rather than widens global disparities.

Speculative but grounded futures

Looking further ahead, one can imagine scenarios where DP and AI converge with other technologies:

- Radiomics and histomics: combined models predicting therapy response.
- Liquid biopsy correlations: AI linking circulating tumor DNA results with histology.
- Wearables and digital pathology: patient monitoring feeding back into predictive pathology models.
- Federated learning: algorithms trained across continents without moving data, preserving privacy while building global capacity.

These futures are not science fiction; they are emerging in pilot studies. The challenge is to integrate them responsibly, with pathologists leading the effort.

A reflective conclusion

Digital pathology preserves the essence of the discipline while extending its reach into the digital, computational, and global domains. The microscope defined pathology in the 20th century; the digital slide, enriched by computational co-intelligence, is defining it in the 21st. The pathologist of the future is not replaced by algorithms but elevated by them—the diagnostic conductor of an integrative, global orchestra guiding patient care.

Author perspective: Computational co-intelligence and diagnostic responsibility

As digital pathology increasingly intersects with algorithmic and computational tools, careful attention to language becomes essential. Terms such as *artificial intelligence*, while deeply embedded in scientific and public discourse, often imply autonomy, agency, or cognition that exceeds the actual capabilities of current systems. In clinical medicine, these metaphors are not merely semantic; they shape expectations, influence trust, and affect how responsibility is perceived by clinicians, patients, and regulators.

In practice, the technologies integrated into digital pathology workflows perform computational pattern recognition, probabilistic inference, and quantitative analysis. Their clinical value does not arise from independent decision-making but from their integration with human expertise, contextual reasoning, and ethical judgment. Conceptualizing these systems as **computational co-intelligence** more accurately reflects their role: not as artificial minds but as computational partners that extend and augment human diagnostic capability.

This framing reinforces a central principle of modern pathology—that agency and accountability remain with the clinician. Computational co-intelligence emphasizes collaboration rather than replacement, precision rather than mystique, and transparency rather than metaphor. It aligns naturally with the evolution of the pathologist described in this chapter: from isolated microscopist to integrative diagnostic conductor, responsible for harmonizing morphology, biomarkers, imaging, and computation into a coherent clinical narrative.

As digital pathology continues to mature, clarity of language becomes a crucial component of patient safety and professional integrity. By grounding computational tools in an honest conceptual framework, the field can advance innovation while preserving trust, accountability, and the fundamentally human nature of medical diagnosis.

AI disclosure

During the preparation of this work, the author(s) used ChatGPT to polish the text. After using this tool/service, the author(s) reviewed and edited the content as needed and take(s) full responsibility for the content of the publication.

References

Alassiri, A., Almutrafi, A., Alsufiani, F., Al Nehkilan, A., Al Salim, A., Musleh, H., Aziz, M., Khalbuss, W., 2020. Whole slide imaging compared with light microscopy for primary diagnosis in surgical neuropathology: a validation study. Ann. Saudi Med. 0256-4947. 40 (1), 36–41. doi:10.5144/0256-4947.2020.36.

Babawale, M., Gunavardhan, A., Walker, J., Corfield, T., Huey, P., Savage, A., Bansal, A., Atkinson, M., Abdelsalam, H., Raweily, E., Christian, A., Evangelou, I., Thomas, D., Shannon, J., Youd, E., Brumwell, P., Harrison, J., Thompson, I., Rashid, M., Leopold, G., Finall, A., Roberts, S., Housa, D., Nedeva, P., Davies, A., Fletcher, D., Aslam, M., 2021. Verification and validation of digital pathology (whole slide imaging) for primary histopathological diagnosis: all wales experience. J. Pathol. Inform. 21533539. 12 (1), 4. doi:10.4103/jpi.jpi_55_20.

Baidoshvili, A., Bucur, A., van Leeuwen, J., van der Laak, J., Kluin, P., van Diest, P.J., 2018. Evaluating the benefits of digital pathology implementation: time savings in laboratory logistics. Histopathology. 13652559. 73 (5), 784–794. Available from: http://onlinelibrary.wiley.com/journal/10.1111/(ISSN)1365-2559. doi:10.1111/his.13691.

Browning, L., Colling, R., Rakha, E., Rajpoot, N., Rittscher, J., James, J.A., Salto-Tellez, M., Snead, D.R.J., Verrill, C., 2021. Digital pathology and artificial intelligence will be key to supporting clinical and academic cellular pathology through COVID-19 and future crises: the PathLAKE consortium perspective. J. Clin. Pathol. 0021-9746. 74 (7), 443–447. doi:10.1136/jclinpath-2020-206854.

Bruce, C., Prassas, I., Mokhtar, M., Clarke, B., Youssef, E., Wang, C., Yousef, G.M., 2024. Transforming diagnostics: the implementation of digital pathology in clinical laboratories. Histopathology. 0309-0167. 85 (2), 207–214. doi:10.1111/his.15178.

Bueno, G., Fernández-Carrobles, M.M., Deniz, O., García-Rojo, M., 2016. New trends of emerging technologies in digital pathology. Pathobiology. 1015-2008. 83 (2–3), 61–69. doi:10.1159/000443482.

Bulten, W., Pinckaers, H., van Boven, H., Vink, R., de Bel, T., van Ginneken, B., van der Laak, J., Hulsbergen-van de Kaa, C., Litjens, G., 2020. Automated deep-learning system for Gleason grading of prostate cancer using biopsies: a diagnostic study. Lancet Oncol. 14702045. 21 (2), 233–241. doi:10.1016/s1470-2045(19)30739-9.

Chang, L.C.F., Kuo, H.C., Wang, H.M., Kuo, Y.C., Wang, C.T., Chen, L.C., Hsieh, J.C.H., 2025. The use of an integrated digital tool to improve the efficiency of multidisciplinary tumor boards—a prospective trial in Taiwan. Cancers. 20726694. 17 (3). Available from: http://www.mdpi.com/journal/cancers/. doi:10.3390/cancers17030444.

Cho, W.C., 2022. Digital pathology: new initiative in pathology. Biomolecules. 2218273X. 12 (9). Available from: http://www.mdpi.com/journal/biomolecules. doi:10.3390/biom12091314.

CLIA, 1992. Standard: Retention Requirements In. 42 CFR §493.1105. US Government Publishing Office.

Colling, R., Pitman, H., Oien, K., Rajpoot, N., Macklin, P., Snead, D., Sackville, T., Verrill, C., 2019. Artificial intelligence in digital pathology: a roadmap to routine use in clinical practice. J. Pathol. 0022-3417. 249 (2), 143–150. doi:10.1002/path.5310.

Cornish, T.C., 2020. Clinical application of image analysis in pathology. Adv. Anat. Pathol. 15334031. 27 (4), 227–235. Available from: http://www.anatomicpathology.com. doi:10.1097/PAP.0000000000000263.

Coulter, C., McKay, F., Hallowell, N., Browning, L., Colling, R., Macklin, P., Sorell, T., Aslam, M., Bryson, G., Treanor, D., Verrill, C., 2022. Understanding the ethical and legal considerations of digital pathology. J. Pathol. Clin. Res. 2056-4538. 8 (2), 101–115. doi:10.1002/cjp2.251.

Eloy, C., Vale, J., Curado, M., Polónia, A., Campelos, S., Caramelo, A., Sousa, R., Sobrinho-Simões, M., 2021a. Digital pathology workflow implementation at IPATIMUP. Diagnostics. 2075-4418. 11 (11), 2111. doi:10.3390/diagnostics11112111.

Eloy, C., Zerbe, N., Fraggetta, F., 2021b. Europe Unites for the digital transformation of pathology: the role of the new ESDIP. J. Pathol. Inform. 21533539. 12 (1), 10. doi:10.4103/jpi.jpi_80_20.

Evans, A.J., Brown, R.W., Bui, M.M., Chlipala, E.A., Lacchetti, C., Milner, D.A., Pantanowitz, L., Parwani, A.V., Reid, K., Riben, M.W., Reuter, V.E., Stephens, L., Stewart, R.L., Thomas, N.E., 2022. Validating whole slide imaging systems for diagnostic purposes in pathology. Arch. Pathol. Lab. Med. 1543-2165. 146 (4), 440–450. doi:10.5858/arpa.2020-0723-cp.

Fraggetta, F., L'Imperio, V., Ameisen, D., Carvalho, R., Leh, S., Kiehl, T.-R., Serbanescu, M., Racoceanu, D., Mea, V. D., Polonia, A., Zerbe, N., Eloy, C., 2021. Best practice recommendations for the implementation of a digital pathology workflow in the anatomic pathology laboratory by the European Society of Digital and Integrative Pathology (ESDIP). Diagnostics. 2075-4418. 11 (11), 2167. doi:10.3390/diagnostics11112167.

Goacher, E., Randell, R., Williams, B., Treanor, D., 2017. The diagnostic concordance of whole slide imaging and light microscopy: a systematic review. Arch. Pathol. Lab. Med. 0003-9985. 141 (1), 151–161. doi:10.5858/arpa.2016-0025-ra.

Hanna, M.G., Reuter, V.E., Ardon, O., Kim, D., Sirintrapun, S.J., Schüffler, P.J., Busam, K.J., Sauter, J.L., Brogi, E., Tan, L.K., Xu, B., Bale, T., Agaram, N.P., Tang, L.H., Ellenson, L.H., Philip, J., Corsale, L., Stamelos, E., Friedlander, M.A., Ntiamoah, P., Labasin, M., England, C., Klimstra, D.S., Hameed, M., 2020. Validation of a digital pathology system including remote review during the COVID-19 pandemic. Mod. Pathol. 15300285. 33 (11), 2115–2127. Available from: http://www.nature.com/modpathol/index.html. doi:10.1038/s41379-020-0601-5.

Hanna, M.G., Ardon, O., Reuter, V.E., Sirintrapun, S.J., England, C., Klimstra, D.S., Hameed, M.R., 2022. Integrating digital pathology into clinical practice. Mod. Pathol. 15300285. 35 (2), 152–164. Available from: http://www.nature.com/modpathol/index.html. doi:10.1038/s41379-021-00929-0.

He, X., Wang, L., Wang, L., Gao, J., Cui, F., Ma, Q., Zhang, W., Wang, L., Zhai, Y., Zhao, J., 2021. Effectiveness of a cloud-based telepathology system in China: large-sample observational study. J. Med. Internet Res. 1438-8871. 23 (7), e23799. doi:10.2196/23799.

Henriksen, J., Kolognizak, T., Houghton, T., Cherne, S., Zhen, D., Cimino, P.J., Latimer, C.S., Scherpelz, K.P., Yoda, R. A., Alpers, C.E., Chhieng, D.F., Keene, C.D., Gonzalez-Cuyar, L.F., 2020. Rapid validation of telepathology by an academic neuropathology practice during the COVID-19 pandemic. Arch. Pathol. Lab. Med. 1543-2165. 144 (11), 1311–1320. doi:10.5858/arpa.2020-0372-sa.

Huisman, A., Looijen, A., van den Brink, S.M., van Diest, P.J., 2010. Creation of a fully digital pathology slide archive by high-volume tissue slide scanning. Hum. Pathol. 00468177. 41 (5), 751–757. doi:10.1016/j.humpath.2009.08.026.

Leng, H., Deng, R., Bao, S., Fang, D., Millis, B.A., Tang, Y., Yang, H., Wang, X., Peng, Y., Wan, L., Huo, Y., 2024. High-performance data management for whole slide image analysis in digital pathology. In: Progress in Biomedical Optics and Imaging—Proceedings of SPIE. SPIE, United States, p. 12933. 9781510671706. http://spie.org/x1848.xml. doi:10.1117/12.3006273.

Lloyd, M.C., Monaco, J.P., Bui, M.M., 2016. Image analysis in surgical pathology. Surg. Pathol. Clin. 18759157. 9 (2), 329–337. Available from: http://www.elsevier.com/wps/find/journaldescription.cws_home/716205/descriptio. doi:10.1016/j.path.2016.02.001.

Lujan, G., Quigley, J.C., Hartman, D., Parwani, A., Roehmholdt, B., Van Meter, B., Ardon, O., Hanna, M.G., Kelly, D., Sowards, C., Montalto, M., Bui, M., Zarella, M.D., LaRosa, V., Slootweg, G., Retamero, J.A., Lloyd, M.C., Madory, J., Bowman, D., 2021a. Dissecting the business case for adoption and implementation of digital pathology: a white paper from the digital pathology association. J. Pathol. Inform. 21533539. 12 (1), 17. doi:10.4103/jpi.jpi_67_20.

Lujan, G.M., Savage, J., Shana'ah, A., Yearsley, M., Thomas, D., Allenby, P., Otero, J., Limbach, A.L., Cui, X., Scarl, R. T., Hardy, T., Sheldon, J., Plaza, J.A., Whitaker, B., Frankel, W., Parwani, A.V., Li, Z., 2021b. Digital pathology initiatives and experience of a large academic institution during the coronavirus disease 2019 (COVID-19) pandemic. Arch. Pathol. Lab. Med. 15432165. 145 (9), 1051–1061. Available from: https://allen.silverchair-cdn.com. doi:10.5858/arpa.2020-0715-SA.

Milana, F., Famularo, S., Luberto, A., Rimassa, L., Scorsetti, M., Comito, T., Pressiani, T., Franzese, C., Poretti, D., Di Tommaso, L., Personeni, N., Rodari, M., Pedicini, V., Donadon, M., Torzilli, G., 2022. Multidisciplinary tumor board in the management of patients with colorectal liver metastases: a single-center review of 847 patients. Cancers. 2072-6694. 14 (16), 3952. doi:10.3390/cancers14163952.

Mukhopadhyay, S., Feldman, M.D., Abels, E., Ashfaq, R., Beltaifa, S., Cacciabeve, N.G., Cathro, H.P., Cheng, L., Cooper, K., Dickey, G.E., Gill, R.M., Heaton, R.P., Kerstens, R., Lindberg, G.M., Malhotra, R.K., Mandell, J.W., Manlucu, E.D., Mills, A.M., Mills, S.E., Moskaluk, C.A., Nelis, M., Patil, D.T., Przybycin, C.G., Reynolds, J.P., Rubin, B.P., Saboorian, M.H., Salicru, M., Samols, M.A., Sturgis, C.D., Turner, K.O., Wick, M.R., Yoon, J.Y., Zhao, P., Taylor, C.R., 2018. Whole slide imaging versus microscopy for primary diagnosis in surgical pathology: a multicenter blinded randomized noninferiority study of 1992 cases (pivotal study). Am. J. Surg. Pathol. 15320979. 42 (1), 39–52. Available from: http://journals.lww.com/ajsp/pages/default.aspx. doi:10.1097/PAS.0000000000000948.

Nagpal, K., Foote, D., Tan, F., Liu, Y., Chen, P.H.C., Steiner, D.F., Manoj, N., Olson, N., Smith, J.L., Mohtashamian, A., Peterson, B., Amin, M.B., Evans, A.J., Sweet, J.W., Cheung, C., Kwast, T.V.D., Sangoi, A.R., Zhou, M., Allan, R., Humphrey, P.A., Hipp, J.D., Gadepalli, K., Corrado, G.S., Peng, L.H., Stumpe, M.C., Mermel, C.H., 2020. Development and validation of a deep learning algorithm for gleason grading of prostate cancer from biopsy specimens. JAMA Oncol. 23742445. 6 (9), 1372–1380. Available from: http://oncology.jamanetwork.com/journal.aspx. doi:10.1001/jamaoncol.2020.2485.

Pantanowitz, L., Sharma, A., Carter, A.B., Kurc, T., Sussman, A., Saltz, J., 2018. Twenty years of digital pathology: an overview of the road travelled, what is on the horizon, and the emergence of vendor-neutral archives. J. Pathol. Inform. 21533539. 9 (1), 40. doi:10.4103/jpi.jpi_69_18.

Retamero, J.A., Aneiros-Fernandez, J., Aneiros-Fernandez, J., Moral, R.G.D., Moral, R.G.D., 2020a. Microscope? No, thanks: user experience with complete digital pathology for routine diagnosis. Arch. Pathol. Lab. Med. 15432165. 144 (6), 672–673. Available from: https://www.archivesofpathology.org/doi/pdf/10.5858/arpa.2019-0355-LE. doi:10.5858/arpa.2019-0355-LE.

Retamero, J.A., Aneiros-Fernandez, J., Del Moral, R.G., 2020b. Complete digital pathology for routine histopathology diagnosis in a multicenter hospital network. Arch. Pathol. Lab. Med. 15432165. 144 (2), 221–228. Available from: https://www.archivesofpathology.org/doi/pdf/10.5858/arpa.2018-0541-OA. doi:10.5858/arpa.2018-0541-OA.

Rosado, A.M., Vizcarra, J.C., Sharma, S.R.R., Keene, C.D., White, C.L., Kim, A., Forrest, S.L., Kovacs, G.G., Chuah, C., Flanagan, M.E., Pearce, T.M., Dugger, B.N., Gutman, D.A., 2025. Digital neuropathology of neurodegenerative disorders: foundations, research advances, and future directions. Alzheimers Dement. 1552-5260. 21 (11). doi:10.1002/alz.70775.

Sato, S., Maki, S., Yamanaka, T., Hoshino, D., Ota, Y., Yoshioka, E., Kawachi, K., Washimi, K., Suzuki, M., Ohkubo, Y., Yokose, T., Yamashita, T., Ohtori, S., Miyagi, Y., 2021. Machine learning-based image analysis for accelerating the diagnosis of complicated preneoplastic and neoplastic ductal lesions in breast biopsy tissues. Breast Cancer Res. Treat. 0167-6806. 188 (3), 649–659. doi:10.1007/s10549-021-06243-2.

Shean, R.C., Rets, A.V., 2025. Digital pathology in hematopathology: from vision to deployment. Int. J. Lab. Hematol. 1751553X. Available from: http://onlinelibrary.wiley.com/journal/10.1111/(ISSN)1751-553X. doi:10.1111/ijlh.14533.

Stathonikos, N., Veta, M., Huisman, A., van Diest, P.J., 2013. Going fully digital: perspective of a Dutch academic pathology lab. J. Pathol. Inform. 21533539. 4 (1), 15. doi:10.4103/2153-3539.114206.

Subramanian, H., Subramanian, S., 2022. Improving diagnosis through digital pathology: proof-of-concept implementation using smart contracts and decentralized file storage. J. Med. Internet Res. 1438-8871. 24 (3), e34207. doi:10.2196/34207.

Teo, L., Koh, M.S., Chua, K.L.M., Leow, L., Lim, C.H., Lim, D.W.T., Lim, H.L., Kong, H.L., Ooi, G., Samol, J., Tan, C.S., Soo, R.A., 2025. Early lung cancer patient journey in singapore: challenges and recommendations. Asia Pac. J. Clin. Oncol. 17437563. 21 (6), 586–596. Available from: http://onlinelibrary.wiley.com/journal/10.1111/(ISSN)1743-7563. doi:10.1111/ajco.14180.

Têtu, B., Perron, É., Louahlia, S., Paré, G., Trudel, M.C., Meyer, J., 2014. The Eastern Québec Telepathology Network: a three-year experience of clinical diagnostic services. Diagn. Pathol. 17461596. 9 (1). Available from: http://www.diagnosticpathology.org/. doi:10.1186/1746-1596-9-S1-S1.

Tsuchihashi, Y., Takamatsu, T., Hashimoto, Y., Takashima, T., Nakano, K., Fujita, S., 2008. Use of virtual slide system for quick frozen intra-operative telepathology diagnosis in Kyoto, Japan. Diagn. Pathol. 1746-1596. 3 (Suppl 1), S6. doi:10.1186/1746-1596-3-S1-S6.

Tsuchihashi, Y., 2011. Expanding application of digital pathology in Japan—from education, telepathology to autodiagnosis. Diagn. Pathol. 1746-1596. 6 (S1). doi:10.1186/1746-1596-6-s1-s19.

Wang, C.W., Huang, S.C., Lee, Y.C., Shen, Y.J., Meng, S.I., Gaol, J.L., 2022. Deep learning for bone marrow cell detection and classification on whole-slide images. Med. Image Anal. 13618423. 75. Available from: http://www.elsevier.com/inca/publications/store/6/2/0/9/8/3/index.htt. doi:10.1016/j.media.2021.102270.

Weinstein, R.S., Bloom, K.J., Rozek, L.S., 1987. Telepathology and the networking of pathology diagnostic services. Arch. Pathol. Lab. Med. 111 (7), 646–652.

Wilkinson, S.R., Aloqalaa, M., Belhajjame, K., Crusoe, M.R., de Paula Kinoshita, B., Gadelha, L., Garijo, D., Gustafsson, O.J.R., Juty, N., Kanwal, S., Khan, F.Z., Köster, J., Gehlen, K.P.-V., Pouchard, L., Rannow, R.K., Soiland-Reyes, S., Soranzo, N., Sufi, S., Sun, Z., Vilne, B., Wouters, M.A., Yuen, D., Goble, C., 2025. Applying the FAIR Principles to computational workflows. Sci. Data. 20524463. 12 (1). Available from: https://www.nature.com/sdata/. doi:10.1038/s41597-025-04451-9.

Zhao, C., Wu, T., Ding, X., Parwani, A.V., Chen, H., McHugh, J., Piccoli, A., Xie, Q., Lauro, G.R., Feng, X., Hartman, D.J., Seethala, R.R., Shangwei, W., Yousem, S., Liang, Y., Pantanowitz, L., 2015. International telepathology consultation: three years of experience between the University of Pittsburgh Medical Center and KingMed Diagnostics in China. J. Pathol. Inform. 21533539. 6 (1), 63. doi:10.4103/2153-3539.170650.

CHAPTER

5

Other applications of digital pathology

Ghulam Rasool[a], *Rajendra Singh*[b], *and Marilyn Bui*[c]

[a]Department of Machine Learning, Moffitt Cancer Center and Research Institute, Tampa, FL, United States

[b]Department of Pathology and Laboratory Medicine, Perelman School of Medicine at the University of Pennsylvania, Philadelphia, PA, United States

[c]Department of Pathology, Moffitt Cancer Center, Tampa, FL, United States

Introduction

Digital pathology has transformed pathology practice beyond primary diagnosis, encompassing education, research, and quality improvement by removing constraints imposed by glass slides, fixed laboratories, and conventional microscopes. By converting traditional slides into high-resolution whole-slide images (WSI), digital pathology enables flexible, remote, and collaborative learning that is difficult to achieve with conventional methods. Students can learn anytime, anywhere, and educators can teach using tools ranging from virtual microscopy and annotated slide repositories to virtual and augmented reality platforms and learning management systems (Hassell et al., 2023; Yin et al., 2016; Kumar et al., 2004). As pathology moves toward a more connected and technology-driven future, digital pathology prepares learners for modern practice, expands access across institutions and regions, and supports innovative teaching models such as flipped classrooms, adaptive learning, and interactive online sessions. Rather than simply replacing glass slides, it creates an integrated educational ecosystem that enhances learning, fosters collaboration, and aligns with contemporary educational principles (Pantanowitz et al., 2018; Pantanowitz, 2010; Waqas et al., 2023, 2024b; Mukhopadhyay et al., 2018).

Beyond education, the impact of digital pathology extends well beyond primary diagnostic sign-out. As a digital infrastructure, it supports a broad range of applications in biomedical research, quality improvement, quality control, and other emerging use cases that leverage WSIs, computational analysis, and data integration. This chapter focuses on these nonprimary diagnostic applications, highlighting the roles of digital pathology in education, research, quality assurance, and additional domains that collectively underscore its value as

Advances in Digital Pathology and Artificial Intelligence in Pathology
https://doi.org/10.1016/B978-0-443-40383-5.00009-1

a foundational platform for innovation, standardization, and transformation across the pathology enterprise (Parwani, 2019; Bera et al., 2019; Hanna et al., 2024; 2019b).

Digital pathology in education

The advantages of digital pathology for education over conventional methods

Digital pathology provides unprecedented accessibility to high-quality pathology educational materials. Learners in remote or resource-limited settings can access the same WSIs, lectures, and assessments as trainees at major academic centers. Cloud-based sharing of digital slides further facilitates cross-institutional teaching collaborations and the development of global pathology education communities (Yin et al., 2016; Kumar et al., 2004; Parker et al., 2020).

Standardization and quality assurance represent another major advantage of digital pathology in education. Unlike glass slides, which can fade, break, or vary in staining quality, digital slides preserve image integrity indefinitely. Annotated cases, standardized teaching sets, and assessment materials can be reused consistently over time and across institutions, ensuring uniform educational experiences and evaluation standards (Yin et al., 2016; Kumar et al., 2004).

Digital platforms also offer significant flexibility and interactivity. Learners can study at their own pace and from any location, navigating slides dynamically to focus on key histological features. Educators can incorporate annotations, quizzes, hyperlinks, and supplementary resources directly into digital slides, creating immersive learning environments. When integrated with adaptive learning systems, digital pathology enables personalized educational pathways tailored to learners' performance and experience (Yin et al., 2016; Kumar et al., 2004).

From an economic perspective, while initial investments in glass slide scanners and data storage may be substantial, digital pathology offers long-term cost savings by reducing expenses for slide preparation, storage, maintenance, and microscope infrastructure. Importantly, digital platforms allow educational programs to scale without proportional increases in physical resources and can support revenue-generating activities such as paid virtual courses, conferences, and continuing medical education (CME) programs (Hanna et al., 2019a; Jahn et al., 2020).

Finally, digital pathology serves as a foundation for innovation through integration with emerging technologies. Artificial intelligence-assisted slide interpretation can function as an educational aid, while virtual and augmented reality applications can simulate grossing procedures and tissue handling. These technologies not only modernize pathology education but also prepare trainees for an increasingly data-driven diagnostic environment (Waqas et al., 2025; Hassell et al., 2025b; Madrigal et al., 2016).

Key use cases of digital pathology in education

Undergraduate and allied health education

Digital microscopy has steadily replaced traditional teaching laboratories since the early 2000s, enabling remote access to virtual histology and pathology laboratories.

The rapid transition to online education during the COVID-19 pandemic highlighted the effectiveness of this approach. For example, the University of Washington's online pathology course leveraged curated digital slides, case-based discussions, and live collaborative tools to engage learners with limited pathology exposure. Multiple studies have demonstrated that digital slides can support learning outcomes equivalent to or better than those of conventional microscopy, particularly when combined with flipped classrooms, game-based learning, and augmented reality. Platforms such as PathElective (https://www.pathelective.com/) provide interactive, self-guided educational resources that introduce pathology concepts to medical students and early trainees (Hassell et al., 2023).

Graduate medical education (residency and fellowship)

In residency and fellowship training, digital pathology enhances exposure to diverse diagnostic scenarios across surgical pathology, cytology, hematopathology, and molecular pathology. Digital platforms support case preview, real-time case review, virtual tumor boards, and remote sign-out, facilitating hybrid educational models that became essential during the COVID-19 pandemic and now persist as standard practice (Hassell et al., 2023). The Digital Anatomic Pathology Academy (DAPA) (https://digitalpathologyassociation.org/digital-anatomic-pathology-academy), developed by the Digital Pathology Association (DPA), exemplifies this approach by offering a curated global repository of thousands of cases with expert annotations and integrated didactic content. This "residency in the cloud" model democratizes access to high-quality educational resources, including for programs without local scanning infrastructure. In subspecialties such as cytopathology and hematopathology, digital pathology enables the simulation of diagnostic workflows through interactive z-stacking, improving pattern recognition while standardizing training and assessment across institutions.

Continuing medical education and peer learning

CME presents unique challenges in a rapidly evolving field marked by advances in precision medicine, biomarker testing, digital pathology, and augmented intelligence. Many practicing pathologists lack access to frequent in-person educational opportunities, creating an ongoing need for accessible, high-quality, CME-accredited programs. Digital pathology extends beyond formal training into lifelong professional development through virtual conferences, digital case discussions, and self-assessment modules. WSI-based CME activities allow participants to review cases asynchronously, assess diagnostic accuracy, and receive immediate feedback. Virtual tumor boards and peer-to-peer discussions involving multidisciplinary participants further promote global collaboration and expert consultation (Hassell et al., 2023).

The Precision Medicine Academy (PMA) (https://flpath.org/), supported by the Florida Society of Pathologists' digital learning center, illustrates this model by integrating biomarker testing, molecular pathology, digital pathology, and artificial intelligence into outcome-driven, CME-accredited curricula. Although designed for Florida-based pathologists, PMA activities are accessible both nationally and internationally.

Patient and public education

An emerging educational application of digital pathology is patient and public engagement. Digitized biopsy slides can be shared with patients during consultations, allowing clinicians to visually explain disease processes and pathological findings. This approach promotes transparency, improves patient understanding, and enhances trust. Early studies suggest that patients value viewing their own histopathology slides. Digital platforms, such as HER2Know.com, further support education by helping healthcare professionals and the public understand how pathologists evaluate biomarkers, such as HER2 immunohistochemistry.

Current challenges and the path forward

Despite its advantages, the adoption of digital pathology in education faces challenges, including financial costs, interoperability limitations, data privacy concerns, and institutional resistance to change. Nevertheless, initiatives such as the Digital Anatomic Pathology Academy and collaborations among international pathology societies are addressing these barriers by creating shared repositories, promoting open standards such as Digital Imaging and Communications in Medicine (DICOM) for pathology, and developing best-practice frameworks (Mildenberger et al., 2002). To address disparities in low-resource settings, the Open Pathology Education Network (OPEN) aims to provide free and globally accessible pathology education, underscoring digital pathology's potential to reduce educational inequities worldwide (Hassell et al., 2023, 2025a; Williams et al., 2020).

Digital pathology in research

Introduction and evolution of digital pathology in research

Digital pathology has become a cornerstone of modern biomedical research, evolving in tandem with advances in computational imaging and artificial intelligence (Waqas et al., 2023, 2024b; Hassell et al., 2025b; Kim et al., 2024). Historically, pathology research relied on manual microscopy, limited sample throughput, and subjective interpretation—constraints that slowed discovery and limited reproducibility. Early quantitative approaches, based on handcrafted image features such as region annotations and morphometric measurements, were foundational but lacked scalability, robustness, and cross-institutional generalizability (Waqas et al., 2023).

The introduction of convolutional neural networks (CNNs) in the mid-2010s marked a transformative step, enabling automated feature learning directly from WSIs. CNN-based approaches advanced tumor classification, segmentation, and prognostic modeling but often required large, labeled datasets and remained sensitive to domain shifts (Bera et al., 2019). More recently, foundation models—large-scale, pre-trained AI architectures developed using weakly labeled, multiinstitutional datasets—have ushered in a new era. Built using self-supervised, contrastive, or vision-language paradigms, these models learn generalizable image representations that can be fine-tuned for specific research tasks using relatively small, labeled datasets. This shift has democratized computational pathology and reduced dependence on extensive computational infrastructure (Chen et al., 2024).

The latest frontier integrates digital pathology with multiomics data, creating multimodal systems that link tissue morphology with molecular and clinical features. Such approaches enable the inference of gene expression from histology, identification of molecular subtypes, characterization of tumor-immune interactions, and prediction of prognosis and therapeutic response. In this context, digital pathology functions not merely as an imaging modality but as a foundational research infrastructure supporting high-throughput, data-driven discovery across academia and industry (Waqas et al., 2024b; Chen et al., 2025; Huang et al., 2025; Tripathi et al., 2026).

Why digital pathology is transformative for research

Digital pathology enables scalable, high-throughput analysis by allowing thousands of slides to be evaluated efficiently and repetitively. Unlike glass slides, digital images remain stable over time, supporting reproducibility and standardization across institutions and longitudinal studies. AI-driven feature extraction introduces quantitative, objective measurements of tissue architecture and cellular composition, reducing the subjectivity inherent to manual interpretation (Bera et al., 2019; Madabhushi and Lee, 2016).

As the substrate for modern computational pathology, digital pathology integrates seamlessly with advanced analytics, including CNNs, transformer-based architectures, and multiomics AI systems (Waqas et al., 2023, 2024b). Cloud-native platforms further facilitate global collaboration, federated research models, and remote expert participation. Although initial infrastructure costs may be substantial, digital workflows ultimately reduce labor, shipping, recuts, and manual review burdens, improving long-term economic efficiency in research environments (Waqas et al., 2023).

Key use cases of digital pathology in research

Computational pathology and AI research

AI-driven digital pathology research encompasses automated tissue segmentation, large-scale cell detection and quantification, prognostic and predictive modeling, unsupervised phenotyping, and algorithmic quality control for detecting staining or scanning artifacts. These approaches transform histology into a measurable, analyzable, and computable research modality (Bera et al., 2019; Lu et al., 2021b; Jurgas et al., 2024).

Biomarker discovery and validation

Digital pathology supports standardized quantification of protein biomarkers, deep learning-based discovery of novel morphological phenotypes, and predictive modeling linking image-derived features with genomic, proteomic, and clinical outcomes. These capabilities are central to patient stratification in immunotherapy and to the development of targeted treatments (Ilie and Hofman, 2017; Kather et al., 2019; Echle et al., 2021).

Clinical trials and regulatory science

In clinical research, digital pathology enables centralized pathology review, real-time collaboration among trial pathologists, regulatory-grade image archiving with complete metadata, and reduced logistical delays by eliminating physical slide shipment. Automated image analysis supports exploratory endpoints such as immune cell density and tumor burden assessment (Niazi et al., 2019; Baxi et al., 2022; Saltz et al., 2018).

Multiomics and systems biology

Integration of WSI with genomics, transcriptomics, spatial proteomics, and radiology has redefined digital pathology research. Image-omics fusion enables risk prediction, treatment forecasting, and systems-level modeling of tissue ecosystems, supporting precision oncology and translational science (Chen et al., 2025; Echle et al., 2021; Jackson et al., 2020).

Foundation models and fine-tuned applications in digital pathology research

Foundation models trained using large-scale pathology and medical image datasets learn transferable representations of tissue morphology that generalize across institutions and technical variability (Waqas et al., 2024a; Tripathi et al., 2024, 2025). Their advantages include robustness to domain shifts, reduced annotation requirements, rich feature embeddings, and accelerated research cycles (Chen et al., 2024, 2025). Fine-tuning these models enables specialized applications such as rare tumor subtype detection, tumor-immune mapping, automated grading, recurrence prediction, and modeling of spatial biomarker heterogeneity (Tripathi et al., 2025). Emerging multimodal foundation models that integrate histology with genomics, radiology, clinical text, and spatial omics generate unified biological representations, supporting histology-to-gene-expression prediction, integrated patient stratification, and multiscale disease modeling (Brodsky et al., 2025; Lu et al., 2021a).

Digital pathology in quality improvement and quality control

Digital pathology plays a critical role in quality improvement (QI) and quality control (QC), both of which are essential for maintaining diagnostic confidence, research integrity, and operational efficiency. By transforming histopathology slides into durable, analyzable digital assets, digital pathology enables systematic, scalable, and objective approaches to quality assurance that are difficult or impossible to achieve with conventional glass-slide workflows (Tripathi et al., 2026; Madabhushi and Lee, 2016; Jurgas et al., 2024; Pantanowitz et al., 2024).

Quality control across the digital pathology workflow

Quality control in digital pathology spans the entire workflow, from slide preparation and image acquisition to downstream analysis and data reuse. Unlike traditional microscopy, digital systems allow continuous monitoring and automated assessment of image quality at multiple stages.

Automated artifact detection

Computational pathology models can be trained to function as QC algorithms that automatically detect and flag imaging and preparation artifacts, including out-of-focus regions, tissue folds, staining irregularities, and scanner stitching errors. These tools ensure that only high-quality WSIs are used for clinical interpretation, research analysis, or educational purposes, reducing downstream error propagation and improving overall reliability (Jurgas et al., 2024; Tripathi et al., 2025; Janowczyk and Madabhushi, 2016; Janowczyk et al., 2019).

Normalization of variability

A major challenge in multicenter studies and routine practice is the variability introduced by differences in tissue processing, staining protocols, and scanner hardware. Digital pathology addresses this challenge by developing normalization pipelines that computationally adjust color, intensity, and contrast across images. By reducing technical variability, these approaches support consistent feature extraction, fair algorithm performance, and reliable cross-institutional comparisons, which are critical for both research reproducibility and quality assurance (Tripathi et al., 2025; Reinhard et al., 2001; Tellez et al., 2019).

Data integrity and traceability

Unlike physical glass slides, which may degrade, fade, or be lost over time, digital slides preserve image quality indefinitely. Digital pathology systems maintain regulatory-grade archives that capture complete metadata, timestamps, version histories, and audit trails for each image (Tripathi et al., 2024). This level of traceability enables comprehensive monitoring of the digital workflow, supports internal and external audits, and facilitates compliance with regulatory and accreditation requirements (Hanna et al., 2019b).

AI-enabled quantitative quality assurance of immunohistochemistry staining

The staining quality of immunohistochemistry (IHC) is an important preanalytical factor that has a significant impact on staining results. This is particularly important for immunohistochemical biomarkers that are used as companion diagnostics. There is a great need for AI-driven solutions to quantify and monitor staining quality in pathology laboratories, addressing variability that impacts diagnostic assays and research reproducibility. There are commercially available AI-enabled tools to assess IHC staining consistency across key prognostic and predictive biomarkers, which demonstrate the practical use of AI-driven QC metrics in a surgical pathology quality context (https://imageanalysis.visiopharm.com/qualitopix-scanner?utm_source=chatgpt.com).

Quality improvement enabled by digital pathology

Beyond static quality control checks, digital pathology enables continuous quality improvement by providing structured data streams that can be analyzed longitudinally.

Aggregated QC metrics, such as scan failure rates, artifact frequency, turnaround times, and re-scan rates, can be used to identify systemic issues, optimize laboratory processes, and benchmark performance across sites or over time. These capabilities support proactive interventions that improve efficiency, reduce waste, and enhance patient care (Pantanowitz, 2010; Hanna et al., 2019a; Schwen et al., 2023).

Quality assurance in research and education

Quality assurance is equally important in research and educational contexts. In computational pathology research, QC algorithms ensure that training datasets are free of artifacts and biases that could compromise model performance or generalizability. In educational settings, digital QC supports the creation of standardized, high-quality teaching sets that remain consistent across cohorts and institutions, reinforcing fairness and reproducibility in assessment (Hassell et al., 2023; Jurgas et al., 2024; Pantanowitz et al., 2024). By enabling robust QC and continuous QI across clinical, research, and educational domains, digital pathology establishes a foundation for standardized, scalable, and reliable pathology workflows. These capabilities ultimately translate into improved efficiency, higher-quality data, and greater confidence in both diagnostic and research outcomes (Pantanowitz et al., 2024).

Other and emerging applications of digital pathology

Beyond its established roles in education, research, and quality assurance, digital pathology supports a range of additional applications that are increasingly relevant to contemporary pathology practice. These other applications leverage WSI and associated metadata as shared, reusable digital assets that enhance communication, operational insight, capacity building, and long-term data stewardship. Importantly, many of these use cases are already in active use and deliver immediate practical value to pathology departments (Waqas et al., 2023; Hanna et al., 2019a).

Support for multidisciplinary care and tumor boards

Digital pathology has become an effective tool for multidisciplinary team (MDT) meetings and tumor boards, where clear visualization and shared understanding of pathologic findings are essential. WSIs allow pathologists to present representative regions of interest dynamically, facilitating discussion with oncologists, surgeons, radiologists, and other specialists. Unlike static photomicrographs, WSIs enable real-time navigation, annotation, and contextual explanation of histologic features, improving communication and consensus building (Hanna et al., 2020; Randell et al., 2012). Because digital slides can be accessed asynchronously and remotely, they also support pre-meeting review and post-meeting documentation. Annotated slides used during tumor boards can be archived and revisited for education, quality review, or follow-up discussions. In this context, digital pathology functions not as a diagnostic replacement but as a communication and collaboration platform that strengthens multidisciplinary care.

Operational analytics and workflow optimization

Digital pathology generates rich metadata that can be repurposed for operational analytics and workflow optimization. Scan timestamps, file sizes, case volumes, subspecialty distribution, rescan rates, and turnaround times can be aggregated to provide insights into laboratory performance and resource utilization. These data enable pathology leaders to identify bottlenecks, balance workloads, and benchmark performance across services or institutions (Hanna et al., 2019b, 2020). When combined with quality metrics, operational analytics support data-driven management decisions, such as optimizing scanner deployment, staffing models, or case routing strategies. This secondary use of digital pathology data extends beyond traditional quality control by focusing on system-level efficiency and continuous operational improvement.

Global pathology and capacity building

Digital pathology also plays an important role in global pathology and capacity-building initiatives. By enabling the creation and sharing of curated digital slide libraries, educational content, and case-based discussions, digital platforms support training and mentorship in regions with limited access to subspecialty expertise (Williams et al., 2018). These activities often combine elements of education and research but are distinct in their emphasis on sustainable skill development rather than remote diagnosis. In practice, digital pathology facilitates international collaborations, exchange programs, and shared educational resources, strengthening pathological capacity without requiring continuous physical slide shipment or on-site presence. This model is particularly valuable for building long-term partnerships and fostering equitable access to pathology knowledge worldwide (Hassell et al., 2023; Walsh and Orsi, 2024).

Biobanking, data reuse, and longitudinal cohort development

Digitization of histopathology slides enables new approaches to biobanking and data reuse. WSIs can be linked to biospecimens, clinical data, and outcomes, creating durable digital cohorts that support retrospective studies, longitudinal analyses, and future research initiatives (Tripathi et al., 2024, 2025). Unlike glass slides, which may degrade or become unavailable over time, digital slides remain accessible and analyzable indefinitely (Pantanowitz et al., 2018, 2024). These digital cohorts are increasingly used to support translational research, biomarker validation, and the development and evaluation of computational pathology tools. From a practical perspective, digitized archives enhance institutional memory, reduce redundancy in slide handling, and maximize the long-term value of existing tissue resources (Bera et al., 2019; Kothari et al., 2013).

Challenges and the path forward

Despite its substantial promise, digital pathology faces several important barriers that affect its broader adoption across research, quality improvement, and other nonprimary diagnostic applications. High capital costs associated with slide scanners, data storage, and computational infrastructure remain a significant obstacle, particularly for smaller institutions and resource-limited settings. In parallel, data interoperability challenges persist, driven by uneven adoption of standards such as DICOM for pathology and variability in vendor-specific file formats and metadata structures (Pantanowitz, 2010, Pantanowitz et al., 2018, 2024; Tellez et al., 2019; Clunie, 2021). Technical variability introduced by differences in tissue processing, staining protocols, and scanner platforms further complicates large-scale deployment, necessitating robust normalization and quality control pipelines. Regulatory considerations also present challenges, especially as digital pathology increasingly intersects with artificial intelligence. Requirements for model validation, transparency, and explainability continue to evolve and must be addressed to ensure safe and responsible implementation (Topol, 2019; Sheller et al., 2020).

Human factors remain equally important. Gaps in digital literacy and computational expertise within pathology and research teams can limit the effective use of digital tools, while data privacy and security concerns, particularly in multiinstitutional and cloud-based collaborations, require careful governance and oversight. Encouragingly, many of these challenges are being actively addressed through global initiatives, including the broader adoption of DICOM standards, the development of open and curated datasets, the implementation of federated learning frameworks, and collaboration within international pathology and AI consortia. As technical infrastructure matures, standards converge, and AI systems become more interpretable and robust, digital pathology is poised to continue expanding its impact across biomedical research, quality improvement, education, and precision medicine, reinforcing its role as a foundational platform for the future of pathology (Sheller et al., 2020; Koutsoubis et al., 2025).

Conclusion

Digital pathology has evolved beyond a tool for slide digitization into a foundational platform that supports a wide range of nonprimary diagnostic applications across the pathology enterprise. As highlighted in this chapter, its impact extends through education, biomedical research, quality improvement, and emerging domains that collectively reshape how pathological knowledge is generated, disseminated, and applied. In education, digital pathology enables scalable, standardized, and interactive learning environments that prepare trainees and practicing pathologists for a technology-driven future. In research, it serves as the digital substrate for computational pathology, biomarker discovery, clinical trials, and multiomics integration, accelerating discovery while improving reproducibility and collaboration. In quality improvement and quality control, digital workflows introduce objective, automated, and traceable mechanisms that enhance reliability, efficiency, and confidence across clinical, research, and educational settings.

Importantly, these applications underscore digital pathology's role as infrastructure rather than a standalone technology. By enabling data reuse, integration with artificial intelligence, and global collaboration, digital pathology supports continuous innovation while addressing long-standing challenges related to variability, access, and standardization. Although barriers remain, including cost, interoperability, regulatory complexity, and workforce readiness, ongoing advances in standards, automation, and collaborative frameworks continue to lower these thresholds. As digital pathology matures, its value will increasingly be measured not only by diagnostic efficiency but also by its ability to support learning, discovery, quality assurance, and system-wide improvement. Together, these other applications position digital pathology as a central pillar of modern pathology practice and research, with enduring implications for precision medicine and the future of the discipline.

AI Disclosure

During the preparation of this work, the author(s) used ChatGPT to polish the writing. After using this tool/service, the author(s) reviewed and edited the content as needed and take(s) full responsibility for the content of the publication.

References

Baxi, V., Edwards, R., Montalto, M., Saha, S., 2022. Digital pathology and artificial intelligence in translational medicine and clinical practice. Mod. Pathol. 15300285. 35 (1), 23–32. Available from: http://www.nature.com/modpathol/index.html. doi:10.1038/s41379-021-00919-2.

Bera, K., Schalper, K.A., Rimm, D.L., Velcheti, V., Madabhushi, A., 2019. Artificial intelligence in digital pathology—new tools for diagnosis and precision oncology. Nat. Rev. Clin. Oncol. 17594782. 16 (11), 703–715. Available from: http://www.nature.com/nrclinonc/archive/index.html. doi:10.1038/s41571-019-0252-y.

Brodsky, V., Ullah, E., Bychkov, A., Song, A.H., Walk, E.E., Louis, P., Rasool, G., Singh, R.S., Mahmood, F., Bui, M.M., Parwani, A.V., 2025. Generative artificial intelligence in anatomic pathology. Arch. Pathol. Lab. Med. 1543-2165. 149 (4), 298–318. doi:10.5858/arpa.2024-0215-ra.

Chen, R.J., Ding, T., Lu, M.Y., Williamson, D.F.K., Jaume, G., Song, A.H., Chen, B., Zhang, A., Shao, D., Shaban, M., Williams, M., Oldenburg, L., Weishaupt, L.L., Wang, J.J., Vaidya, A., Le, L.P., Gerber, G., Sahai, S., Williams, W., Mahmood, F., 2024. Towards a general-purpose foundation model for computational pathology. Nat. Med. 1546170X. 30 (3), 850–862. Available from: https://www.nature.com/nm/. doi:10.1038/s41591-024-02857-3.

Chen, W., Zhang, P., Tran, T.N., Xiao, Y., Li, S., Shah, V.V., Cheng, H., Brannan, K.W., Youker, K., Lai, L., Fang, L., Yang, Y., Le, N.T., Abe, J.I., Chen, S.H., Ma, Q., Chen, K., Song, Q., Cooke, J.P., Wang, G., 2025. A visual-omics foundation model to bridge histopathology with spatial transcriptomics. Nat. Methods. 15487105. 22 (7), 1568–1582. Available from: https://www.nature.com/nmeth/. doi:10.1038/s41592-025-02707-1.

Clunie, D.A., 2021. DICOM Format and protocol standardization—a core requirement for digital pathology success. Toxicol. Pathol. 15331601. 49 (4), 738–749. Available from: http://tpx.sagepub.com/archive/. doi:10.1177/0192623320965893.

Echle, A., Rindtorff, N.T., Brinker, T.J., Luedde, T., Pearson, A.T., Kather, J.N., 2021. Deep learning in cancer pathology: a new generation of clinical biomarkers. Br. J. Cancer. 15321827. 124 (4), 686–696. Available from: http://www.nature.com/bjc/index.html. doi:10.1038/s41416-020-01122-x.

Hanna, M.G., Olson, N.H., Zarella, M., Dash, R.C., Herrmann, M.D., Furtado, L.V., Stram, M.N., Raciti, P.M., Hassell, L., Mays, A., Pantanowitz, L., Sirintrapun, J.S., Krishnamurthy, S., Parwani, A., Lujan, G., Evans, A., Glassy, E.F., Bui, M.M., Singh, R., Souers, R.J., de Baca, M.E., Seheult, J.N., 2024. Recommendations for performance evaluation of machine learning in pathology: a concept paper from the College of American Pathologists. Arch. Pathol. Lab. Med. 1543-2165. 148 (10), e335. doi:10.5858/arpa.2023-0042-cp.

Hanna, M.G., Reuter, V.E., Samboy, J., England, C., Corsale, L., Fine, S.W., Agaram, N.P., Stamelos, E., Yagi, Y., Hameed, M., Klimstra, D.S., Sirintrapun, S.J., 2019a. Implementation of digital pathology offers clinical and operational increase in efficiency and cost savings. Arch. Pathol. Lab. Med. 15432165. 143 (12), 1545–1555. Available from: https://www.archivesofpathology.org/doi/pdf/10.5858/arpa.2018-0514-OA. doi:10.5858/arpa.2018-0514-OA.

Hanna, M.G., Reuter, V.E., Hameed, M.R., Tan, L.K., Chiang, S., Sigel, C., Hollmann, T., Giri, D., Samboy, J., Moradel, C., Rosado, A., Otilano, J.R., England, C., Corsale, L., Stamelos, E., Yagi, Y., Schüffler, P.J., Fuchs, T., Klimstra, D.S., Sirintrapun, S.J., 2019b. Whole slide imaging equivalency and efficiency study: experience at a large academic center. Mod. Pathol. 15300285. 32 (7), 916–928. Available from: http://www.nature.com/modpathol/index.html. doi:10.1038/s41379-019-0205-0.

Hanna, M.G., Reuter, V.E., Ardon, O., Kim, D., Sirintrapun, S.J., Schüffler, P.J., Busam, K.J., Sauter, J.L., Brogi, E., Tan, L.K., Xu, B., Bale, T., Agaram, N.P., Tang, L.H., Ellenson, L.H., Philip, J., Corsale, L., Stamelos, E., Friedlander, M.A., Ntiamoah, P., Labasin, M., England, C., Klimstra, D.S., Hameed, M., 2020. Validation of a digital pathology system including remote review during the COVID-19 pandemic. Mod. Pathol. 15300285. 33 (11), 2115–2127. Available from: http://www.nature.com/modpathol/index.html. doi:10.1038/s41379-020-0601-5.

Hassell, L.A., Absar, S.F., Chauhan, C., Dintzis, S., Farver, C.F., Fathima, S., Glassy, E.F., Goldstein, J.A., Gullapalli, R., Ho, J., Koch, L.K., Madory, J.E., Mirza, K.M., Nguyen, P.N., Pantanowitz, L., Parwani, A., Rojansky, R., Seifert, R.P., Singh, R., ElGabry, E.A., Bui, M., 2023. Pathology education powered by virtual and digital transformation now and the future. Arch. Pathol. Lab. Med. 15432165. 147 (4), 474–491. Available from: https://meridian.allenpress.com/aplm/article/147/4/474/484251/Pathology-Education-Powered-by-Virtual-and-Digital. doi:10.5858/ARPA.2021-0473-RA.

Hassell, L.A., Wong, A., Parkash, V., Ng, J.S., Tran, N.T., Huynh, L., Han, T.N., Quynh, T.T.N., Ngoc, P.T.H., Quy, T. T., 2025a. Virtual remote pathology education in support of virtual remote gynecologic-oncology training. Arch. Pathol. Lab. Med. 15432165. 149 (7), 597–601. doi:10.5858/arpa.2023-0449-EP.

Hassell, L.A., Forsythe, M.L., Bhalodia, A., Lan, T., Rashid, T., Powers, A., Bui, M.M., Brickman, A., Qiangqiang, G., Bychkov, A., Cree, I., Pantanowitz, L., 2025b. Toward optimizing the impact of digital pathology and augmented intelligence on issues of diagnosis, grading, staging, and classification. Mod. Pathol. 08933952. 38 (7), 100765. doi:10.1016/j.modpat.2025.100765.

Huang, Q., Shen'ao, W., Zhenkai, O., Gao, Y., 2025. Computational pathology: a comprehensive review of recent developments in digital and intelligent pathology. Intell. Oncol. 29502616. 1 (2), 139–159. doi:10.1016/j.intonc.2025.03.004.

Ilie, M., Hofman, P., 2017. Reproducibility of PD-L1 assessment in non-small cell lung cancer-know your limits but never stop trying to exceed them. Transl. Lung Cancer Res. 22264477. 6, S51. Available from: http://tlcr.amegroups.com/article/download/16898/13530. doi:10.21037/tlcr.2017.10.13.

Jackson, H.W., Fischer, J.R., Zanotelli, V.R.T., Ali, H.R., Mechera, R., Soysal, S.D., Moch, H., Muenst, S., Varga, Z., Weber, W.P., Bodenmiller, B., 2020. The single-cell pathology landscape of breast cancer. Nature. 14764687. 578 (7796), 615–620. Available from: http://www.nature.com/nature/index.html. doi:10.1038/s41586-019-1876-x.

Jahn, S.W., Plass, M., Moinfar, F., 2020. Digital pathology: advantages, limitations and emerging perspectives. J. Clin. Med. 20770383. 9 (11). Available from: https://www.mdpi.com/2077-0383/9/11/3697/pdf. doi:10.3390/jcm9113697.

Janowczyk, A., Zuo, R., Gilmore, H., Feldman, M., Madabhushi, A., 2019. HistoQC: an open-source quality control tool for digital pathology slides. JCO Clin. Cancer Inform. 2473-4276. 3 (3), 1–7. doi:10.1200/cci.18.00157.

Janowczyk, A., Madabhushi, A., 2016. Deep learning for digital pathology image analysis: a comprehensive tutorial with selected use cases. J. Pathol. Inform. 21533539. 7 (1), 29. doi:10.4103/2153-3539.186902.

Jurgas, A., Wodzinski, M., D'Amato, M., van der Laak, J., Atzori, M., Müller, H., 2024. Improving quality control of whole slide images by explicit artifact augmentation. Sci. Rep. 20452322. 14 (1). Available from: https://www.nature.com/srep/. doi:10.1038/s41598-024-68667-2.

Kather, J.N., Pearson, A.T., Halama, N., Jäger, D., Krause, J., Loosen, S.H., Marx, A., Boor, P., Tacke, F., Neumann, U. P., Grabsch, H.I., Yoshikawa, T., Brenner, H., Chang-Claude, J., Hoffmeister, M., Trautwein, C., Luedde, T., 2019. Deep learning can predict microsatellite instability directly from histology in gastrointestinal cancer. Nat. Med. 1546170X. 25 (7), 1054–1056. Available from: http://www.nature.com/nm/index.html. doi:10.1038/s41591-019-0462-y.

Kim, D., Thrall, M.J., Michelow, P., Schmitt, F.C., Vielh, P.R., Siddiqui, M.T., Sundling, K.E., Virk, R., Alperstein, S., Bui, M.M., Chen-Yost, H., Donnelly, A.D., Lin, O., Liu, X., Madrigal, E., Zakowski, M.F., Parwani, A.V., Jenkins, E., Pantanowitz, L., Li, Z., 2024. The current state of digital cytology and artificial intelligence (AI): global survey results from the American Society of Cytopathology Digital Cytology Task Force. J. Am. Soc.

Cytopathol. 22132945. 13 (5), 319–328. Available from: https://www.sciencedirect.com/science/journal/22132945. doi:10.1016/j.jasc.2024.04.003.

Kothari, S., Phan, J.H., Wang, M.D., 2013. Eliminating tissue-fold artifacts in histopathological whole-slide images for improved image-based prediction of cancer grade. J. Pathol. Inform. 21533539. 4 (1), 22. doi:10.4103/2153-3539.117448.

Koutsoubis, N., Waqas, A., Yilmaz, Y., Ramachandran, R.P., Schabath, M.B., Rasool, G., 2025. Privacy-preserving federated learning and uncertainty quantification in medical imaging. Radiol. Artif. Intell. 26386100. 7 (4). Available from: https://pubs.rsna.org/doi/epdf/10.1148/ryai.240637. doi:10.1148/ryai.240637.

Kumar, R.K., Velan, G.M., Korell, S.O., Kandara, M., Dee, F.R., Wakefield, D., 2004. Virtual microscopy for learning and assessment in pathology. J. Pathol. 204 (5), 613–618. doi:10.1002/path.1658.

Lu, M.Y., Chen, T.Y., Williamson, D.F.K., Zhao, M., Shady, M., Lipkova, J., Mahmood, F., 2021a. AI-based pathology predicts origins for cancers of unknown primary. Nature. 14764687. 594 (7861), 106–110. Available from: http://www.nature.com/nature/index.html. doi:10.1038/s41586-021-03512-4.

Lu, M.Y., Williamson, D.F.K., Chen, T.Y., Chen, R.J., Barbieri, M., Mahmood, F., 2021b. Data-efficient and weakly supervised computational pathology on whole-slide images. Nat. Biomed. Eng. 2157846X. 5 (6), 555–570. Available from: www.nature.com/natbiomedeng/. doi:10.1038/s41551-020-00682-w.

Madabhushi, A., Lee, G., 2016. Image analysis and machine learning in digital pathology: challenges and opportunities. Med. Image Anal. 13618423. 33, 170–175. Available from: http://www.elsevier.com/inca/publications/store/6/2/0/9/8/3/index.htt. doi:10.1016/j.media.2016.06.037.

Madrigal, E., Prajapati, S., Hernandez-Prera, J.C., 2016. Introducing a virtual reality experience in anatomic pathology education. Am. J. Clin. Pathol. 19437722. 146 (4), 462–468. Available from: http://ajcp.oxfordjournals.org/content/by/year. doi:10.1093/ajcp/aqw133.

Mildenberger, P., Eichelberg, M., Martin, E., 2002. Introduction to the DICOM standard. Eur. Radiol. 12 (4), 920–927. doi:10.1007/s003300101100.

Mukhopadhyay, S., Feldman, M.D., Abels, E., Ashfaq, R., Beltaifa, S., Cacciabeve, N.G., Cathro, H.P., Cheng, L., Cooper, K., Dickey, G.E., Gill, R.M., Heaton, R.P., Kerstens, R., Lindberg, G.M., Malhotra, R.K., Mandell, J.W., Manlucu, E.D., Mills, A.M., Mills, S.E., Moskaluk, C.A., Nelis, M., Patil, D.T., Przybycin, C.G., Reynolds, J.P., Rubin, B.P., Saboorian, M.H., Salicru, M., Samols, M.A., Sturgis, C.D., Turner, K.O., Wick, M.R., Yoon, J.Y., Zhao, P., Taylor, C.R., 2018. Whole slide imaging versus microscopy for primary diagnosis in surgical pathology: a multicenter blinded randomized noninferiority study of 1992 cases (pivotal study). Am. J. Surg. Pathol. 15320979. 42 (1), 39–52. Available from: http://journals.lww.com/ajsp/pages/default.aspx. doi:10.1097/PAS.0000000000000948.

Niazi, M.K.K., Parwani, A.V., Gurcan, M.N., 2019. Digital pathology and artificial intelligence. Lancet Oncol. 14745488. 20 (5), e253. Available from: http://www.journals.elsevier.com/the-lancet-oncology/. doi:10.1016/S1470-2045(19)30154-8.

Pantanowitz, L., 2010. Digital images and the future of digital pathology. J. Pathol. Inform. 21533539. 1 (1), 15. doi:10.4103/2153-3539.68332.

Pantanowitz, L., Sharma, A., Carter, A.B., Kurc, T., Sussman, A., Saltz, J., 2018. Twenty years of digital pathology: an overview of the road travelled, what is on the horizon, and the emergence of vendor-neutral archives. J. Pathol. Inform. 21533539. 9 (1), 40. doi:10.4103/jpi.jpi_69_18.

Pantanowitz, L., Hanna, M., Pantanowitz, J., Lennerz, J., Henricks, W.H., Shen, P., Quinn, B., Bennet, S., Rashidi, H. H., 2024. Regulatory aspects of artificial intelligence and machine learning. Mod. Pathol. 08933952. 37 (12), 100609. doi:10.1016/j.modpat.2024.100609.

Parker, E.U., Chang, O., Koch, L., 2020. Remote anatomic pathology medical student education in Washington state. Am. J. Clin. Pathol. 19437722. 154 (5), 585–591. Available from: http://ajcp.oxfordjournals.org/content/by/year. doi:10.1093/ajcp/aqaa154.

Parwani, A.V., 2019. Next generation diagnostic pathology: use of digital pathology and artificial intelligence tools to augment a pathological diagnosis. Diagn. Pathol. 17461596. 14 (1). Available from: http://www.diagnosticpathology.org/. doi:10.1186/s13000-019-0921-2.

Randell, R., Ruddle, R.A., Thomas, R., Treanor, D., 2012. Diagnosis at the microscope: a workplace study of histopathology. Cognition, Technology and Work. 14355566. 14 (4), 319–335. doi:10.1007/s10111-011-0182-7.

Reinhard, E., Adhikhmin, M., Gooch, B., Shirley, P., 2001. Color transfer between images. IEEE Comput. Graph. Appl. 02721716. 21 (4), 34–41. doi:10.1109/38.946629.

Saltz, J., Gupta, R., Hou, L., Kurc, T., Singh, P., Nguyen, V., Samaras, D., Shroyer, K.R., Zhao, T., Batiste, R.,, Van Arnam, J., Caesar-Johnson, S.J., Demchok, J.A., Felau, I., Kasapi, M., Ferguson, M.L., Hutter, C.M., Sofia, H.J., Tarnuzzer, R., Wang, Z., Yang, L., Zenklusen, J.C., Zhang, J. (Julia), Chudamani, S., Liu, J., Lolla, L., Naresh, R., Pihl, T., Sun, Q., Wan, Y., Wu, Y., Cho, J., De Freitas, T., Frazer, S., Gehlenborg, N., Getz, G., Heiman, D.I., Kim, J., Lawrence, M.S., Lin, P., Meier, S., Noble, M.S., Saksena, G., Voet, D., Zhang, H., Bernard, B., Chambwe, N., Dhankani, V., Knijnenburg, T., Kramer, R., Leinonen, K., Liu, Y., Miller, M., Reynolds, S., Shmulevich, I., Thorsson, V., Zhang, W., Akbani, R., Broom, B.M., Hegde, A.M., Zhenlin, J., Kanchi, R.S., Korkut, A., Jun, L., Liang, H., Ling, S., Liu, W., Yiling, L., Mills, G.B., Kwok-Shing, N., Rao, A., Ryan, M., Wang, J., Weinstein, J.N., Zhang, J., Abeshouse, A., Armenia, J., Chakravarty, D., Chatila, W.K., de Bruijn, I., Gao, J., Gross, B.E., Heins, Z.J., Kundra, R., Konnor, L., Ladanyi, M., Luna, A., Nissan, M.G., Ochoa, A., Phillips, S.M., Reznik, E., Sanchez-Vega, F., Sander, C., Schultz, N., Sheridan, R., Sumer, S.O., Sun, Y., Taylor, B.S., Wang, J., Zhang, H., Anur, P., Peto, M., Spellman, P., Benz, C., Stuart, J.M., Wong, C.K., Yau, C., Hayes, D.N., Parker, J.S., Wilkerson, M.D., Ally, A., Balasundaram, M., Bowlby, R., Brooks, D., Carlsen, R., Chuah, E., Dhalla, N., Holt, R., Jones, S.J.M., Kasaian, K., Lee, D., Yussanne, M., Marra, M.A., Mayo, M., Moore, R.A., Mungall, A.J., Mungall, K., Robertson, A.G., Sadeghi, S., Schein, J.E., Sipahimalani, P., Tam, A., Thiessen, N., Tse, K., Wong, T., Berger, A.C., Beroukhim, R., Cherniack, A.D., Cibulskis, C., Gabriel, S.B., Gao, G.F., Gavin, H., Meyerson, M., Schumacher, S.E., Shih, J., Kucherlapati, M.H., Kucherlapati, R.S., Baylin, S., Cope, L., Danilova, L., Bootwalla, M.S., Lai, P.H., Maglinte, D. T., Berg, D.J.V.D., Weisenberger, D.J., Auman, J.T., Balu, S., Bodenheimer, T., Fan, C., Hoadley, K.A., Hoyle, A.P., Jefferys, S.R., Jones, C.D., Meng, S., Mieczkowski, P.A., Mose, L.E., Perou, A.H., Perou, C.M., Roach, J., Shi, Y., Simons, J.V., Skelly, T., Soloway, M.G., Tan, D., Veluvolu, U., Fan, H., Hinoue, T., Laird, P.W., Shen, H., Zhou, W., Bellair, M., Chang, K., Covington, K., Creighton, C.J., Dinh, H., Doddapaneni, H., Donehower, L.A., Drummond, J., Gibbs, R.A., Glenn, R., Hale, W., Han, Y., Jianhong, H., Korchina, V., Lee, S., Lewis, L., Wei, L., Liu, X., Morgan, M., Morton, D., Muzny, D., Santibanez, J., Sheth, M., Shinbrot, E., Wang, L., Wang, M., Wheeler, D.A., Liu, X., Zhao, F., Hess, J., Appelbaum, E.L., Bailey, M., Cordes, M.G., Ding, L., Fronick, C.C., Fulton, L.A., Fulton, R.S., Kandoth, C., Mardis, E.R., McLellan, M.D., Miller, C.A., Schmidt, H.K., Wilson, R.K., Crain, D., Curley, E., Gardner, J., Lau, K., Mallery, D., Morris, S., Paulauskis, J., Penny, R., Shelton, C., Shelton, T., Sherman, M., Thompson, E., Yena, P., Bowen, J., Gastier-Foster, J.M., Gerken, M., Leraas, K.M., Lichtenberg, T.M., Ramirez, N.C., Wise, L., Zmuda, E., Corcoran, N., Costello, T., Hovens, C., Carvalho, A.L., de Carvalho, A.C., Fregnani, J.H., Longatto-Filho, A., Reis, R.M., Scapulatempo-Neto, C., Silveira, H.C.S., Vidal, D.O., Burnette, A., Eschbacher, J., Hermes, B., Noss, A., Singh, R., Anderson, M.L., Castro, P.D., Ittmann, M., Huntsman, D., Kohl, B., Xuan, L., Thorp, R., Andry, C., Duffy, E.R., Lyadov, V., Paklina, O., Setdikova, G., Shabunin, A., Tavobilov, M., McPherson, C., Warnick, R., Berkowitz, R., Cramer, D., Feltmate, C., Horowitz, N., Kibel, A., Muto, M., Raut, C.P., Malykh, A., Barnholtz-Sloan, J.S., Barrett, W., Devine, K., Fulop, J., Ostrom, Q.T., Shimmel, K., Wolinsky, Y., Sloan, A.E., De Rose, A., Giuliante, F., Goodman, M., Karlan, B.Y., Hagedorn, C.H., Eckman, J., Harr, J., Myers, J., Tucker, K., Zach, L.A., Deyarmin, B., Hai, H., Kvecher, L., Larson, C., Mural, R.J., Somiari, S., Vicha, A., Zelinka, T., Bennett, J., Iacocca, M., Rabeno, B., Swanson, P., Latour, M., Lacombe, L., Têtu, B., Bergeron, A., McGraw, M., Staugaitis, S.M., Chabot, J., Hibshoosh, H., Sepulveda, A., Tao, S., Wang, T., Potapova, O., Voronina, O., Desjardins, L., Mariani, O., Roman-Roman, S., Sastre, X., Stern, M.-H., Cheng, F., Signoretti, S., Berchuck, A., Bigner, D., Lipp, E., Marks, J., McCall, S., McLendon, R., Secord, A., Sharp, A., Behera, M., Brat, D.J., Chen, A., Delman, K., Force, S., Khuri, F., Magliocca, K., Maithel, S., Olson, J.J., Owonikoko, T., Pickens, A., Ramalingam, S., Shin, D.M., Sica, G., Van Meir, E.G., Zhang, H., Eijckenboom, W., Gillis, A., Korpershoek, E., Looijenga, L., Oosterhuis, W., Stoop, H., van Kessel, K.E., Zwarthoff, E.C., Calatozzolo, C., Cuppini, L., Cuzzubbo, S., DiMeco, F., Finocchiaro, G., Mattei, L., Perin, A., Pollo, B., Chen, C., Houck, J., Lohavanichbutr, P., Hartmann, A., Stoehr, C., Stoehr, R., Taubert, H., Wach, S., Wullich, B., Kycler, W., Murawa, D., Wiznerowicz, M., Chung, K., Jeffrey Edenfield, W., Martin, J., Baudin, E., Bubley, G., Bueno, R., De Rienzo, A., Richards, W.G., Kalkanis, S., Mikkelsen, T., Noushmehr, H., Scarpace, L., Girard, N., Aymerich, M., Campo, E., Giné, E., López Guillermo, A., Van Bang, N., Hanh, P.T., Phu, B.D., Tang, Y., Colman, H., Evason, K., Dottino, P.R., Martignetti, J.A., Gabra, H., Juhl, H., Akeredolu, T., Stepa, S., Hoon, D., Ahn, K., Kang, K.J., Beuschlein, F., Breggia, A., Birrer, M., Bell, D., Borad, M., Bryce, A.H., Castle, E., Chandan, V., Cheville, J., Copland, J.A., Farnell, M., Flotte, T., Giama, N., Thai, H., Kendrick, M., Kocher, J.-P., Kopp, K., Moser, C., Nagorney, D., O'Brien, D., O'Neill, B.P., Patel, T., Petersen, G., Que, F., Rivera, M., Roberts, L., Smallridge, R., Smyrk, T., Stanton, M., Houston Thompson, R., Torbenson, M., Yang, J.D., Zhang, L., Brimo, F., Ajani, J.A., Gonzalez, A.M.A., Behrens, C., Bondaruk, J., Broaddus, R., Czerniak, B., Esmaeli, B., Fujimoto, J.,

Gershenwald, J., Guo, C., Lazar, A.J., Logothetis, C., Meric-Bernstam, F., Moran, C., Ramondetta, L., Rice, D., Sood, A., Tamboli, P., Thompson, T., Troncoso, P., Tsao, A., Wistuba, I., Carter, C., Haydu, L., Hersey, P., Jakrot, V., Kakavand, H., Kefford, R., Lee, K., Long, G., Mann, G., Quinn, M., Saw, R., Scolyer, R., Shannon, K., Spillane, A., Stretch, O., Synott, M., Thompson, J., Wilmott, J., Al-Ahmadie, H., Chan, T.A., Ghossein, R., Gopalan, A., Levine, D.A., Reuter, V., Singer, S., Singh, B., Tien, N.V., Broudy, T., Mirsaidi, C., Nair, P., Drwiega, P., Miller, J., Smith, J., Zaren, H., Park, J.-W., Hung, N.P., Kebebew, E., Marston Linehan, W., Metwalli, A.R., Pacak, K., Pinto, P.A., Schiffman, M., Schmidt, L.S., Vocke, C.D., Wentzensen, N., Worrell, R., Yang, H., Moncrieff, M., Goparaju, C., Melamed, J., Pass, H., Botnariuc, N., Caraman, I., Cernat, M., Chemencedji, I., Clipca, A., Doruc, S., Gorincioi, G., Mura, S., Pirtac, M., Stancul, I., Tcaciuc, D., Albert, M., Alexopoulou, I., Arnaout, A., Bartlett, J., Engel, J., Gilbert, S., Parfitt, J., Sekhon, H., Thomas, G., Rassl, D.M., Rintoul, R.C., Bifulco, C., Tamakawa, R., Urba, W., Hayward, N., Timmers, H., Antenucci, A., Facciolo, F., Grazi, G., Marino, M., Merola, R., de Krijger, R., Gimenez-Roqueplo, A.-P., Piché, A., Chevalier, S., McKercher, G., Birsoy, K., Barnett, G., Brewer, C., Farver, C., Naska, T., Pennell, N.A., Raymond, D., Schilero, C., Smolenski, K., Williams, F., Morrison, C., Borgia, J.A., Liptay, M.J., Pool, M., Seder, C.W., Junker, K., Omberg, L., Dinkin, M., Manikhas, G., Alvaro, D., Bragazzi, M.C., Cardinale, V., Carpino, G., Gaudio, E., Chesla, D., Cottingham, S., Dubina, M., Moiseenko, F., Dhanasekaran, R., Becker, K.-F., Janssen, K.-P., Slotta-Huspenina, J., Abdel-Rahman, M.H., Aziz, D., Bell, S., Cebulla, C.M., Davis, A., Duell, R., Bradley Elder, J., Hilty, J., Kumar, B., Lang, J., Lehman, N.L., Mandt, R., Nguyen, P., Pilarski, R., Rai, K., Schoenfield, L., Senecal, K., Wakely, P., Hansen, P., Lechan, R., Powers, J., Tischler, A., Grizzle, W.E., Sexton, K.C., Kastl, A., Henderson, J., Porten, S., Waldmann, J., Fassnacht, M., Asa, S.L., Schadendorf, D., Couce, M., Graefen, M., Huland, H., Sauter, G., Schlomm, T., Simon, R., Tennstedt, P., Olabode, O., Nelson, M., Bathe, O., Carroll, P.R., Chan, J.M., Disaia, P., Glenn, P., Kelley, R.K., Landen, C.N., Phillips, J., Prados, M., Simko, J., Smith-McCune, K., VandenBerg, S., Roggin, K., Fehrenbach, A., Kendler, A., Sifri, S., Steele, R., Jimeno, A., Carey, F., Forgie, I., Mannelli, M., Carney, M., Hernandez, B., Campos, B., Herold-Mende, C., Jungk, C., Unterberg, A., von Deimling, A., Bossler, A., Galbraith, J., Jacobus, L., Knudson, M., Knutson, T., Deqin, M., Milhem, M., Sigmund, R., Godwin, A.K., Madan, R., Rosenthal, H.G., Adebamowo, C., Adebamowo, S.N., Boussioutas, A., Beer, D., Giordano, T., Mes-Masson, A.-M., Saad, F., Bocklage, T., Landrum, L., Mannel, R., Moore, K., Moxley, K., Postier, R., Walker, J., Zuna, R., Feldman, M., Valdivieso, F., Dhir, R., Luketich, J., Pinero, E.M.M., Quintero-Aguilo, M., Carlotti, C.G., Sebastião Dos Santos, J., Kemp, R., Sankarankuty, A., Tirapelli, D., Catto, J., Agnew, K., Swisher, E., Creaney, J., Robinson, B., Shelley, C.S., Godwin, E.M., Kendall, S., Shipman, C., Bradford, C., Carey, T., Haddad, A., Moyer, J., Peterson, L., Prince, M., Rozek, L., Wolf, G., Bowman, R., Fong, K.M., Yang, I., Korst, R., Kimryn Rathmell, W., Fantacone-Campbell, J.L., Hooke, J.A., Kovatich, A.J., Shriver, C.D., DiPersio, J., Drake, B., Govindan, R., Heath, S., Ley, T., Van Tine, B., Westervelt, P., Rubin, M.A., Lee, J.I., Aredes, N.D., Mariamidze, A., Shmulevich, I., Rao, A.U.K., Lazar, A.J., Sharma, A., Thorsson, V., 2018. Spatial organization and molecular correlation of tumor-infiltrating lymphocytes using deep learning on pathology images. Cell Rep. 22111247. 23 (1), 181. doi:10.1016/j.celrep.2018.03.086.

Schwen, L.O., Kiehl, T.-R., Carvalho, R., Zerbe, N., Homeyer, A., 2023. Digitization of pathology labs: a review of lessons learned. Lab. Investig. 00236837. 103 (11), 100244. doi:10.1016/j.labinv.2023.100244.

Sheller, M.J., Edwards, B., Reina, G.A., Martin, J., Pati, S., Kotrotsou, A., Milchenko, M., Xu, W., Marcus, D., Colen, R. R., Bakas, S., 2020. Federated learning in medicine: facilitating multi-institutional collaborations without sharing patient data. Sci. Rep. 20452322. 10 (1). Available from: www.nature.com/srep/index.html. doi:10.1038/s41598-020-69250-1.

Tellez, D., Litjens, G., Bándi, P., Bulten, W., Bokhorst, J.-M., Ciompi, F., van der Laak, J., 2019. Quantifying the effects of data augmentation and stain color normalization in convolutional neural networks for computational pathology. Med. Image Anal. 13618415. 58, 101544. doi:10.1016/j.media.2019.101544.

Topol, E.J., 2019. High-performance medicine: the convergence of human and artificial intelligence. Nat. Med. 1546170X. 25 (1), 44–56. Available from: http://www.nature.com/nm/index.html. doi:10.1038/s41591-018-0300-7.

Tripathi, A., Waqas, A., Venkatesan, K., Yilmaz, Y., Rasool, G., 2024. Building flexible, scalable, and machine learning-ready multimodal oncology datasets. Sensors. 1424-8220. 24 (5), 1634. doi:10.3390/s24051634.

Tripathi, A., Waqas, A., Schabath, M.B., Yilmaz, Y., Rasool, G., 2025. HONeYBEE: enabling scalable multimodal AI in oncology through foundation model-driven embeddings. Npj Digit. Med. 2398-6352. 8 (1). doi:10.1038/s41746-025-02003-4.

Tripathi, A., Waqas, A., Venkatesan, K., Ullah, E., Khan, A., Khalil, F., Chen, W.-S., Ozturk, Z.G., Saeed-Vafa, D., Bui, M.M., Schabath, M.B., Rasool, G., 2026. Using consensus-based reasoning and large language models to extract structured data from surgical pathology reports. Lab. Investig. 00236837. 106 (2), 104272. doi:10.1016/j.labinv.2025.104272.

Walsh, E., Orsi, N.M., 2024. The current troubled state of the global pathology workforce: a concise review. Diagn. Pathol. 17461596. 19 (1). Available from: https://diagnosticpathology.biomedcentral.com. doi:10.1186/s13000-024-01590-2.

Waqas, A., Tripathi, A., Ramachandran, R.P., Stewart, P.A., Rasool, G., 2024a. Multimodal data integration for oncology in the era of deep neural networks: a review. Front. Artif. Intell. 26248212. 7. Available from: www.frontiersin.org/journals/artificial-intelligence#. doi:10.3389/frai.2024.1408843.

Waqas, A., Ullah, E., Khan, A., Khalil, F., Ozturk, Z.G., Chumbalkar, V., Saeed-Vafa, D., Jameel, Z., Chen, W., Bittar, H.T., Dhillon, J., Singh, R.S., Bychkov, A., Parwani, A.V., Bui, M.M., Schabath, M.B., Rasool, G., 2025. Reasoning beyond accuracy: expert evaluation of large language models in diagnostic pathology. medRxiv Available from https://www.medrxiv.org/. doi:10.1101/2025.04.11.25325686.

Waqas, A., Bui, M.M., Glassy, E.F., Naqa, I.E., Borkowski, P., Borkowski, A.A., Rasool, G., 2023. Revolutionizing digital pathology with the power of generative artificial intelligence and foundation models. Lab. Investig. 00236837. 103 (11), 100255. doi:10.1016/j.labinv.2023.100255.

Waqas, A., Naveed, J., Shahnawaz, W., Asghar, S., Bui, M.M., Rasool, G., 2024b. Digital pathology and multimodal learning on oncology data. BJR | Artif. Intell. 2976-8705. 1 (1). doi:10.1093/bjrai/ubae014.

Williams, B.J., Lee, J., Oien, K.A., Treanor, D., 2018. Digital pathology access and usage in the UK: results from a national survey on behalf of the National Cancer Research Institute's CM-Path initiative. J. Clin. Pathol. 14724146. 71 (5), 463–466. Available from: http://jcp.bmj.com/. doi:10.1136/jclinpath-2017-204808.

Williams, B.J., Williams, B.J., Treanor, D., Treanor, D., 2020. Practical guide to training and validation for primary diagnosis with digital pathology. J. Clin. Pathol. 14724146. 73 (7), 418–422. Available from: http://jcp.bmj.com/. doi:10.1136/jclinpath-2019-206319.

Yin, F., Han, G., Bui, M.M., Gibbs, J., Martin, I., Sundharkrishnan, L., King, L., Jabcuga, C., Stuart, L.N., Hassell, L.A., 2016. Educational value of digital whole slides accompanying published online pathology journal articles a multi-institutional study. Arch. Pathol. Lab. Med. 15432165. 140 (7), 694–697. Available from: http://www.archivesofpathology.org/doi/pdf/10.5858/arpa.2015-0366-OA. doi:10.5858/arpa.2015-0366-OA.

CHAPTER

6

Digital cytology

Swati Satturwar[a], *Liron Pantanowitz*[b], *and Zaibo Li*[a]

[a]Department of Pathology, The Ohio State University, Columbus, OH, United States
[b]Department of Pathology, University of Pittsburgh Medical Center, Pittsburgh, PA, United States

Introduction

Digital pathology (DP) is an image-based environment that integrates image acquisition using whole-slide imaging (WSI), data management, and interpretation of pathology from digitized glass slides. Digital cytology is a specialized field of pathology involving the use of similar principles of DP for cytology. Technical advancements in DP have transformed the practice of surgical pathology over the past decade. However, WSI in cytopathology has not been as pervasive; therefore the adoption of digital cytology is still in the experimental stage (Kim et al., 2024a, 2024b; Pantanowitz et al., 2013; El-Garby et al., 2014; Zhao et al., 2015; Amin et al., 2014). Validation studies have proven that WSI is not inferior to the brightfield light microscope for diagnosis in surgical pathology (Bauer et al., 2013; Snead et al., 2016) but comparative studies in cytopathology have demonstrated that glass slides are more accurate and faster than WSI (House et al., 2013; Evered and Dudding, 2011). This chapter will review topics related to establishing a digital cytology workflow, such as scanners currently available for digital cytology, technical challenges related to cytology, the current state of implementation of digital cytology, practical tips for adopting this technology, applications of digital cytology, and the future of digital cytology.

Unique barriers for digital cytology

Differences in cytology and surgical pathology specimens and preparations

With the growing interest in the digitization of the cytology workflow, it is essential to discuss the uniqueness of cytology specimen types, preparation techniques, stains unique to cytology, and the cytology workflow in comparison to surgical pathology. Cytology is the study of diseases at the cellular level. The primary benefit of cytologic evaluation is the fast turn-around time due to the rapid collection, processing, and evaluation of samples.

Advances in Digital Pathology and Artificial Intelligence in Pathology
https://doi.org/10.1016/B978-0-443-40383-5.00007-8

TABLE 6.1 Differences between cytology and surgical pathology specimens.

	Cytology	Surgical pathology
Specimen types	Exfoliative, abrasive, fine needle aspiration	Biopsy, resections
Preparation types	Smears, LBC, cytospin, touch preparation, cell-blocks	Formalin fixation and paraffin-embedded tissue
Commonly used Stains	Diff-Quik, Pap, H&E, IHC	H&E, IHC
Differences in procured material	3-D clusters, uneven distribution of cells, material extending beyond cover slips, frequent scant cellularity	Uniform thickness of tissue sections on a glass slide and simple topography
Case level differences	Single cases can have multiple preparation types, such as smears, cytospin, LBC, and cell-block	NA
Workflow	All specimens are screened by cytologists, and areas of interest are marked using colored markers	No screening
Automated barcode	Only subset of cases	Yes, all cases

The application of WSI to cytopathology is less pervasive than surgical pathology because of some unique challenges in cytology image acquisition and the complex cytology workflow (Jhala and Jhala, 2014; Fan and Bradley, 2016; Pantanowitz et al., 2011; Donnelly et al., 2013, 2014; Wright et al., 2013; Tawfik et al., 2015) (Table 6.1).

Types of acquisition of cytology specimens include exfoliative cytology, abrasive cytology, and aspiration cytology. Cytology preparation types include smear (the specimen is thinly spread directly on a glass slide), liquid-based cytology (LBC) (the specimen is preserved in a liquid suspension medium, mechanically filtered, concentrated, and deposited as a thin monolayer on a glass slide), cytocentrifuge (using centrifugal force, cytology specimens such as body fluids are concentrated and deposited onto a glass slide), touch preparation/imprint cytology (a piece of tissue, such as a biopsy, is touched onto a glass slide), and cell-block (tissue is collected, fixed in formalin, embedded in paraffin, cut to place onto a glass slide, and stained using Hematoxylin and Eosin, similar to conventional surgical pathology) (Jhala and Jhala, 2014) (Fig. 6.1).

Two most used stains in cytology include Diff-Quik stain and Papanicolaou (Pap) stain, which complement each other for optimal cytologic evaluation. Diff-Quik stain is performed on air-dried smears and enables better evaluation of cytoplasmic details and stroma, though it is prone to producing air-dry artifacts. Pap stain is performed on wet-fixed smears, thus allowing better evaluation of nuclear details and highlighting keratinization, but it is less useful for assessing cytoplasmic details (Fig. 6.2).

Technical challenges of digital cytology

Ideal characteristics of a glass slide to acquire high-quality WSIs are summarized in Table 6.2. Cytology slides, both smears as well as LBC preparations, are often characterized

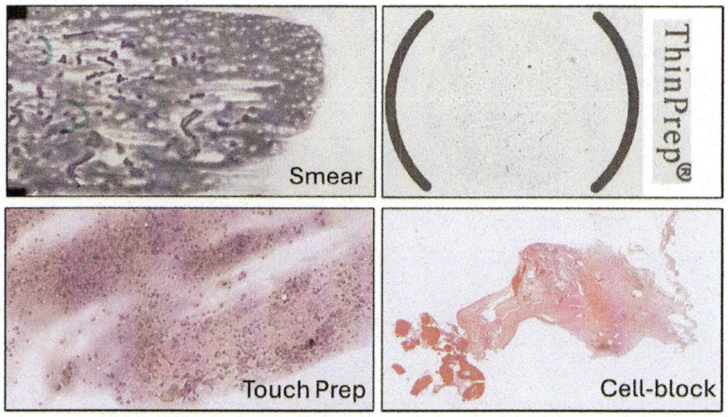

FIG. 6.1 Cytology preparation types include Smear, ThinPrep, Touch preparation, and cell-block.

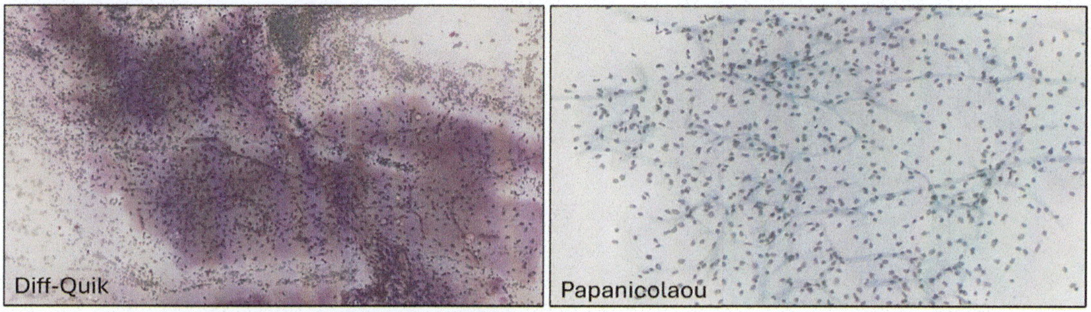

FIG. 6.2 Two commonly used stains in cytology—Diff-Quik stain highlighting myxoid matrix/stroma in a case of myxoid liposarcoma, and Papanicolaou stain highlighting nuclei.

TABLE 6.2 Ideal characteristics of a cytology glass slide to acquire high-quality WSIs.

Required characteristics	Need to avoid
Evenly spread material	Poorly spread material, dense tissue fragments, thick smears
Good cellular preservation	Overstained slides
Optimal staining	Plastic coverslip
Glass coverslip	Scantly cellular specimen
No air bubbles under coverslip	Material beyond coverslip

by unevenly distributed cellular material and variable thickness, 3D crowded clusters. Not infrequently, cytology cases with smear slides may have cellular material extending beyond the coverslip (Figs. 6.3 and 6.4). Cytology samples require pathologists to evaluate and interpret dispersed cells across multiple focal planes. Digitizing such cytology slides with all materials in focus can be very challenging and time-consuming, and often needs human intervention, such as manual manipulation for optimal scanning, or technology modification, such as Z-stacking.

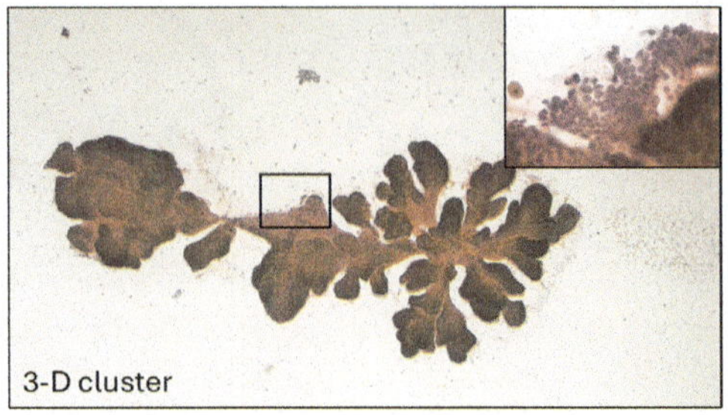

FIG. 6.3 3-D cluster in a smear of papillary thyroid carcinoma. The inset shows a higher magnification view of one edge, highlighting the nuclear features of papillary thyroid carcinoma.

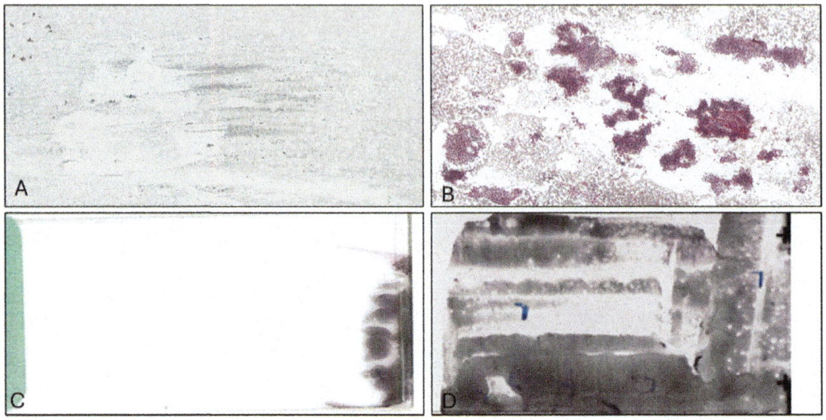

FIG. 6.4 (A–B) Aspirate smears with variable distribution of cells and variable thickness of cell clusters. (C) Scantily cellular aspirate smear with cellular material spread beyond the coverslip. (D) Uneven thickness of the smear with scant cellular material.

Surgical pathology slides with tissue sections have little variation in thickness topologically. Therefore scanning surgical pathology slides with one Z-focus to generate a two-dimensional WSI is generally sufficient for interpretation. In contrast, cytology slides often contain thick and uneven smears, as well as 3-D cell clusters (Fig. 6.3) especially in cases with highly cellular smears. Cytopathologists and cytologists often focus up and down the vertical or Z-axis to visualize these clusters when using conventional light microscopes. Such clusters, if scanned with only one Z-plane, are suboptimal for interpretation. Therefore scanning in both horizontal (X and Y) and vertical (Z) axes is necessary to ensure focus in cytology preparations (Bauer et al., 2013; Snead et al., 2016) (Fig. 6.5). Scanners with Z-stacking technology scan slides at multiple Z planes/layers and generate a final composite WSI to enable visualization of these 3-D clusters or uneven smears, just like a conventional light microscope. Automated smearing techniques produce slides with evenly distributed materials in a monolayer, which would potentially reduce the necessity for Z-stack scanning. In addition, certain preparations, like LBC, are more conducive to digitization due to less 3D material on slides. A handful of studies evaluating z-layer requirements used 3, 7, and 10–15 z-layers in their designs. A variety of specimen types and differences in scanners used

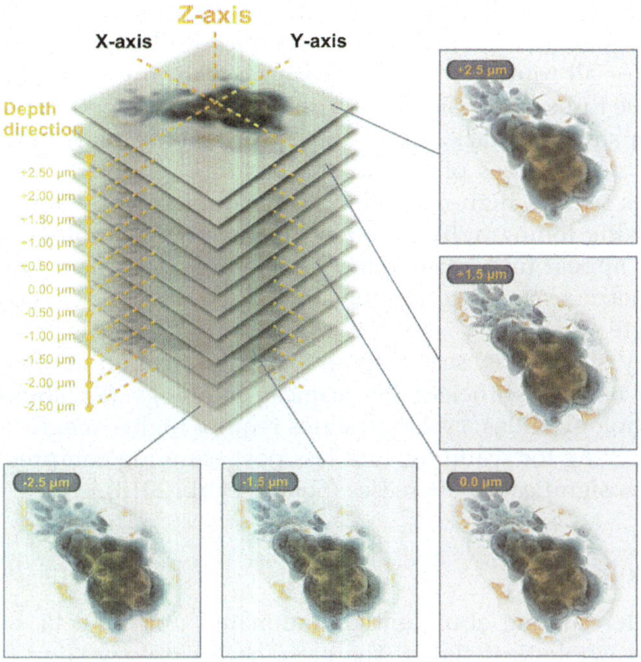

FIG. 6.5 Z-stacking images are acquired at multiple planes along the z-axis above and below the "optimal" focal plane.

preclude meaningful comparison (Fan and Bradley, 2016; Pantanowitz et al., 2011; Donnelly et al., 2013). LBC required fewer z-layers compared to smears. The exact number of z planes and the distance between z planes for digitizing cytology slides have not yet been standardized.

However, Z-stacking with multiple layers significantly increases scanning time, WSI file size, cost, and bandwidth for data transmission (Zarella and Jakubowski, 2019). Z-stacking typically increases file size in parallel with the number of levels scanned. Cytology slides with abundant cellular material may not require Z-stacking if the cellular material in focus is sufficient to render a diagnosis. A recent study by Hang et al. (2024) evaluated a novel solution using AI to simulate z-layers. The heuristic scan simulation calculated the total number of suspicious cells for high-grade urothelial carcinoma from 21 z-layer scan data. The heuristic scan reduced scanning time and file size and achieved higher accuracy (62.5% versus 37.5%) compared to single z-layer scans.

Emerging technologies aim to mitigate the need for deep Z-stacks. **Extended focus** (also referred to as *extended depth of field*, *z-stack flattening*, or *focus stacking*) is a technique that creates a single, fully in-focus whole-slide image by combining the sharpest regions from multiple z-level images. Because tissue sections and cytology preparations often vary in thickness, pathologists normally adjust fine focus on a microscope to visualize structures at different depths. Whole-slide scanners, however, capture a fixed focal plane unless z-stacking is used. Extended focus compensates for this by algorithmically merging focal planes, producing an image in which diagnostically important features across varying depths appear sharply defined. This approach improves clarity and diagnostic confidence for thick sections, smears, and uneven slides, though it increases scan time and file size and is therefore typically used selectively rather than for all routine slides. **Volumetric scanning** in digital pathology refers to the acquisition of true three-dimensional (3D) data from tissue specimens, going beyond the traditional 2D WSI or simple z-stack. Instead of capturing only a few focal planes, volumetric scanners acquire a dense series of optical sections throughout the full thickness of the tissue, using modalities such as high-resolution confocal microscopy, light-sheet microscopy, micro-optical sectioning tomography, or other 3D imaging technologies. This makes uneven or thick specimens appear uniformly in focus but eliminates true depth information. WSI scanners using volumetric scanning technology include the Hologic Genius System, Evident/Pramana, and Techcyte scanners.

For optimal examination of cytology specimens, scanning at 40× magnification is often necessary, while 20× magnification is usually sufficient for surgical pathology (Khalbuss et al., 2011). Compared to surgical pathology slides, cytology slides require higher magnification, a higher number of pixels, z-stacking for scanning, and less post-scanning compression; therefore the size of cytology WSIs significantly increases (Sellaro et al., 2013).

Workflow challenges of digital cytology

Diagnostic materials are often hidden among abundant contaminated normal cells or buried within background material on cytological slides. Therefore each cytology slide is typically screened by cytologists in a conventional workflow. In addition, diagnostic cells/clusters are annotated by cytologists using markers (Fig. 6.6). Khalbuss et al. (2013) reported that screening of cytology WSIs was tedious and time-consuming compared to manual conventional screening using light microscopes. Navigating cytology WSIs can be challenging, given the fact that the whole slide area is often covered with materials. However, technical advances allow the use of several approaches, such as keyboard-controlled

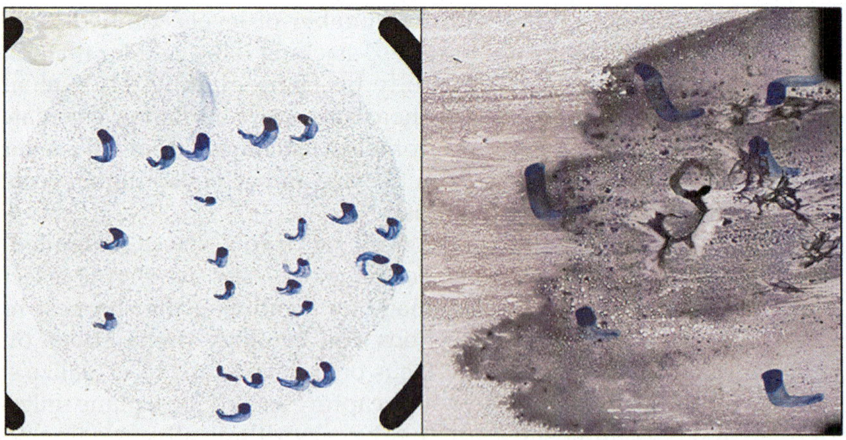

FIG. 6.6 ThinPrep and cytology smear with manual annotations by a cytologist using a blue marker.

navigation together with displaying a thumbnail image, in-built digital tracking tools, track-balls, touch pads, and touchscreens, to facilitate the navigation process. In addition, digital cytology annotation tools are much easier to use and offer the end-user the ability to annotate images using different shapes and colors, as well as in-built functions such as incorporating text, measurement, and alignment between H&E and immunostains (Yagi et al., 2012). Initial studies exploring diagnostic accuracy using digital cytology show conflicting results (Eccher and Girolami, 2020; Pantanowitz, 2020; Bongaerts et al., 2018; Gerhard et al., 2013). Despite these challenges, there is a clear trend toward the adoption of a digital cytology workflow (Bychkov et al., 2023). Furthermore, AI-generated galleries would remove the need for this difficult navigation, screening, and annotation.

Applications for digital cytology

Various applications of digital cytology include primary diagnosis, second-opinion consultations, rapid onsite evaluation (ROSE) for fine-needle aspiration (FNA) and touch preparations of small biopsies, virtual education, quality assurance, proficiency testing, archiving, research, and development of AI models (Antonini et al., 2023; Yao and Li, 2020).

Primary diagnosis using digital cytology

Primary diagnosis refers to the final interpretation rendered on a digital slide or slides without previewing the glass slides of the corresponding cytology slide. Digital cytology workflow streamlines processes, allows easy distribution, enables remote reviewing, and saves the time needed for the transportation of slides associated with conventional workflows. Despite these advancements, only a handful of laboratories utilize digital cytology for primary diagnosis (Bychkov et al., 2023). This is attributed to the initial cost of establishing a digital workflow, technical issues unique to cytology, as well as storage and regulatory challenges. Given the lack of published literature, validation guidelines for primary digital cytology diagnosis are excluded from CAP's guidelines (Antonini et al., 2023; Evans et al., 2022).

Recently, the Hologic Genius Digital Diagnostic System (HGDDS) received FDA approval for primary screening/diagnosis of cervical Pap smears using ThinPrep slides. The HGDDS WSI scanner is optimized to scan cytology slides with volumetric scanning technology. When coupled with deep learning algorithms, this digital cytology platform will likely transform the practice of cytology. The output of the newer Hologic Pap test image analysis algorithm no longer directs end users to use a microscope to examine particular fields of view but will instead allow them to view images on a computer display. Moreover, cytologists will now only need to examine a gallery of images (30 FOVs arranged in 5 labeled rows with 6 FOVs in each row) instead of screening through thousands of cells. If necessary, the cytologist will still be given the option to review 30 additional FOVs or the WSI. The ability to compress multiple Z-stacks into a single image makes it possible to now easily examine 3D cell groups. Since this new digital cytology platform has the potential for users to review many more cases in a shorter timeframe, perhaps new workload limits will need to be established.

Secondary diagnosis using digital cytology

Telecytology for second opinion consultations has been evaluated in some studies (Khurana, 2012; Satturwar et al., 2020; Archondakis et al., 2021). A study of telecytology using only cell-blocks for nongynecologic cases by Satturwar et al. has shown the feasibility of digital cytology to provide a meaningful second opinion interpretation in the majority of cases (Satturwar et al., 2020). Cell-block-only consultations solve focus issues associated with conventional cytology whole slide imaging. Interobserver concordance for salivary gland telecytology second opinions showed substantial to perfect K-values, ranging from 0.71 to 0.89 (Archondakis et al., 2021). Remote second opinion cytology consultations have the potential to address healthcare disparities across the globe.

Rapid onsite evaluation (ROSE)

ROSE for fine-needle aspiration (FNA) and touch preparations of small biopsies significantly improves diagnostic quality and enables appropriate specimen triage. Conventional methods of ROSE are rapidly being replaced by modern telecytology implementations across multiple academic as well as private hospitals. A telecytology system comprises onsite equipment with a camera, hardware, and software to capture high-quality images and transmit them to a remote location using a telecommunication link, and a remote workstation with a viewer to assess transmitted images (Yao and Li, 2020; Sirintrapun et al., 2017; Yao et al., 2018; Xing et al., 2020; Perez et al., 2021; Lin et al., 2024; Russell, 2022). Three types of telecytology systems are currently available, including microscopes with mounted cameras, robotic microscopes, and whole slide scanners. Table 6.3 summarizes the benefits and limitations of different telecytology systems. Currently, robotic and WSI-based systems are the most popular in the United States. The American Society of Cytopathology (ASC), in association with the College of American Pathologists (CAP), recommends validation of

TABLE 6.3 Pros and cons of different telecytology systems.

	Advantages	Limitations
Microscope with mounted camera (portable cart or fixed set up)	Simple and most cost-effective; can adjust focus	Needs an operator with cytology experience to navigate different areas of interest on the glass slide; high-width internet
Robotic microscopes	Remote users can navigate the images themselves	No need for an onsite operator; slower image review; no ability to adjust focus; high-width internet; costly
Whole slide scanners (e.g. Grundium)	Remote users can navigate the images themselves	Time-consuming (the time needed for creating WSI is proportional to the amount of material on the slide); no ability to adjust focus; costly

telecytology systems for all facilities where telecytology will be implemented (Lin et al., 2024; Russell, 2022).

Proficiency testing and quality assurance

The Clinical Laboratory Improvement Amendments Act of 1988 (CLIA ’88) requires cytology laboratories to participate in and pass an annual national gynecologic cytology proficiency testing (PT) as part of a quality assurance program. These PTs are available from the College of American Pathologists (CAP) or the American Society for Clinical Pathology (ASCP) (Eversole et al., 2010; Gagnon et al., 2004) and require many high-quality glass slides for distribution to thousands of laboratories across the nation. Ideally, all participants should examine the same slides to fairly assess their diagnostic skills. However, this is impossible using conventional cytology glass slides. Since WSIs can be replicated and made accessible to an unlimited number of participants, employing WSIs in PT programs is very attractive and has been recommended (Pantanowitz et al., 2009; Wilbur et al., 2009). Studies have demonstrated that digital WSIs are equivalent to glass slides for PT (Gagnon et al., 2004; Marchevsky et al., 2006). CLIA requires 10% prospective re-screening of gynecologic Pap tests before releasing final results, a 5-year retrospective screen of any new diagnosis that is a high-grade squamous intraepithelial lesion (HSIL) or carcinoma, and cytology-histopathology correlations. These QC/QA processes require slide retrieval, which can be a time-consuming and labor-intensive process. Digital cytology could allow significant time savings and streamline these lengthy processes.

Education and research

WSIs offer significant advantages over conventional glass slides in medical education. Cytology WSIs can be replicated and shared simultaneously by multiple students virtually (Stewart et al., 2008; Wilbur, 2016; Capitanio et al., 2018; Chiou, 2020). WSIs with annotations enhance the learning experience and allow the dissemination of knowledge without boundaries. Several online education programs now offer WSIs, including the International Academy of Cytology (IAC), the Digital Pathology Association (DPA), and the ASC.

Digital cytology WSIs offer powerful opportunities for research by enabling large-scale, reproducible analyses of cytologic specimens. High-resolution WSIs facilitate quantitative image assessment, AI algorithm development, and the exploration of new biomarkers using computational and AI-driven approaches. Because WSIs can be easily stored, shared, and annotated, they support multicenter collaborations and accelerate the generation of robust datasets, particularly for rare entities. As digital cytology expands, its research applications will continue to grow, fostering innovation in diagnostic accuracy and precision cytopathology.

Cytology slide archiving

Archiving cytology slides using WSIs can save physical storage space, allow instant review of old cases, and enable laboratories to retain a permanent digital record for consulting cases from outside institutions. WSI data can be easily deidentified for research purposes. Curated datasets can be used for developing machine learning and deep learning algorithms.

Digital cytology implementation

Implementing digital cytology demands a carefully structured, phased approach that reflects the inherent complexities of cytologic specimens, the need for high-resolution imaging, and the extensive integration required with existing laboratory workflows and clinical practice. The rollout of a digital cytology platform typically involves several stages.

Planning and system selection

Before deploying a digital cytology system, a laboratory should engage in thorough planning that addresses use-case definition, system evaluation, budget considerations, facility modifications (if needed), integration strategy, data storage requirements, and a rollout plan. It is essential to review multiple digital cytology platforms and gather input from all relevant stakeholders—cytologists, pathologists, informatics personnel, IT staff, and administrators. This process often involves issuing a formal request for proposal (RFP), request for information (RFI), or request for quote (RFQ). Inviting vendors to demonstrate their systems, ideally by scanning a set of the laboratory's own cytology slides, can offer valuable real-world insight. The budget must account not only for hardware (scanners, monitors) and software but also for data storage, ongoing maintenance, and necessary personnel such as scanning technologists or a digital pathology manager. For laboratories migrating from an existing platform, maintaining an overlap period for parallel validation is highly advisable to ensure continuity of clinical operations and avoid workflow disruptions.

The first step involves the analysis of existing digital cytology WSI platforms and obtaining quotes. Selecting WSI platforms requires in-depth knowledge of cytology case volume, use cases, and a list of potential users. A dedicated team should obtain input from stakeholders such as cytologists, cytopathologists, cytology prep technologists, IT staff, and administrators. Currently available scanners with z-stacking capabilities include the following vendors (Pernick, n.d.) (Table 6.4).

Deployment and validation

Validation of digital cytology solutions is mandatory, irrespective of FDA approval status. ASC's digital cytology task force white paper (Kim et al., 2024a) adopted CAP's guidelines for surgical pathology and recommends a validation set of at least 60 cases (not slides) per use case, with >95% intraobserver concordance between digital and glass slides. These recommendations may be subject to change based on evolving literature. The validation set should include a variety of sample types, preparation types, and stain types encountered in the practice setting to be validated. The number of slides may vary per the requirements of individual laboratories; however, at least one slide per specimen/preparation/stain type should be included in the validation set. An additional 20 cases should be included with additional IHC/special stains. A wash-out period of at least 2 weeks is recommended for establishing concordance. Formal training of cytology staff is necessary for using digital cytology for screening or primary diagnosis. The task force also recommends developing

TABLE 6.4 List of commercially available scanners with Z-stacking capabilities.

Vendor	Scanner model
Bionovation Biotech	CSFA800
3DHISTECH	Panoramic Flash Desk DX, Panoramic 1000 DX
Evident/Pramana	VS-M1, VS200
Grundium	Ocus40, Ocus20
Hamamatsu	Nanozoomer S360, Nanozoomer S20
Huron	TissueScope iQ, TissueScope LE
KFBIO	KF-PRO-002
Leica (Aperio)	Aperio GT 450
Mikroscan	SL5-20
Motic	Motic EasyScan NEW Infinity. Motic EasyScan One
Objective imaging	Glissando 20SL
OptraSCAN	OS-Ultra, OS-Lite Brightfield Scanner
Precipoint	M8, O8
Roche (Ventana)	VENTANA DP 600
Sakura	VisionTek, VisionTek M6
Visiopharm	Oncotopix Scan Nanozoomer S360

a digital cytology QC/QA program and revalidation of the system as needed. Scanning performance metrics should be monitored (Table 6.5).

Quality control and maintenance

A comprehensive quality control program is essential to ensure the reliability of digital cytology operations. Ongoing monitoring should include scan failure rates, image quality metrics, turnaround time, and user competency. Major hardware or software changes require repeat validation to confirm system performance. A robust QC/QA framework should also be in place to detect errors and address technical issues promptly. Key performance indicators—such as the number and types of cytology slides scanned, scanning failure rates, and the frequency of poor-quality WSI requiring rescanning—should be routinely recorded, trended, and reviewed.

Routine maintenance of digital cytology systems should encompass all components, including scanner hardware, monitors, software updates, data storage, and LIS or other system integrations. Ongoing competency maintenance for personnel involved in the digital cytology workflow—IT staff, scanning technologists, and cytology laboratory staff—is also recommended, either through proficiency testing or structured QC/QA programs. Although

TABLE 6.5 ASC recommendations for validating a digital cytology system for clinical practice.

Category	ASC Task force recommendation	Notes/rationale
Validation set size	Minimum of 60 cytology cases (not slides)	Parallels CAP surgical pathology guidance; the number may evolve as more evidence emerges
Concordance requirement	≥95% concordance between WSI and glass-slide interpretation	Should reflect primary diagnostic or screening use case(s)
Case selection	Cases should represent the expected clinical workload: various stains, preparation types (e.g., LBC, smears), benign and neoplastic diagnoses, and some nondiagnostic cases	Ensures generalizability and real-world performance
Use case definition	Each laboratory must specify its intended use (primary diagnosis, primary screening, specific preparation types)	Validation must match the actual clinical application
Expansion to new sample types	Twenty additional cases are recommended for adding new stains or IHC; a full 60-case revalidation is recommended for different preparation types (smears, LBC, cytospins)	Sample-preparation significantly affects WSI appearance and scanning behavior
Intraobserver comparison	Required; ≥2-week washout between reads	Ensures reproducible interpretation and avoids recall bias

the CAP found insufficient evidence to define specific competency requirements for digital pathology, it emphasized that validation should not substitute for training (Evans et al., 2022). A practical approach is to establish a competency set of at least 20 cases representing a spectrum of diagnoses and to define an appropriate passing threshold.

Digital image retention

If digital slides are used for primary diagnosis, and if storage capacity and cost permit, they should ideally be retained for at least the same duration required for glass slide retention. Institutional, state, federal, and accrediting agency regulations will ultimately determine the cytology laboratory's specific retention policy. Under current CAP requirements, Pap test slides must be stored for 5 years, whereas nongynecologic slides, including FNA slides, must be retained for 10 years. When a digitized glass slide is to be destroyed, for example, when material is scraped for ancillary testing, it becomes particularly important to preserve the corresponding digital slide. Digital slides may also warrant long-term retention for medicolegal, documentation, or quality assurance purposes.

Return on investment

Financial considerations play a significant role in decisions to adopt digital cytology. Across digital pathology more broadly, investments have been associated with meaningful cost savings driven by improved productivity and workflow efficiency, enhanced diagnostic accuracy with reduced unnecessary treatment, shorter turnaround times, and fewer ancillary tests (Ho et al., 2014; Hanna et al., 2019). Similar or potentially greater financial benefits are anticipated for digital cytology; however, dedicated cost-benefit analyses are still needed to fully quantify its economic impact.

Conclusion

In summary, the adoption of digital cytology in routine clinical practice has progressed more slowly than in surgical pathology. However, emerging digital solutions tailored specifically to cytologic specimens have the potential to transform traditional workflows into streamlined, fully digital processes. Such workflows enable faster case review, integration of AI-assisted tools, improved turnaround times, enhanced collaboration, and ultimately better patient care. As the evidence base continues to grow, validation guidelines for digital cytology will continue to evolve. The authors encourage the cytology community to publish their experiences with validation, implementation, and the return on investment associated with digital cytology to help inform and strengthen future best practices.

AI Disclosure

During the preparation of this work, the author(s) used ChatGPT to polish the writing. After using this tool/service, the author(s) reviewed and edited the content as needed and take(s) full responsibility for the content of the publication.

References

Antonini, P., Santonicco, N., Pantanowitz, L., Girolami, I., Rizzo, P.C., Brunelli, M., Bellevicine, C., Vigliar, E., Negri, G., Troncone, G., Fadda, G., Parwani, A., Marletta, S., Eccher, A., 2023. Relevance of the College of American Pathologists guideline for validating whole slide imaging for diagnostic purposes to cytopathology. Cytopathology. 13652303. 34 (1), 5–14. Available from: http://onlinelibrary.wiley.com/journal/10.1111/(ISSN)1365-2303. doi:10.1111/cyt.13178.

Archondakis, S., Roma, M., Kaladelfou, E., 2021. Remote cytological diagnosis of salivary gland lesions by means of precaptured videos. J. Am. Soc. Cytopathol. 22132945. 10 (4), 435–443. Available from: http://www.elsevier.com/journals/journal-of-the-american-society-of-cytopathology/2213-2945. doi:10.1016/j.jasc.2021.02.003.

Bauer, T.W., Schoenfield, L., Slaw, R.J., Yerian, L., Sun, Z., Henricks, W.H., 2013. Validation of whole slide imaging for primary diagnosis in surgical pathology. Arch. Pathol. Lab. Med. 15432165. 137 (4), 518–524. Available from: http://www.archivesofpathology.org/doi/pdf/10.5858/arpa.2011-0678-OA. doi:10.5858/arpa.2011-0678-OA.

Bongaerts, O., Clevers, C., Debets, M., Paffen, D., Senden, L., Rijks, K., Ruiten, L., Sie-Go, D., Van Diest, P.J., Nap, M., 2018. Conventional microscopical versus digital whole-slide imaging-based diagnosis of thin-layer cervical specimens: a validation study. J. Pathol. Inform. 21533539. 9 (1), 29. doi:10.4103/jpi.jpi_28_18.

Bychkov, A., Yoshikawa, A., Munkhdelger, J., Hori, T., Fukuoka, J., 2023. Integrating cytology into routine digital pathology workflow: a 5-year journey. Virchows Arch. 14322307. 483 (4), 555–559. Available from: https://www.springer.com/journal/428. doi:10.1007/s00428-023-03547-0.

Capitanio, A., Dina, R.E., Treanor, D., 2018. Digital cytology: a short review of technical and methodological approaches and applications. Cytopathology. 13652303. 29 (4), 317–325. Available from: http://onlinelibrary.wiley.com/journal/10.1111/(ISSN)1365-2303. doi:10.1111/cyt.12554.

Chiou, P.Z., 2020. Learning cytology in times of pandemic: an educational institutional experience with remote teaching. J. Am. Soc. Cytopathol. 22132945. 9 (6), 579–585. Available from: http://www.elsevier.com/journals/journal-of-the-american-society-of-cytopathology/2213-2945. doi:10.1016/j.jasc.2020.05.004.

Donnelly, A.D., Mukherjee, M.S., Lyden, E.R., Bridge, J.A., Lele, S.M., Wright, N., McGaughey, M.F., Culberson, A.M., Horn, A.J., Wedel, W.R., Radio, S.J., 2013. Optimal z-axis scanning parameters for gynecologic cytology specimens. J. Pathol. Inform. 21533539. 4 (1), 38. doi:10.4103/2153-3539.124015.

Donnelly, A., Mukherjee, M., Schneider, J., Lyden, E., Lele, S., McGaughey, M., Radio, S., 2014. Z-Axis scanning parameters: an investigation of consistency. J. Am. Soc. Cytopathol. 22132945. 3 (5), S79. doi:10.1016/j.jasc.2014.09.187.

Eccher, A., Girolami, I., 2020. Current state of whole slide imaging use in cytopathology: pros and pitfalls. Cytopathology. 13652303. 31 (5), 372–378. Available from: http://onlinelibrary.wiley.com/journal/10.1111/(ISSN)1365-2303. doi:10.1111/cyt.12806.

El-Garby, E.A., Parwani, A., Pantanowitz, L., 2014. Whole slide imaging: widening the scope of cytopathology. Diagn. Histopathol. 20, 456–461.

Evans, A.J., Brown, R.W., Bui, M.M., Chlipala, E.A., Lacchetti, C., Milner, D.A., Pantanowitz, L., Parwani, A.V., Reid, K., Riben, M.W., Reuter, V.E., Stephens, L., Stewart, R.L., Thomas, N.E., 2022. Validating whole slide imaging systems for diagnostic purposes in pathology. Arch. Pathol. Lab. Med. 1543-2165. 146 (4), 440–450. doi:10.5858/arpa.2020-0723-cp.

Evered, A., Dudding, N., 2011. Accuracy and perceptions of virtual microscopy compared with glass slide microscopy in cervical cytology. Cytopathology. 09565507. 22 (2), 82–87. doi:10.1111/j.1365-2303.2010.00758.x.

Eversole, G.M., Moriarty, A.T., Schwartz, M.R., Clayton, A.C., Souers, R., Fatheree, L.A., Chmara, B.A., Tench, W.D., Henry, M.R., Wilbur, D.C., 2010. Practices of participants in the College of American Pathologists interlaboratory comparison program in cervicovaginal cytology, 2006. Arch. Pathol. Lab. Med. 15432165. 134 (3), 331–335. Available from . doi:10.1043/1543-2165-134.3.331.

Fan, Y., Bradley, A.P., 2016. A method for quantitative analysis of clump thickness in cervical cytology slides. Micron. 09684328. 80, 73–82. Available from: www.elsevier.com/inca/publications/store/4/7/5/index.htt. doi:10.1016/j.micron.2015.09.002.

Gagnon, M.B., Inhorn, S., Hancock, J., Keller, B., Carpenter, D., Merlin, T., Hearn, T., Thompson, P., Whalen, R., 2004. Comparison of cytology proficiency testing: glass slides vs. virtual slides. Acta Cytol. 19382650. 48 (6), 788–794. Available from: http://content.karger.com/ProdukteDB/produkte.asp?Aktion=JournalHome&ProduktNr=254338&ContentOnly=false. doi:10.1159/000326447.

Gerhard, R., Teixeira, S., Rocha, A.G.D., Schmitt, F., 2013. Thyroid fine-needle aspiration cytology: is there a place to virtual cytology? Diagn. Cytopathol. 10970339. 41 (9), 793–798. doi:10.1002/dc.22958.

Hang, J.F., Ou, Y.C., Yang, W.L., Tsao, T.Y., Yeh, C.H., Li, C.B., Hsu, E.Y., Hung, P.Y., Lin, M.Y., Hwang, Y.T., Liu, T. J., Tung, M.C., 2024. Evaluating urine cytology slide digitization efficiency: a comparative study using an artificial intelligence-based heuristic scanning simulation and multiple Z-plane scanning. Acta Cytol. 19382650. 68 (4), 342–350. Available from: http://content.karger.com/ProdukteDB/produkte.asp?Aktion=JournalHome&ProduktNr=254338&ContentOnly=false. doi:10.1159/000538985.

Hanna, M.G., Reuter, V.E., Samboy, J., England, C., Corsale, L., Fine, S.W., Agaram, N.P., Stamelos, E., Yagi, Y., Hameed, M., Klimstra, D.S., Sirintrapun, S.J., 2019. Implementation of digital pathology offers clinical and operational increase in efficiency and cost savings. Arch. Pathol. Lab. Med. 15432165. 143 (12), 1545–1555. Available from: https://www.archivesofpathology.org/doi/pdf/10.5858/arpa.2018-0514-OA. doi:10.5858/arpa.2018-0514-OA.

Ho, J., Ahlers, S.M., Stratman, C., Aridor, O., Pantanowitz, L., Fine, J.L., Kuzmishin, J.A., Montalto, M.C., Parwani, A.V., 2014. Can digital pathology result in cost savings? A financial projection for digital pathology implementation at a large integrated health care organization. J. Pathol. Inform. 21533539. 5 (1), 33. doi:10.4103/2153-3539.139714.

House, J.C., Henderson-Jackson, E.B., Johnson, J.O., Lloyd, M.C., Dhillon, J., Ahmad, N., Hakam, A., Khalbuss, W.E., Leon, M.E., Chhieng, D., Zhang, X., Centeno, B.A., Bui, M.M., 2013. Diagnostic digital cytopathology: are we ready yet? J. Pathol. Inform. 21533539. 4 (1), 28. doi:10.4103/2153-3539.120727.

Jhala, N., Jhala, D., 2014. Definitions in tissue acquisition. core biopsy, cell block, and beyond. Gastrointest. Endosc. Clin. N. Am. 15581950. 24 (1), 19–27. doi:10.1016/j.giec.2013.08.005.

Khalbuss, W.E., Cuda, J., Cucoranu, I.C., 2013. Screening and dotting virtual slides: a new challenge for cytotechnologists. CytoJournal. 0974-5963. 10 (1), 22. doi:10.4103/1742-6413.120790.

Khalbuss, W.E., Pantanowitz, L., Parwani, A.V., 2011. Digital imaging in cytopathology. Pathol. Res. Int. 2011, 264683. doi:10.4061/2011/264683.

Khurana, K., 2012. Telecytology and its evolving role in cytopathology. Diagn. Cytopathol. 10970339. 40 (6), 498–502. doi:10.1002/dc.22822.

Kim, D., Sundling, K.E., Virk, R., Thrall, M.J., Alperstein, S., Bui, M.M., Chen-Yost, H., Donnelly, A.D., Lin, O., Liu, X., Madrigal, E., Michelow, P., Schmitt, F.C., Vielh, P.R., Zakowski, M.F., Parwani, A.V., Jenkins, E., Siddiqui, M.T., Pantanowitz, L., Li, Z., 2024a. Digital cytology part 1: digital cytology implementation for practice: a concept paper with review and recommendations from the American Society of Cytopathology Digital Cytology Task Force. J. Am. Soc. Cytopathol. 22132945. 13 (2), 86–96. Available from: https://www.sciencedirect.com/science/journal/22132945. doi:10.1016/j.jasc.2023.11.006.

Kim, D., Sundling, K.E., Virk, R., Thrall, M.J., Alperstein, S., Bui, M.M., Chen-Yost, H., Donnelly, A.D., Lin, O., Liu, X., Madrigal, E., Michelow, P., Schmitt, F.C., Vielh, P.R., Zakowski, M.F., Parwani, A.V., Jenkins, E., Siddiqui, M.T., Pantanowitz, L., Li, Z., 2024b. Digital cytology part 2: artificial intelligence in cytology: a concept paper with review and recommendations from the American Society of Cytopathology Digital Cytology Task Force. J. Am. Soc. Cytopathol. 22132945. 13 (2), 97–110. Available from: https://www.sciencedirect.com/science/journal/22132945. doi:10.1016/j.jasc.2023.11.005.

Lin, O., Alperstein, S., Barkan, G.A., Cuda, J.M., Kezlarian, B., Jhala, D., Jin, X., Mehrotra, S., Monaco, S.E., Rao, J., Saieg, M., Thrall, M., Pantanowitz, L., 2024. American Society of Cytopathology Telecytology validation recommendations for rapid on-site evaluation (ROSE). J. Am. Soc. Cytopathol. 22132945. 13 (2), 111–121. Available from: https://www.sciencedirect.com/science/journal/22132945. doi:10.1016/j.jasc.2023.12.001.

Marchevsky, A.M., Khurana, R., Thomas, P., Scharre, K., Farias, P., Bose, S., 2006. The use of virtual microscopy for proficiency testing in gynecologic cytopathology: a feasibility study using ScanScope. Arch. Pathol. Lab. Med. 00039985. 130 (3), 349–355. Available from: http://www.archivesofpathology.org/loi/arpa. doi:10.5858/2006-130-349-tuovmf.

PA, A.M., Parwani, A., Pantanowitz, L., 2014. In: Parwani, A. (Ed.), Digital Imaging. Springer, New York.

Pantanowitz, L., 2020. Digital cytology: look how much has been achieved. Cytopathology. 13652303. 31 (5), 370–371. Available from: http://onlinelibrary.wiley.com/journal/10.1111/(ISSN)1365-2303. doi:10.1111/cyt.12866.

Pantanowitz, L., Hornish, M., Goulart, R.A., 2009. The impact of digital imaging in the field of cytopathology. CytoJournal. 0974-5963. 6, 6. doi:10.4103/1742-6413.48606.

Pantanowitz, L., Parwani, A.V., Khalbuss, W.E., 2011. Digital imaging for cytopathology: are we there yet? Cytopathology 22 (2), 73–74.

Pantanowitz, L., Sinard, J.H., Henricks, W.H., Fatheree, L.A., Carter, A.B., Contis, L., Beckwith, B.A., Evans, A.J., Otis, C.N., Lal, A., Parwani, A.V., 2013. Validating whole slide imaging for diagnostic purposes in Pathology: guideline from the College of American pathologists Pathology and Laboratory Quality Center. Arch. Pathol. Lab. Med. 15432165. 137 (12), 1710–1722. Available from: http://www.archivesofpathology.org/doi/pdf/10.5858/arpa.2013-0093-CP. doi:10.5858/arpa.2013-0093-CP.

Perez, D., Stemmer, M.N., Khurana, K.K., 2021. Utilization of dynamic telecytopathology for rapid onsite evaluation of touch imprint cytology of needle core biopsy: diagnostic accuracy and pitfalls. Telemed. e-Health. 15563669. 27 (5), 525–531. Available from: http://www.liebertpub.com/publication.aspx?pub_id=54. doi:10.1089/tmj.2020.0117.

Pernick, N. n.d., WSI commercial platforms. PathologyOutlines.com website. https://www.pathologyoutlines.com/topic/informaticswsicommercialplatforms.html. Accessed 28 September 2025.

Russell, D.K., 2022. Cytopathology in focus: know the accreditation requirements for telecytology. CAP Today. https://www.captodayonline.com/cytopathology-in-focus-know-the-accreditation-requirements-for-telecytology/.

Satturwar, S., Monaco, S.E., Xing, J., Pantanowitz, L., 2020. The utility of cell blocks for international cytopathology teleconsultation by whole slide imaging. Cytopathology. 13652303. 31 (5), 419–425. Available from: http://onlinelibrary.wiley.com/journal/10.1111/(ISSN)1365-2303. doi:10.1111/cyt.12800.

Sellaro, T.L., Filkins, R., Hoffman, C., Fine, J.L., Jon, H., Parwani, A.V., Pantanowitz, L., Montalto, M., 2013. Relationship between magnification and resolution in digital pathology systems. J. Pathol. Inform. 21533539. 4 (1), 21. doi:10.4103/2153-3539.116866.

Sirintrapun, S., Rudomina, D., Mazzella, A., Feratovic, R., Alago, W., Siegelbaum, R., Lin, O., 2017. Robotic telecytology for remote cytologic evaluation without an on-site cytotechnologist or cytopathologist: a tale of implementation and review of constraints. J. Pathol. Inform. 21533539. 8 (1). Available from: www.jpathinformatics.org/browse.asp?sabs=n. doi:10.4103/jpi.jpi_26_17.

Snead, D.R.J., Tsang, Y.W., Meskiri, A., Kimani, P.K., Crossman, R., Rajpoot, N.M., Blessing, E., Chen, K., Gopalakrishnan, K., Matthews, P., Momtahan, N., Read-Jones, S., Sah, S., Simmons, E., Sinha, B., Suortamo, S., Yeo, Y., Daly, H.E., Cree, I.A., 2016. Validation of digital pathology imaging for primary histopathological diagnosis. Histopathology. 13652559. 68 (7), 1063–1072. Available from: http://www3.interscience.wiley.com/journal/118511588/home. doi:10.1111/his.12879.

Stewart, J., Bevans-Wilkins, K., Bhattacharya, A., Ye, C., Miyazaki, K., Kurtycz, D.F.I., 2008. Virtual microscopy: an educator's tool for the enhancement of cytotechnology students' locator skills. Diagn. Cytopathol. 10970339. 36 (6), 363–368. doi:10.1002/dc.20821.

Tawfik, O., Davis, M., Dillon, S., Tawfik, L., Diaz, F.J., Amin, K., Fan, F., 2015. Whole-slide imaging of pap cellblock preparations is a potentially valid screening method. Acta Cytol. 19382650. 59 (2), 187–200. Available from: http://content.karger.com/ProdukteDB/produkte.asp?Aktion=JournalHome&ProduktNr=254338&ContentOnly=false. doi:10.1159/000430082.

Wilbur, D.C., 2016. Digital pathology and its role in cytology education. Cytopathology. 13652303. 27 (5), 325–330. Available from: http://onlinelibrary.wiley.com/journal/10.1111/(ISSN)1365-2303. doi:10.1111/cyt.12377.

Wilbur, D.C., Black-Schaffer, W.S., Luff, R.D., Abraham, K.P., Kemper, C., Molina, J.T., Tench, W.D., 2009. The Becton Dickinson focalpoint GS imaging system: clinical trials demonstrate significantly improved sensitivity for the detection of important cervical lesions. Am. J. Clin. Pathol. 00029173. 132 (5), 767–775. Available from: http://ajcp.ascpjournals.org/content/132/5/767.full.pdf. doi:10.1309/AJCP8VE7AWBZCVQT.

Wright, A.M., Smith, D., Dhurandhar, B., Fairley, T., Scheiber-Pacht, M., Chakraborty, S., Gorman, B.K., Mody, D., Coffey, D.M., 2013. Digital slide imaging in cervicovaginal cytology: a pilot study. Arch. Pathol. Lab. Med. 15432165. 137 (5), 618–624. Available from: http://www.archivesofpathology.org/doi/pdf/10.5858/arpa.2012-0430-OA. doi:10.5858/arpa.2012-0430-OA.

Xing, J., Monaco, S.E., Cuda, J., Pantanowitz, L., 2020. Telecytology rapid on-site evaluation: diagnostic challenges, technical issues and lessons learned. Cytopathology. 13652303. 31 (5), 402–410. Available from: http://onlinelibrary.wiley.com/journal/10.1111/(ISSN)1365-2303. doi:10.1111/cyt.12801.

Yagi, Y., Yoshioka, S., Kyusojin, H., Onozato, M., Mizutani, Y., Osato, K., Yada, H., Mark, E.J., Frosch, M.P., Louis, D. N., 2012. An ultra-high speed whole slide image viewing system. Stud. Health Technol. Inform. 18798365. 179, 239–249. Available from: http://www.iospress.nl/bookserie/studies-in-health-technology-and-informatics/. doi:10.3233/978-1-61499-086-4-239.

Yao, K., Li, Z., 2020. Review of different platforms to perform rapid onsite evaluation via telecytology. Cytopathology. 13652303. 31 (5), 379–384. Available from: http://onlinelibrary.wiley.com/journal/10.1111/(ISSN)1365-2303. doi:10.1111/cyt.12871.

Yao, K., Shen, R., Parwani, A., Zaibo, L., 2018. A comprehensive study of telecytology using robotic digital microscope and single Z-stack digital scan for fine-needle aspiration-rapid on-site evaluation. J. Pathol. Inform. 21533539. 9 (1), 49. doi:10.4103/jpi.jpi_75_18.

Zarella, M.D., Jakubowski, J., 2019. Video compression to support the expansion of whole-slide imaging into cytology. J. Med. Imaging (Bellingham) 6 (4), 047502. doi:10.1117/1.JMI.6.4.047502.

Zhao, C., Tao, W., Ding, X., Parwani, A.V., Chen, H., McHugh, J., Piccoli, A., Xie, Q., Lauro, G.R., Feng, X., Hartman, D.J., Seethala, R.R., Shangwei, W., Yousem, S., Liang, Y., Pantanowitz, L., 2015. International telepathology consultation: three years of experience between the University of Pittsburgh Medical Center and KingMed Diagnostics in China. J. Pathol. Inform. 21533539. 6 (1), 63. doi:10.4103/2153-3539.170650.

CHAPTER

7

Telepathology

Mauro Saieg[a] *and Rafael Parra-Medina*[b]

[a]McGill University, Montreal, QC, Canada

[b]Instituto de Investigación, Fundación Universitaria de Ciencias de la Salud (FUCS), Bogotá, Colombia

Introduction and definition

Telepathology, broadly defined as the practice of transmitting digital pathology images for remote interpretation, has evolved into a critical component in contemporary diagnostic workflows. It includes both static-image transmission (store-and-forward) and dynamic modalities (real-time video or robotic microscopy), each offering unique applications in routine and urgent pathology practice. Particularly in intraoperative frozen section analysis and rapid on-site evaluation (ROSE) of cytological specimens, telepathology provides a robust framework for delivering timely, expert consultation when immediate diagnostic input is required. Recent advances in high-resolution whole-slide imaging (WSI), coupled with reliable telecommunication infrastructures, have enabled remote pathologists to render diagnoses with accuracy comparable to conventional microscopy (Weinstein et al., 2009; Pantanowitz et al., 2014).

The clinical utility of telepathology in frozen section and ROSE settings lies in its ability to overcome geographic and logistical barriers while maintaining diagnostic integrity (Al-Janabi et al., 2012; Mukhopadhyay et al., 2018). Collectively, these developments position telepathology not only as a pragmatic solution to workforce distribution challenges but also as a transformative tool in the pursuit of more efficient, equitable, and scalable pathology services. In this chapter, we will discuss the main available commercial platforms and their uses, as well as the primary impediments and avenues for broad implementation.

Available commercial platforms and uses

Digital pathology platforms can generally be categorized into two principal modalities based on image acquisition and transmission mechanisms: static image systems and WSI systems, including those enabling real-time transmission. Static systems involve capturing

Advances in Digital Pathology and Artificial Intelligence in Pathology
https://doi.org/10.1016/B978-0-443-40383-5.00001-7

selected high-resolution still images from specific fields of interest on a glass slide using conventional light microscopes fitted with digital cameras. These static images are commonly saved in formats such as JPEG or TIFF and can be efficiently stored, transmitted, and reviewed asynchronously, making them suitable for teleconsultation and educational applications (Pantanowitz et al., 2014). In contrast, WSI systems digitize the entire glass slide at diagnostic magnifications (typically 20× or 40×), generating a comprehensive virtual slide that can be interactively navigated, zoomed, and annotated (Pantanowitz et al., 2014). Furthermore, real-time transmission systems, often coupled with robotic microscopes or live video streaming technologies, enable remote users to control the microscope dynamically or observe procedures in real time, offering an effective solution (Weinstein et al., 2009; Pantanowitz et al., 2014). Smartphones have also been used to capture and share dynamic images through videoconferencing. However, resolution issues may arise, and maintaining image clarity without lag requires a high-speed internet connection (Tanveer et al., 2022).

The last few years have seen the onset of various DP platforms, which differ significantly in their technical specifications, including slide capacity, continuous loading options, supported glass slide sizes, image magnification, fluorescence capabilities, scan speed, image availability time, Z-stacking support, physical footprint, and ergonomic design (Ardon et al., 2025). While some systems are purpose-built for routine clinical diagnosis, including the regulatory approvals (FDA, CE-IVD), others are optimized for research, teleconsultation, or educational applications. Key distinctions include the mode of image acquisition (static capture vs. real-time streaming vs. full-slide scanning), image resolution, software ecosystem, and degree of interoperability with existing laboratory information systems (LIS) or picture archiving and communication systems (PACS) (Table 7.1).

In the United States, formal regulatory engagement between the FDA and digital pathology stakeholders began in 2009 with an informational panel. In 2011, the FDA classified WSI systems as Class III medical devices, thereby requiring a premarket approval (PMA) process for clinical applications. The Digital Pathology Association (DPA) initiated dialog with the FDA in 2014, leading to an agreement on pivotal trial design in 2015. This culminated in the de novo approval of the Philips IntelliSite Pathology Solution in April 2017—the first WSI platform authorized for primary diagnosis in surgical pathology. The system, which integrates a scanner, image management software, and display, was approved based on a non-inferiority study involving nearly 2000 formalin-fixed, paraffin-embedded (FFPE) slides across various histopathological entities. Notably, the approval excluded frozen sections, cytology, and hematopathology cases lacking FFPE processing. This landmark FDA decision established a regulatory precedent, allowing subsequent WSI systems to seek market clearance through the less stringent 510(k) pathway. As a result, the Leica Aperio AT2 DX received FDA clearance in 2019, followed by the Hamamatsu NanoZoomer S360MDin in 2022. In addition, the Sectra Digital Pathology Module, when used in conjunction with the FDA-cleared Leica AT2 DX scanner, became the second FDA-approved system for primary histopathological diagnosis as of September 2020 (Zia et al., 2025).

Several commercial platforms, including those from Philips, Roche/Ventana, Leica/Aperio, Hamamatsu, and 3DHISTECH, are currently CE-IVD labeled under this regulatory framework. However, with the implementation of the new In Vitro Diagnostic Medical Device Regulation (IVDR) (EU 2017/746) in May 2022, all in vitro diagnostic devices,

TABLE 7.1 Summary of some digital pathology platforms.

Platform	Manufacturer	Type	Scanner throughput/mode	Multilayer support	Image capture magnification	Scan speed	Primary use cases
IntelliSite pathology solution	Philips	WSI	High (≥80 slides/h)	No	20×/40×	~1 min/slide	Primary diagnosis, AI integration, telepathology
Aperio AT2 DX	Leica Biosystems	WSI	High (≥100 slides/h)	Yes (limited)	20×/40×	~1 min/slide	Clinical diagnostics, retrospective review
NanoZoomer S360MD	Hamamatsu	WSI	High (230 slides/h)	Yes	20×/40×	~50 s/slide	Routine diagnostics, research
VENTANA DP 200	Roche/Ventana	WSI	Mid (≤150 slides/h)	Yes	20×/40×	~1 min/slide	Routine diagnostics, AI-ready workflows
VENTANA DP 600	Roche/Ventana	WSI	High (up to 450 slides/day)	Yes	20×/40×	~45 s/slide	High-throughput labs, primary diagnosis, AI workflows
KF-SC-AN Series	KF-Biotech (KFBIO)	WSI	High (150–300 slides/h)	Yes	20×/40×	~1 min/slide	National screening, digital infrastructure in Asia
MoticEasyScan Pro 6	Motic	WSI	Low to Mid (40–60 slides/h)	Yes	20×/40×	~1.5 min/slide	Education, diagnostics in LMICs
3DHISTECH Pannoramic 1000	3DHISTECH	WSI	Very High (1000 slides/day)	Yes	20×/40×	~40 s/slide	Pharma research, high-throughput AI
Genius Digital Diagnostics System	Hologic	WSI (Cytology + AI)	Moderate (60–80 slides/h)	Yes (volumetric scanning)	20×/40×	~1 min/slide	Cervical screening, AI-assisted cytology
TissueScope iQ/LE120/Touch	Huron Digital Pathology	WSI	High (up to 120 slides)	No	20×/40×	~1–2 min/slide	Research, education, large-format scanning
OptraSCAN OS-SiA	OptraSCAN	WSI + AI	Modular/variable	Yes	20×/40×	~1 min/slide	Real-time AI-assisted scanning, research workflows

Continued

TABLE 7.1 Summary of some digital pathology platforms—cont'd

Platform	Manufacturer	Type	Scanner throughput/mode	Multilayer support	Image capture magnification	Scan speed	Primary use cases
Grundium Ocus40/Ocus M	Grundium	Live microscopy/ Static	Real-time streaming	No	20×/40×	Live view	Field consults, portable diagnostics
PathScan Enabler IV	Meyer Instruments	Static imaging	Manual	No	20×/40×	N/A	Documentation, basic imaging
Olympus VS200	Olympus	WSI	High	Yes	20×/40×	~1 min/slide	Research, academic settings
TelePi	Open-source (Yan Yin et al.)	Live microscopy +Static	Real-time stream + stills	No	Up to 40×	Live view	Education, frozen sections, low-resource, remote diagnosis
GT450	Leica	WSI	High (81 slides/h)	Yes	20×/40×	~45 s/slide	Clinical diagnostics
SpectralHT	Pramana	WSI	Low (24 slides/h)	Yes (extended focal imaging)	20×/40×	~2.5 min/slide	Extended focal imaging, research

including WSI scanners and digital pathology software, must now undergo a more rigorous assessment. The IVDR requires manufacturers to provide performance evaluations comprising scientific validity, analytical performance, and clinical performance reports (Jahn et al., 2020).

An easy and cost-effective option that has become increasingly popular is the use of smartphones, thanks to advances in smartphone cameras, faster internet connections, and software applications that allow the transmission of images despite large file sizes. These improvements have also helped address important issues such as pixelation. Using these devices is very simple and requires little practice; however, they are not suitable for capturing multiple images in rapid succession. To overcome this limitation, smartphone scope adapters have been developed, which can be attached to telescopes, binoculars, and microscopes, enabling smartphones to capture high-quality images (Tanveer and Naz, 2025). In addition, custom applications have been designed specifically for smartphone-based telepathology to support second opinions (Hartman et al., 2014).

Issues, impediments of broad usage

The implementation of digital pathology is accompanied by a complex array of challenges that institutions must carefully navigate. Among the foremost considerations is cost, encompassing not only the acquisition of hardware and software but also long-term investments in interoperability, digital storage, and human resources. The technological landscape further complicates this process, requiring robust information technology (IT) infrastructure capable of integrating seamlessly with existing systems such as Electronic Medical Records (EMR), Laboratory Information Systems (LIS), Image Management Systems, AI tools, middleware, and scalable storage solutions. Transitioning to a fully digital workflow is inherently gradual and logistically demanding, especially for laboratories attempting to operate within a hybrid analog-digital model during the interim. With a growing number of available commercial platforms, each tailored to different use cases, the selection process can feel overwhelming, particularly without a clear understanding of institutional needs and operational priorities (Eloy et al., 2025; Lujan et al., 2022).

Moreover, in addition to the infrastructure (scanners, monitors, image management systems, laboratory information systems, data center infrastructure), digital transformation also entails substantial organizational change management, a steep learning curve for personnel—which demands proper training—and the need to address new paradigms such as remote case sign-out, while simultaneously ensuring the ethical and secure management of sensitive patient data, including compliance with privacy regulations and data protection standards (Lujan et al., 2022; Coulter et al., 2022). The American Telehealth Association (ATA) mandates that all telepathology data must be protected using encryption that complies with recognized standards. It further recommends that confidential data be stored only in secure locations and explicitly advises against the use of noncompliant cloud services for data storage (Tanveer and Naz, 2025).

Cultural and adoption-related obstacles are reflected in professional resistance due to limited trust in technology, along with the need for continuous training of both technical and medical personnel (Lujan et al., 2022). Furthermore, subspecialties like cytopathology, hematopathology, and renal pathology pose unique challenges due to their specific diagnostic workflows; while tailored solutions do exist for these areas, the present focus remains on establishing a foundational digital workflow within general surgical pathology (Lujan et al., 2022). Specifically for cytopathology, validation protocols have been published in a white paper (Lin et al., 2024).

One of the key benefits of telepathology is its ability to facilitate expert consultation in underserved regions. Within the GLORIA's project (Mosquera-Zamudio et al., 2024), which aims to establish a nationwide telepathology network in Colombia to reduce diagnostic delays in cancer care, the primary obstacle identified was the development of the complete infrastructure and the limited quality and reliability of available bandwidth. However, in low- and middle-income countries, as in Latin America, the most significant challenges extend beyond the acquisition of hardware and software, with infrastructure limitations playing a critical role. Additional barriers include regulatory constraints, limited training opportunities, shortages of essential inputs, concerns about cost and return on investment, and, importantly, low sample quality resulting from preanalytical factors (Coudry et al., 2024).

Validation of telepathology systems is a critical requirement. The College of American Pathologists (CAP) Pathology and Laboratory Quality Center guideline for WSI specifies that laboratories must validate all specimen types intended for digital diagnosis, including hematoxylin and eosin sections, cytology, frozen sections, and immunohistochemistry. Validation requires a minimum of 60 cases for the primary application, with at least 20 additional cases for each subsequent application. The study must demonstrate diagnostic concordance of at least 95% between digital and glass slides, assessed by the same observer. If concordance falls below this threshold, laboratories are expected to investigate and resolve the discrepancy. Importantly, validation must evaluate intraobserver, rather than interobserver, variability, with a minimum washout period of 2 weeks between digital and glass slide review to minimize recall bias (Evans et al., 2022).

With the rapid advancement of digital pathology, reimbursement has emerged as one of the most pressing challenges both now and in the future. Existing payment models frequently fail to cover the substantial costs associated with implementation, including slide scanners, data storage, and information technology infrastructure, which particularly hinders adoption (Eloy et al., 2025). Moreover, the absence of standardized reimbursement mechanisms exacerbates inequities in access, thereby compromising one of the central promises of telepathology (Sulaieva et al., 2024). While recent policy initiatives, such as the introduction of Category III CPT add-on codes in the United States, approved by the American Medical Association (AMA) with strong support from the CAP, represent an encouraging step toward formalizing billing for digital pathology services, these codes remain temporary and are primarily intended for data collection. Thus long-term progress will depend on establishing permanent reimbursement pathways to ensure equitable and sustainable integration of digital pathology into clinical practice (Evans et al., 2022).

Telecytology and teleROSE

Telecytology has become a cornerstone for delivering rapid on-site evaluation (ROSE) in fine-needle aspiration (FNA) and core biopsy settings, with a spectrum of approaches ranging from static image transmission to dynamic video streaming and robotic microscopy. Dynamic modalities, especially when integrated into secure enterprise videoconferencing platforms, have shown high concordance for adequacy calls compared with conventional ROSE, while low-cost alternatives, such as smartphone-based streaming, have demonstrated feasibility in budget-restrained settings (Agarwal et al., 2016; Dudas et al., 2014). These options offer institutions flexibility: high-resource centers can deploy dedicated telecytology carts with robust IT integration, while resource-limited facilities can leverage affordable solutions without compromising adequacy assessment (Kouanda et al., 2023; Lin et al., 2019). Importantly, dynamic video transmission has generally been favored over static image capture, as it better replicates real-time slide navigation and interactive diagnostic decision-making.

Implementation strategies, however, must be accompanied by rigorous validation. The newly released 2024 American Society of Cytopathology consensus recommendations provide the most comprehensive framework to date, emphasizing that validation must test the entire end-to-end "pixel pathway" (from camera to remote monitor) rather than isolated components (Lin et al., 2024). These recommendations stress the inclusion of both retrospective and prospective cases, competency assessment of operators, integration into quality management systems, and prespecified discrepancy monitoring, particularly for malignant-to-benign downgrades. They also highlight that any material changes in hardware, software, or workflow—such as switching capture devices, updating conferencing platforms, or introducing robotic microscopes—constitute a new system requiring revalidation (Lin et al., 2024).

The 2024 ASC task force further contextualizes its proposals by referencing key validation and feasibility studies across the telecytology literature. For example, Perez et al. (2021) underscored diagnostic accuracy and pitfalls in touch imprint cytology via dynamic telecytology, while Agarwal et al. (2016) and Costa et al. (2018) validated FaceTime as a low-cost adequacy assessment tool. Larger series, such as Kouanda et al. (2023), confirmed the equivalence of telecytology and in-room cytopathologist evaluations for pancreatic FNAs, and Lin et al. (2019) documented successful high-definition streaming workflows in a cancer center environment. Collectively, these data illustrate that while implementation models differ—from enterprise-grade systems to economical smartphone setups—the key determinant of safety and reliability is adherence to robust, transparent validation. The ASC recommendations, therefore, provide a consensus-based roadmap for harmonizing practice, ensuring diagnostic integrity, and paving the way for regulatory alignment and eventual reimbursement.

Another important consideration when dealing with telecytology is choosing between static and dynamic transmission of images. Static telecytology—where an on-site operator captures representative photomicrographs or WSIs and transmits them for remote review—offers multiple pragmatic advantages for ROSE workflows. It requires lower sustained bandwidth, is relatively inexpensive to deploy, and produces image files that are straightforward to archive, audit, and integrate into a laboratory quality-management system. These attributes make static workflows attractive for geographically dispersed or resource-constrained sites focused primarily on adequacy assessment, and several studies report acceptable adequacy concordance when field selection and image quality are adequate (Agarwal et al., 2016; Dudas et al., 2014; Lin et al., 2019). However, static approaches are highly dependent on the skill of the on-site operator to select diagnostically relevant fields and can introduce sampling bias when important material is not photographed.

On the other hand, live (dynamic) telecytology—encompassing real-time video streaming—more closely replicates the interactivity of in-room microscopy and therefore reduces the risk that diagnostically important regions will be missed. Dynamic systems permit the remote cytopathologist/cytotechnologist to direct field selection, alter magnification and focus, and request immediate additional passes for sample adequacy (Kouanda et al., 2023; Perez et al., 2021; Trabzonlu et al., 2022). However, while it is less dependent on the skill of the operator, it also requires training, since the whole slide is usually not transmitted (due to the time restrictions of the procedure), and preselecting suspicious areas is vital for proper clinical correlation (Costa et al., 2018).

Therefore choosing between static and live models should be driven by the intended clinical use and local resources (https://www.cap.org/advocacy/payments-for-pathology-services/digital-pathology-codes, n.d.; Agarwal et al., 2016; Al-Janabi et al., 2012; Ardon et al., 2025; Costa et al., 2018; Coudry et al., 2024; Coulter et al., 2022; Dudas et al., 2014; Eloy et al., 2025; Evans et al., 2022; Hartman et al., 2014; Jahn et al., 2020; Kouanda et al., 2023; Lin et al., 2024, 2019; Lujan et al., 2022; Meyer and Pare, 2014; Mosquera-Zamudio et al., 2024; Mukhopadhyay et al., 2018; Pantanowitz et al., 2014; Perez et al., 2021; Sulaieva et al., 2024; Tanveer and Naz, 2025; Tanveer et al., 2022; Trabzonlu et al., 2022; Weinstein et al., 2009; Zia et al., 2025). For small centers, where the primary objective is adequacy calls, well-validated static image workflows can deliver immediate clinical value. By contrast, tertiary centers or high-volume ROSE programs that require interactive diagnostic triage might benefit more from dynamic platforms.

Conclusion

In summary, telepathology and telecytology have advanced significantly in recent years, driven by the integration of WSI, robotic microscopy, and smartphone-based technologies. Widespread adoption continues to face significant barriers, including the high cost of hardware and software, stringent validation mandates, and complex regulatory frameworks. Data security and ethical concerns, particularly regarding compliance with privacy regulations in different institutions, further compound the challenges. Nevertheless, recent advances demonstrate that telepathology has the potential not merely to augment but to fundamentally transform diagnostic practice, bridging geographic disparities in access while safeguarding diagnostic integrity worldwide.

AI Disclosure

During the preparation of this work, the author(s) used ChatGPT for language editing. After using this tool/service, the author(s) reviewed and edited the content as needed and take(s) full responsibility for the content of the publication.

References

https://www.cap.org/advocacy/payments-for-pathology-services/digital-pathology-codes, n.d.

Agarwal, S., Zhao, L., Zhang, R., Hassell, L., 2016. FaceTime validation study: low-cost streaming video for cytology adequacy assessment. Cancer Cytopathol. 19346638. 124 (3), 213–220. Available from: http://onlinelibrary.wiley.com/journal/10.1002/(ISSN)1934-6638. doi:10.1002/cncy.21636.

Al-Janabi, S., Huisman, A., Vink, A., Leguit, R.J., Offerhaus, G.J.A., Kate, F.J.W.T., Van Dijk, M.R., Van Diest, P.J., 2012. Whole slide images for primary diagnostics in dermatopathology: a feasibility study. J. Clin. Pathol. 14724146. 65 (2), 152–158. Available from: http://jcp.bmj.com/content/65/2/152.full.pdf+html. doi:10.1136/jclinpath-2011-200277Netherlands.

Ardon, O., Manzo, A., Spencer, J., Reuter, V.E., Hameed, M., Hanna, M.G., 2025. Digital slide scanning at scale: comparison of whole slide imaging devices in a clinical setting. J. Pathol. Inform. 21533539. 18. Available from: https://www.sciencedirect.com/science/journal/21533539. doi:10.1016/j.jpi.2025.100446.

Costa, C., Pastorello, R.G., Mendonça, A., Tamaro, C., Morais, C., Barbosa, B., Ribeiro, K.B., Caivano, A., Saieg, M.A., 2018. Use of a low-cost telecytopathology method for remote assessment of thyroid FNAs. Cancer Cytopathol. 19346638. 126 (9), 767–772. Available from: http://onlinelibrary.wiley.com/journal/10.1002/(ISSN)1934-6638. doi:10.1002/cncy.22026.

Coudry, R.A., Assis, E.A.C.P., Frassetto, F.P., Jansen, A.M., da Silva, L.M., Parra-Medina, R., Saieg, M., 2024. Crossing the Andes: challenges and opportunities for digital pathology in Latin America. J. Pathol. Inform. 21533539. 15. Available from: https://www.sciencedirect.com/science/journal/21533539. doi:10.1016/j.jpi.2024.100369.

Coulter, C., McKay, F., Hallowell, N., Browning, L., Colling, R., Macklin, P., Sorell, T., Aslam, M., Bryson, G., Treanor, D., Verrill, C., 2022. Understanding the ethical and legal considerations of digital pathology. J. Pathol. Clin. Res. 20564538. 8 (2), 101–115. Available from: http://onlinelibrary.wiley.com/journal/10.1002/(ISSN)2056-4538. doi:10.1002/cjp2.251.

Dudas, R., VandenBussche, C., Baras, A., Ali, S.Z., Olson, M.T., 2014. Inexpensive telecytology solutions that use the Raspberry Pi and the iPhone. J. Am. Soc. Cytopathol. 22132945. 3 (1), 49–55. doi:10.1016/j.jasc.2013.09.005.

Eloy, C., Fraggetta, F., van Diest, P.J., Polónia, A., Curado, M., Temprana-Salvador, J., Zlobec, I., Purqueras, E., Weis, C.A., Matias-Guiu, X., Schirmacher, P., Ryška, A., 2025. Digital transformation of pathology—the European Society of Pathology expert opinion paper. Virchows Arch. 14322307. 487 (5), 971–981. Available from: https://www.springer.com/journal/428. doi:10.1007/s00428-025-04090-w.

Evans, A.J., Brown, R.W., Bui, M.M., Chlipala, E.A., Lacchetti, C., Milner, D.A., Pantanowitz, L., Parwani, A.V., Reid, K., Riben, M.W., Reuter, V.E., Stephens, L., Stewart, R.L., Thomas, N.E., 2022. Validating whole slide imaging systems for diagnostic purposes in pathology. Arch. Pathol. Lab. Med. 1543-2165. 146 (4), 440–450. doi:10.5858/arpa.2020-0723-cp.

Hartman, D.J., Parwani, A.V., Cable, B., Cucoranu, I.C., McHugh, J.S., Kolowitz, B.J., Yousem, S.A., Palat, V., Reden, A.V., Sloka, S., Lauro, G.R., Ahmed, I., Pantanowitz, L., 2014. Pocket pathologist: a mobile application for rapid diagnostic surgical pathology consultation. J. Pathol. Inform. 21533539. 5 (1), 10. doi:10.4103/2153-3539.129443.

Jahn, S.W., Plass, M., Moinfar, F., 2020. Digital pathology: advantages, limitations and emerging perspectives. J. Clin. Med. 20770383. 9 (11). Available from: https://www.mdpi.com/2077-0383/9/11/3697/pdf. doi:10.3390/jcm9113697.

Kouanda, A., Mclean, R., Faggen, A., Demissie, E., Balassanian, R., Kamal, F., Avila, P., Arain, M., Dai, S.C., Munroe, C., 2023. Telecytology versus in-room cytopathologist for EUS-guided FNA or fine-needle biopsy sampling of solid pancreatic lesions. Gastrointest. Endosc. 10976779. 97 (3), 466–471. Available from: http://www.elsevier.com/inca/publications/store/6/2/3/1/3/7/index.htt. doi:10.1016/j.gie.2022.10.015.

Lin, O., Rudomina, D., Feratovic, R., Sirintrapun, S.J., 2019. Rapid on-site evaluation using telecytology: a major cancer center experience. Diagn. Cytopathol. 10970339. 47 (1), 15–19. Available from: http://onlinelibrary.wiley.com/journal/10.1002/(ISSN)1097-0339. doi:10.1002/dc.23925.

Lin, O., Alperstein, S., Barkan, G.A., Cuda, J.M., Kezlarian, B., Jhala, D., Jin, X., Mehrotra, S., Monaco, S.E., Rao, J., Saieg, M., Thrall, M., Pantanowitz, L., 2024. American Society of Cytopathology Telecytology validation recommendations for rapid on-site evaluation (ROSE). J. Am. Soc. Cytopathol. 22132945. 13 (2), 111–121. Available from: https://www.sciencedirect.com/science/journal/22132945. doi:10.1016/j.jasc.2023.12.001.

Lujan, G., Li, Z., Parwani, A.V., 2022. Challenges in implementing a digital pathology workflow in surgical pathology. Hum. Pathol. Rep. 2772736X. 29. Available from: https://www.journals.elsevier.com/human-pathology-reports. doi:10.1016/j.hpr.2022.300673.

Meyer, J., Pare, G., 2014. Telepathology implementation challenges and benefits: a scoping review. In: 47th Hawaii International Conference on System Sciences. IEEE, pp. 2838–2847.

Mosquera-Zamudio, A., Gomez-Suarez, M., Sprockel, J., Riaño-Moreno, J.C., Janssen, E.A.M., Pantanowitz, L., Parra-Medina, R., 2024. Globalization of a telepathology network with artificial intelligence applications in Colombia: the GLORIA program study protocol. J. Pathol. Inform. 21533539. 15. Available from: https://www.sciencedirect.com/science/journal/21533539. doi:10.1016/j.jpi.2024.100394.

Mukhopadhyay, S., Feldman, M.D., Abels, E., Ashfaq, R., Beltaifa, S., Cacciabeve, N.G., Cathro, H.P., Cheng, L., Cooper, K., Dickey, G.E., Gill, R.M., Heaton, R.P., Kerstens, R., Lindberg, G.M., Malhotra, R.K., Mandell, J.W., Manlucu, E.D., Mills, A.M., Mills, S.E., Moskaluk, C.A., Nelis, M., Patil, D.T., Przybycin, C.G., Reynolds, J.P., Rubin, B.P., Saboorian, M.H., Salicru, M., Samols, M.A., Sturgis, C.D., Turner, K.O., Wick, M.R., Yoon, J.Y., Zhao, P., Taylor, C.R., 2018. Whole slide imaging versus microscopy for primary diagnosis in surgical pathology: a multicenter blinded randomized noninferiority study of 1992 cases (pivotal study). Am. J. Surg. Pathol. 15320979. 42 (1), 39–52. Available from: http://journals.lww.com/ajsp/pages/default.aspx. doi:10.1097/PAS.0000000000000948.

Pantanowitz, L., Dickinson, K., Evans, A.J., Hassell, L.A., Henricks, W.H., Lennerz, J.K., Lowe, A., Parwani, A.V., Michael Riben, C.O.L.D.S., Tuthill, J.M., Weinstein, R.S., Wilbur, D.C., Krupinski, E.A., Bernard, J., 2014. American Telemedicine Association clinical guidelines for telepathology. J. Pathol. Inform. 21533539. 5 (1), 39. doi:10.4103/2153-3539.143329.

Perez, D., Stemmer, M.N., Khurana, K.K., 2021. Utilization of dynamic telecytopathology for rapid onsite evaluation of touch imprint cytology of needle core biopsy: diagnostic accuracy and pitfalls. Telemed. E Health. 15563669. 27 (5), 525–531. Available from: http://www.liebertpub.com/publication.aspx?pub_id=54. doi:10.1089/tmj.2020.0117.

Sulaieva, O., Dudin, O., Koshyk, O., Panko, M., Kobyliak, N., 2024. Digital pathology implementation in cancer diagnostics: towards informed decision-making. Front. Digit. Health. 2673-253X. 6. doi:10.3389/fdgth.2024.1358305.

Tanveer, N., Dixit, S., Jain, P., Rani, P., Gautum, H.V., Naskar, S., 2022. Static smartphone images supplemented by videoconferencing for breast telepathology. Clin. Breast Cancer. 19380666. 22 (5), 455–461. Available from: http://www.journals.elsevier.com/clinical-breast-cancer. doi:10.1016/j.clbc.2022.02.010.

Tanveer, N., Naz, F., 2025. Smartphone assisted telepathology: a review. Ann. Diagn. Pathol. 10929134. 78, 152493. doi:10.1016/j.anndiagpath.2025.152493.

Trabzonlu, L., Chatt, G., McIntire, P.J., Eshraghi, R., Lapadat, R., Atieh, M., Pambuccian, S.E., Wojcik, E.M., Mehrotra, S., Barkan, G.A., 2022. Telecytology validation: is there a recipe for everybody? J. Am. Soc. Cytopathol. 22132945. 11 (4), 218–225. Available from: http://www.elsevier.com/journals/journal-of-the-american-society-of-cytopathology/2213-2945. doi:10.1016/j.jasc.2022.03.001.

Weinstein, R.S., Graham, A.R., Richter, L.C., Barker, G.P., Krupinski, E.A., Lopez, A.M., Erps, K.A., Bhattacharyya, A.K., Yagi, Y., Gilbertson, J.R., 2009. Overview of telepathology, virtual microscopy, and whole slide imaging: prospects for the future. Hum. Pathol. 00468177. 40 (8), 1057–1069. doi:10.1016/j.humpath.2009.04.006.

Zia, S., Yildiz-Aktas, I.Z., Zia, F., Parwani, A.V., 2025. An update on applications of digital pathology: primary diagnosis; telepathology, education and research. Diagn. Pathol. 17461596. 20 (1). Available from: https://diagnosticpathology.biomedcentral.com. doi:10.1186/s13000-025-01610-9.

CHAPTER

8

Artificial intelligence (imaging analysis) fundamentals

Abdul Rehman Akbar[a] *and Muhammad Khalid Khan Niazi*[b]

[a]Department of Pathology, School of Medicine, The Ohio State University Wexner Medical Center, Columbus, OH, United States

[b]Department of Pathology, The Ohio State University, Columbus, OH, United States

Introduction to artificial intelligence in pathology

Artificial Intelligence (AI) has emerged as one of the most transformative technologies in modern medicine, and its impact on diagnostic pathology is profound (Niazi et al., 2019b; Coudray et al., 2018). Pathology, the cornerstone of disease diagnosis, relies heavily on the interpretation of complex tissue structures and cellular morphology. Traditionally, this process has been manual, requiring years of training and experience to achieve proficiency. However, the increasing complexity of cases, the growing demand for precision medicine, and the global shortage of pathologists have created an urgent need for computational augmentation. AI offers a solution by enabling machines to perform tasks that mimic human intelligence—such as pattern recognition, decision making, and adaptive learning—at unprecedented speed and scale (Niazi et al., 2019b; Campanella et al., 2019).

Pathology images as unstructured data

Unlike structured data, such as spreadsheets or laboratory values, histopathology images are inherently unstructured (Niazi et al., 2019a). A whole slide image (WSI) is essentially a grid of millions of pixels, each representing color intensity values but without any inherent semantic meaning. The computer does not "see" a nucleus, a gland, or a tumor—it only sees pixel arrays. There are no rows, columns, or predefined labels that indicate diagnostic features. This lack of structure poses a significant challenge for computational analysis because algorithms cannot directly interpret biological context from raw pixel data (Fig. 8.1).

Advances in Digital Pathology and Artificial Intelligence in Pathology
https://doi.org/10.1016/B978-0-443-40383-5.00006-6

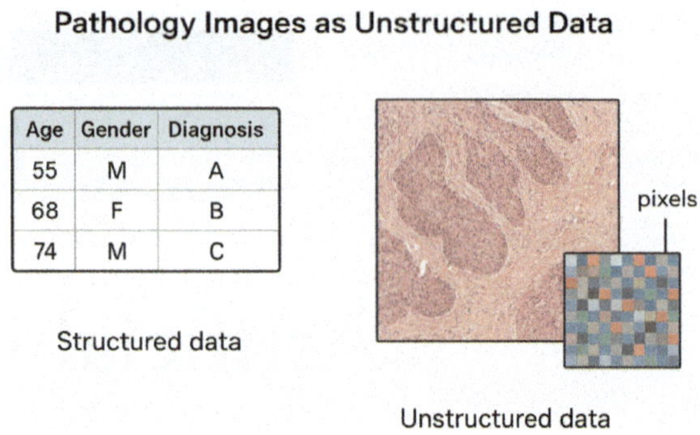

FIG. 8.1 Comparison of structured and unstructured data in pathology. Structured data (e.g., patient demographics in a table) contains predefined semantics, making it easy for computers to interpret. In contrast, histopathology WSIs are unstructured, consisting of millions of pixel values without inherent biological meaning, posing challenges for computational analysis.

For example:

- **A spreadsheet of patient demographics is structured:** Each column has a defined meaning—gender, diagnosis, treatment history. Every row corresponds to a patient, and the relationships between data points are explicit. This structure allows computers to easily sort, filter, and analyze the information because the semantics are embedded in the format.
- **A histology slide, in contrast, is fundamentally different:** It contains rich visual information but no explicit markers for "tumor region," "mitotic figure," or "necrosis." To a computer, the slide is simply an array of millions of pixel values representing color intensities. There is no inherent concept of a cell, a nucleus, or a gland. Without additional processing, the image is just a collection of numbers—meaningless in isolation.

For a pathologist, identifying a tumor region involves recognizing complex spatial patterns—clusters of atypical cells, disrupted glandular architecture, or stromal invasion. For a computer, these patterns must be translated into measurable features (Senaras et al., 2018). This requires segmentation (dividing the image into meaningful regions), feature extraction (quantifying attributes like nuclear size or chromatin texture), and classification (assigning biological meaning to those features).

AI bridges this gap by converting unstructured pixel data into structured representations. Through machine learning and deep learning, algorithms learn to detect and label regions of interest, compute quantitative metrics, and organize them into structured outputs such as tumor grade, biomarker expression levels, or prognostic scores. This transformation is what enables downstream tasks like automated diagnosis, risk prediction, and treatment planning.

How AI converts unstructured data into structured insights

Every pathology image begins as a silent canvas—a vast ocean of pixels rich in color and texture but devoid of meaning. To the human eye, these pixels form recognizable structures: nuclei, glands, stroma, and tumor regions. To a computer, however, they are nothing more than numerical values arranged in a grid. The challenge is monumental: how do we teach

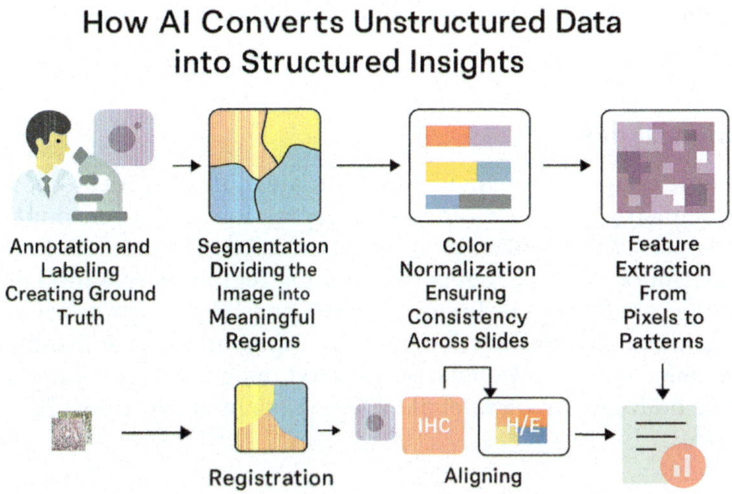

FIG. 8.2 AI pipeline for transforming unstructured pathology images into structured insights. The process includes annotation and labeling to create ground truth, segmentation of biologically meaningful regions, color normalization for consistency, registration to align multi modal slides (H&E and IHC), feature extraction from pixels to patterns, and generation of structured outputs for clinical interpretation.

a machine to see what a pathologist sees, and beyond that, to quantify and interpret these patterns in ways that support clinical decisions? This transformation from raw pixels to actionable insights is not a single leap; it unfolds through a series of deliberate steps, each mirroring—and often surpassing—the reasoning process of a trained pathologist (Fig. 8.2).

Annotation and labeling: Creating ground truth

The journey begins with human expertise. Pathologists provide annotations—marking mitotic figures, outlining tumor boundaries, and identifying regions of necrosis. These labeled examples serve as the foundation for supervised learning, acting as a bridge between biological knowledge and computational logic. Without this ground truth, AI would wander blindly through a sea of pixels, unable to distinguish a malignant cell from benign stroma. Annotation is more than a technical step; it is a dialog between human and machine, where clinical wisdom is distilled into data that algorithms can learn from, as noted by Niazi et al. (2018).

Segmentation: Dividing the image into meaningful regions

Once annotated, images are often segmented—a process akin to carving a complex landscape into distinct territories. Segmentation partitions the WSI into biologically relevant regions: tumor, stroma, necrosis, and lymphocytes. This step is critical because quantitative analysis depends on context. For example, tumor segmentation enables accurate calculation of the tumor to stroma ratio, a prognostic marker linked to patient outcomes. Similarly, cell segmentation allows precise counting of mitotic figures or measurement of nuclear morphology. By transforming a continuous pixel grid into discrete, interpretable units, segmentation lays the groundwork for downstream tasks such as grading and biomarker quantification

(Su et al., 2024b, 2024c, 2024e; Camalan et al., 2024; Tavolara et al., 2020b, 2021c; Binol et al., 2020; Niazi et al., 2020a, 2020b).

Registration: Aligning multimodal data

Pathology rarely relies on a single image. A specimen may generate multiple slides—H&E for morphology, immunohistochemistry (IHC) for biomarker expression, and even multiplexed imaging for spatial proteomics. These slides often represent adjacent tissue sections, making alignment essential. Registration ensures that corresponding regions across modalities line up perfectly, enabling integrated analysis. Imagine correlating HER2 expression on an IHC slide with tumor architecture on an H&E section; without registration, such insights would be fragmented and unreliable. This step transforms isolated images into a coherent, multidimensional view of disease biology (Tavolara et al., 2020a; Khan and Nyström, 2010; Niazi et al., 2010; Chen et al., 2025).

Color normalization: Ensuring consistency across slides

Even within a single modality, variability lurks. Differences in staining protocols, scanner settings, and laboratory conditions can alter color profiles, confusing AI models that rely on color cues for feature extraction. Color normalization addresses this challenge by standardizing stain appearance across slides. By harmonizing hematoxylin and eosin intensity, normalization ensures that nuclei and cytoplasm appear consistent regardless of origin. This seemingly simple adjustment enhances model robustness, reducing the risk of misclassification due to technical artifacts rather than biological differences (Boschman et al., 2022; Salvi et al., 2024; Tellez et al., 2019; Vahadane et al., 2016).

Feature extraction: From pixels to patterns

With segmentation and normalization complete, the stage is set for feature extraction—the heart of computational pathology. In traditional machine learning, this involves handcrafted features such as nuclear size, chromatin texture, and glandular architecture, computed from annotated regions. These features reflect decades of pathological insight, translated into numbers that machines can process. Deep learning takes a different approach. Convolutional neural networks learn hierarchical features directly from raw pixels, uncovering patterns that may elude even expert eyes. Whether through human guided engineering or autonomous discovery, feature extraction converts visual complexity into structured representations (Tavolara et al., 2021b; Niazi et al., 2017; Lu et al., 2021).

Structured outputs: Turning numbers into knowledge

The final act in this transformation is interpretation. AI models use learned patterns to generate structured outputs: tumor grade or subtype classification, quantitative metrics like mitotic count or Ki-67 positivity, and predictive scores for prognosis or therapy response.

These outputs are not mere numbers; they are clinically actionable insights that inform treatment decisions and patient care. By translating the pathologist's visual expertise into reproducible, scalable computations, AI enables large scale analysis, supports diagnostic consistency, and opens new frontiers in biomarker discovery and precision medicine (Afzaal et al., 2025; Akbar et al., 2025; Sajjad et al., 2023, 2025; She et al., 2020; Su et al., 2022, 2023a, 2023c, 2024a, 2024d, 2025a; Tavolara et al., 2023).

In essence, this pipeline transforms chaos into clarity. It takes the silent language of pixels and converts it into structured knowledge—a process that not only mirrors human reasoning but amplifies it, revealing patterns invisible to the naked eye and unlocking insights that redefine the future of pathology.

Why AI is relevant to pathology

Imagine standing before a digital slide of a breast biopsy. The image spans an enormous canvas—hundreds of thousands of pixels wide and tall—capturing every nuance of tissue architecture. Within this expanse lie thousands of cells, each with subtle variations in size, shape, and chromatin texture. For a pathologist, interpreting these details requires meticulous scanning, mental pattern recognition, and years of experience. Now multiply this effort across dozens of slides for a single patient, and then across hundreds of patients in a busy clinical setting. The sheer scale of this task illustrates why pathology is both art and science—and why it is ripe for computational augmentation.

Histopathology images are not just large; they are complex and unstructured. Unlike a spreadsheet, where each column has a defined meaning, a WSI is a grid of color values without inherent semantic labels. The computer does not know where the tumor begins or ends, nor can it distinguish a mitotic figure from background stroma without guidance. Manual review of such data is labor intensive and prone to variability between observers, especially when diagnostic criteria involve subjective thresholds.

This is where AI becomes indispensable. AI systems can process these massive datasets rapidly and consistently, reducing turnaround times and improving diagnostic reproducibility. But efficiency is only part of the story. AI introduces capabilities that extend human vision—detecting subtle textural changes linked to molecular alterations, quantifying features with mathematical precision, and even predicting patient outcomes directly from hematoxylin and eosin (H&E) stained slides. These insights are not limited to mimicking human tasks; they often reveal patterns invisible to the naked eye, bridging morphology with genomics and enabling a new era of precision pathology (Kather et al., 2020; Li et al., 2025).

Core functions of AI in pathology

To appreciate the role of AI in pathology, imagine the journey of a digital slide through an AI system. A WSI begins as a vast ocean of pixel color values without meaning. For a human pathologist, these pixels form recognizable structures: nuclei, glands, stroma, and tumor regions. For a computer, however, they are just numbers. The challenge is to teach the machine to see what the pathologist sees and, beyond that, to quantify and interpret these patterns in ways that support clinical decisions.

The first step in this journey is feature extraction. In the context of pathology, extracting meaningful features means converting visual patterns into measurable attributes. For example, when analyzing breast cancer slides, features might include nuclear size, chromatin texture, or glandular architecture. These features are not arbitrary—they represent biological signals that correlate with disease severity or prognosis. AI systems learn to compute these features either through handcrafted algorithms (machine learning) or automatically through deep learning models (Echle et al., 2021).

Next, AI must identify cellular and tissue level patterns indicative of disease. This is more than detecting individual cells; it involves understanding spatial relationships and tissue architecture. For instance, distinguishing invasive carcinoma from in situ disease requires recognizing glandular disruption and stromal invasion—patterns that emerge only when cells are considered in context. AI models achieve this by segmenting tissue regions, classifying cell types, and modeling their spatial organization.

Finally, these learned features and patterns become the foundation for assisting in diagnostic workflows. Once the system understands what constitutes a mitotic figure or a tumor boundary, it can perform downstream tasks such as:

- Counting mitotic figures to support tumor grading.
- Quantifying biomarker expression in immunohistochemistry (IHC) slides.
- Predicting molecular subtypes or recurrence risk directly from H&E images.

These capabilities extend beyond mimicking human tasks. AI often uncovers patterns invisible to the naked eye—subtle textural cues linked to genetic mutations or prognostic signatures embedded in tissue morphology. By transforming raw pixel data into structured, clinically relevant insights, AI not only accelerates diagnosis but also opens new frontiers in precision pathology.

In summary, AI algorithms are designed to:

- Analyze WSIs and extract meaningful features from raw pixel data.
- Identify cellular and tissue level patterns indicative of disease.
- Assist in diagnostic workflows by providing quantitative measurements, preliminary classifications, or prognostic predictions.

Applications for AI in pathology

To understand the practical impact of AI in pathology, imagine the daily challenges faced in a busy diagnostic lab. Each case brings unique complexities—some subtle, others glaring—and every decision carries clinical consequences. AI steps in as a digital assistant, not replacing the pathologist but amplifying their capabilities. Below are key areas where AI is making a difference, illustrated through real world scenarios (Fig. 8.3).

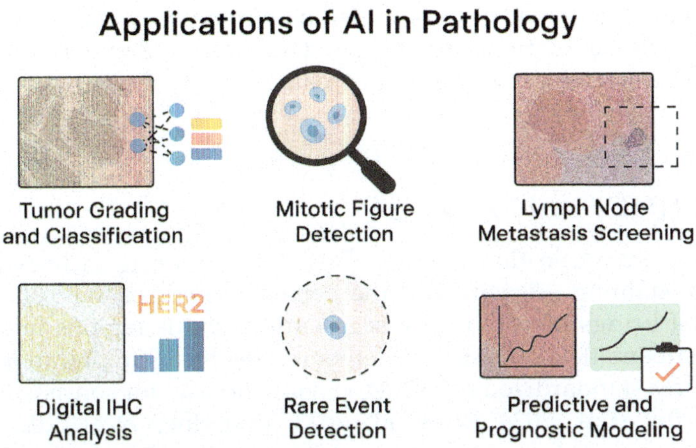

FIG. 8.3 Key applications of AI in pathology: (1) Tumor grading and classification for consistent diagnosis, (2) Mitotic figure detection to accelerate workflow, (3) Lymph node metastasis screening for accurate staging, (4) Digital IHC analysis for biomarker quantification, (5) Rare event detection for early identification of uncommon patterns, and (6) Predictive and prognostic modeling to forecast outcomes and therapy response.

Tumor grading and classification

Picture a prostate biopsy slide under the microscope. The pathologist scans for glandular architecture, assessing whether the glands are well formed or fused—a critical determinant of the Gleason grade. This task demands precision, yet even experts can disagree on borderline cases. AI models, trained on thousands of annotated slides, can analyze glandular patterns at scale, offering consistent grading. In breast cancer, convolutional neural networks (CNNs) learn to interpret nuclear morphology and mitotic activity, classifying subtypes with accuracy comparable to seasoned pathologists. These systems act as a second set of eyes, reducing variability and supporting evidence based decisions (Niazi et al., 2017; Su et al., 2022; Goldstraw et al., 2016; Tavolara et al., 2021a; Borczuk, 2022; Niazi et al., 2014; Lu et al., 2024; Wang et al., 2016).

Mitotic figure detection

Counting mitotic figures is a cornerstone of tumor grading, but it is notoriously tedious. Imagine scanning a large section of sarcoma tissue, hunting for tiny mitotic spindles among thousands of cells. Fatigue and time pressure can lead to missed figures. AI algorithms excel here—they can sweep across entire slides, pinpointing mitotic cells with remarkable speed and consistency. This automation not only accelerates workflow but also ensures reproducibility, a critical factor in high stakes diagnoses (Shen et al., 2024; Bertram et al., 2022; Jahanifar et al., 2024; Ganz et al., 2024).

Lymph node metastasis screening

Consider a sentinel lymph node from a breast cancer patient. The pathologist searches for micro metastases—clusters of malignant cells that may be only a few microns wide.

Missing these can alter staging and treatment. AI systems trained on vast datasets can detect these elusive deposits, flagging suspicious regions for review. This partnership between human and machine minimizes false negatives and enhances patient safety (Sajjad et al., 2023; Su et al., 2023c, 2024f).

Digital immunohistochemistry (IHC) analysis

In precision oncology, biomarker quantification matters. Take HER2 scoring in breast cancer: pathologists must evaluate staining intensity and the percentage of positive cells, a process prone to subjectivity. AI brings objectivity by segmenting tumor regions and calculating staining metrics with mathematical precision. Whether it is the Ki-67 proliferation index or PD-L1 expression, AI ensures standardized reporting, paving the way for consistent therapeutic decisions (Krishnamurthy et al., 2024; Wu et al., 2023; Vesterinen et al., 2022; Zehra et al., 2023; Wu et al., 2022).

Rare event detection

Some tasks push the limits of human perception. Detecting circulating tumor cells or rare infectious organisms in tissue sections is like finding a needle in a haystack. AI thrives in these scenarios. By learning from millions of examples, algorithms can identify rare patterns that even expert eyes might overlook. This capability is invaluable for early detection and research into uncommon disease presentations (Afzaal et al., 2025; Akbar et al., 2025; Sajjad et al., 2025).

Predictive and prognostic modeling

Perhaps the most exciting frontier is prediction. Imagine an H&E slide that not only confirms a diagnosis but also forecasts the patient's future—risk of recurrence, likelihood of therapy response, or underlying genetic mutations. AI models can bridge morphology and genomics, extracting hidden signals from tissue architecture to generate prognostic insights. This transforms pathology from a descriptive science into a predictive one, aligning perfectly with the goals of precision medicine (Afzaal et al., 2025; Akbar et al., 2025; Sajjad et al., 2023, 2025; She et al., 2020; Su et al., 2022, Su et al., 2023b, 2023c, 2024d, 2024f, 2025a, 2025b; Tavolara et al., 2023; Wang et al., 2016; Wang et al., 2025).

Subfields of AI relevant to pathology

Overview of machine learning (ML) and deep learning (DL)

AI encompasses a broad spectrum of computational techniques, but two key subfields dominate imaging analysis (Fig. 8.4).

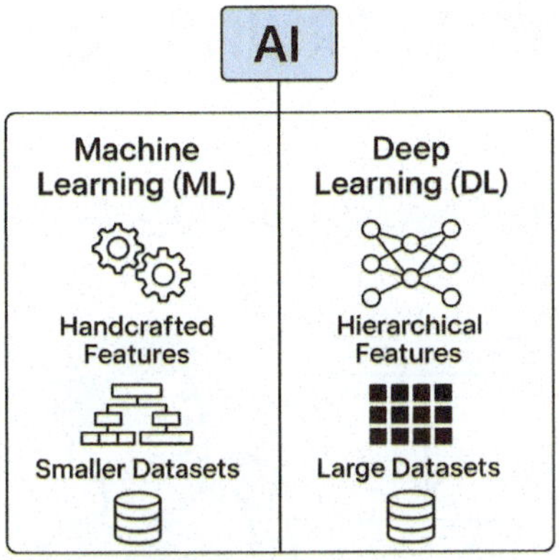

FIG. 8.4 Comparison of Machine Learning (ML) and Deep Learning (DL) within Artificial Intelligence (AI). Machine Learning relies on handcrafted features and smaller datasets, emphasizing interpretability, while Deep Learning uses neural networks to automatically learn hierarchical features from large datasets, often functioning as a "black box." Both collectively power modern AI applications in computational pathology.

- Machine Learning (ML) relies on handcrafted features such as nuclear size, chromatin texture, and cell shape. Pathologists or data scientists define these features, and models learn decision boundaries based on labeled examples.
- Deep Learning (DL) uses neural networks to automatically learn hierarchical features from raw pixel data, eliminating the need for manual feature engineering. DL models can discover patterns beyond human perception, such as subtle textural cues linked to genetic mutations.

The distinction between ML and DL is not merely technical; it influences the following:

- Data requirements: DL typically requires large, annotated datasets.
- Interpretability: ML models are often more explainable, whereas DL models can be "black boxes."
- Clinical integration: Regulatory approval and pathologist trust hinge on transparency and reproducibility.

Machine learning: Handcrafted intelligence

Machine learning equips computers with the ability to learn from data without being explicitly programmed. In pathology, ML models are often trained using annotated datasets—such as images labeled with mitotic figures or tumor grades. The process begins with feature engineering, where domain experts or data scientists extract measurable characteristics from images. These features might include nuclear size, shape, chromatin texture, or intensity (Fig. 8.5).

This transformation of raw image data into structured numerical inputs allows the model to learn decision boundaries. For example, a classifier might learn to distinguish mitotic from nonmitotic cells based on circularity and texture. These features are handcrafted, meaning they are designed by humans to reflect biologically relevant or statistically useful properties.

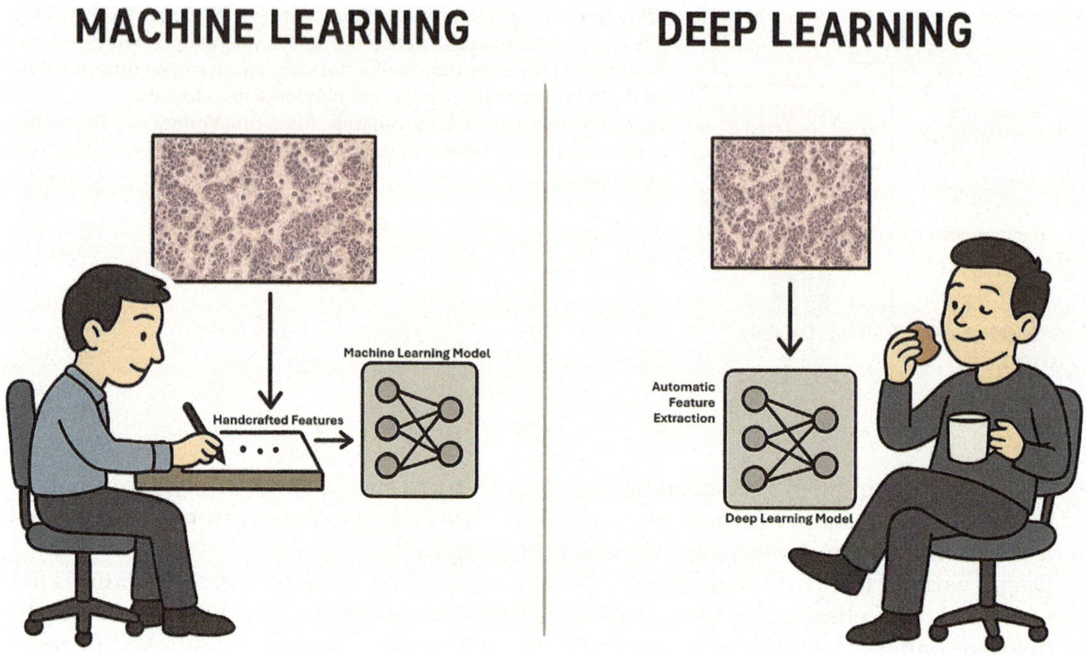

FIG. 8.5 Comparison of traditional machine learning and deep learning workflows for histopathology images (H&E). In machine learning, a human manually extracts features from the image before feeding them into the model. In deep learning, the image is processed end to end by the model.

Importantly, ML models are often interpretable. Because the features are explicitly defined and familiar to pathologists, the model's decisions can be traced and understood. This transparency makes ML particularly attractive for clinical applications, where explainability is critical.

*Supervised **versus** unsupervised learning: What does learning mean in pathology?*

When we talk about "learning" in artificial intelligence, it is tempting to imagine a machine memorizing facts like a student preparing for an exam. But in pathology, learning is far more sophisticated. It means teaching a computer to recognize biological structures—nuclei, glands, mitotic figures—and to interpret these patterns in clinically meaningful ways. This process mirrors how a pathologist learns: through repeated exposure to examples, feedback, and pattern recognition.

The concept of learning in AI is fundamentally about mapping patterns to meaning. For a computer, an image is just a grid of pixel values. It does not inherently know what a tumor looks like or what defines a mitotic figure. Learning is the process by which the machine moves from raw pixels to structured knowledge, guided either by human provided labels or by its own ability to discover patterns.

Supervised learning: Guided by ground truth

Supervised learning is akin to training a pathology resident with a labeled atlas. The model is given examples where the correct answer is known—H&E slides annotated by experts marking mitotic figures, tumor boundaries, or necrotic regions. These annotations serve as ground truth, enabling the algorithm to learn associations between visual features and diagnostic labels.

For instance:

- When detecting mitotic figures, the model learns that condensed chromatin and spindle formation are key indicators.
- When grading prostate cancer, it is understood that glandular architecture and nuclear atypia correlate with Gleason patterns.

This guided approach ensures the model focuses on clinically relevant features. Over time, just like a resident who improves with feedback, the algorithm refines its decision boundaries, reducing errors and improving accuracy. Supervised learning excels in tasks where diagnostic criteria are well defined—mitosis detection, tumor grading, and biomarker quantification.

Unsupervised learning: Discovering hidden structures

Unsupervised learning tells a different story. Here, the model explores data without labels—like a resident wandering through cases without guidance. It looks for patterns and groupings on its own:

- Clustering cells with similar morphology.
- Identifying tissue regions that share textural characteristics.
- Detecting anomalies that might represent rare subtypes or artifacts.

This approach is powerful for discovery. For example, unsupervised clustering of tumor samples might reveal previously unrecognized histologic subgroups that correlate with genetic mutations or treatment response (Long et al., 2024). It is akin to a pathologist noticing subtle variations across cases and hypothesizing new disease categories.

Bridging the two approaches

Supervised learning provides precision and reliability for routine diagnostic tasks, while unsupervised learning opens doors to innovation and hypothesis generation. Together, they form a continuum of computational pathology—one focused on replicating known expertise, the other on expanding the boundaries of knowledge.

Deep learning: Learning without explicit instruction

Imagine training a junior pathologist. Instead of giving them a checklist—"look for nuclear size, chromatin texture, glandular architecture"—you hand them thousands of slides and diagnoses and let them learn by exposure. Over time, they begin to recognize patterns: what invasive carcinoma looks like, how mitotic figures appear, and which stromal changes signal aggressive disease. They may even notice subtle features that seasoned experts overlook. This is the essence of deep learning (Fig. 8.5).

Deep learning represents a paradigm shift in computational pathology. Inspired by the structure of the human brain, deep learning models use artificial neural networks to learn directly from raw data. Unlike traditional machine learning, which depends on handcrafted features, deep learning automatically discovers features through experience. The model does not need explicit instructions—it learns what matters by analyzing vast amounts of annotated examples (Bulten et al., 2022; Srinidhi et al., 2021).

In pathology, this means feeding thousands of WSIs into a neural network and allowing it to uncover patterns on its own. The network might learn to:

- Detect mitotic figures without being told what chromatin condensation looks like.
- Recognize tumor architecture and stromal invasion without predefined rules.
- Identify subtle textural cues linked to molecular alterations—features even experts had not considered.

The output of this process is a feature vector—a numerical representation of learned patterns. But unlike traditional pathology descriptors, these features are not labeled or easily interpretable. They resemble a table without headers: each column captures something meaningful, but we do not always know what. This opacity is both a strength and a challenge—deep learning can discover novel biomarkers, but explaining its reasoning remains difficult.

Models vs. Architectures: What is the difference?

In the world of AI, the terms "model" and "architecture" are often used interchangeably, but they represent different concepts:

- Architecture refers to the blueprint—the design of the neural network. For example, a Convolutional Neural Network (CNN) architecture specifies layers for convolution, pooling, and activation functions. A Vision Transformer (ViT) architecture uses attention mechanisms to capture global context.
- Model refers to an architecture that has been trained on data. For instance, ResNet-50 is a CNN architecture, but when trained on millions of pathology tiles, it becomes a model capable of feature extraction for histology tasks.

Understanding this distinction matters because architecture defines capability, while models embody learned knowledge. A single architecture can produce multiple models depending on the dataset and training objectives.

Flavors of deep learning in pathology

Deep learning is not monolithic; it comes in several flavors, each suited to different challenges in pathology (Fig. 8.6):

Convolutional neural networks (CNNs)

The workhorse for image analysis, CNNs excel at detecting local patterns such as nuclei, glandular structures, and mitotic figures. They are widely used for segmentation, classification, and patch level analysis (Khosravi et al., 2018; Dörrich et al., 2023).

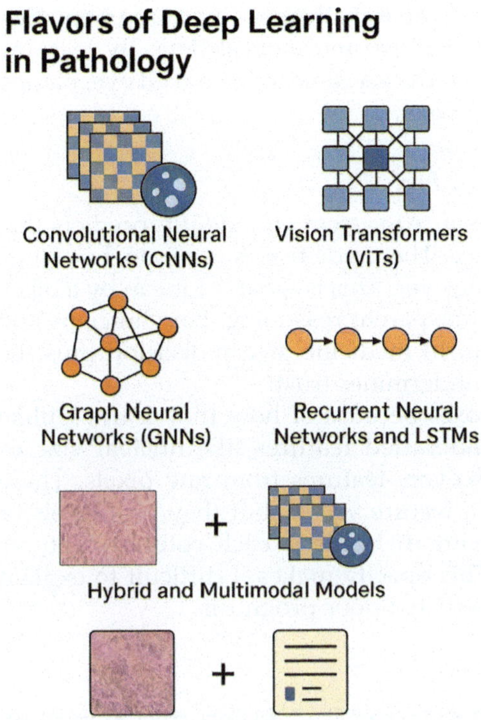

FIG. 8.6 Different deep learning architectures applied in pathology. Common approaches include Convolutional Neural Networks (CNNs), Vision Transformers (ViTs), Graph Neural Networks (GNNs), and Recurrent Neural Networks (RNNs) with LSTMs. Hybrid and multimodal models combine image data with other modalities, such as clinical text or structured data, for comprehensive analysis.

Vision transformers (ViTs)

Emerging as a powerful alternative to CNNs, ViTs leverage attention mechanisms to capture global context across entire slides. This is critical for tasks like tumor grading, where spatial relationships matter (Park et al., 2025; Wessels et al., 2023).

Graph neural networks (GNNs)

Ideal for modeling tissue architecture at the cellular level, GNNs represent cells as nodes and their spatial relationships as edges, enabling analysis of microenvironments and tumor stroma interactions (Nair et al., 2022; Abbas et al., 2023).

Recurrent neural networks (RNNs) and LSTMs

Useful for sequential data, such as modeling temporal changes in serial biopsies or integrating text from pathology reports with image features (Kim et al., 2023; Pham, 2021).

Hybrid and multimodal models

Combining CNNs or ViTs with natural language processing (NLP) models to integrate histology images with pathology reports, genomic data, and clinical notes—paving the way for comprehensive diagnostic systems.

These architectures enable deep learning to tackle diverse challenges—from segmentation and classification to multimodal integration with genomics and clinical data. By learning without explicit instruction, these models uncover patterns invisible to the naked eye, paving the way for predictive diagnostics and precision oncology.

Explainability in deep learning: Opening the black box

Imagine sitting in a tumor board meeting where an AI system confidently predicts that a breast cancer case has an elevated risk of recurrence. The prediction is accurate, but when you ask, "Why?" the system cannot provide a clear answer. This is what we mean by a black box in AI—a model that produces results without transparent reasoning. For clinicians and regulators, this lack of interpretability is unsettling. In medicine, every decision must be justified, documented, and defensible. A black box undermines trust.

Deep learning models often function as black boxes because of how they learn. Unlike traditional machine learning, which relies on handcrafted features like nuclear size or chromatin texture, deep learning automatically discovers features from raw pixels. These learned features are stored as numerical values in a feature vector, but they lack labels or biological meaning. It is like looking at a table without headers: each column captures something important, but we do not know what. This opacity makes it difficult to explain why the model classified a slide as high grade or predicted poor prognosis.

Why explainability matters in pathology

Pathology is a high stakes domain. Misclassification can lead to incorrect treatment, delayed care, or unnecessary procedures. Regulatory bodies require transparency for clinical approval, and pathologists need interpretability to trust AI recommendations. Without explainability, even the most accurate model may fail to gain adoption in real world workflows.

Techniques to open the black box

Researchers have developed tools to shed light on deep learning decisions:

- Saliency Maps: Highlight regions of the image that most influenced the prediction (Simonyan et al., 2014).
- GradCAM (Gradient-weighted Class Activation Mapping): Visualizes heatmaps over tissue areas, showing where the model "looked" when making a decision (Selvaraju et al., 2020; Zhang et al., 2025).
- Attention Mechanisms: Used in transformer based models to indicate which image patches receive the most focus (Su et al., 2024f; Ilse et al., 2018; Mao et al., 2025; Vaswani et al., 2017).

These techniques do not fully solve the interpretability challenge, but they provide a window into the model's reasoning. For example, if a Grad-CAM heatmap highlights tumor nuclei when predicting high grade, the pathologist can validate that the model's focus aligns with clinical logic.

The impact of explainability

Explainability is not just a technical requirement; it is a clinical necessity. It builds trust, facilitates regulatory approval, and enables collaborative decision making between AI systems and human experts. In pathology, where every diagnosis carries life altering implications, opening the black box is essential for safe and ethical AI integration.

Foundation models in pathology: The era of scalable intelligence

The emergence of foundation models marks a turning point in computational pathology. Traditionally, AI systems were built for narrow tasks—detecting mitotic figures, grading tumors, or quantifying biomarkers—each requiring its own dataset and training pipeline. Foundation models change this paradigm by learning from millions of image tiles drawn from thousands of WSIs across diverse tissue types and staining protocols. Models such as CTransPath, PLIP, UNI, Virchow, and CHIEF are trained at an unprecedented scale, enabling them to capture generalizable representations of histology (Su et al., 2025b; Chen et al., 2024; Vorontsov et al., 2023; Alfasly et al., 2024; Wang et al., 2022, 2024; Xu et al., 2024). These representations serve as universal building blocks that can be fine tuned for specialized tasks with minimal labeled data, dramatically reducing the annotation burden that has long constrained AI development in pathology (Fig. 8.7).

Beyond scalability, foundation models introduce adaptability. Through transfer learning, knowledge gained from one domain—such as breast cancer morphology—can accelerate learning in another, like prostate cancer or rare sarcomas. This flexibility is essential for deploying AI across the heterogeneous landscape of pathology. Furthermore, multimodal foundation models like PLIP extend this capability by integrating image and text, linking WSIs with pathology reports and clinical narratives. This integration moves us closer to holistic diagnostic systems that combine visual and contextual information, paving the way for intelligent platforms that support precision medicine.

FIG. 8.7 Overview of recent foundation models in computational pathology.

Commonalities between ML and DL: Different roads, same destination

Although machine learning (ML) and deep learning (DL) differ in architecture and methodology, their ultimate goals in pathology converge by transforming unstructured visual data into structured insights that inform clinical decisions. Both approaches tackle core tasks such as segmentation, classification, and biomarker discovery. Segmentation is foundational—whether achieved through handcrafted algorithms or learned representations—because delineating tumor boundaries, identifying cell types, and quantifying tissue compartments are prerequisites for meaningful analysis. Classification follows naturally, enabling models to assign diagnostic categories, predict tumor grades, or identify molecular subtypes. Biomarker discovery represents the frontier, where both ML and DL seek patterns correlated with prognosis or therapeutic response, though DL often uncovers subtle features invisible to human experts.

Increasingly, hybrid strategies combine the interpretability of ML with the representational power of DL. For example, a system might employ deep learning for feature extraction while relying on traditional ML classifiers for decision making. This constructive collaboration offers a balance between accuracy and transparency, addressing the clinical need for high performing models that remain explainable. In essence, ML and DL represent different roads leading to the same destination: computational tools that augment pathologists' expertise and enable data driven diagnostics.

Limitations and challenges: The roadblocks ahead

Despite remarkable progress, AI in pathology faces significant hurdles that must be overcome for widespread clinical adoption. Hardware requirements remain a major barrier; training deep learning models on gigapixel WSIs demands high performance GPUs and large scale computing infrastructure, which may be inaccessible to smaller institutions. Data scarcity and annotation burden compound the challenge. While pathology archives contain millions of slides, obtaining high quality annotations is labor intensive and costly, limiting the development of robust supervised models. Variability in staining and scanning introduces another layer of complexity. Differences in color profiles, section thickness, and image resolution across laboratories necessitate sophisticated color normalization and domain adaptation techniques to ensure model generalizability (Alfasly et al., 2024; Tizhoosh and Pantanowitz, 2018).

Regulatory and workflow integration present additional obstacles. AI systems must undergo rigorous validation to meet safety and efficacy standards, and their deployment requires seamless interoperability with laboratory information systems. Finally, interpretability remains a critical concern. Deep learning models often function as "black boxes," making it difficult to explain predictions—a barrier to clinical trust and regulatory approval. Addressing these challenges will require collaborative efforts among pathologists, data scientists, engineers, and policymakers to create solutions that are technically sound, clinically relevant, and ethically responsible (Bera et al., 2019; van der Laak et al., 2021).

Future directions: Toward intelligent, integrated pathology

The future of AI in pathology lies in building systems that are not only accurate but also integrated, interpretable, and clinically impactful. One promising direction is the development of multimodal models that combine histology with genomics, proteomics, radiology, and clinical data. By synthesizing these diverse sources of information, AI can provide a comprehensive view of disease biology, enabling precision diagnostics and personalized treatment strategies. Another critical frontier is interpretability. Future models must go beyond prediction to provide transparent reasoning, using techniques such as attention maps, concept based explanations, and interactive visualization tools. These capabilities will foster trust among pathologists and facilitate regulatory approval.

Foundation models will continue to evolve, forming the backbone of scalable AI solutions. Expanding these models to include rare diseases, special stains, and global datasets will democratize access to high quality AI tools, ensuring equitable benefits across healthcare systems. Clinical deployment will require robust, user friendly platforms that integrate seamlessly into existing workflows, reducing cognitive load rather than adding complexity. Ultimately, the success of AI in pathology will depend on collaborative innovation. Pathologists and data scientists must work together to define clinically relevant problems, curate datasets, and validate solutions. In this future, AI will not replace pathologists—it will empower them, augmenting human expertise with computational precision and unlocking insights that drive the next generation of precision medicine.

References

Abbas, S.F., Vuong, T.T.L., Kim, K., Song, B., Kwak, J.T., 2023. Multi-cell type and multi-level graph aggregation network for cancer grading in pathology images. Med. Image Anal. 1361-8415. 90, 102936. doi:10.1016/j.media.2023.102936.

Afzaal, U., Su, Z., Sajjad, U., Stack, T., Lu, H., Niu, S., Akbar, A.R., Gurcan, M.N., Niazi, M.K.K., 2025. HistoChat: instruction-tuning multimodal vision language assistant for colorectal histopathology on limited data. Patterns. 2666-3899. 6 (8), 101284. doi:10.1016/j.patter.2025.101284.

Akbar, A.R., Sajjad, U., Su, Z., Li, W., Xing, F., Ruiz, J., Chen, W., Niazi, M.K.K., 2025. CellEcoNet: decoding the cellular language of pathology with deep learning for invasive lung adenocarcinoma recurrence prediction. arXiv 23318422. Available from: https://arxiv.org. doi:10.48550/arXiv.2508.16742.

Alfasly, S., Alabtah, G., Hemati, S., Kalari, K.R., Tizhoosh, H.R., 2024. Zero-shot whole slide image retrieval in histopathology using embeddings of foundation models. arXiv 23318422. Available from: https://arxiv.org. doi:10.48550/arXiv.2409.04631.

Bera, K., Schalper, K.A., Rimm, D.L., Velcheti, V., Madabhushi, A., 2019. Artificial intelligence in digital pathology—new tools for diagnosis and precision oncology. Nat. Rev. Clin. Oncol. 1759-4782. 16 (11), 703–715. Available from: http://www.nature.com/nrclinonc/archive/index.html. doi:10.1038/s41571-019-0252-y.

Bertram, C.A., Aubreville, M., Donovan, T.A., Bartel, A., Wilm, F., Marzahl, C., Assenmacher, C.A., Becker, K., Bennett, M., Corner, S., Cossic, B., Denk, D., Dettwiler, M., Gonzalez, B.G., Gurtner, C., Haverkamp, A.K., Heier, A., Lehmbecker, A., Merz, S., Noland, E.L., Plog, S., Schmidt, A., Sebastian, F., Sledge, D.G., Smedley, R. C., Tecilla, M., Thaiwong, T., Fuchs-Baumgartinger, A., Meuten, D.J., Breininger, K., Kiupel, M., Maier, A., Klopfleisch, R., 2022. Computer-assisted mitotic count using a deep learning–based algorithm improves interobserver reproducibility and accuracy. Vet. Pathol. 1544-2217. 59 (2), 211–226. Available from: https://journals.sagepub.com/home/VET. doi:10.1177/03009858211067478.

Binol, H., Moberly, A.C., Niazi, M.K.K., Essig, G., Shah, J., Elmaraghy, C., Teknos, T., Taj-Schaal, N., Yu, L., Gurcan, M.N., 2020. SelectStitch: automated frame segmentation and stitching to create composite images from otoscope video clips. Appl. Sci. 2076-3417. 10 (17), 5894. doi:10.3390/app10175894.

Borczuk, A.C., 2022. Updates in grading and invasion assessment in lung adenocarcinoma. Mod. Pathol. 0893-3952. 35 (Suppl 1), 28–35. doi:10.1038/s41379-021-00934-3.

Boschman, J., Farahani, H., Darbandsari, A., Ahmadvand, P., Van Spankeren, A., Farnell, D., Levine, A.B., Naso, J.R., Churg, A., Jones, S.J.M., Yip, S., Köbel, M., Huntsman, D.G., Gilks, C.B., Bashashati, A., 2022. The utility of color normalization for AI -based diagnosis of hematoxylin and eosin-stained pathology images. J. Pathol. 0022-3417. 256 (1), 15–24. doi:10.1002/path.5797.

Bulten, W., Kartasalo, K., Chen, P.H.C., Ström, P., Pinckaers, H., Nagpal, K., Cai, Y., Steiner, D.F., van Boven, H., Vink, R., Hulsbergen-van de Kaa, C., van der Laak, J., Amin, M.B., Evans, A.J., van der Kwast, T., Allan, R., Humphrey, P.A., Grönberg, H., Samaratunga, H., Delahunt, B., Tsuzuki, T., Häkkinen, T., Egevad, L., Demkin, M., Dane, S., Tan, F., Valkonen, M., Corrado, G.S., Peng, L., Mermel, C.H., Ruusuvuori, P., Litjens, G., Eklund, M., Brilhante, A., Çakır, A., Farré, X., Geronatsiou, K., Molinié, V., Pereira, G., Roy, P., Saile, G., Salles, P. G.O., Schaafsma, E., Tschui, J., Billoch-Lima, J., Pereira, E.M., Zhou, M., He, S., Song, S., Sun, Q., Yoshihara, H., Yamaguchi, T., Ono, K., Shen, T., Ji, J., Roussel, A., Zhou, K., Chai, T., Weng, N., Grechka, D., Shugaev, M.V., Kiminya, R., Kovalev, V., Voynov, D., Malyshev, V., Lapo, E., Campos, M., Ota, N., Yamaoka, S., Fujimoto, Y., Yoshioka, K., Juvonen, J., Tukiainen, M., Karlsson, A., Guo, R., Hsieh, C.L., Zubarev, I., Bukhar, H.S.T., Li, W., Li, J., Speier, W., Arnold, C., Kim, K., Bae, B., Kim, Y.W., Lee, H.S., Park, J., 2022. Artificial intelligence for diagnosis and Gleason grading of prostate cancer: the PANDA challenge. Nat. Med. 1546-170X. 28 (1), 154–163. Available from: http://www.nature.com/nm/index.html. doi:10.1038/s41591-021-01620-2.

Camalan, S., Niazi, M.K.K., Elmaraghy, C., Moberly, A.C., Gurcan, M.N., 2024. Tympanic membrane segmentation of video frames to create composite images using SAM. In: Paper presented at: Medical Imaging 2024: Computer-Aided Diagnosis.

Campanella, G., Hanna, M.G., Geneslaw, L., Miraflor, A., Werneck Krauss Silva, V., Busam, K.J., Brogi, E., Reuter, V. E., Klimstra, D.S., Fuchs, T.J., 2019. Clinical-grade computational pathology using weakly supervised deep learning on whole slide images. Nat. Med. 1546-170X. 25 (8), 1301–1309. Available from: http://www.nature.com/nm/index.html. doi:10.1038/s41591-019-0508-1.

Chen, M., Ma, H., Sun, X., Schwartz, M., Brand, R.E., Xu, J., Gotsis, D.S., Nguyen, P., Moore, B.A., Snyder, L., Brand, R.M., Liu, Y., 2025. Multimodal whole slide image processing pipeline for quantitative mapping of tissue architecture and tissue microenvironment. Npj Imaging. 2948-197X. 3 (1). doi:10.1038/s44303-025-00088-w.

Chen, R.J., Ding, T., Lu, M.Y., Williamson, D.F.K., Jaume, G., Song, A.H., Chen, B., Zhang, A., Shao, D., Shaban, M., Williams, M., Oldenburg, L., Weishaupt, L.L., Wang, J.J., Vaidya, A., Le, L.P., Gerber, G., Sahai, S., Williams, W., Mahmood, F., 2024. Towards a general-purpose foundation model for computational pathology. Nat. Med. 1546-170X. 30 (3), 850–862. Available from: https://www.nature.com/nm/. doi:10.1038/s41591-024-02857-3.

Coudray, N., Ocampo, P.S., Sakellaropoulos, T., Narula, N., Snuderl, M., Fenyö, D., Moreira, A.L., Razavian, N., Tsirigos, A., 2018. Classification and mutation prediction from non–small cell lung cancer histopathology images using deep learning. Nat. Med. 1546-170X. 24 (10), 1559–1567. Available from: http://www.nature.com/nm/index.html. doi:10.1038/s41591-018-0177-5.

Dörrich, M., Hecht, M., Fietkau, R., Hartmann, A., Iro, H., Gostian, A.O., Eckstein, M., Kist, A.M., 2023. Explainable convolutional neural networks for assessing head and neck cancer histopathology. Diagn. Pathol. 1746-1596. 18 (1). Available from: https://diagnosticpathology.biomedcentral.com. doi:10.1186/s13000-023-01407-8.

Echle, A., Rindtorff, N.T., Brinker, T.J., Luedde, T., Pearson, A.T., Kather, J.N., 2021. Deep learning in cancer pathology: a new generation of clinical biomarkers. Br. J. Cancer 1532-1827. 124 (4), 686–696. Available from: http://www.nature.com/bjc/index.html. doi:10.1038/s41416-020-01122-x.

Ganz, J., Marzahl, C., Ammeling, J., et al., 2024. On the value of PHH3 for mitotic figure detection on H&E-stained images. arXiv 240619899.

Goldstraw, P., Chansky, K., Crowley, J., Rami-Porta, R., Asamura, H., Eberhardt, W.E.E., Nicholson, A.G., Groome, P., Mitchell, A., Bolejack, V., Goldstraw, P., Rami-Porta, R., Asamura, H., Ball, D., Beer, D.G., Beyruti, R., Bolejack, V., Chansky, K., Crowley, J., Detterbeck, F., Eberhardt, W.E.E., Edwards, J., Galateau-Sall é, F., Giroux, D., Gleeson, F., Groome, P., Huang, J., Kennedy, C., Kim, J., Kim, Y.T., Kingsbury, L., Kondo, H., Krasnik, M., Kubota, K., Lerut, A., Lyons, G., Marino, M., Marom, E.M., van Meerbeeck, J., Mitchell, A., Nakano, T., Nicholson, A.G., Nowak, A., Peake, M., Rice, T., Rosenzweig, K., Ruffini, E., Rusch, V., Saijo, N.,

Van Schil, P., Sculier, J.-P., Shemanski, L., Stratton, K., Suzuki, K., Tachimori, Y., Thomas, C.F., Travis, W., Tsao, M.S., Turrisi, A., Vansteenkiste, J., Watanabe, H., Wu, Y.-L., Baas, P., Erasmus, J., Hasegawa, S., Inai, K., Kernstine, K., Kindler, H., Krug, L., Nackaerts, K., Pass, H., Rice, D., Falkson, C., Filosso, P.L., Giaccone, G., Kondo, K., Lucchi, M., Okumura, M., Blackstone, E., Abad Cavaco, F., Ansótegui Barrera, E., Abal Arca, J., Lamelas, I.P., Arnau Obrer, A., Guijarro Jorge, R., Ball, D., Bascom, G.K., Blanco Orozco, A.I., González Castro, M. A., Blum, M.G., Chimondeguy, D., Cvijanovic, V., Defranchi, S., de Olaiz Navarro, B., Escobar Campuzano, I., Macía Vidueira, I., Fernández Araujo, E., Andreo García, F., Fong, K.M., Francisco Corral, G., Cerezo González, S., Freixinet Gilart, J., García Arangüena, L., García Barajas, S., Girard, P., Goksel, T., González Budiño, M.T., González Casaurrán, G., Gullón Blanco, J.A., Hernández Hernández, J., Hernández Rodríguez, H., Herrero Collantes, J., Iglesias Heras, M., Izquierdo Elena, J.M., Jakobsen, E., Kostas, S., León Atance, P., Núñez Ares, A., Liao, M., Losanovscky, M., Lyons, G., Magaroles, R., De Esteban Júlvez, L., Mariñán Gorospe, M., McCaughan, B., Kennedy, C., Íñiguez, R.M., Miravet Sorribes, L., Naranjo Gozalo, S., Álvarez de Arriba, C., Núñez Delgado, M., Padilla Alarcón, J., Peñalver Cuesta, J.C., Park, J.S., Pass, H., Pavón Fernández, M. J., Rosenberg, M., Ruffini, E., Rusch, V., Sánchez de Cos Escuín, J., Saura Vinuesa, A., Serra Mitjans, M., Strand, T. E., Subotic, D., Swisher, S., Terra, R., Thomas, C., Tournoy, K., Van Schil, P., Velasquez, M., Wu, Y.L., Yokoi, K., 2016. The IASLC lung cancer staging project: proposals for revision of the TNM stage groupings in the forthcoming (Eighth) edition of the TNM classification for lung cancer. J. Thorac. Oncol. 1556-0864. 11 (1), 39–51. doi:10.1016/j.jtho.2015.09.009.

Ilse, M., Tomczak, J., Welling, M., 2018. Attention-based deep multiple instance learning. In: Paper presented at: International Conference on Machine Learning.

Jahanifar, M., Shephard, A., Zamanitajeddin, N., Graham, S., Raza, S.E.A., Minhas, F., Rajpoot, N., 2024. Mitosis detection, fast and slow: robust and efficient detection of mitotic figures. Med. Image Anal. 1361-8415. 94, 103132. doi:10.1016/j.media.2024.103132.

Kather, J.N., Heij, L.R., Grabsch, H.I., Loeffler, C., Echle, A., Muti, H.S., Krause, J., Niehues, J.M., Sommer, K.A.J., Bankhead, P., Kooreman, L.F.S., Schulte, J.J., Cipriani, N.A., Buelow, R.D., Boor, P., Ortiz-Brüchle, N., Hanby, A.M., Speirs, V., Kochanny, S., Patnaik, A., Srisuwananukorn, A., Brenner, H., Hoffmeister, M., van den Brandt, P.A., Jäger, D., Trautwein, C., Pearson, A.T., Luedde, T., 2020. Pan-cancer image-based detection of clinically actionable genetic alterations. Nat. Cancer. 2662-1347. 1 (8), 789–799. Available from: https://www.nature.com/natcancer/. doi:10.1038/s43018-020-0087-6.

Khan, M.K., Nyström, I., 2010. A modified particle swarm optimization applied in image registration. In: Paper presented at: 2010 20th International Conference on Pattern Recognition.

Khosravi, P., Kazemi, E., Imielinski, M., Elemento, O., Hajirasouliha, I., 2018. Deep convolutional neural networks enable discrimination of heterogeneous digital pathology images. EBioMedicine. 2352-3964. 27, 317–328. doi:10.1016/j.ebiom.2017.12.026.

Kim, S., Lee, E., Liu, J., 2023. A deep attention LSTM embedded aggregation network for multiple histopathological images. PLOS ONE. 1932-6203. 18 (6), e0287301. doi:10.1371/journal.pone.0287301.

Krishnamurthy, S., Schnitt, S.J., Vincent-Salomon, A., Canas-Marques, R., Colon, E., Kantekure, K., Maklakovski, M., Finck, W., Thomassin, J., Globerson, Y., Bien, L., Mallel, G., Grinwald, M., Linhart, C., Sandbank, J., Vecsler, M., 2024. Fully automated artificial intelligence solution for human epidermal growth factor receptor 2 immunohistochemistry scoring in breast cancer: a multireader study. JCO Precis. Oncol. 2473-4284. 8. Available from: https://ascopubs.org/journal/po. doi:10.1200/PO.24.00353.

Li, Z., Mirjahanmardi, S.H., Sali, R., Eweje, F., Gopaulchan, M., Kloker, L., Zhang, X., Li, G., Jiang, Y., Li, R., 2025. Automated cell annotation and classification on histopathology for spatial biomarker discovery. Nat. Commun. 2041-1723. 16 (1). Available from: https://www.nature.com/ncomms/. doi:10.1038/s41467-025-61349-1.

Long, X., Wang, T., Kan, Y., et al., 2024. Pseudo training data generation for unsupervised cell membrane segmentation in immunohistochemistry images. In: Paper presented at: 2024 IEEE International Conference on Bioinformatics and Biomedicine (BIBM).

Lu, H., Rezapour, M., Baha, H., Niazi, M.K.K., Narayanan, A., Gurcan, M.N., 2024. Gene pointNet for tumor classification. Neural Comput. Appl. 1433-3058. 36 (33), 21107–21121. Available from: https://www.springer.com/journal/521. doi:10.1007/s00521-024-10307-x.

Lu, M.Y., Williamson, D.F.K., Chen, T.Y., Chen, R.J., Barbieri, M., Mahmood, F., 2021. Data-efficient and weakly supervised computational pathology on whole-slide images. Nat. Biomed. Eng. 2157-846X. 5 (6), 555–570. Available from: www.nature.com/natbiomedeng/. doi:10.1038/s41551-020-00682-w.

Mao, J., Xu, J., Tang, X., Liu, Y., Zhao, H., Tian, G., Yang, J., Gao, X., 2025. CAMIL: channel attention-based multiple instance learning for whole slide image classification. Bioinformatics. 1367-4811. 41 (2). doi:10.1093/bioinformatics/btaf024.

Nair, A., Arvidsson, H., Gatica V, J.E., Tudzarovski, N., Meinke, K., Sugars, R.V., 2022. A graph neural network framework for mapping histological topology in oral mucosal tissue. BMC Bioinform. 1471-2105. 23 (1). Available from: https://bmcbioinformatics.biomedcentral.com/. doi:10.1186/s12859-022-05063-5.

Niazi, M., Hedrich, J., Ingela, N., 2010. Image registration using particle swarm optimization approach. In: Paper presented at: SSBA 2010.

Niazi, M.K.K., Abas, F.S., Senaras, C., Pennell, M., Sahiner, B., Chen, W., Opfer, J., Hasserjian, R., Louissaint, A., Shana'ah, A., Lozanski, G., Gurcan, M.N., Maitland, K.C., 2018. Nuclear IHC enumeration: a digital phantom to evaluate the performance of automated algorithms in digital pathology. PLOS ONE. 1932-6203. 13 (5), e0196547. doi:10.1371/journal.pone.0196547.

Niazi, M.K.K., Hemminger, J., Kurt, H., Lozanski, G., Gurcan, M., 2014. Grading vascularity from histopathological images based on traveling salesman distance and vessel size 9041. In: Progress in Biomedical Optics and Imaging—Proceedings of SPIE. SPIE, United States. 16057422. doi:10.1117/12.2043808.

Niazi, M.K.K., Lin, Y., Liu, F., Ashok, A., Marcellin, M.W., Tozbikian, G., Gurcan, M.N., Bilgin, A., 2019a. Pathological image compression for big data image analysis: application to hotspot detection in breast cancer. Artif. Intell. Med. 1873-2860. 95, 82–87. Available from: www.elsevier.com/locate/artmed. doi:10.1016/j.artmed.2018.09.002.

Niazi, M.K.K., Parwani, A.V., Gurcan, M.N., 2019b. Digital pathology and artificial intelligence. Lancet Oncol. 1474-5488. 20 (5), e253. Available from: http://www.journals.elsevier.com/the-lancet-oncology/. doi:10.1016/S1470-2045(19)30154-8.

Niazi, M.K.K., Yao, K., Zynger, D.L., Clinton, S.K., Chen, J., Koyuturk, M., Laframboise, T., Gurcan, M., 2017. Visually meaningful histopathological features for automatic grading of prostate cancer. IEEE J. Biomed. Health Inform. 2168-2208. 21 (4), 1027–1038. Available from: http://ieeexplore.ieee.org/xpl/RecentIssue.jsp?punumber=6221020. doi:10.1109/JBHI.2016.2565515.

Niazi, M.K.K., Yazgan, E., Lee, C., Parwani, A., Gurcan, M.N., 2020b. Identifying bladder layers from H and E images using U-Net image segmentation. In: Paper presented at: Medical Imaging 2020: Digital Pathology.

Niazi, M.K.K., Yazgan, E., Tavolara, T.E., Li, W., Lee, C.T., Parwani, A., Gurcan, M.N., 2020a. Semantic segmentation to identify bladder layers from H&E images. Diagn. Pathol. 1746-1596. 15 (1). Available from: http://www.diagnosticpathology.org/. doi:10.1186/s13000-020-01002-1.

Park, S.Y., Ayana, G., Wako, B.D., Jeong, K.C., Yoon, S.D., Choe, S.W., 2025. Vision transformers for low-quality histopathological images: a case study on squamous cell carcinoma margin classification. Diagnostics. 2075-4418. 15 (3). Available from: http://www.mdpi.com/journal/diagnostics/. doi:10.3390/diagnostics15030260.

Pham, T.D., 2021. Time-frequency time-space long short-term memory networks for image classification of histopathological tissue. Sci. Rep. 2045-2322. 11 (1). Available from: www.nature.com/srep/index.html. doi:10.1038/s41598-021-93160-5.

Sajjad, U., Akbar, A.R., Su, Z., Knight, D., Frankel, W.L., Gurcan, M.N., Chen, W., Niazi, M.K.K., 2025. Morphology-aware prognostic model for five-year survival prediction in colorectal cancer from H&E whole slide images. arXiv 23318422. Available from: https://arxiv.org. doi:10.48550/arXiv.2510.14800.

Sajjad, U., Rezapour, M., Su, Z., Tozbikian, G.H., Gurcan, M.N., Niazi, M.K.K., 2023. NRK-ABMIL: subtle metastatic deposits detection for predicting lymph node metastasis in breast cancer whole-slide images. Cancers. 2072-6694. 15 (13). Available from: http://www.mdpi.com/journal/cancers/. doi:10.3390/cancers15133428.

Salvi, M., Branciforti, F., Molinari, F., Meiburger, K.M., 2024. Generative models for color normalization in digital pathology and dermatology: advancing the learning paradigm. Expert Syst. Appl. 0957-4174. 245, 123105. doi:10.1016/j.eswa.2023.123105.

Selvaraju, R.R., Cogswell, M., Das, A., Vedantam, R., Parikh, D., Batra, D., 2020. Grad-CAM: visual explanations from deep networks via gradient-based localization. Int. J. Comput. Vis. 1573-1405. 128 (2), 336–359. Available from: http://www.kluweronline.com/issn/0920-5691/. doi:10.1007/s11263-019-01228-7.

Senaras, C., Niazi, M.K.K., Lozanski, G., Gurcan, M.N., Lo, C.-M., 2018. DeepFocus: detection of out-of-focus regions in whole slide digital images using deep learning. PLOS ONE. 1932-6203. 13 (10), e0205387. doi:10.1371/journal.pone.0205387.

She, Y., Jin, Z., Wu, J., Deng, J., Zhang, L., Su, H., Jiang, G., Liu, H., Xie, D., Cao, N., Ren, Y., Chen, C., 2020. Development and validation of a deep learning model for non–small cell lung cancer survival. JAMA Netw. Open. 2574-3805. 3 (6), e205842. doi:10.1001/jamanetworkopen.2020.5842.

Shen, Z., Simard, M., Brand, D., Andrei, V., Al-Khader, A., Oumlil, F., Trevers, K., Butters, T., Haefliger, S., Kara, E., Amary, F., Tirabosco, R., Cool, P., Royle, G., Hawkins, M.A., Flanagan, A.M., Collins-Fekete, C.-A., 2024. A deep learning framework deploying segment anything to detect pan-cancer mitotic figures from haematoxylin and eosin-stained slides. Commun. Biol. 2399-3642. 7 (1). doi:10.1038/s42003-024-07398-6.

Simonyan, K., Vedaldi, A., Zisserman, A. 2014 Deep inside convolutional networks: visualising image classification models and saliency maps. In: 2nd International Conference on Learning Representations, ICLR 2014—Workshop Track Proceedings International Conference on Learning Representations. ICLR, United Kingdom. Available from: https://dblp.org/db/conf/iclr/iclr2014w.html.

Srinidhi, C.L., Ciga, O., Martel, A.L., 2021. Deep neural network models for computational histopathology: a survey. Med. Image Anal. 1361-8415. 67, 101813. doi:10.1016/j.media.2020.101813.

Su, Z., Afzaal, U., Niu, S., de Toro, M.M., Xing, F., Ruiz, J., Gurcan, M.N., Li, W., Niazi, M.K.K., 2024f. Deep learning model for predicting lung adenocarcinoma recurrence from whole slide images. Cancers. 2072-6694. 16 (17), 3097. doi:10.3390/cancers16173097.

Su, Z., Akbar, A.R., Sajjad, U., Parwani, A.V., Niazi, M.K.K., 2025a. Streamline pathology foundation model by cross-magnification distillation. arXiv 23318422. Available from: https://arxiv.org. doi:10.48550/arXiv.2509.23097.

Su, Z., Chen, W., Annem, S., et al., 2024b. Adapting SAM to histopathology images for tumor bud segmentation in colorectal cancer. In: Paper presented at: Proceedings of SPIE–the International Society for Optical Engineering.

Su, Z., Chen, W., Leigh, P.J., et al., 2024c. Few-shot tumor bud segmentation using generative model in colorectal carcinoma. In: Paper presented at: Medical Imaging 2024: Digital and Computational Pathology.

Su, Z., Guo, Y., Wesolowski, R., Tozbikian, G., O'Connell, N.S., Niazi, M.K.K., Gurcan, M.N., 2025b. Computational pathology for accurate prediction of breast cancer recurrence: development and validation of a deep learning–based tool. Mod. Pathol. 08933952. 38 (12), 100847. doi:10.1016/j.modpat.2025.100847.

Su, Z., Niazi, M.K.K., Tavolara, T.E., Niu, S., Tozbikian, G.H., Wesolowski, R., Gurcan, M.N., Xu, C., 2023b. BCR-Net: a deep learning framework to predict breast cancer recurrence from histopathology images. PLOS ONE. 1932-6203. 18 (4), e0283562. doi:10.1371/journal.pone.0283562.

Su, Z., Rezapour, M., Sajjad, U., Gurcan, M.N., Niazi, M.K.K., 2023a. Attention2Minority: a salient instance inference-based multiple instance learning for classifying small lesions in whole slide images. Comput. Biol. Med. 0010-4825. 167, 107607. doi:10.1016/j.compbiomed.2023.107607.

Su, Z., Rezapour, M., Sajjad, U., Gurcan, M.N., Niazi, M.K.K., 2023c. Attention2Minority: a salient instance inference-based multiple instance learning for classifying small lesions in whole slide images. arXiv 23318422. Available from: https://arxiv.org. doi:10.48550/arXiv.2301.07700.

Su, Z., Rezapour, M., Sajjad, U., Niu, S., Gurcan, M.N., Niazi, M.K.K., 2024a. Cross-attention-based saliency inference for predicting cancer metastasis on whole slide images. IEEE J. Biomed. Health Inform. 21682208. 28 (12), 7206–7216. Available from: http://ieeexplore.ieee.org/xpl/RecentIssue.jsp?punumber=6221020. doi:10.1109/JBHI.2024.3439499.

Su, Z., Rosen, A., Wesolowski, R., Tozbikian, G., Niazi, M.K.K., Gurcan, M.N., 2024d. Deep-ODX: an efficient deep learning tool to risk stratify breast cancer patients from histopathology images. In: Paper presented at: Medical Imaging 2024: Digital and Computational Pathology.

Su, Z., Tavolara, T.E., Carreno-Galeano, G., Lee, S.J., Gurcan, M.N., Niazi, M.K.K., 2022. Attention2majority: weak multiple instance learning for regenerative kidney grading on whole slide images. Med. Image Anal. 1361-8423. 79. Available from: http://www.elsevier.com/inca/publications/store/6/2/0/9/8/3/index.htt. doi:10.1016/j.media.2022.102462.

Su, Z., Wei Chen, M., Sajjad, U., 2024e. Adapting segment anything model for tumor bud segmentation on hematoxylin and eosin images of colorectal cancer. Arch. Pathol. Lab. Med. 148 (9), E183.

Tavolara, T.E., Carreno-Galeano, G., Gurcan, M.N., Lee, S.J., Niazi, M.K.K., 2021a. Grading and localization of histological features for bioengineered kidney constructs. In: Progress in Biomedical Optics and Imaging—Proceedings of SPIE, vol. 11603. SPIE, United States. 9781510640351. Available from: http://spie.org/x1848.xml. doi:10.1117/12.2581949.

Tavolara, T.E., Carreno-Galeano, G., Gurcan, M.N., Lee, S.J., Niazi, M.K.K., 2021b. Grading and localization of histological features for bioengineered kidney constructs. In: Paper presented at: Medical Imaging 2021: Digital Pathology.

Tavolara, T.E., Jorgensen, A.M., Gurcan, M.N., Murphy, S.V., Niazi, M.K.K., 2021c. Panoptic segmentation of wounds in a pig model. In: Paper presented at: Medical Imaging 2021: Computer-Aided Diagnosis.

Tavolara, T.E., Niazi, M.K.K., Beamer, G., Gurcan, M.N., 2020b. Segmentation of Mycobacterium tuberculosis bacilli clusters from acid-fast stained lung biopsies: a deep learning approach. In: Progress in Biomedical Optics and Imaging—Proceedings of SPIE, vol. 11320. SPIE, United States. 9781510634077. Available from: http://spie.org/x1848.xml. doi:10.1117/12.2549016.

Tavolara, T.E., Niazi, M.K.K., Ginese, M., Piedra-Mora, C., Gatti, D.M., Beamer, G., Gurcan, M.N., 2020a. Automatic discovery of clinically interpretable imaging biomarkers for Mycobacterium tuberculosis supersusceptibility using deep learning. EBioMedicine. 2352-3964. 62. Available from: http://www.journals.elsevier.com/ebiomedicine/. doi:10.1016/j.ebiom.2020.103094.

Tavolara, T.E., Su, Z., Gurcan, M.N., Niazi, M.K.K., 2023. One label is all you need: interpretable AI-enhanced histopathology for oncology. In: Paper presented at: Seminars in Cancer Biology.

Tellez, D., Litjens, G., Bándi, P., Bulten, W., Bokhorst, J.-M., Ciompi, F., van der Laak, J., 2019. Quantifying the effects of data augmentation and stain color normalization in convolutional neural networks for computational pathology. Med. Image Anal. 1361-8415. 58, 101544. doi:10.1016/j.media.2019.101544.

Tizhoosh, H.R., Pantanowitz, L., 2018. Artificial intelligence and digital pathology: challenges and opportunities. J. Pathol. Inform. 2153-3539. 9 (1), 38. doi:10.4103/jpi.jpi_53_18.

Vahadane, A., Peng, T., Sethi, A., Albarqouni, S., Wang, L., Baust, M., Steiger, K., Schlitter, A.M., Esposito, I., Navab, N., 2016. Structure-preserving color normalization and sparse stain separation for histological images. IEEE Trans. Med. Imaging. 0278-0062. 35 (8), 1962–1971. doi:10.1109/tmi.2016.2529665.

van der Laak, J., Litjens, G., Ciompi, F., 2021. Deep learning in histopathology: the path to the clinic. Nat. Med. 1078-8956. 27 (5), 775–784. doi:10.1038/s41591-021-01343-4.

Vaswani, A., Shazeer, N., Parmar, N., Uszkoreit, J., Jones, L., Gomez, A.N., Kaiser, Ł., Polosukhin, I., 2017. Attention is all you need. Adv. Neural Inf. Process. Syst. 30, 1049-5258. 5999–6009.

Vesterinen, T., Säilä, J., Blom, S., Pennanen, M., Leijon, H., Arola, J., 2022. Automated assessment of Ki-67 proliferation index in neuroendocrine tumors by deep learning. APMIS. 0903-4641. 130 (1), 11–20. doi:10.1111/apm.13190.

Vorontsov, E., Bozkurt, A., Casson, A., Shaikovski, G., Zelechowski, M., Liu, S., Severson, K., Zimmermann, E., Hall, J., Tenenholtz, N., Fusi, N., Mathieu, P., van Eck, A., Lee, D., Viret, J., Robert, E., Wang, Y.K., Kunz, J.D., Lee, M.C.H., Bernhard, J.H., Godrich, R.A., Oakley, G., Millar, E., Hanna, M., Retamero, J., Moye, W.A., Yousfi, R., Kanan, C., Klimstra, D., Rothrock, B., Fuchs, T.J., 2023. Virchow: a million-slide digital pathology foundation model. arXiv 23318422. Available from: https://arxiv.org. doi:10.48550/arXiv.2309.07778.

Wang, D., Khosla, A., Gargeya, R., Irshad, H., Beck, A.H., 2016. Deep learning for identifying metastatic breast cancer. arXiv 160605718.

Wang, X., Yang, S., Zhang, J., Wang, M., Zhang, J., Yang, W., Huang, J., Han, X., 2022. Transformer-based unsupervised contrastive learning for histopathological image classification. Med. Image Anal. 1361-8415. 81, 102559. doi:10.1016/j.media.2022.102559.

Wang, X., Zhao, J., Marostica, E., Yuan, W., Jin, J., Zhang, J., Li, R., Tang, H., Wang, K., Li, Y., Wang, F., Peng, Y., Zhu, J., Zhang, J., Jackson, C.R., Zhang, J., Dillon, D., Lin, N.U., Sholl, L., Denize, T., Meredith, D., Ligon, K.L., Signoretti, S., Ogino, S., Golden, J.A., Nasrallah, M.P., Han, X., Yang, S., Kun-Hsing, Y., 2024. A pathology foundation model for cancer diagnosis and prognosis prediction. Nature. 0028-0836. 634 (8035), 970–978. doi:10.1038/s41586-024-07894-z.

Wang, Y., Smith, M.R., Dixon, C.B., D'Agostino, R., Liu, Y., Ruiz, J., Chan, M.D., Su, J., Mileham, K.F., Lycan, T., Green, M.E., Hassan, O.A., Jiang, Y., Khan Niazi, M.K., Li, W., Xing, F., 2025. IASLC grading system predicts distant metastases for resected lung adenocarcinoma. J. Clin. Pathol. 1472-4146. 78 (6), 409–415. Available from: http://jcp.bmj.com/. doi:10.1136/jcp-2024-209649.

Wessels, F., Schmitt, M., Krieghoff-Henning, E., Nientiedt, M., Waldbillig, F., Neuberger, M., Kriegmair, M.C., Kowalewski, K.-F., Worst, T.S., Steeg, M., Popovic, Z.V., Gaiser, T., von Kalle, C., Utikal, J.S., Fröhling, S., Michel, M.S., Nuhn, P., Brinker, T.J., 2023. A self-supervised vision transformer to predict survival from histopathology in renal cell carcinoma. World J. Urol. 1433-8726. 41 (8), 2233–2241. doi:10.1007/s00345-023-04489-7.

Wu, J., Liu, C., Liu, X., Sun, W., Li, L., Gao, N., Zhang, Y., Yang, X., Zhang, J., Wang, H., Liu, X., Huang, X., Zhang, Y., Cheng, R., Chi, K., Mao, L., Zhou, L., Lin, D., Ling, S., 2022. Artificial intelligence-assisted system for precision diagnosis of PD-L1 expression in non-small cell lung cancer. Mod. Pathol. 0893-3952. 35 (3), 403–411. doi:10.1038/s41379-021-00904-9.

Wu, S., Yue, M., Zhang, J., Li, X., Li, Z., Zhang, H., Wang, X., Han, X., Cai, L., Shang, J., Jia, Z., Wang, X., Li, J., Liu, Y., 2023. The role of artificial intelligence in accurate interpretation of HER2 immunohistochemical scores 0 and 1+ in breast cancer. Mod. Pathol. 0893-3952. 36 (3), 100054. doi:10.1016/j.modpat.2022.100054.

Xu, H., Usuyama, N., Bagga, J., Zhang, S., Rao, R., Naumann, T., Wong, C., Gero, Z., González, J., Gu, Y., Xu, Y., Wei, M., Wang, W., Ma, S., Wei, F., Yang, J., Li, C., Gao, J., Rosemon, J., Bower, T., Lee, S., Weerasinghe, R., Wright, B.J., Robicsek, A., Piening, B., Bifulco, C., Wang, S., Poon, H., 2024. A whole-slide foundation model for digital pathology from real-world data. Nature. 0028-0836. 630 (8015), 181–188. doi:10.1038/s41586-024-07441-w.

Zehra, T., Jaffar, N., Shams, M., Chundriger, Q., Ahmed, A., Anum, F., Alsubaie, N., Ahmad, Z., 2023. Use of a novel deep learning open-source model for quantification of Ki-67 in breast cancer patients in Pakistan: a comparative study between the manual and automated methods. Diagnostics. 2075-4418. 13 (19), 3105. doi:10.3390/diagnostics13193105.

Zhang, Z., Gu, J., Chowdhury, A., Mai, Z., Carlyn, D., Berger-Wolf, T., Su, Y., Chao, W.L., 2025. Finer-CAM: spotting the difference reveals finer details for visual explanation. In: Proceedings of the IEEE Computer Society Conference on Computer Vision and Pattern Recognition. IEEE Computer Society, United States, pp. 9611–9620. 10636919. doi:10.1109/CVPR52734.2025.00898.

CHAPTER

9

Applications of artificial intelligence in surgical pathology

Zaibo Li[a], *Yueping Liu*[b], *William MacDonald*[a], *Shaoli Sun*[a], *Wei Chen*[a], *Ankush U. Patel*[a], *Anil V. Parwani*[a], *and Swati Satturwar*[a]

[a]Department of Pathology, The Ohio State University, Columbus, OH, United States
[b]Department of Pathology, Hebei Medical University Fourth Hospital, Shijiazhuang, China

Applications of artificial intelligence in breast pathology

Conventional breast pathology relies on the expertise of pathologists to interpret microscopic images derived from tissue biopsies and surgical resections. However, rising case volumes, inter-observer variability, and the growing complexity of precision pathology have underscored the need for innovative diagnostic approaches. The emergence of digital pathology, driven by whole slide imaging (WSI), has created new opportunities for integrating artificial intelligence (AI) into this field. To facilitate this transition, the College of American Pathologists (CAP) has published validation guidelines to support laboratories in the clinical implementation of digital pathology technologies (Evans et al., 2022; Pantanowitz et al., 2013). Validation studies focused on breast pathology have confirmed that digital evaluation of tissue is diagnostically equivalent to traditional microscopy, ensuring safe and effective patient care (Williams et al., 2018; Williams et al., 2020; Mastrosimini et al., 2022).

AI is being actively developed and applied to a broad range of diagnostic tasks in breast pathology, including lesion classification, tumor grading, and lymph node metastasis detection. Some commercially available AI tools for breast pathology are summarized in Table 9.1.

This section reviews key advances in the application of AI across different domains of breast pathology, its potential to improve diagnostic accuracy and efficiency, and the challenges that remain for clinical adoption.

Advances in Digital Pathology and Artificial Intelligence in Pathology
https://doi.org/10.1016/B978-0-443-40383-5.00008-X

TABLE 9.1 Summary of some commercially available AI tools for breast pathology.

Company	Country	Product	Applications	Regulatory approval
Aiforia	Finland	Aiforia Clinical Suites: AI models	Biomarker quantification, grading	CE-IVD marked
AIRA Matrix	India	AIRA clinical products	Biomarker quantification, grading	
Artera	Unites States	ArteraAI breast cancer test	Prediction of recurrent risk	FDA approval
IBEX	Israel	Galen Breast	Diagnosis/classification	FDA Breakthrough device
Indica Labs	Unites States	Breast IHC AI, Slide QC	Diagnosis/classification, biomarker quantification and QC	CE-IVD certified
Mindpeak	Germany	Breast Ki-67, HER2, ER/PR, metastasis	Biomarker quantification, detection	CE-IVD marked
nucleai	Israel	Nucleai Atom	Biomarker quantification, spatial analysis	
OWKIN	France	RlapsRisk	Prediction of recurrent risk	CE-IVD marked
Paige AI	Unites States	Breast Suite (Detect and Neoplasm, Mitosis, Lymph Node, HER2Complete)	Diagnosis/classification, biomarker quantification	FDA Breakthrough device
Qritive	Singapore	Lymph Node Dx, Ki67 Quant, Breast HER2 Quant	Diagnosis/classification, quantification, detection, prediction	
Stratipath	Sweden	Stratipath Breast	Prediction of recurrent risk	CE-IVD marked
Visiopharm	Denmark	Biomarker quantification, lymph node detection	Quantification, detection	CE-IVD certified

Classification and diagnosis of breast lesions

Precise classification of breast lesions is critical for appropriate treatment selection. AI algorithms are increasingly being designed to support pathologists in the primary diagnosis of breast lesions, including differentiating benign from malignant entities. Abdolahi et al. showed that AI models could identify invasive ductal carcinoma with 81%–85% accuracy after training on up to 277,524 images (Abdolahi et al., 2020). Han et al. proposed a deep learning (DL) model for multiclass classification of breast cancer subtypes, including ductal, lobular, mucinous, and papillary carcinoma, achieving an average accuracy of 93.2% on a large dataset (Han et al., 2017). Mercan et al. reported 70% accuracy in distinguishing benign tissue from ductal carcinoma in situ (DCIS) and demonstrated performance comparable to expert pathologists in differentiating atypia from DCIS (Mercan et al., 2019). Cruz-Roa et al. (2017) developed a convolutional neural network (CNN) to classify image patches from breast cancer WSIs containing invasive ductal carcinoma and then applied a ConvNet classifier to estimate tumor extent and infiltration across whole slides. Trained on 400 manually annotated slides from multiple institutions and validated on 200 TCGA slides, their model achieved a pixel-level F1 score of 75.86%. It maintained high accuracy in detecting invasive regions on independent validation, although mixed invasive and in situ lesions were challenging. Improved representation of in situ carcinomas and more complex algorithms enhanced performance.

Screening tools such as the Galen Breast (IBEX system) have been designed to streamline diagnostic workflows by recognizing benign and malignant lesions. This algorithm employs an ensemble of CNNs trained on over two million labeled image patches from 2153 hematoxylin and eosin (H&E)-stained slides (Sandbank et al., 2022). In a recent study evaluating the IBEX AI solution, the algorithm was benchmarked against pathologists for lesion classification using WSIs. For invasive carcinoma, AI achieved an area under the curve (AUC) of 0.976 with 91.7% sensitivity and 95% specificity. For DCIS, the AUC was 0.976 with 93.3% sensitivity and 96.6% specificity, while for lobular neoplasia, the AUC reached 0.953 with 94.1% sensitivity and 95.8% specificity. With AI support, pathologists' diagnostic accuracy improved from 97.1% to 100%. Moreover, AI assistance enhanced efficiency, reducing review time by an average of 16.5% across three pathologists and decreasing immunohistochemistry (IHC) utilization by 33% (Tahir et al., 2025).

Overall, AI-assisted diagnostic classification has matured rapidly, with algorithms approaching or even exceeding expert-level performance in specific tasks. The greatest near-term utility lies in workflow augmentation, including triage of negative or high-probability malignant cases, pre-screening, and decision support for diagnostically challenging lesions.

Detection of lymph node metastasis and other findings

Lymph node metastasis

Accurate detection of axillary lymph node metastasis is essential for breast cancer staging and prognosis. Identifying macrometastases is typically straightforward, but detecting micrometastases and isolated tumor cells can be laborious and time-consuming. Recent studies have demonstrated that AI tools can markedly improve both the sensitivity and efficiency of lymph node assessments (Liu et al., 2019; Zheng et al., 2020; Steiner et al., 2018). The CAMELYON16 challenge was a landmark event that stimulated the development of automated algorithms for metastasis detection (Ehteshami Bejnordi et al., 2017). It provided a dataset of WSIs comprising 110 metastatic and 160 negative lymph nodes confirmed by IHC. Algorithm performance on an independent test set of 129 WSIs yielded AUCs from 0.556 to 0.994, with the best model (AUC = 0.994) surpassing pathologists in accuracy. Liu et al. (2019) trained their model on 270 lymph node images and validated it on 129, reaching an AUC of 0.99—significantly outperforming pathologists (AUC = 0.88). Steiner et al. (2018) conducted a study in which six pathologists reviewed 70 lymph node WSIs with and without AI support. AI-assisted evaluation significantly improved sensitivity for detecting micrometastases (91% vs. 83%, P = .02) and reduced interpretation time, demonstrating clear benefits in accuracy and efficiency.

Commercial AI solutions are now available, such as the Visiopharm metastasis detection app, which automates tissue detection, metastasis identification, measurement, and calculation, generating annotated WSIs with maximum metastasis size and area. One recent work using the Visiopharm Integrator System (VIS) AI algorithm demonstrated perfect sensitivity (100%) and negative predictive value for detecting metastases in both sentinel and

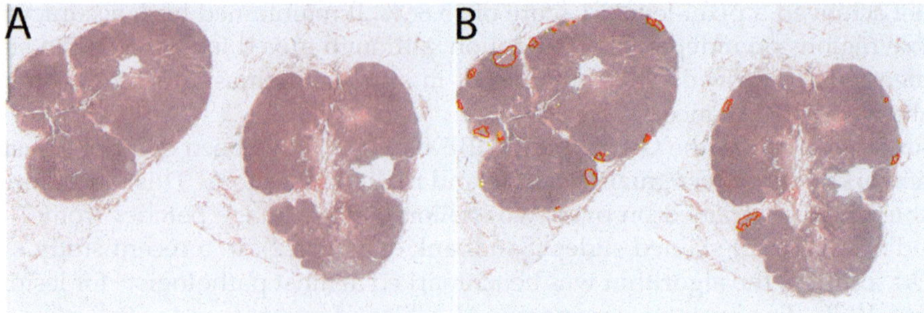

FIG. 9.1 Detection of metastatic tumor in lymph node specimen. (A) Lymph node specimen, H&E stain. (B) Metastatic tumors were successfully detected and annotated in red.

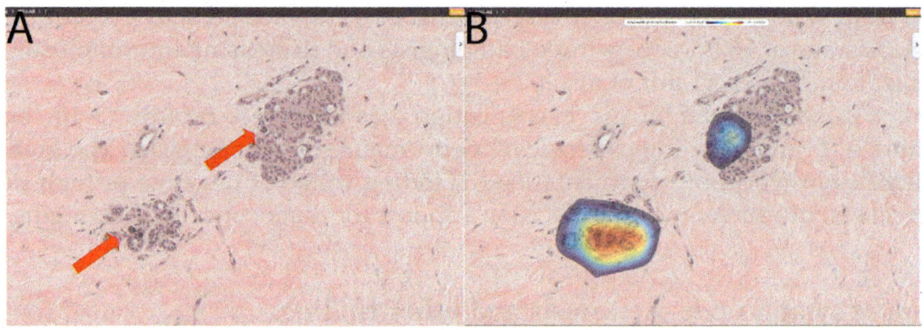

FIG. 9.2 Detection of microcalcification in breast core biopsy by the AI solution. (A) Breast core biopsy specimen with microcalcification, H&E stain. (B) Microcalcification was successfully detected and marked with high probability (orange-red color in heatmap).

nonsentinel nodes, including lobular carcinoma and post-neoadjuvant therapy cases (Challa et al., 2023) (Fig. 9.1). Pathologists reviewing AI-annotated slides required less time than when using traditional IHC slides. These findings emphasize AI's role in enhancing both diagnostic precision and workflow efficiency.

Microcalcifications

It is challenging for pathologists to identify mammary microcalcifications on H&E-stained slides of stereotactic biopsy from screening mammography, which are associated with a wide range of benign, premalignant, and malignant conditions. Besides breast malignant and atypical lesions, the IBEX AI solution was also designed to identify microcalcifications in breast biopsy WSIs. One study has demonstrated that the algorithm achieved an AUC of 0.925 with 95% sensitivity, indicating strong potential for supporting microcalcification detection in routine practice (Tahir et al., 2025) (Fig. 9.2).

AI-driven quantification of biomarkers and grading

Tissue biomarkers are central to diagnosis, prognosis, and treatment selection. Traditionally, protein markers are evaluated by IHC, and pathologists have historically interpreted these manually. With the advancement of digital pathology, AI technology is now being used to provide more objective and standardized assessments. Key biomarkers in breast cancer—including estrogen receptor (ER), progesterone receptor (PR), HER2/neu, Ki-67, and PD-L1—have been the focus of many AI studies.

Estrogen receptor and progesterone receptor

ER and PR are predictive of breast cancer patient prognosis and treatment responses. ER and PR are routinely assessed by IHC, typically reported as the percentage of tumor nuclei showing positive staining. Accurate scoring near the 1% threshold is critical, as even low-level ER positivity can influence therapeutic decisions. The American Society of Clinical Oncology (ASCO)/CAP guidelines recommend testing ER and PR in primary and recurrent breast cancers to guide prognosis and therapy (Allred et al., 1998; Hahnel et al., 1979; Fitzgibbons et al., 2000; Hammond et al., 2010). Manual ER/PR scoring can vary among observers (Nadji et al., 2005; Mann et al., 2005; Hede, 2008; Collins et al., 2005; Badve et al., 2008; Ciocca and Elledge, 2000; Gelber and Gelber, 2009; Reisenbichler et al., 2013; Viale et al., 2007). AI algorithms have shown strong agreement with manual scoring and can segment cells by staining intensity to calculate H-scores, enhancing reproducibility (Thomsen et al., 2019; Bolton et al., 2010; Diaz et al., 2004; Faratian et al., 2009; Rizzardi et al., 2012; Turbin et al., 2008; Gokhale et al., 2007; Rexhepaj et al., 2008; Lujan et al., 2021; Ahern et al., 2017). Challenges remain, including false positives from benign tissues and difficulty with faint staining, necessitating pathologist oversight (Shafi et al., 2022).

HER2

HER2 is overexpressed in approximately 20% of breast cancers and is a critical biomarker in breast cancer. ASCO/CAP guidelines recommend assessing HER2 status via IHC and/or in situ hybridization (ISH) for all invasive carcinomas (Slamon et al., 1987; Tandon et al., 1989; Press et al., 1993; Park et al., 2008). Manual HER2 scoring suffers from variability (Gancberg et al., 2002; Jacobs et al., 1999; Wolff et al., 2018). AI-based methods have demonstrated superior reproducibility and accuracy (Brugmann et al., 2012; Dobson et al., 2010; Helin et al., 2016; Laurinaviciene et al., 2011; Skaland et al., 2008; Holten-Rossing et al., 2015). Algorithms evaluate membrane staining patterns or staining intensity, while AI-assisted microscopy incorporating augmented reality is further improving reliability (Brugmann et al., 2012; Koopman et al., 2019; Koopman et al., 2018; Hartage et al., 2020; Yue et al., 2021). Emerging research suggests that AI may help refine classification, such as distinguishing HER2-low subgroups (Zhang et al., 2020; Modi et al., 2020; Modi et al., 2022).

HER2 ISH testing, which evaluates gene amplification, is more complex. AI algorithms have been developed to automate ISH signal counting—both for HER2 and chromosome 17 centromere probes—within tumor nuclei. These algorithms can accelerate analysis and reduce variability in equivocal cases, though careful quality control remains essential.

Ki-67

Ki-67 is a marker of cell proliferation and prognostic significance (Yerushalmi et al., 2010; Pathmanathan et al., 2014; Cheang et al., 2009; Petrelli et al., 2015). High Ki-67 expression indicates aggressive disease, particularly in ER-positive, HER2-negative tumors (Harbeck et al., 2021; Johnston et al., 2020). Manual Ki-67 scoring is inconsistent, relying on variable counting methods (Yerushalmi et al., 2010; Pollack et al., 2004; Dowsett et al., 2011; Christgen et al., 2015; Leung et al., 2019; Shui et al., 2015). AI tools can automate nuclei detection and positivity calculation, providing more consistent results (Stalhammar et al., 2016; Grala et al., 2009; Remes et al., 2012; Ács et al., 2018; Stålhammar et al., 2018; Klauschen et al., 2015; Koopman et al., 2018; Røge et al., 2016). While some challenges persist, such as distinguishing nontumor cells, AI represents a promising avenue for improving Ki-67 assessment (Goodell et al., 2012). Automated scoring can also process entire tumor areas rather than a few selected fields, providing more comprehensive proliferation estimates.

Other biomarkers

PD-L1 IHC is increasingly used to guide immune checkpoint inhibitor therapy, especially in triple-negative breast cancer. Scoring relies on the combined positive score (CPS), which considers both tumor and immune cells. Manual CPS assessment is challenging due to the need to identify and count multiple cell types across large tumor areas. Early AI models for PD-L1 scoring have demonstrated promising results, accurately distinguishing immune cells from tumor cells and computing CPS values that closely match expert assessments (Bankhead et al., 2017; Humphries et al., 2018).

Other emerging biomarkers, including multiplex IHC panels and novel therapeutic targets, are likely to benefit from AI-based quantification. The ability of AI to handle complex staining patterns, multiplexed markers, and spatial relationships positions it as a powerful tool for biomarker discovery and clinical implementation.

Tumor grading

Consistent tumor histologic grading in breast cancer is critical for prognosis and therapy planning. The Nottingham grading system comprises three components: tubule formation, nuclear pleomorphism, and mitotic activity. Traditionally, pathologists manually evaluate these components, which can lead to interobserver variability. DL models have improved the reproducibility of tubule formation, nuclear pleomorphism, and mitotic activity assessments (Das et al., 2020; Veta et al., 2015; Mantrala et al., 2022; Balkenhol et al., 2019). For example, a DL classifier was developed to automatically identify tubule-forming tumor cell nuclei from WSIs and calculate the ratio of tubule-forming tumor cell nuclei to the total number of tumor cell nuclei—a tubule formation indicator (Romo-Bucheli et al., 2016). Mitosis detection has been enhanced through automated DL pipelines that identify hotspots and mitotic figures, improving proliferation assessment accuracy (Nateghi et al., 2021; Ibrahim et al., 2024; Azam et al., 2024; Li et al., 2018; Sebai et al., 2020; Mahmood et al., 2020). AI models for nuclear pleomorphism have also demonstrated superior reproducibility and prognostic stratification compared with traditional grading (Wang et al., 2022; Elsharawy et al., 2021).

Tumor microenvironment

The tumor microenvironment, particularly tumor-infiltrating lymphocytes (TILs), plays an important prognostic role, especially in TNBC and HER2-positive cancers (Maley et al., 2015; Heindl et al., 2018). High TIL levels are associated with improved disease-free survival and a better response to adjuvant/neoadjuvant chemotherapy. Despite their clinical relevance, manual TIL assessment is challenging. The International TILs Working Group has published guidelines recommending the visual estimation of stromal TIL density in predefined regions. However, manual evaluation is inherently subjective, and interobserver agreement is often modest. The complex spatial distribution of immune cells within tumors further complicates accurate and reproducible assessment. These challenges make AI-based TME analysis an attractive solution.

DL-based algorithms have greatly improved TIL detection and quantification. CNN models trained in pathologist-annotated regions can distinguish lymphocytes with high precision and segment tumor versus stromal areas, enabling automated calculation of TIL density according to standardized guidelines. Indeed, AI-derived TIL scores correlate strongly with manual pathologist assessments but offer far superior reproducibility (Loi et al., 2019; Hendry et al., 2017). Automated TIL quantification also allows for continuous measurements rather than subjective categorical bins, enabling finer stratification of immune infiltration. This is particularly valuable in research settings and clinical trials, where TIL levels may serve as biomarkers for patient selection or therapeutic response monitoring.

Beyond simple density measurements, the spatial organization of immune cells relative to tumor nests carries important biological and clinical information. AI and computational pathology methods enable systematic spatial profiling of lymphocytes and other immune cells within the TME. Using DL-based cell segmentation and classification, algorithms can map the precise coordinates of thousands of cells across entire slides, allowing the calculation of spatial statistics such as nearest-neighbor distances, cell clustering, and immune infiltration gradients (Le et al., 2020). These spatial biomarkers have been linked to prognosis and therapeutic response in breast cancer and are increasingly being explored in translational research.

Prognosis, risk stratification, and prediction of treatment response

Histopathologic features of breast cancer contain a wealth of prognostic information. Traditional prognostic factors include histologic type, grade, tumor size, lymph node status, lymphovascular invasion, and biomarker expression (Jaroensri et al., 2022; Lu et al., 2018; Fernandez et al., 2022; Albusayli et al., 2023; Ivanova et al., 2024). In recent decades, genomic assays such as Oncotype DX, MammaPrint, and Prosigna (PAM50) have become integral to risk stratification, guiding decisions regarding adjuvant chemotherapy and endocrine therapy in early-stage, hormone receptor–positive breast cancer. However, these assays are costly, time-consuming, and not universally available, particularly in resource-limited settings. By analyzing histopathologic features, AI provides accurate, cost-effective risk stratification, reducing reliance on expensive molecular assays. This emerging field leverages DL to identify complex morphologic patterns associated with tumor behavior, many of which are imperceptible to the human eye.

Prognosis

AI algorithms trained on digital histopathology have demonstrated the ability to predict patient outcomes directly from H&E-stained slides, sometimes approaching or even matching the performance of genomic assays (Fernandez et al., 2022; Ahn et al., 2023; McCaffrey et al., 2024; Chan et al., 2023). Automated AI models can score key features from H&E slides with accuracy comparable to pathologists and link morphological patterns with clinical outcomes (Jaroensri et al., 2022). Su et al. developed BCR-Net, a DL model predicting recurrence from WSIs, achieving AUCs of 0.775 on H&E and 0.811 on Ki-67 slides (Su et al., 2023). A subsequent model, Deep-BCR-Auto, improved performance to an AUC of 0.83, demonstrating the potential of AI as a scalable solution for recurrence prediction (Su et al., 2024).

Risk stratification

A particularly exciting direction is the use of AI to predict molecular assay results directly from histology. Molecular assays, such as Oncotype DX (ODX), are routinely used to assess prognosis in early-stage ER-positive/HER2-negative breast cancer (Klein et al., 2013). Romo-Bucheli et al. showed that quantitative morphologic features, such as tubule and nuclei characteristics, correlated strongly with Oncotype DX recurrence risk categories (Romo-Bucheli et al., 2016). More recent DL models have been trained to predict Oncotype DX recurrence scores or PAM50 subtypes from WSIs (Romo-Bucheli et al., 2017). For example, AI-based Deep-ODX, employing cross-attention networks on H&E slides (AUC 0.862), offers more consistent predictions (Su et al., 2024).

Several AI algorithms for breast cancer patient risk stratification are commercially available, including Stratipath Breast (Stratipath AB) and RlapsRisk® BC (Owkin). These algorithms use routine H&E whole-slide images to generate risk scores for patients with ER-positive/HER2-negative breast cancer.

Prediction of treatment response

Predicting treatment response is a major clinical objective in breast cancer management, especially for patients receiving neoadjuvant chemotherapy. In TNBC, only 40%–50% of patients achieve pathologic complete response (pCR), and those with residual disease face significantly higher recurrence risk (van den Ende et al., 2023; Zhu et al., 2021; Abuhadra et al., 2022; Schmid et al., 2020; Mittendorf et al., 2020; Zhao et al., 2020). Early pCR prediction could guide therapy selection and avoid unnecessary toxicity (da Costa Re et al., 2023; Biswas et al., 2017). Histopathologic slides contain abundant information about tumor biology, microenvironment, and aggressiveness—factors closely linked to treatment response. Recent advances in AI have demonstrated that DL models can extract predictive features directly from routine H&E slides, offering a potentially low-cost and widely accessible alternative or complement to molecular and imaging biomarkers (Ahn et al., 2023; Liu et al., 2023). Multimodal models that combine imaging, clinical variables, and histology have improved predictive accuracy (Soliman et al., 2024; Xu et al., 2025; Jimenez et al., 2022). For example, Krishnamurthy et al. (2023) demonstrated that H&E-based AI prediction achieved an AUC of 0.75, offering a cost-effective alternative to imaging. Huang et al. introduced IMPRESS, an AI pipeline integrating H&E and multiplex IHC, achieving AUCs of 0.8975 for HER2+and 0.7674 for TNBC, outperforming manual assessments (Huang et al., 2023). These studies illustrate that routine histopathologic slides can contain sufficient predictive information to stratify response, potentially guiding therapy before treatment initiation.

Response prediction can be improved by multimodal integration, combining histologic AI models with other data streams, including biomarkers (ER, PR, HER2, TILs, Ki-67, and PD-L1), genomics, and imaging features. Multimodal models are likely to be the most clinically powerful, reflecting the multifactorial nature of treatment response. Future work should focus on integrating multiomics data, validating AI models in large prospective trials, and developing interpretable frameworks to facilitate clinical adoption. By leveraging AI to decode complex biological patterns, oncology is advancing toward truly personalized treatment strategies for breast cancer.

Prediction of molecular alterations

Breast cancer is a heterogeneous disease characterized by a spectrum of molecular alterations that influence tumor biology, prognosis, and therapeutic response. Genomic assays, gene expression profiling, and mutational analyses provide critical information for tumor classification and precision oncology. However, these tests require specialized infrastructure, are costly, and may be unavailable in many settings.

AI applied to digital histopathology has emerged as a promising approach to infer molecular alterations directly from routine H&E slides, enabling "virtual molecular testing." DL models can detect subtle morphologic correlates of molecular phenotypes, many of which are imperceptible to human observers. This capability has far-reaching implications for both research and clinical practice.

Hormone receptor and HER2 status

ER, PR, and HER2 status are routinely determined by IHC and/or ISH. Recent studies have shown that DL models can predict their status directly from H&E slides with high accuracy. For example, Couture et al. trained CNNs on large breast cancer cohorts to classify ER, PR, and HER2 status based solely on morphology, achieving AUC values of 0.90 for ER, 0.86 for PR, and 0.88 for HER2 prediction (Couture et al., 2018). Farahmand et al. developed a DL model that predicted HER2 status from H&E slides and achieved AUCs of 0.81–0.89 across multiple independent cohorts, approaching the performance of IHC and ISH testing (Farahmand et al., 2022). Visualization techniques revealed that the model focused on cell membrane morphology and growth patterns, consistent with known HER2-associated morphologic traits. Although these models are not yet replacements for IHC/ISH testing, they highlight the latent molecular information encoded in routine histology and may serve as adjuncts or quality assurance tools, especially in settings where confirmatory testing is limited.

Intrinsic molecular subtypes

Gene expression profiling classifies breast cancers into intrinsic molecular subtypes—Luminal A, Luminal B, HER2-enriched, and Basal-like—each with distinct prognoses and therapeutic responses (Perou et al., 2000). DL models have been trained to predict these subtypes directly from H&E slides. For example, Bychkov et al. trained a CNN on thousands of WSIs linked to gene expression data, demonstrating accurate classification of intrinsic subtypes with AUC values in the range of 0.80–0.90 (Bychkov et al., 2021). These models likely learn complex combinations of morphologic features, including nuclear atypia, glandular architecture, mitotic activity, and immune infiltration, that correlate with underlying transcriptional programs. Predicting intrinsic subtype from histology could provide a rapid and low-cost surrogate for gene expression profiling, enabling earlier risk stratification and treatment planning, particularly in resource-limited environments.

Mutation status prediction

Some studies have extended AI-based inference to somatic and germline mutation status. DL models have been trained to predict BRCA mutation status from H&E slides, achieving AUCs of approximately 0.80 in some cohorts (Wang et al., 2021). Similarly, TP53 mutation status has been inferred from histology in breast and other cancers. Coudray et al. demonstrated that DL models could predict common driver mutations in lung adenocarcinoma with high accuracy (Coudray et al., 2018) and similar methods have since been applied to breast cancer. A key question in molecular prediction from histology is what the AI models are learning. Unlike IHC, where staining intensity is visually interpretable, DL-based molecular prediction relies on patterns that may be subtle or high-dimensional. Explainable AI techniques, such as Class Activation Mapping (CAM) and attention heatmaps, have revealed that models often focus on regions containing tumor epithelium, nuclear morphology, stromal context, and lymphocytic infiltration (Farahmand et al., 2022; Wang et al., 2021). These findings support the biological plausibility of AI-based molecular inference and may also yield new insights into genotype–phenotype relationships, potentially identifying novel morphologic biomarkers of molecular alterations.

Applications of AI in gastrointestinal pathology

Introduction

Gastrointestinal (GI) pathology is a vital component of any surgical pathology department or clinical practice. As awareness of GI diseases grows, endoscopic technologies advance, and clinical guidelines evolve, the demands on GI pathologists continue to increase. In response, AI is emerging as a transformative tool in the field, offering enhanced diagnostic accuracy and improved workflow efficiency. As the GI system encompasses a diverse array of organs and is vulnerable to a wide range of both neoplastic and nonneoplastic pathologies, AI-powered tools have the potential to support pathologists across multiple diagnostic and analytical tasks. These applications can be broadly grouped into five key areas in GI pathology currently: disease recognition and classification, quantification of cellular features on routine H&E stains, identification of microbial infections, quantification of IHC biomarkers, and disease risk stratification and survival outcome prediction. In this review, we explore the current landscape of AI tools in GI pathology and examine future directions for their development and integration. While some platforms have received regulatory approval for clinical use in select regions, many remain investigational or limited to research settings, particularly in the United States. Nonetheless, ongoing validation efforts and technological advancements indicate a clear trajectory toward wider clinical adoption. This review aims to highlight key AI applications in GI pathology. With numerous AI products already available and new ones emerging daily, the examples provided here represent only a small portion of the broader landscape (Table 9.2).

Histological pattern recognition and classification

A primary application of AI in GI pathology is the detection and classification of a wide array of diseases, both nonneoplastic and neoplastic, across the diverse organs of the GI system. The GI tract is a long, continuous tube designed for complex and coordinated functions, including digestion, nutrient absorption, and waste elimination, and accordingly features a broad range of anatomical and histological variations (Yoshida and Kiyuna, 2021). In addition to the tubular GI tract, accessory organs such as the liver, pancreas, and gallbladder each display distinct histomorphologic patterns.

Organ-specific versus Pan-GI models

To address the diversity of tissue types and diseases in GI pathology, AI models have been developed with two major training strategies. Some systems are trained for organ-specific analysis, allowing them to focus on the unique histologic features of a particular organ. For example, IBEX's Galen Gastric Suite is based on fully supervised, multilayered CNNs, designed specifically for gastric biopsy analysis (Kris Lami et al., 2025). In contrast, other platforms, such as Paige AI's GI Suite, which was developed on the Virchow2 foundation model, are trained on a range of GI specimens, including those from the esophagus, stomach, small intestine, colon, rectum, and pancreas (Paige.ai, n.d.).

Beyond carcinoma detection: Multiclassification capabilities

Advanced AI models can classify dozens of distinct histologic entities, extending well beyond simple binary assessments. While carcinoma detection remains a central focus, many tools are now trained to identify a wide variety of histologic entities. These include lymphomas, neuroendocrine tumors, and a range of nonneoplastic conditions such as inflammatory and reactive changes. The expansion of classification capabilities enables AI tools to provide more nuanced and clinically relevant interpretations, truly serving as a detection and classification aid in GI pathology, particularly in cases where overlapping features or subtle findings may complicate diagnosis.

Detecting lymph node metastases for staging

In addition to primary tumor detection, AI is also being employed in cancer staging through the identification and quantification of metastatic disease. Tools such as Paige's PanCancer Detect, though not specific to GI pathology, have been trained to recognize metastatic lesions, including those originating from GI primaries. Both PanCancer Detect and the Aiforia Colon Suite are capable of identifying metastatic deposits in lymph nodes, which play a crucial role in accurate tumor staging (Paige.ai, n.d.; Inc. A, n.d.). By automating this labor-intensive process, AI has the potential to improve staging accuracy and efficiency and support better-informed clinical decisions.

TABLE 9.2 Summary of AI-based tools relevant to GI pathology.

Vendor		Histopathologic classification	H&E Cell counting	Microbial Identification	IHC Biomarkers	Predictive Modeling
Aiforia	Gastric Suite	Gastric cancer, lymph node metastases		*H. pylori*		
	Colorectal Cancer QuantCRC					Risk stratification of colorectal adenocarcinoma
DeepLIIF					Ki-67	
DoMore Platform						Stratifies stage II/III CRC patients by prognosis
GastroMIL/MIL-GC						Prognostic prediction in gastric cancer
IBEX Galen Gastric		Gastric cancer, dysplasia		*H. pylori*		
MSINet						Predicts MSI status directly from H&E-stained slides
Owkin MSIntuit						Pre-screening MSI in colorectal carcinoma
PaigeAI	GI Suite	>40 GI pathologies, metastases	Mitotic count			MSI, HER2
	PanCancer Detect	Detects metastatic disease, including GI metastases				
PathAI IBD Explore and AIM-HI UC						Quantify histologic indices and predict the clinical course in IBD mucosal biopsies
PictorLabs DeepStain and ReStain					Virtual IHC and special stains	
Visiopharm (APPs)		Framework not exclusive to GI but adaptable			Modular APPs for PD-L1, Ki-67, and others (currently validated in non-GI cancers but adaptable to GI pathology)	
		APP for lymph node metastasis detection.				

Note: Most algorithms do not have FDA approval for clinical use in the United States and remain intended for research and educational use only at this time.

Improving diagnostic consistency in challenging GI cases

Importantly, AI has shown promise in addressing diagnostic variability and under-detection associated with certain subtle or borderline lesions. Grading of epithelial dysplasia in the GI tract is well known to suffer from interobserver variability. Early neoplastic changes, such as low-grade dysplasia in Barrett's esophagus, may be small, focal, and obscured by inflammation or reactive epithelial changes, and the interobserver agreement is often low, as shown in a study by Vennalaganti et al., with a κ value of only 0.11 for low-grade dysplasia (Vennalaganti et al., 2017). Part of the reason is that the criteria for diagnosing dysplasia are somewhat subjective. In such scenarios, AI models with high sensitivity can aid by localizing suspicious regions and suggesting consistent interpretations based on more quantitative, and therefore more objective, evaluation of subtle changes on the pixel level; thereby reducing subjectivity and improving the standardization of dysplasia grading (Faghani et al., 2022). Similarly, in gastric biopsies, AI can support the classification and grading of both acute and chronic inflammation, as well as identify premalignant changes such as dysplasia, intestinal metaplasia, and atrophy (Shi et al., 2022).

AI also contributes to the detection and grading of diagnostically subtle neoplasms. For instance, neuroendocrine neoplasms, which can occur throughout the GI tract, may be well differentiated and subtle in appearance, hiding in the cellular lamina propria or mimicking granulation tissue, especially on the edges of small biopsies. AI can assist in their detection and perform classification and grading by mitotic figure counting, as discussed further below.

In hepatic pathology, AI tools have also been used to differentiate hepatocellular lesions, which can be diagnostically challenging, especially in differentiating high-grade dysplastic nodules from well-differentiated hepatocellular carcinoma (International Consensus Group for Hepatocellular Neoplasia, 2009).

Enhancing interpretability with visual overlays

To support interpretability and clinical usability, many AI tools in pathology now incorporate visual overlays, such as heatmaps, on WSIs (Fig. 9.1). These overlays indicate regions of interest by presenting the model's estimated likelihood or probability that certain histologic features are present, rather than issuing a definitive diagnosis. This probabilistic output aligns with regulatory and clinical guidelines that emphasize AI as a decision-support tool rather than a diagnostic authority. AI models highlight areas based on learned patterns associated with specific outcomes, but the final interpretation remains the responsibility of the pathologist. Such overlays can assist in navigating large or complex specimens and are especially valuable in screening tasks or small biopsies, where diagnostically significant but subtle features might otherwise be overlooked. In practice, these tools can improve diagnostic efficiency and consistency by directing attention to regions with higher model confidence. However, careful validation and pathologist oversight remain essential to mitigate risks, such as overreliance or misinterpretation of model outputs.

Workflow integration and efficiency gains

Beyond diagnostic accuracy, using AI for pathological classification offers tangible benefits in workflow optimization. The growing volume of endoscopic biopsies, many of which yield normal or only minor findings, places a significant burden on GI pathologists. AI algorithms with high sensitivity can function as triage tools, rapidly screening slides and flagging abnormal findings for closer review. Such triage tools are already widely used in the automated analysis of hematologic specimens, as in automated hematology analyzers (Zhang et al., 2021; Gedefaw et al., 2023), and more recently in cervical cytology (Pap smear) screening (Zeng et al., 2024). This prioritization helps allocate pathologists' attention more efficiently, potentially reducing turnaround time and increasing productivity without compromising diagnostic quality.

Quantitative cellular feature analysis on routine H&E stain

The GI pathologist faces increasing diagnostic demands, both in volume and complexity. One area in which AI has begun to make inroads is the quantification of cellular features on routine H&E-stained slides, a task often essential for diagnosis, disease monitoring, or risk stratification in GI pathology. These assessments are frequently time-consuming, tedious, and prone to interobserver variability, making them ideal targets for AI-assisted workflows. Emerging algorithms now offer support for several key tasks, ranging from cell counting to steatosis quantification. Below, we review select use cases where AI shows potential to enhance accuracy, efficiency, and reproducibility in GI pathology practice.

Eosinophilic esophagitis: counting eosinophils

Eosinophilic esophagitis (EoE) has been gaining prevalence and, while once considered rare, is now a commonplace disease and an important consideration on routine esophageal biopsies (Straumann and Katzka, 2018). The most important histologic feature of EoE is the quantity of eosinophilic infiltrate in the esophageal squamous epithelium, also known as the peak eosinophil count per high-power field (HPF). The Eosinophilic Esophagitis Histologic Scoring System (EoEHSS) extends beyond simple eosinophil quantification to include features such as lamina propria fibrosis and basal hyperplasia (Ma et al., 2022). Manual quantification of eosinophils is labor-intensive, and the implementation of the full EoEHSS can be prohibitively time-consuming in routine practice. Recently, an AI model developed by Mayo Clinic has demonstrated the ability to quantify peak eosinophil count and score other features of the EoEHSS with performance similar to that of GI pathologists (Ricaurte Archila et al., 2023). This approach promises to increase throughput and consistency, particularly in high-volume settings.

GIST and neuroendocrine neoplasms: Mitotic count

While mitotic activity is an important consideration in the histologic evaluation of any neoplasm, there are several scenarios in which mitotic count has direct implications for risk stratification and treatment decisions in the GI tract. Mitotic activity in gastrointestinal stromal tumor (GIST) is an indicator of aggressive behavior and is one of the features used in risk stratification, along with tumor size and location (Campora et al., 2023). Accurate assessment, however, has several challenges. First, the area of highest proliferative activity (the "hotspot") must be identified, which is often subjective and varies between observers. Even within hotspots, mitotic figures can be rare and difficult to distinguish from apoptotic bodies or pyknotic nuclei, especially in inflamed or poorly fixed tissues (Alkhasawneh et al., 2015).

Similarly, for GI neuroendocrine neoplasms (GI-NEN), mitotic count is a key component of the WHO grading system and requires scanning of at least 50 HPFs (Hirabayashi et al., 2013) and counting mitotic figures in 10 HPFs within a 2 mm area (Pantanowitz et al., 2013) at the hotspot, further adding to workload and variability (Malon et al., 2012). To the best of our knowledge, while there are no AI models currently validated specifically for mitotic counting in GIST or GI-NEN in the United States, several generalizable models initially developed for breast and other cancers have demonstrated strong performance in mitosis detection across tissue types (Shen et al., 2024). In practice, AI may be most valuable not only as an automated counter but as a decision-supporting tool, assisting with hotspot detection, prescreening candidate mitotic figures for pathologist review, and standardizing field selection across cases to reduce variability.

Intraepithelial lymphocytosis: Counting lymphocytes

Several GI pathologies are characterized by increased intraepithelial lymphocytes (IELs), often within an otherwise normal-appearing mucosa. Lymphocytic esophagitis is defined entirely by histologic findings, with key features including intraepithelial lymphocytosis and associated mucosal injury (Patil et al., 2018). One of the histologic hallmarks of celiac disease is increased IELs (>25 per 100 enterocytes), in addition to other findings such as altered crypt architecture and villus blunting (Gruver et al., 2023). Lymphocytic colitis, a subtype of microscopic colitis, is classically defined by $\geq$20 IELs per 100 surface epithelial cells with minimal crypt distortion (Langner et al., 2015). In routine practice, formal IEL counting is often unnecessary due to the obvious nature of lymphocytosis in most cases. However, in borderline cases, poorly oriented biopsies, or where histologic overlap exists with conditions like reflux esophagitis, early celiac disease, or infectious colitis, objective IEL quantification can aid in diagnostic clarity. Although no AI models have yet been validated specifically for lymphocytic esophagitis or colitis, the task is well suited to AI-assisted analysis. Potential benefits include automated identification of surface epithelium, precise distinction between intraepithelial and lamina propria lymphocytes, and consistent quantification over larger tissue areas—features that could reduce observer variability and enhance reproducibility. AI tools under development for celiac disease, which similarly depend on IEL counts for Marsh classification, demonstrate the feasibility of this approach (Gruver et al., 2023).

Tumor buds: detection and quantification

Tumor budding is a well-established histopathologic feature in colorectal carcinoma (CRC) associated with adverse outcomes, including lymph node metastasis, local recurrence, and reduced survival. It is defined as single tumor cells or clusters of fewer than five cells at the invasive front of the tumor, representing active epithelial–mesenchymal transition and local invasion (Shah et al., 2022). Tumor budding is especially critical in early-stage (pT1 or pT2) CRC, where a high tumor bud score may indicate the need for more aggressive surgical intervention or adjuvant therapy (Petrelli et al., 2015; Cappellesso et al., 2017).

However, the reproducible identification and quantification of tumor buds remain challenging. Tumor buds are often subtle and can be obscured by inflammation, mucin, necrosis, or crush artifacts. Interobserver variability is common, even when using consensus-based methods such as those outlined by the International Tumor Budding Consensus Conference (ITBCC) (Lugli et al., 2017). While cytokeratin IHC can enhance the visibility of buds, it may also overestimate their number by highlighting nonbudding epithelial fragments (Kai et al., 2016).

AI offers promising solutions by enabling the automated detection of tumor budding at the invasive front using either H&E or cytokeratin-stained slides. DL algorithms can assist by accurately identifying the tumor–stroma interface, segmenting tumor buds, and generating reproducible bud counts across entire tumor regions. Studies have demonstrated that AI-based budding assessment improves consistency and correlates well with pathologist scoring (Bokhorst et al., 2023; Sajjad et al., 2024; Su et al., 2024). These models could be particularly valuable in borderline pT1 lesions, where treatment decisions hinge on accurate risk stratification. As digital pathology becomes more widely adopted, AI-assisted tumor budding evaluation has the potential to improve prognostication and guide therapeutic decision-making in CRC by reducing variability and improving standardization.

Hepatic steatosis: Quantification

Quantifying hepatic steatosis is a key component in evaluating both chronic liver disease and donor liver suitability. In the setting of metabolic dysfunction-associated fatty liver disease (MAFLD), liver biopsies are commonly assessed using the Nonalcoholic Steatohepatitis Clinical Research Network (NASH CRN) Scoring System. This system grades histologic features, including steatosis, lobular inflammation, and hepatocyte ballooning, to generate a NAFLD Activity Score (NAS) and separately stages fibrosis (Forlano et al., 2020). However, visual estimation of steatosis by pathologists has been shown to have limited reproducibility and frequent discrepancies compared to objective, quantitative assessments (El-Badry et al., 2009; Hall et al., 2013).

AI-based algorithms trained on routine H&E-stained liver biopsies have demonstrated strong performance in quantifying steatosis, with improved agreement between observers and better reproducibility than manual assessment (Forlano et al., 2020). In addition to steatosis, these tools can also measure lobular inflammation, hepatocyte ballooning, and even fibrosis using H&E-based morphology, potentially reducing the need for special stains. Quantitative outputs from such models may improve consistency in longitudinal studies and clinical trials, where subtle changes in disease activity are critical to assess.

Hepatic steatosis also plays a pivotal role in liver transplantation, where excessive fat content is associated with an increased risk of primary graft dysfunction and ischemia-reperfusion injury (El-Badry et al., 2009; Jiao et al., 2024). In 2022, the Banff Working Group on Liver Allograft Pathology issued consensus guidelines emphasizing the standardized assessment of both fat droplet size and the percentage of hepatocytes involved (Neil et al., 2022). AI models have shown promise in aligning with these recommendations by enabling consistent, quantitative evaluation of macrovesicular steatosis on donor liver biopsies (Jiao et al., 2024).

Microbial infection identification

Helicobacter pylori

Similar to the challenges associated with cell counting in histopathology, the accurate identification of microbial infections can be a tedious and time-consuming task, yet it is critical for the proper and timely treatment of the patient. For example, *Helicobacter pylori* is a prevalent bacterial infection in the United States and across the world, which is a known risk factor for malignancies such as gastric mucosa-associated lymphoid tissue (MALT) lymphoma and gastric carcinoma (Teng et al., 2025). However, identifying *H. pylori* on H&E-stained slides can be time-consuming and challenging. The bacteria are small, may be obscured by mucus or staining artifacts, and can be present in low numbers following prior treatment. While IHC staining is an alternative with superior sensitivity and specificity, it is also expensive and significantly increases turnaround times, limiting its use in high-throughput clinical settings (Wan et al., 2023). Several AI-based tools have emerged for research use in the AI-based detection of *H. pylori* in histopathology (Fig. 9.3). Models trained on both H&E-stained slides as well as IHC have demonstrated high performance in identifying *H. pylori,* in some cases outperforming traditional microscopic review (Teng et al., 2025; Klein et al., 2020).

Cytomegalovirus

Another focus has been the use of AI to detect viral cytopathic effects from histologic images. The identification of cytomegalovirus (CMV) is an important consideration in any GI

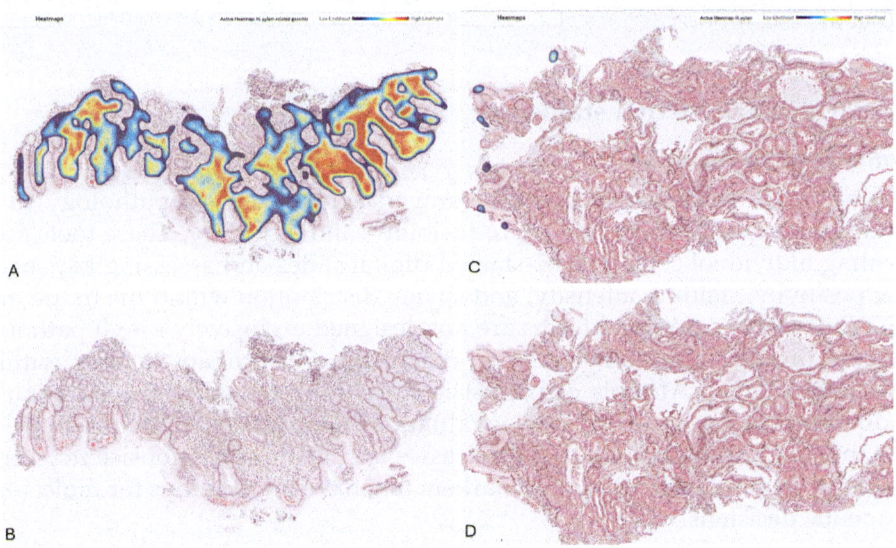

FIG. 9.3 Example of visual overlay (heatmap) on digital images. Heatmaps highlight regions with varying probabilities of *Helicobacter pylori* gastritis, with red indicating high likelihood and blue indicating low likelihood. (A, B) Gastric biopsy positive for *H. pylori* infection, shown with and without heatmap overlay, respectively. (C, D) Gastric biopsy negative for *H. pylori* infection, shown with and without heatmap overlay, respectively. Original magnification, ×5.

tract biopsy from immunocompromised patients, including those with inflammatory bowel disease (IBD), a history of organ transplantation, or those receiving oncology treatment. While serology can be a useful tool, local infection (e.g., CMV colitis) may occur without positive viremia, and histologic examination remains the gold standard (McCurdy et al., 2015). CMV inclusions can be difficult to identify on H&E slides and are often obscured by inflammation. AI algorithms can aid the pathologist by screening biopsies for CMV infection. A recent study demonstrated multiple DL models with high performance in detecting CMV inclusions in H&E slides, with the highest performance from the EfficientNetV2BO model (Post et al., 2023). Given this success, histopathology-based CMV detection is a likely area of future development for AI.

Other common GI infections

Beyond histopathology, AI-powered tools offer promising utility in the diagnosis of other GI infections. Intestinal parasitic infections, such as *Giardia lamblia*, *Entamoeba histolytica*, and *Cryptosporidium,* can be difficult to identify due to their low abundance, subtle morphological features, and the rarity of these infections, all of which may contribute to missed diagnoses. DL models have demonstrated high accuracy in detecting *G. lamblia* cysts and trophozoites in stool microscopy (Yarahmadi et al., 2025). DL-enabled imaging flow cytometry, combined with models like MCellNet, has also achieved high accuracy and sensitivity for both *Giardia* and *Cryptosporidium* in water samples (Luo et al., 2021). Fungal infections, particularly in immunocompromised hosts, can be difficult to differentiate due to morphologic similarity. DL frameworks have shown promise in classifying various fungi on microscopic images, potentially expediting diagnostic turnaround by bypassing extended biochemical testing (Zielinski et al., 2020).

IHC biomarkers and virtual stains

IHC biomarker quantification

AI platforms are increasingly being leveraged across surgical pathology to quantify IHC markers with greater precision, reproducibility, and efficiency. These tools are capable of segmenting individual cells on IHC-stained digital slides and assessing key metrics such as marker positivity, staining intensity, and spatial distribution within the tissue microenvironment. While most current platforms are not designed exclusively for GI pathology, their functions are directly applicable to several diagnostic and prognostic tasks within the GI workflow. For example, AI tools can assist in measuring the Ki-67 proliferation index in neuroendocrine tumor grading and evaluating abnormal p53 expression patterns (Ghahremani et al., 2022). Automating these assessments improves consistency across cases and between observers, particularly for markers that have implications for molecular testing and therapeutic decisions.

Some AI platforms offer modular, app-based frameworks tailored for specific diagnostic tasks within IHC quantification. While many of these modules have been validated in other tumor types, such as Visiopharm's PD-L1 quantification app in non-small cell lung cancer and Ki-67 app for breast cancer, they represent flexible and adaptable infrastructures that could be readily extended to GI pathology (Rewcastle et al., 2024; Plass et al., 2025).

This approach holds promise for developing dedicated tools targeting biomarkers commonly used in colorectal and upper GI cancers, including HER2, Claudin18, or PD-L1.

Virtual staining

A particularly innovative application of AI in pathology is the development of virtual staining technologies. These tools utilize digital images of unstained formalin-fixed, paraffin-embedded (FFPE) tissue sections and apply DL models to generate realistic simulations of traditional stains such as H&E, IHC, and special stains. By utilizing tissue autofluorescence (Rivenson et al., 2019), quantitative phase imaging (Rivenson et al., 2019), or cycle-consistent generative adversarial networks (CycleGAN) (Yoon et al., 2024) these models can reconstruct a "virtual stain" that mimics the appearance and diagnostic utility of conventional histochemical techniques. These tools aim to reduce tissue consumption and improve staining standardization. This technology can potentially support pathologists by enabling multiple virtual stains from a single slide, which is particularly useful in limited tissue samples.

In the context of liver pathology, generating a virtual trichrome stain, which is commonly used in liver biopsies to evaluate fibrosis, can allow for efficient assessment of collagen architecture without consuming additional tissue. PictorLabs' Deep Stain application is one such tool that uses DL to generate virtual stained images from WSIs of unlabeled tissue (Zhang et al., 2020). As these tools evolve, they may enable multiplexed virtual staining from a single slide, offering comprehensive tissue evaluation while conserving material and resources.

AI predictive modeling: From gene mutations to disease prognostication

An essential component of the GI pathologist's report in cancer cases is the integration of histologic features, TNM staging, and molecular characteristics to inform prognosis and guide treatment decisions. Increasingly, AI tools are being developed to predict clinical outcomes and reveal underlying molecular alterations from histological images, using information hidden in H&E pixels that is not readily detectable by visual inspection. These technologies have the potential to streamline complex diagnostic workflows, provide prognostic insights, and enhance clinical decision-making.

CRC is the third most common malignancy in the United States. It is a histologically and molecularly heterogeneous disease, as noted by McHugh and Pai (2023), with many of its variations linked to distinct tumor behaviors. As our understanding of these histologic correlations has expanded, synoptic reporting has become increasingly complex. Features such as mucinous differentiation, extramural large venous invasion, immune cell infiltration (Idos et al., 2020) tumor budding (Lugli et al., 2017), poorly differentiated clusters (Shivji et al., 2020) and immature desmoplastic stroma (Huijbers et al., 2013) all show established prognostic or therapeutic significance. However, the growing number of reportable features has introduced added complexity, posing challenges for both pathologists in how to communicate these elements in their reports and for oncologists in interpreting their clinical relevance. With recent advances in machine learning tools, several algorithms have focused on integrating these histopathological features to generate automated predictions of clinical outcomes.

Molecular mutation prediction

Predicting microsatellite instability

Microsatellite instability (MSI) is a well-established molecular pathway in the development of CRC, driven by deficient DNA mismatch repair (Boland and Goel, 2010; Kawakami et al., 2015). Determining MSI status is clinically important, as it carries both prognostic and therapeutic implications. Although most cases of MSI CRC occur sporadically, a subset develops because of an inherited germline mutation in DNA mismatch repair genes, as seen in Lynch syndrome. Testing for MSI is typically performed algorithmically with IHC and/or molecular testing. These methods are well validated but can be resource-intensive. Histologically, several morphologic features that are visible to the human eye on routine H&E-stained slides have long been established to have an association with MSI CRC, including mucinous/signet ring cell differentiation, tumor-infiltrating lymphocytes, and Crohn's-like reaction at the tumor invasive front (Greenson et al., 2003). While suggestive, these features lack the specificity and reproducibility needed for diagnostic certainty. Several DL models, including the MSINet model, have demonstrated high accuracy in identifying MSI CRCs based solely on histopathologic patterns (Yamashita et al., 2021). These AI tools potentially streamline traditional testing, particularly in settings where IHC/molecular resources are limited.

Predicting other key mutations

Beyond MSI, additional molecular alterations such as mutations in *KRAS*, *BRAF*, and *TP53* as well as the CpG island methylator phenotype (CIMP), are increasingly recognized as clinically significant in CRC. These biomarkers play critical roles in determining prognosis and predicting response to targeted therapies. Emerging AI-based tools have begun to explore the feasibility of inferring these molecular features from digitized histopathology slides. While predicting single-gene mutations such as *KRAS* and *BRAF* from histological morphology presents greater challenges due to the absence of overt, distinctive features, some studies have reported moderate to high predictive performance using CNNs (Bilal et al., 2021; Guitton et al., 2023; Li et al., 2024). Similarly, DL approaches have been applied to infer CIMP status, often in combination with other epigenetic or transcriptomic signatures (Bilal et al., 2021). These tools are still in the early stages of validation but represent a promising adjunct to traditional molecular testing.

Prognostication prediction

Colorectal carcinoma

Clinically effective biomarkers enable personalized therapy by aligning treatment decisions to the specific molecular profile of the tumor. In the setting of CRC, decisions regarding adjuvant treatment have historically been made based on the tumor stage rather than clinically validated biomarkers. However, several groups are now using AI-driven tools to refine prognostic models to further stratify subgroups for more targeted therapeutic decisions. The DoMore platform employs DL to assess morphologic features and classify patients into distinct prognostic categories (Skrede et al., 2020). The clinical decision support system based on the model has shown potential in identifying patients with stage II and III CRC who have a prognosis similar to low-risk groups that may safely forego adjuvant chemotherapy (Kleppe et al., 2022).

Gastric carcinoma

Clinically relevant biomarkers for gastric cancer remain limited, with prognosis traditionally guided by tumor staging rather than nuanced morphological insights (Cao et al., 2019). However, recent AI-driven histomorphological tools are poised to transform prognostic stratification in GC. For instance, the GastroMIL and MIL-GC models employ DL on H&E-stained whole-slide images to stratify patients into high- and low-risk groups and have been demonstrated to be strong predictors of outcomes in gastric cancer (Huang et al., 2021). These AI platforms thus enable refined risk stratification from histomorphology that may ultimately guide adjuvant treatment decisions.

Inflammatory bowel disease

Beyond neoplastic diseases, AI is now being applied to chronic inflammatory diseases, such as IBD, for disease severity evaluation and outcome prediction. Traditionally, histologic assessment of colonic mucosal biopsies is essential in confirming ulcerative colitis (UC) diagnosis, excluding infection, and monitoring for dysplasia (Najdawi et al., 2023). Furthermore, histologic assessment of inflammation has been shown to be important for predicting IBD disease severity and prognosis (Marchal Bressenot et al., 2015; Gupta et al., 2021). However, quantitative assessment of complex features on histology can be difficult and time-consuming for the pathologist to discern unassisted.

Two leading platforms developed by PathAI (IBD Explore and AIM-HI UC) leverage AI to quantify histologic features in IBD and predict clinical course. The PathAI AIM-HI UC was trained using pathologist-labeled histological features of UC to extract "human interpretable features" (HIF) correlating with pathologist scoring. Generated HIF include areas of erosion/ulceration, crypt abscesses, infiltrated epithelium, and density of neutrophils, and these HIF showed strong correlation with pathologist scoring using the National Histological Index (NHI) (Najdawi et al., 2023). These tools have been deployed in collaborative research initiatives, such as PathAI's partnership with the Crohn's & Colitis Foundation to leverage IBD Explore and AIM-HI UC on large-scale pathology datasets for biomarker discovery, subtype identification, and therapeutic target development (PathAI, 2024). Together, IBD Explore and AIM-HI UC represent a growing paradigm in AI-powered histopathology: transforming labor-intensive, subjective scoring systems into automated, quantitative, and reproducible analyses with the ability for longitudinal comparison, which can drive precision medicine in IBD.

Miscellaneous considerations

Explainable AI

While DL models have proven adept at recognizing complex patterns and generating accurate predictions, a common critique lies in their lack of interpretability. Despite their accuracy, the internal reasoning process is not easily interpretable by humans. Most current algorithms do not explicitly disclose which histological features they use in decision-making, creating challenges in adapting to evolving classification systems or communicating results to clinicians and patients. This has prompted an increasing interest in "explainable AI" models that leverage DL in a manner with more transparency in their reasoning processes.

The Aiforia CRC QuantCRC is an example of this next generation of AI tools. The model segments CRC into layered morphologic features, which are then validated against interpretations by expert GI pathologists. After segmentation, quantitative parameters such as tumor-stroma ratio, percentage of tumor buds/poorly differentiated clusters within the tumor bed, and percentage of immature stroma are measured. These metrics, when calculated manually using routine microscopy, are prohibitively time-consuming and subject to observer variability. Segmentation algorithms, by contrast, offer speed, consistency, and reproducibility in reporting (Pai et al., 2021). Once computed, these morphologic metrics can be integrated with other clinical parameters to deliver a recurrence risk score, providing a comprehensive risk assessment for individual patients (Pai et al., 2022). As segmentation technology evolves, its integration into digital pathology platforms may become a central component of standardized CRC reporting.

ROSE DL models

Rapid on-site evaluation (ROSE) for minimally invasive sampling by endoscopic ultrasound is another potential target for AI development. It is important to ensure adequate tissue sampling in real time, but having this service always readily available can be resource-intensive. In a recent study by Lin et al., the ROSE-AI model demonstrated levels of accuracy approaching those of well-trained cytopathologists and may also improve the speed at which results can be reported (Euscher et al., 2023). As AI continues to evolve, its role in real-time decision support during endoscopic procedures is likely to expand, potentially improving diagnostic accuracy, reducing inter-operator variability, and optimizing patient outcomes.

AI advancements in GI endoscopy and imaging

While AI applications have been developed for GI pathology, similar technological transformations are occurring in GI endoscopy and imaging. In endoscopy, conventional colonoscopy often misses adenomas and struggles to distinguish lesion types, but AI-driven systems now improve detection of subtle lesions and assess tumor invasion depth to support treatment planning (Zhu et al., 2019; Misawa and Kudo, 2025; Gao et al., 2025). Similarly, in GI imaging, AI radiomics enhances CT utility by improving lesion detection, staging, and prognostication—for example, predicting invasion depth and recurrence-free survival in gastric cancer (Niu et al., 2020) and differentiating malignant from benign disease (Chidambaram et al., 2022). These advances parallel GI pathology, where AI helps reduce variability and improve diagnostic accuracy.

Conclusion

AI applications in the GI pathology workflow can fulfill a wide variety of roles, ranging from routine, mundane detection and quantification tasks to offering insights into disease prognostication not visible to the human eye alone. As challenges in daily practice are identified, AI applications are developed to fulfill clinical needs. While there is a theoretically limitless number of directions that AI can take, pathologists must work with the relevant stakeholders to determine which AI tools should be implemented into real-life practice to improve workflow in a cost-effective manner. As diagnostic criteria and terminology evolve, continuous updates to pathology AI tools are necessary, as with any other software product.

Applications of AI in urologic pathology

Introduction

AI has transitioned from experimental technology to routine assistance in urologic surgical pathology, with prostate showing the highest clinical maturity (Verghese et al., 2023; Parwani et al., 2023; Patel et al., 2022; Patel et al., 2022). The practical value of AI in genitourinary (GU) pathology lies in the augmentation of three workflow phases: pre-sign-out (specimen intake, quality control, and case prioritization), sign-out (region of interest identification, grading assistance, and quantification), and post-sign-out (quality assurance, performance monitoring, and tumor board preparation) (Parwani et al., 2023; Patel et al., 2022).

Two success factors resonate across implementations: First, pre-analytic quality control determines downstream accuracy, i.e., ensuring adequate focus and tissue coverage in whole slide images and sufficient cellularity in cytology specimens (Patel et al., n.d.; Vorontsov et al., 2024; Chen et al., 2024; Amann et al., 2020; Diao et al., 2021). Second, explainable AI ("xAI") features that provide region-level grounding through heatmaps and confidence indicators enable pathologists to verify AI assessments and maintain diagnostic control (Vorontsov et al., 2024; Chen et al., 2024; Amann et al., 2020; Diao et al., 2021).

Across GU pathology, AI performs best where case volumes are high, labeling is consistent, and datasets span multiple centers (Patel et al., 2022; Bulten, 2022; Bulten et al., 2022; Ström et al., 2020; Challenge, n.d.). Real-world performance depends on local variables, including scanner variations, staining protocols, and case mix, necessitating site-specific validation and continuous monitoring, even for well-studied algorithms (Ramwala et al., 2024; Froń et al., 2023; Park and Kressel, 2018; Fehr et al., 2022; Patel et al., 2024; Ennab and McHeick, 2024).

AI for prostate pathology

Prostate cancer represents approximately 40% of the GU histopathology workload in Western laboratories (Bulten, 2022; Challenge, n.d.) The subspecialty characteristics of prostate pathology, including high biopsy volumes, standardized pattern-based grading systems, and diagnostic complexity, create compelling conditions for AI implementation (Bulten et al., 2022; Ström et al., 2020; Challenge, n.d.; Bulten et al., 2020; Busby et al., 2024; Faryna et al., 2024; Marletta et al., 2024; Mulliqi et al., 2025; Steiner et al., 2020). A synergistic cycle, where clinical need fuels innovation and innovation enhances clinical capability, has accelerated AI adoption in prostate pathology while simultaneously driving technological advancement, as seen in the development of extensive prostate-specific annotated datasets such as PANDA, CAMELYON-PCam, and STHLM3 (Bulten, 2022; Challenge, n.d.).

Workflow improvements are documented in prospective studies and live implementation, with reports of 22%–30% reduction in prostate biopsy review time and pre-screening paradigms achieving higher efficiency gains (Eloy et al., 2023; Eloy et al., 2023; Flach et al., 2025; Hagland, 2024). Many diagnostic accuracy studies report AUCs exceeding 0.95 for cancer detection, supporting adjunctive use in clinical practice (Bulten et al., 2022; Bulten et al., 2020; Steiner et al., 2020; Raciti et al., 2023; Olsson et al., 2022). A 13%–14% discordance rate with pathologist diagnosis also supports the necessity of local validation and appropriate guardrails (Bulten et al., 2022; Faryna et al., 2024; Wang et al., 2025; Yeung et al., 2023; Zhu et al., 2024; Rienda et al., 2024).

Some commercially available AI platforms for prostate pathology are summarized in Table 9.3, and one of them is exemplified in Fig. 9.4.

AI for bladder pathology

AI applications for bladder pathology span both cytology and histology workflows. For lymph-node assessment, AI triage has safely excluded 80%–92% of true-negative slides while maintaining high sensitivity for metastasis detection (Froń et al., 2023). In urine cytology, automated prescreening aligned with Paris System quality gates improved reliability. Slides with **≥2644 analyzable cells** met a practical QC threshold associated with **consistent downstream AI accuracy** (Patel et al., n.d.). Commercial cytology platforms, including TechCyte, CytoBay, and VISIOCYT, are undergoing multisite validation with

TABLE 9.3 Summary of commercial prostate AI platforms.

Company	Product	Core Functions	Key Evidence	Distinguishing Features
Paige	Paige Prostate Suite	Cancer detection, grading, quantification	FDA De Novo clearance; multisite validation (U.S. Food and Drug Administration, 2021; Paige, n.d.; Smith and Belanger, 2024)	First FDA-cleared prostate AI; comprehensive suite approach.
Ibex	Galen™ Prostate	Detection, grading, structured reporting	22%–30% efficiency improvement; prospective studies (Eloy et al., 2023; Eloy et al., 2023; Flach et al., 2025; Chatrian et al., 2021)	Integrated triage and sign-out workflows
Aiforia	Aiforia® Clinical	CE-IVD detection, high-risk features	CE-marked; cloud deployment (Shafi et al., 2022; Wolff et al., 2023).	Cloud-native platform; multiorgan capabilities
Deep Bio	DeepDx® Prostate	Detection, Gleason grading support	Asian market studies (Marletta et al., 2024; Paik et al., 2025)	Strong grading focus; regional deployment
Indica Labs	HALO® Prostate	Detection ± grading modules	CE-IVD certification (Sandbank et al., 2022; Patel et al., 2022)	Integration with HALO analytics ecosystem
Inify Labs	INIFY® Prostate	Suspicious region outlining	Service model deployment (Anon., 2022)	Laboratory service approach

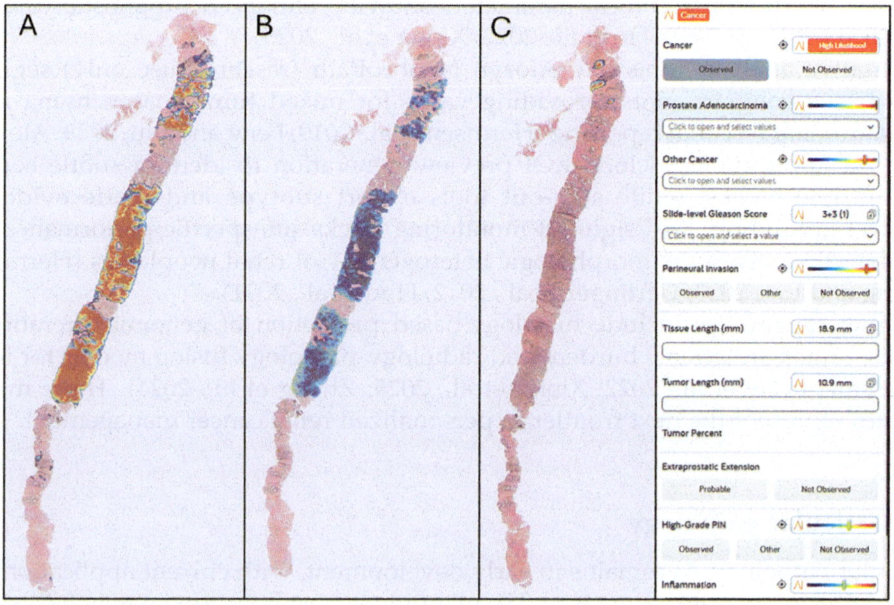

FIG. 9.4 A: Red heatmap generated by the Ibex AI algorithm highlighting cancer foci on a prostate needle core biopsy. B: Blue-green color-coded heatmap displaying color-coded Gleason score. C: Heatmap highlighting perineural invasion focus. The right column displays AI outputs.

promising sensitivity and specificity profiles (Patel et al., 2022; Scientific, 2025; Anon., 2024; Lebret et al., 2023).

For transurethral resection specimens (TURBT), AI assists with grade assessment and invasion detection through compartment segmentation (urothelium, lamina propria, muscularis propria) (Khoraminia et al., 2023; Pan et al., 2023). Artifact-aware models incorporate abstention rules to prevent false-positive invasion calls caused by cautery or crush artifacts (Khoraminia et al., 2023). Emerging applications include molecular surrogate prediction from H&E images, such as FGFR3 mutation status, to pre-select cases for confirmatory molecular testing (Bannier et al., 2024; Loeffler et al., 2022; Woerl et al., 2020). Risk stratification algorithms for BCG-treated nonmuscle invasive bladder cancer are entering prospective multicenter evaluation (Lotan et al., 2024; Lotan et al., 2025; Abbas et al., 2024).

AI for renal pathology (neoplastic)

Renal cell carcinoma (RCC) subtyping is the most mature AI application in kidney pathology, with algorithms distinguishing clear cell, papillary, and chromophobe subtypes with AUCs exceeding 0.90 across multiple validation cohorts (Chandramohan et al., 2024; Hermsen et al., 2019; Zheng et al., 2025; Tian et al., 2019; Li et al., 2025; Feng and Liu, 2023). Clear cell RCC grading achieves comparable performance (AUC: 0.89–0.96), with multimodal approaches incorporating clinical parameters showing enhanced prognostic value (Zheng et al., 2025; Tian et al., 2019; He et al., 2022; Xiong et al., 2025).

Quantitative analysis tools like Aiosyn NephroPath (research use only) segment and quantify tissue compartments, providing value for mixed tumor-parenchyma resections and standardizing research reporting (Hermsen et al., 2019; Feng and Liu, 2023; Aiosin, 2025). Pre-sign-out applications include ROI preview generation to identify subtle components across multiple blocks, while sign-out tools export subtype and grade evidence with region-level grounding. Post-sign-out monitoring tracks site-specific performance and rare variant detection, given the morphologic heterogeneity of renal neoplasms (Hermsen et al., 2019; Feng and Liu, 2023; Gerlinger et al., 2012; Huo et al., 2021).

Emerging capabilities include histology-based prediction of genomic alterations (VHL mutations, tumor mutational burden) and radiology-pathology fusion models for integrated risk assessment (He et al., 2022; Xiong et al., 2025; Zheng et al., 2024). These multimodal approaches represent the next frontier in personalized renal cancer management.

AI for testicular pathology

Testicular pathology AI remains in early development, with current applications focused on screening and quantification tasks. TIL mapping achieves approximately 89% sensitivity, with F1 scores around 0.88, providing standardized immune assessment (Ghosh et al., 2021; Abousamra et al., 2021; Medvedev et al., 2024; Chuluunbaatar et al., 2023). Lymphovascular invasion detection shows modest precision, requiring liberal abstention thresholds and mandatory region-level evidence for clinical use (Medvedev et al., 2024; Chuluunbaatar et al., 2023). Germ cell neoplasia in situ (GCNIS) quantification standardizes margin assessment, though limited case volumes necessitate transfer learning from higher-volume organs and multiinstitutional data pooling for model development (Patel et al., 2022; Patel et al., 2021).

The relative rarity of testicular neoplasms and their morphologic diversity present unique challenges for AI development, with most progress occurring through the adaptation of models trained on other organ systems rather than de novo development.

AI for penile pathology

No commercially available or research-stage AI applications currently exist for penile pathology. The combination of low case volumes, morphologic overlap with other squamous lesions, and limited research funding has precluded meaningful AI development in this domain. Future progress will likely require multiinstitutional consortia and transfer learning from cutaneous and mucosal squamous neoplasia models.

Explainability in urologic pathology

Core explainability mechanisms

For pathology, xAI predicates an AI solution's capacity to "show" pathologists contextually supportive analytical factors for algorithmic outcomes that will be used in practice or research via three general methods: where AI "looked", what AI measured, and how certain the AI is of its final output. Essential components of these methods include:

Region-level grounding (proving the AI "looked" at the right place)

The AI links its predictions to specific areas on the slide (Vorontsov et al., 2024; Lu et al., 2021). This typically appears as:

- A colored heatmap overlay, showing areas of concern
- A gallery of thumbnail images showing the most suspicious regions
- Each thumbnail clickable to navigate directly to that location at full magnification

These visual indicators strengthen pathologists' confidence by confirming that the AI has identified the same subtle diagnostic features they would examine (e.g., tiny cribriform glands, early invasion), while also revealing when the AI may be inappropriately focused on artifacts like pen marks or tissue folds. Best practice favors native attention mechanisms over post-hoc explanations, which can be less stable (Table 9.4) (Fitzgibbons et al., 2000; Vorontsov et al., 2024; Amann et al., 2020; Diao et al., 2021; Ström et al., 2020; Steiner et al., 2020; Raciti et al., 2023; Lu et al., 2021; Rudin, 2019).

Feature-level quantification (supporting "what" with "how much")

The AI provides measurements that map directly to routine reporting (Diao et al., 2021):

- Tumor percentage
- Gleason pattern percentages (3, 4, 5)
- Linear extent of tumor
- TIL density
- Mitotic counts

These appear in a measurement panel with clear units, confidence intervals, and one-click export to reports. This standardization reduces inter-observer variability and eliminates transcription errors (Diao et al., 2021; Bulten et al., 2022; Tolkach et al., 2023).

Uncertainty and abstention (showing when not to trust)

Well-designed AI systems communicate analytical limitations clearly (Olsson et al., 2022):

- An "Abstain" badge appears when confidence is too low. Badges appear differently across various AI solutions and may include color-coded confidence indicators (e.g., circular badges or a rectangular "traffic light system" bar in the corner of the viewing interface), text-based warnings appearing above or beside the heatmap overlay, icon-based indicators (e.g., warning triangle, stop sign/"X" symbol, question mark [?], eye with a slash), confidence meters, and heatmap overlay modifications.
- Reason codes explain why (poor staining, out-of-focus, unusual pattern).
- Quality dashboards show abstention rates.

This selective abstention improves safety more than showing confidence scores alone, as it actively prevents over-reliance on uncertain predictions (Ramwala et al., 2024; Olsson et al., 2022).

TABLE 9.4 xAI mechanisms for GU pathology: what they are, what the pathologist sees, and how to use them.

Method	Description	What the pathologist sees	Strengths/Limitations in practice
Native attention/ MIL (patch-ranking, transformer attention)	Attention weights learned within the model during training show which patches/regions contributed most to the decision—where the model actually "looked" during the analysis.	Heatmap overlay with warm colors highlighting important regions, plus a ranked patch gallery (e.g., top 5–12 tiles) showing the most diagnostic areas. Each tile is clickable, allowing users to navigate directly to that location in the WSI.	**Strengths:** Most faithful and stable evidence; computationally efficient; available for every case. **Limitations:** May highlight correlated rather than causal features if training data are biased; can appear diffuse if tile size is too large. **Best use:** Primary explanation method for prostate, bladder, and kidney tasks.
Post-hoc saliency (Grad-CAM, occlusion analysis)	Reverse-engineered explanations are generated after prediction. **Grad-CAM:** Uses gradients to estimate importance in one pass. **Occlusion:** Systematically masks regions and re-runs the model to identify where predictions change the most.	**Grad-CAM:** Colored heatmap showing sensitivity regions. **Occlusion:** Grid showing prediction changes when different regions are masked or displaying importance scores for each region.	**Strengths:** Works when native attention is unavailable; model-agnostic (occlusion). **Limitations:** Grad-CAM can be unstable with low resolution; occlusion is computationally expensive with potential grid artifacts; both are less reliable on modern architectures. **Best use:** Supplementary evidence only, never as the sole explanation (Froń et al., 2023).
Prototype retrieval/concept activation (for foundation models)	Model retrieves similar cases from a reference database or scores the presence of learned concepts (e.g., "cribriform pattern") based on text-image training.	**Prototype gallery:** "Similar cases" with percentage match scores. **Concept scorecard:** Named features with confidence scores (e.g., "Cribriform: 83%," "Comedonecrosis: 12%").	**Strengths:** Intuitive for pathologists; excellent for education and rare entity verification. **Limitations:** Must still provide pixel-level evidence; risk of hallucinated concepts if unconstrained. **Best use:** Only with region-grounded evidence and clear abstention rules.

Sources: Vorontsov, E., Bozkurt, A., Casson, A., et al., 2024. A foundation model for clinical-grade computational pathology and rare cancers detection. Nat. Med. 30 (10), 2924–2935; Amann, J., Blasimme, A., Vayena, E., Frey, D., Madai VI, T.P.Q., 2020. Explainability for artificial intelligence in healthcare: a multidisciplinary perspective. BMC Med. Inform. Decis. Mak. 20 (1), 310; Diao, J.A., Wang, J.K., Chui, W.F., et al., 2021. Human-interpretable image features derived from densely mapped cancer pathology slides predict diverse molecular phenotypes. Nat. Commun. 12 (1), 1613; Froń, A., Semianiuk, A., Lazuk, U., et al., 2023. Artificial intelligence in urooncology: what we have and what we expect. Cancers 15 (17); Lu, M.Y., Williamson, D.F.K., Chen, T.Y., Chen, R.J., Barbieri, M., Mahmood, F., 2021. Data-efficient and weakly supervised computational pathology on whole-slide images. Nat. Biomed. Eng. 5 (6), 555–570.

Bias and drift detection (preventing shortcuts and catching changes)

AI systems monitor their performance using (Vorontsov et al., 2024):

- Site-specific performance metrics
- Verification that AI focuses on tissue, not artifacts.
- Alerts when performance metrics shift
- Regular reports stratified by scanner and staining batch.

This monitoring detects when AI might be using spurious correlations (e.g., scanner watermarks) or when input distributions have shifted (Vorontsov et al., 2024; Ramwala et al., 2024).

Audit trail (making decisions reproducible)

Every AI-assisted diagnosis should preserve (Vorontsov et al., 2024; Amann et al., 2020):

- Exact coordinates of analyzed regions
- Threshold values used
- Model and dataset versions
- Screenshots of what the pathologist saw

This documentation enables quality review, troubleshooting, and fair evaluation of discordant cases (Vorontsov et al., 2024; Amann et al., 2020; Ramwala et al., 2024).

How these features change daily practice

Explainability features for AI-enhanced GU workflows offer improved diagnostic outcomes:

- Heatmaps and image galleries reduce time to initial assessment, especially for small suspicious foci, because pathologists immediately see the same visual cues they would search for manually (Ström et al., 2020; Ennab and McHeick, 2024; Steiner et al., 2020; Raciti et al., 2023).
- Automated measurements eliminate tedious manual counting and improve consistency across pathologists and over time (Bulten et al., 2022; Tolkach et al., 2023).
- Selective abstention prevents over-reliance by keeping low-confidence cases in fully manual review mode (Olsson et al., 2022).
- Audit trails make it straightforward to review discordant cases, improving trust and enabling rapid corrective action (Vorontsov et al., 2024; Amann et al., 2020; Ramwala et al., 2024).

Real-world examples: How evidence changes clinical decisions

Prostate small cribriform focus: The AI heatmap highlights a 0.3 mm glandular nest. Top-ranked patches show punched-out lumina and rigid architectural patterns. The system updates the diagnosis from Gleason 3+3 to 3+4, with pattern 4 comprising 12%, which exports directly to the report. Trust increases because the visual evidence matches what pathologists look for (Bulten et al., 2022; Ström et al., 2020; Ennab and McHeick, 2024; Raciti et al., 2023; Tolkach et al., 2023).

Bladder TURBT with cautery artifact: The AI suggests possible muscle invasion but abstains due to poor focus and unusual tissue appearance. The pathologist proceeds without AI assistance and requests a collegial review, avoiding a false-positive invasive diagnosis (Olsson et al., 2022; Khoraminia et al., 2023).

Lymph node screening: A set of lymph nodes moves to a "likely negative" queue after AI detects no suspicious regions and raises no quality concerns. The pathologist performs a quick verification review and efficiently signs out the case (Froń et al., 2023).

These explainability features enable specific workflow enhancements across the diagnostic process.

Workflow applications

Pre-sign-out applications: AI systems perform automated quality control, flagging out-of-focus regions and inadequate tissue coverage before pathologist review (Patel et al., 2022; Patel et al., 2021). Case prioritization algorithms rank specimens by cancer probability, generating region-of-interest thumbnails attached to digital worklists (Eloy et al., 2023; Rienda et al., 2024). These pre-screening functions enable efficient batch processing, with benign-likely cases grouped for rapid review, while suspicious cases receive prioritized attention.

During sign-out applications, real-time AI assistance provides heatmap overlays highlighting suspicious regions, complemented by galleries of the most concerning areas at high magnification (Ström et al., 2020; Ennab and McHeick, 2024; Steiner et al., 2020; Raciti et al., 2023). Quantitative features include automated Gleason pattern percentages, tumor volume measurements, and linear extent calculations that export directly to reporting systems with single-click functionality (Bulten et al., 2022; Ennab and McHeick, 2024; Steiner et al., 2020; Tolkach et al., 2023). Confidence scores and abstention indicators help pathologists gauge when to rely on AI assistance versus proceeding with traditional review (Olsson et al., 2022).

Post-Sign Out Applications: Quality assurance dashboards track concordance between AI predictions and final diagnoses, stratified by site, scanner, and staining batch to detect performance drift (Ramwala et al., 2024; Park and Kressel, 2018; Fehr et al., 2022; Patel et al., 2024; Ennab and McHeick, 2024). AI-generated structured data facilitates tumor board packet preparation, combining histologic parameters with molecular and imaging findings when available (Zhou et al., 2024; Hu et al., 2024; Riaz et al., 2024; Parker et al., 2025; Tward et al., 2025). Systematic monitoring of abstention rates and error patterns enables continuous improvement and targeted retraining when performance metrics shift.

Future directions and conclusion

Foundation models trained on diverse pathology datasets offer improved generalization across laboratories and staining conditions, though rare entities and specialized tasks still require targeted fine-tuning and local validation (Chen et al., 2024; Xu et al., 2024; Hacking, 2025; Peng et al., 2025; Lee et al., 2025). Vision-language systems like CONCH enable more intuitive query-based retrieval and automated report generation, though careful validation remains essential to prevent hallucinations and ensure appropriate diagnostic grounding (Lu et al., 2024; Lu et al., 2024).

The trajectory of AI in urologic pathology points toward comprehensive integration across the diagnostic workflow, with multimodal fusion of histology, radiology, and molecular data enabling personalized cancer management. Success depends on maintaining appropriate human oversight while leveraging AI's strengths in pattern recognition, quantification, and consistency. As these tools mature from experimental technologies to clinical standards, the key challenge shifts from technical development to thoughtful implementation that enhances rather than replaces pathologists' expertise.

The clinical deployment curve for AI in urologic pathology is in crescendo, with increased volume in prostate pathology, where multiple commercial solutions demonstrate real-world value. The technology's impact spans the entire diagnostic workflow, from pre-analytical quality control through final report generation, with documented improvements in efficiency and consistency. While bladder and renal applications show promise, and testicular pathology remains nascent, the field's trajectory toward comprehensive AI integration appears inevitable. The institutions best positioned for this transformation are those building experience today with available tools while maintaining rigorous validation standards, appropriate skepticism, and realistic considerations for implementation. Ultimate success lies not in autonomous AI diagnosis but in human-AI collaboration that combines computational precision with pathologist judgment, ensuring that technological advancement translates to improved patient care.

Applications of AI in other subspecialties

Applications of AI in lung pathology

The field of lung pathology presents unique opportunities and challenges for AI integration due to the high prevalence and histologic diversity of lung diseases, particularly lung cancer, which remains the leading cause of cancer mortality worldwide (Russell et al., 2011; Sung et al., 2021; Rudin et al., 2021). AI has demonstrated strong performance across key diagnostic and prognostic tasks, such as tumor classification, grading, mutation prediction, and treatment response stratification.

One of the earliest and most clinically promising applications of AI in lung pathology is tumor detection and classification. Automated algorithms can distinguish between benign and malignant lung lesions, differentiate histologic subtypes of lung carcinoma, and identify subtle morphologic changes associated with early disease (Coudray et al., 2018; Le Page et al., 2021; Kriegsmann et al., 2020). CNN-based models trained on large annotated datasets have achieved pathologist-level accuracy for differentiating adenocarcinoma, squamous cell carcinoma, and small cell carcinoma, even in challenging biopsy samples with limited tissue (Coudray et al., 2018; Wei et al., 2019; Yu et al., 2020; Lami et al., 2023; Sheikh et al., 2022). Importantly, these algorithms can highlight diagnostically relevant regions of interest, functioning as decision-support tools that improve efficiency and reduce oversight errors. AI-assisted classification could be particularly impactful in community and resource-limited settings, where subspecialty thoracic pathology expertise may be scarce. Beyond routine classification, AI models are being explored for more granular histologic tasks, such as tumor grading, detection of lymphovascular invasion, or assessment of tumor budding—parameters that are prognostically important but often suffer from interobserver variability.

Another key area of innovation is the use of AI for lung cancer biomarker prediction from H&E-stained slides. Several landmark studies have demonstrated that DL algorithms can predict oncogenic driver mutations such as *EGFR*, *KRAS*, *ALK*, and *TP53* directly from routine histopathology images, with accuracies approaching or exceeding 80% for some mutations (Coudray et al., 2018; Chen et al., 2020). Such approaches could complement or even partially substitute molecular testing, particularly in low-resource environments or as rapid triage tools to prioritize confirmatory testing. In addition, AI models have shown promise in predicting PD-L1 expression status and tumor mutational burden, two key biomarkers guiding immunotherapy decisions (Chen et al., 2020; Tschuchnig et al., 2020; Kapil et al., 2018; Wu et al., 2022). These "virtual biomarkers" leverage subtle morphological correlates of molecular alterations that are often imperceptible to the human eye, illustrating AI's potential to bridge histology and genomics. While these algorithms are not yet ready to replace molecular assays, they offer exciting opportunities for cost-effective, scalable biomarker screening and could accelerate personalized medicine in lung cancer care.

AI has also demonstrated considerable potential for prognostic and predictive modeling in lung pathology. By learning complex spatial relationships among tumor cells, stroma, and immune infiltrates, algorithms can generate risk scores associated with survival and treatment response. For example, AI-based histologic features outperform traditional clinical staging in predicting patient outcomes for non-small cell lung carcinoma (NSCLC) (Wu et al., 2022; Luo et al., 2017; Wang et al., 2017). Spatial analysis of the tumor microenvironment, including quantification of TILs and tertiary lymphoid structures, has been particularly informative for immunotherapy response prediction (Corredor et al., 2019). AI can also identify high-risk morphologic phenotypes within early-stage lung adenocarcinomas that are not captured by conventional grading systems, supporting more individualized prognostication (Wang et al., 2017). These prognostic models exemplify how AI can move beyond mimicking human diagnostic tasks to uncover novel, clinically relevant tissue phenotypes.

In conclusion, AI tools now demonstrate robust performance in tumor detection, classification, biomarker prediction, and prognostic modeling, with the potential to reshape clinical workflows and accelerate precision medicine. However, realizing this promise requires rigorous validation, interpretability, regulatory oversight, and thoughtful integration into pathology practice.

Applications of AI in hematopathology

Hematopathology was among the earliest fields in pathology to explore computer-assisted image analysis, driven by the inherently quantitative nature of hematologic specimens and the potential to automate labor-intensive tasks such as differential counting and blast identification (Kratz et al., 2019; Cornet et al., 2008). With the advent of high-resolution WSI, CNNs, and other DL architectures, AI applications have expanded rapidly from basic cell classification to comprehensive diagnostic support systems capable of integrating morphology, immunophenotyping, cytogenetics, and molecular data in hematopathology.

Peripheral blood smear analysis represents one of the most mature areas of AI application in hematopathology. Early machine learning algorithms used handcrafted features to classify leukocytes, erythrocytes, and platelets, but modern CNN-based models now achieve human-level accuracy in differential counts and abnormal cell detection (Pohlkamp et al., 2020). These systems can identify subtle morphological changes, flag rare or abnormal cells, and perform rapid quantitative analysis, offering significant advantages in efficiency and reproducibility. Several algorithms have demonstrated high performance in detecting blasts, dysplastic cells, and parasitic infections such as malaria from digitized blood smears (Nanaa et al., 2021; Shouval et al., 2021; Muhsen et al., 2021; Syed-Abdul et al., 2020). Automated blood cell analysis is especially valuable in resource-limited settings and high-volume laboratories, where workforce shortages and variability in expertise can affect diagnostic quality.

Bone marrow aspirate and biopsy interpretation present distinct challenges and opportunities for AI. Bone marrow smears contain densely packed, morphologically diverse cells that are difficult to analyze using traditional rule-based algorithms. DL has overcome many of these limitations by learning complex morphological patterns directly from data. CNNs can be trained to detect blasts, plasma cells, or dysplastic lineages, as well as to assess overall cellularity and lineage distribution from bone marrow specimens (Fu et al., 2020; Tang et al., 2021; Chen et al., 2021; Chandradevan et al., 2020). Automated quantification of hematopoietic elements can reduce observer variability in assessments of marrow cellularity and differential counts—key components in diagnosing acute leukemias, myelodysplastic syndromes, and marrow failure states (Fu et al., 2020; Tang et al., 2021; Chen et al., 2021). Bone marrow core biopsies, meanwhile, have benefited from AI-driven segmentation algorithms that identify and quantify key structures such as trabecular bone, adipocytes, hematopoietic tissue, and fibrotic regions (Reta et al., 2015). These methods enable standardized cellularity estimation, fibrosis grading, and topographic analysis of neoplastic infiltrates, improving both diagnostic accuracy and reproducibility.

Lymph node and extranodal tissue evaluation, critical for lymphoma diagnosis, is another frontier for AI application in hematopathology. Unlike peripheral blood or marrow smears, lymphoid tissues require integration of architectural, cytologic, and immunophenotypic features, making them particularly complex. AI tools for lymphoma classification have made significant strides by combining digital histopathology with IHC and, increasingly, molecular data (Achi et al., 2019; Mohlman et al., 2020; Syrykh et al., 2020). DL algorithms can classify lymphoma subtypes based on morphologic patterns alone, achieving impressive accuracies for certain entities, such as diffuse large B-cell lymphoma versus Hodgkin lymphoma (Achi et al., 2019; Mohlman et al., 2020; Syrykh et al., 2020). Moreover, AI-assisted IHC quantifications, such as for CD20, Ki-67, or PD-L1, enhance precision in biomarker interpretation, which is critical for diagnosis, prognostication, and therapy selection (Achi et al., 2019; Mohlman et al., 2020; Syrykh et al., 2020). Importantly, AI can uncover subtle spatial patterns in cellular organization and marker expression that may correlate with disease subtype or outcome, thus contributing new insights beyond human visual assessment.

An exciting and rapidly evolving application of AI in hematopathology is the integration of digital morphology with high-dimensional ancillary data, including flow cytometry, cytogenetics, and next-generation sequencing. AI algorithms have been demonstrated to be capable of cross-referencing morphological findings with immunophenotypic signatures or mutational profiles, enabling more comprehensive and efficient diagnostic classification. For example, DL models trained on bone marrow images and flow cytometry data have shown promise in predicting specific leukemia subtypes, potentially reducing the time and cost associated with extensive laboratory testing (Biehl et al., 2013; Lakoumentas et al., 2009; Angeletti, 2018). AI tools are also being explored for automated karyotype interpretation and FISH signal quantification, areas traditionally requiring specialized expertise (Schnegg-Kaufmann et al., 2025). By bridging these multiple data modalities, AI has the potential to support fully integrated hematopathology diagnostics that mirror expert decision-making but with greater speed and scalability.

Nonetheless, several challenges must be addressed to realize the full potential of AI in hematopathology. Hematopathology slides present unique imaging challenges, such as variable smear thickness and staining heterogeneity, which can complicate image analysis (Schnegg-Kaufmann et al., 2025). In addition, hematopathology encompasses a wide spectrum of diseases, many of which are rare, making data aggregation and standardization essential. Differences in staining protocols, scanner hardware, and case mix across institutions can reduce algorithm generalizability.

In conclusion, current AI applications already demonstrate high accuracy in peripheral blood smear analysis, bone marrow interpretation, and lymphoma classification, while emerging tools are beginning to integrate complex ancillary data for comprehensive diagnostic support.

Applications of AI in dermatopathology

Dermatopathology relies heavily on nuanced visual interpretation, where diagnostic accuracy can be challenged by overlapping histologic patterns, rare entities, and increasing specimen volumes. Early applications of AI in dermatopathology built upon advances in digital pathology and computer vision, aiming to replicate expert interpretation of H&E slides through machine learning and DL methods. These models excel at pattern recognition, leveraging pixel-level data to learn complex morphologic signatures often imperceptible to the human eye. AI models trained on large, well-annotated datasets have achieved diagnostic performance comparable to, and sometimes exceeding, that of experienced dermatopathologists for narrowly defined tasks such as classification of skin cancers (Esteva et al., 2017; Cruz-Roa et al., 2013; Xie et al., 2019). CNN-based classifiers have shown high accuracy for distinguishing benign nevi, dysplastic nevi, and melanomas (Kulkarni et al., 2020; Hekler et al., 2019; Cazzato et al., 2023). Such models could assist in early melanoma detection. Another area is **inflammatory dermatoses**, where AI can cluster cases based on histopathologic similarity, potentially revealing novel disease subgroups or supporting pattern-based diagnosis (Doeleman et al., 2022). **Quantitative image analysis** of features such as tumor thickness, mitotic activity, or immune infiltrates offers new prognostic and predictive biomarkers that complement traditional histology (Zheng et al., n.d.).

However, several limitations temper these strengths. First, dermatopathology encompasses a broad and heterogeneous disease spectrum; most current algorithms are "narrow AI" systems optimized for single diagnostic questions rather than the complex differential diagnoses encountered in daily practice. Similar to other subspecialties, many models for dermatopathology lack generalizability, performing well only on data from the institutions that generated the training set. Only a minority of studies evaluate algorithm performance across the full dermatopathologic spectrum, assess clinical utility, or examine downstream patient outcomes.

The most immediate utility of real-world applications of AI tools in dermatopathology may lie in triage, quality control, and decision support rather than autonomous diagnosis. Algorithms can pre-screen digitized slides to flag suspicious regions, prioritize cases for review, or detect technical artifacts. Such applications have shown promise in improving efficiency, reducing turnaround times, and enabling pathologists to focus their expertise on challenging cases.

In conclusion, current AI applications in dermatopathology demonstrate impressive technical capabilities but also highlight the importance of rigorous validation, explainability, and integration into human-centered workflows. AI's greatest potential lies not in replacing dermatopathologists but in empowering them with enhanced tools for diagnosis, quantification, discovery, and education.

Applications of AI in gynecologic pathology

Gynecologic pathology encompasses a heterogeneous group of neoplasms and nonneoplastic conditions affecting the cervix, endometrium, ovary, vulva, vagina, and placenta, many of which present diagnostic challenges due to overlapping morphologic features, tumor heterogeneity, and the need for integration of molecular and clinical data (Sung et al., 2021). Recent advances in digital pathology and computational power have led to the development of AI applications that can assist pathologists in making more accurate, reproducible, and efficient diagnoses.

One of the AI applications in gynecologic pathology is in the diagnosis and classification of gynecologic malignancies. AI algorithms have been trained to identify cervical intraepithelial neoplasia and invasive carcinoma on digitized biopsy and excision specimens, achieving accuracy comparable to expert pathologists (Habtemariam et al., 2022; Y-x et al., 2023). For endometrial carcinoma, AI tools have been developed to classify histologic subtypes (e.g., endometrioid, serous, clear cell) and to grade tumors, tasks that are often subject to interobserver variation (Zhang et al., 2022; Mohammadi et al., 2022; Fremond et al., 2023; Song et al., 2022; Hong et al., 2021). In ovarian cancer, DL approaches can assist in distinguishing primary ovarian tumors from metastases and in recognizing borderline tumors, areas that frequently present diagnostic difficulty (BenTaieb et al., 2017; Jiang et al., 2021; Wu et al., 2018; Farahani et al., 2022).

AI has also demonstrated significant promise in biomarker prediction and prognostication for gynecologic cancers. Several studies have demonstrated that DL models applied to H&E slides successfully predicted key molecular alterations, including TP53 mutations, POLE mutations, mismatch repair deficiency (MMRd), BRCA mutations, and homologous recombination deficiency, particularly in endometrial and ovarian carcinomas (Nero et al., 2022; Hong et al., 2021; Veeraraghavan et al., 2020; Liu et al., 2025; Umemoto et al., 2024; Shen et al., 2025; Zhang et al., 2025; Bergstrom et al., 2024). These molecular features are increasingly used to guide risk stratification and therapeutic decisions, including immunotherapy eligibility and PARP inhibitor use. Furthermore, AI-driven histomorphometric analyses have been shown to predict clinical outcomes such as recurrence risk and survival, often outperforming conventional staging systems (Fremond et al., 2023; Zeng et al., 2021; Laury et al., 2021; Sengupta et al., 2022; Volinsky-Fremond et al., 2024; Marmé et al., 2025). For example, spatial analysis of tumor–stroma interactions and immune cell infiltration using DL has yielded prognostic signatures that could complement or refine existing clinicopathologic risk models (Suzuki et al., 2025).

The use of AI in placental pathology represents a rapidly evolving and distinctive subfield within gynecologic pathology. AI tools have been developed to analyze placental villous morphology, vascular patterns, and inflammatory changes with high precision. CNN-based models can classify lesions such as villitis, infarction, and vascular malperfusion, and can quantify morphometric features like villous size, shape, and density, which are relevant to conditions such as preeclampsia and intrauterine growth restriction (Faure et al., 2025; Kidron et al., 2017; Clymer et al., 2020).

In conclusion, AI tools now demonstrate strong capabilities in automated classification, biomarker prediction, prognostication, and morphometric analysis, with significant potential to improve diagnostic accuracy, standardization, and efficiency. The integration of AI into gynecologic pathology promises not only to enhance current diagnostic practices but also to open new avenues for discovery and personalized medicine, ultimately improving outcomes for women's health.

References

Abbas, S., Shafik, R., Soomro, N., Heer, R., Adhikari, K., 2024. AI predicting recurrence in non-muscle-invasive bladder cancer: systematic review with study strengths and weaknesses. Front. Oncol. 14, 1509362.

Abdolahi, M., Salehi, M., Shokatian, I., Reiazi, R., 2020. Artificial intelligence in automatic classification of invasive ductal carcinoma breast cancer in digital pathology images. Med. J. Islam. Repub. Iran 34, 140.

Abousamra, S., Gupta, R., Hou, L., et al, 2021. Deep learning-based mapping of tumor infiltrating lymphocytes in whole slide images of 23 types of cancer. Front. Oncol. 11, 806603.

Abuhadra, N., Stecklein, S., Sharma, P., Moulder, S., 2022. Early-stage triple-negative breast cancer: time to optimize personalized strategies. Oncologist 27 (1), 30–39.

Achi, H.E., Belousova, T., Chen, L., et al, 2019. Automated diagnosis of lymphoma with digital pathology images using deep learning. Ann. Clin. Lab. Sci. 49 (2), 153–160.

Ács, B., Madaras, L., Kovács, K.A., et al, 2018. Reproducibility and prognostic potential of Ki-67 proliferation index when comparing digital-image analysis with standard semi-quantitative evaluation in breast cancer. Pathol. Oncol. Res. 24 (1), 115–127.

Ahern, T.P., Beck, A.H., Rosner, B.A., et al, 2017. Continuous measurement of breast tumour hormone receptor expression: a comparison of two computational pathology platforms. J. Clin. Pathol. 70 (5), 428–434.

Ahn, J.S., Shin, S., Yang, S.A., et al, 2023. Artificial intelligence in breast cancer diagnosis and personalized medicine. J. Breast Cancer 26 (5), 405–435.

Aiosin. NephroPath AI kidney pathology demo images. *Aiosyn Image Gallery* 2025; https://www.aiosyn.com/nephropath-platform/. Accessed 10 Aug 2025.

Albusayli, R., Graham, J.D., Pathmanathan, N., et al, 2023. Artificial intelligence-based digital scores of stromal tumour-infiltrating lymphocytes and tumour-associated stroma predict disease-specific survival in triple-negative breast cancer. The Journal of Pathology 260 (1), 32–42.

Alkhasawneh, A., Reith, J.D., Toro, T.Z., et al, 2015. Interobserver variability of mitotic index and utility of PHH3 for risk stratification in gastrointestinal stromal tumors. Am. J. Clin. Pathol. 143 (3), 385–392.

Allred, D.C., Harvey, J.M., Berardo, M., Clark, G.M., 1998. Prognostic and predictive factors in breast cancer by immunohistochemical analysis. Modern Pathology: an Official Journal of the United States and Canadian Academy of Pathology, Inc 11 (2), 155–168.

Amann, J., Blasimme, A., Vayena, E., Frey, D., Madai VI, T.P.Q., 2020. Explainability for artificial intelligence in healthcare: a multidisciplinary perspective. BMC Med. Inform. Decis. Mak. 20 (1), 310.

Angeletti, C., 2018. A method for the interpretation of flow cytometry data using genetic algorithms. J. Pathol. Inform. 9, 16.

Anon., 2022. Aiforia releases another CE-IVD marked clinical AI Model — expanding its diagnostic pathology offering to prostate cancer; https://tissuepathology.com/2022/05/27/aiforia-releases-another-ce-ivd-marked-clinical-ai-model-expanding-its-diagnostic-pathology-offering-to-prostate-cancer/.

Anon., 2024. TechCyte and CytoBay complete initial validation studies for AI-powered bladder cancer screening solution; https://techcyte.com/techcyte-and-cytobay-complete-validation-for-ai-powered-bladder-cancer-screening-solution/.

Azam, A.S., Tsang, Y.W., Thirlwall, J., et al, 2024. Digital pathology for reporting histopathology samples, including cancer screening samples - definitive evidence from a multisite study. Histopathology 84 (5), 847–862.

Badve, S.S., Baehner, F.L., Gray, R.P., et al, 2008. Estrogen- and progesterone-receptor status in ECOG 2197: comparison of immunohistochemistry by local and central laboratories and quantitative reverse transcription polymerase chain reaction by central laboratory. J. Clin. Oncol. 26 (15), 2473–2481.

Balkenhol, M.C.A., Tellez, D., Vreuls, W., et al, 2019. Deep learning assisted mitotic counting for breast cancer. Lab. Invest. 99 (11), 1596–1606.

Bankhead, P., Loughrey, M.B., Fernández, J.A., et al, 2017. QuPath: open source software for digital pathology image analysis. Sci. Rep. 7 (1), 16878.

Bannier, P.A., Saillard, C., Mann, P., et al, 2024. AI allows pre-screening of FGFR3 mutational status using routine histology slides of muscle-invasive bladder cancer. Nat. Commun. 15 (1), 10914.

BenTaieb, A., Li-Chang, H., Huntsman, D., Hamarneh, G., 2017. A structured latent model for ovarian carcinoma subtyping from histopathology slides. Med. Image Anal. 39, 194–205.

Bergstrom, E.N., Abbasi, A., Díaz-Gay, M., et al, 2024. Deep learning artificial intelligence predicts homologous recombination deficiency and platinum response from histologic slides. J. Clin. Oncol. 42 (30), 3550–3560.

Biehl, M., Bunte, K., Schneider, P., 2013. Analysis of flow cytometry data by matrix relevance learning vector quantization. PLoS One 8 (3), e59401.

Bilal, M., Raza, S.E.A., Azam, A., et al, 2021. Development and validation of a weakly supervised deep learning framework to predict the status of molecular pathways and key mutations in colorectal cancer from routine histology images: a retrospective study. Lancet Digit. Health 3 (12), e763–e772.

Biswas, T., Efird, J.T., Prasad, S., Jindal, C., Walker, P.R., 2017. The survival benefit of neoadjuvant chemotherapy and pCR among patients with advanced stage triple negative breast cancer. Oncotarget 8 (68), 112712–112719.

Bokhorst, J.M., Ciompi, F., Ozturk, S.K., et al, 2023. Fully automated tumor bud assessment in hematoxylin and eosin-stained whole slide images of colorectal cancer. Mod. Pathol. 36 (9), 100233.

Boland, C.R., Goel, A., 2010. Microsatellite instability in colorectal cancer. Gastroenterology 138 (6), 2073–2087e2073.

Bolton, K.L., Garcia-Closas, M., Pfeiffer, R.M., et al, 2010. Assessment of automated image analysis of breast cancer tissue microarrays for epidemiologic studies. Cancer Epidemiol. Biomark. Prev. 19 (4), 992–999.

Brugmann, A., Eld, M., Lelkaitis, G., et al, 2012. Digital image analysis of membrane connectivity is a robust measure of HER2 immunostains. Breast Cancer Res. Treat. 132 (1), 41–49.

Bulten, W. AI for diagnosis of prostate cancer: the PANDA challenge. 2022; https://www.wouterbulten.nl/posts/panda-challenge/.

Bulten, W., Kartasalo, K., Chen, P.C., et al, 2022. Artificial intelligence for diagnosis and Gleason grading of prostate cancer: the PANDA challenge. Nat. Med. 28 (1), 154–163.

Bulten, W., Pinckaers, H., van Boven, H., et al, 2020. Automated deep-learning system for Gleason grading of prostate cancer using biopsies: a diagnostic study. Lancet Oncol. 21 (2), 233–241.

Busby, D., Grauer, R., Pandav, K., et al, 2024. Applications of artificial intelligence in prostate cancer histopathology. Urologic Oncology 42 (3), 37–47.

Bychkov, D., Linder, N., Tiulpin, A., et al, 2021. Deep learning identifies morphological features in breast cancer predictive of cancer ERBB2 status and trastuzumab treatment efficacy. Sci. Rep. 11 (1), 4037.

Campora, M., Paudice, M., Gambella, A., et al, 2023. Counting mitoses in gastrointestinal stromal tumours (GISTs): variable practices in the real-world setting and their clinical implications. Virchows Arch. 482 (3), 589–594.

Cao, L.L., Lu, J., Li, P., et al, 2019. Evaluation of the eighth edition of the American joint committee on cancer TNM staging system for gastric cancer: an analysis of 7371 patients in the SEER database. Gastroenterol. Res. Pract. 2019, 6294382.

Cappellesso, R., Luchini, C., Veronese, N., et al. 2017. Tumor budding as a risk factor for nodal metastasis in pT1 colorectal cancers: a meta-analysis. Hum. Pathol. 65, 62–70.

Cazzato, G., Massaro, A., Colagrande, A., et al, 2023. Artificial intelligence applied to a first screening of naevoid melanoma: a new use of fast random forest algorithm in dermatopathology. Curr. Oncol. 30 (7), 6066–6078.

Challa, B., Tahir, M., Hu, Y., et al, 2023. Artificial intelligence-aided diagnosis of breast cancer lymph node metastasis on histologic slides in a digital workflow. Modern Pathology: an Official Journal of the United States and Canadian Academy of Pathology, Inc 36 (8), 100216.

Challenge, G., n.d. The PANDA Challenge. https://panda.grand-challenge.org. Accessed July 1, 2025.

Chan, R.C., CKC, T., Cheng, K.C.T., Yoshikazu, T., Yan, L.L.A., Tse, G.M., 2023. Artificial intelligence in breast cancer histopathology. Histopathology 82 (1), 198–210.

Chandradevan, R., Aljudi, A.A., Drumheller, B.R., et al, 2020. Machine-based detection and classification for bone marrow aspirate differential counts: initial development focusing on nonneoplastic cells. Lab. Invest. 100 (1), 98–109.

Chandramohan, D., Garapati, H.N., Nangia, U., et al, 2024. Diagnostic accuracy of deep learning in detection and prognostication of renal cell carcinoma: a systematic review and meta-analysis. Front. Med. (Lausanne 11, 1447057.

Chatrian, A., Colling, R.T., Browning, L., et al, 2021. Artificial intelligence for advance requesting of immunohistochemistry in diagnostically uncertain prostate biopsies. Modern Pathology: an Official Journal of the United States and Canadian Academy of Pathology, Inc 34 (9), 1780–1794.

Cheang, M.C., Chia, S.K., Voduc, D., et al, 2009. Ki67 index, HER2 status, and prognosis of patients with luminal B breast cancer. J. Natl. Cancer Inst. 101 (10), 736–750.

Chen, M., Zhang, B., Topatana, W., et al. 2020. Classification and mutation prediction based on histopathology H&E images in liver cancer using deep learning. NPJ Precis. Oncol. 4, 14.

Chen, P., chen Xu, R., Chen, N., et al, 2021. Detection of metastatic tumor cells in the bone marrow aspirate smears by artificial intelligence (AI)-based morphogo system. Front. Oncol. 11, 742395.

Chen, R.J., Ding, T., Lu, M.Y., et al, 2024. Towards a general-purpose foundation model for computational pathology. Nat. Med. 30 (3), 850–862.

Chidambaram, S., Sounderajah, V., Maynard, N., Markar, S.R., 2022. Diagnostic performance of artificial intelligence-centred systems in the diagnosis and postoperative surveillance of upper gastrointestinal malignancies using computed tomography imaging: a systematic review and meta-analysis of diagnostic accuracy. Ann. Surg. Oncol. 29 (3), 1977–1990.

Christgen, M., von Ahsen, S., Christgen, H., Länger, F., Kreipe, H., 2015. The region-of-interest size impacts on Ki67 quantification by computer-assisted image analysis in breast cancer. Human Pathology 46 (9), 1341–1349.

Chuluunbaatar, Y., Bansal, S., Brodie, A., Sharma, A., Vasdev, N., 2023. The current use of artificial intelligence in testicular cancer: a systematic review. Artif. Intell. Surg. 3 (4), 195–206.

Ciocca, D.R., Elledge, R., 2000. Molecular markers for predicting response to tamoxifen in breast cancer patients. Endocrine 13 (1), 1–10.

Clymer, D., Kostadinov, S., Catov, J., et al, 2020. Decidual vasculopathy identification in whole slide images using multiresolution hierarchical convolutional neural networks. Am. J. Pathol. 190 (10), 2111–2122.

Collins, L.C., Botero, M.L., Schnitt, S.J., 2005. Bimodal frequency distribution of estrogen receptor immunohistochemical staining results in breast cancer: an analysis of 825 cases. Am. J. Clin. Pathol. 123 (1), 16–20.

Cornet, E., Perol, J.P., Troussard, X., 2008. Performance evaluation and relevance of the CellaVision DM96 system in routine analysis and in patients with malignant hematological diseases. Int. J. Lab. Hematol. 30 (6), 536–542.

Corredor, G., Wang, X., Zhou, Y., et al, 2019. Spatial architecture and arrangement of tumor-infiltrating lymphocytes for predicting likelihood of recurrence in early-stage non-small cell lung cancer. Clin. Cancer Res. 25 (5), 1526–1534.

Coudray, N., Ocampo, P.S., Sakellaropoulos, T., et al, 2018. Classification and mutation prediction from non-small cell lung cancer histopathology images using deep learning. Nat. Med. 24 (10), 1559–1567.

Couture, H.D., Williams, L.A., Geradts, J., et al. 2018. Image analysis with deep learning to predict breast cancer grade, ER status, histologic subtype, and intrinsic subtype. NPJ Breast Cancer 4, 30.

Cruz-Roa, A.A., Arevalo Ovalle, J.E., Madabhushi, A., González Osorio, F.A. 2013 A deep learning architecture for image representation, visual interpretability and automated basal-cell carcinoma cancer detection. Paper presented at: International conference on medical image computing and computer-assisted intervention.

Cruz-Roa, A., Gilmore, H., Basavanhally, A., et al, 2017. Accurate and reproducible invasive breast cancer detection in whole-slide images: a Deep Learning approach for quantifying tumor extent. Sci. Rep. 7, 46450.

da Costa Re, A.R., de Oliveira Ft, R., Nascimento Araújo, A.L., Vieira, S.C., 2023. Impact of pathologic complete response on the prognosis of triple-negative breast cancer patients: a cohort study. Cureus 15 (4), e37396.

Das, A., Nair, M.S., Peter, D.S., 2020. Batch mode active learning on the riemannian manifold for automated scoring of nuclear pleomorphism in breast cancer. Artif. Intell. Med. 103, 101805.

Diao, J.A., Wang, J.K., Chui, W.F., et al, 2021. Human-interpretable image features derived from densely mapped cancer pathology slides predict diverse molecular phenotypes. Nat. Commun. 12 (1), 1613.

Diaz, L.K., Sahin, A., Sneige, N., 2004. Interobserver agreement for estrogen receptor immunohistochemical analysis in breast cancer: a comparison of manual and computer-assisted scoring methods. Ann. Diagn. Pathol. 8 (1), 23–27.

Dobson, L., Conway, C., Hanley, A., et al, 2010. Image analysis as an adjunct to manual HER-2 immunohistochemical review: a diagnostic tool to standardize interpretation. Histopathology 57 (1), 27–38.

Doeleman, T., Westerbeek, D., Jansen, P., et al, 2022. Artificial intelligence-assisted probability scoring for differentiation of early mycosis fungoides and benign inflammatory dermatoses on H&E stained pathology slides of skin biopsies. Eur. J. Cancer 173.

Dowsett, M., Nielsen, T.O., A'Hern, R., et al, 2011. Assessment of Ki67 in breast cancer: recommendations from the International Ki67 in Breast Cancer working group. J. Natl. Cancer Inst. 103 (22), 1656–1664.

Ehteshami Bejnordi, B., Veta, M., van Diest P, J., et al, 2017. Diagnostic assessment of deep learning algorithms for detection of lymph node metastases in women with breast cancer. Jama 318 (22), 2199–2210.

El-Badry, A.M., Breitenstein, S., Jochum, W., et al, 2009. Assessment of hepatic steatosis by expert pathologists: the end of a gold standard. Ann. Surg. 250 (5), 691–697.

Eloy, C., Marques, A., Pinto, J., et al, 2023a. Artificial intelligence-assisted cancer diagnosis improves the efficiency of pathologists in prostatic biopsies. Virchows Arch. 482 (3), 595–604.

Eloy, C., Marques, A., Pinto, J., et al, 2023b. Artificial intelligence–assisted cancer diagnosis improves the efficiency of pathologists in prostatic biopsies. Virchows Arch. 482 (3), 595–604.

Elsharawy, K.A., Gerds, T.A., Rakha, E.A., Dalton, L.W., 2021. Artificial intelligence grading of breast cancer: a promising method to refine prognostic classification for management precision. Histopathology 79 (2), 187–199.

Ennab, M., McHeick, H., 2024. Enhancing interpretability and accuracy of AI models in healthcare: a comprehensive review on challenges and future directions. Front. Robot. AI 11, 1444763.

Esteva, A., Kuprel, B., Novoa, R.A., et al, 2017. Dermatologist-level classification of skin cancer with deep neural networks. nature 542 (7639).

Euscher, E.D., Marques-Piubelli, M.L., Ramalingam, P., et al, 2023. Extrauterine mesonephric-like carcinoma: a comprehensive single institution study of 33 cases. Am. J. Surg. Pathol. 47 (6), 635–648.

Evans, A.J., Brown, R.W., Bui, M.M., et al, 2022. Validating whole slide imaging systems for diagnostic purposes in pathology. Arch. Pathol. Lab. Med. 146 (4), 440–450.

Faghani, S., Codipilly, D.C., David, V., et al, 2022. Development of a deep learning model for the histologic diagnosis of dysplasia in Barrett's esophagus. Gastrointest. Endosc. 96 (6), 918–925e913.

Farahani, H., Boschman, J., Farnell, D., et al, 2022. Deep learning-based histotype diagnosis of ovarian carcinoma whole-slide pathology images. Mod. Pathol. 35 (12), 1983–1990.

Farahmand, S., Fernandez, A.I., Ahmed, F.S., et al, 2022. Deep learning trained on hematoxylin and eosin tumor region of Interest predicts HER2 status and trastuzumab treatment response in HER2+ breast cancer. Modern Pathology: an Official Journal of the United States and Canadian Academy of Pathology, Inc 35 (1), 44–51.

Faratian, D., Kay, C., Robson, T., et al, 2009. Automated image analysis for high-throughput quantitative detection of ER and PR expression levels in large-scale clinical studies: the TEAM trial experience. Histopathology 55 (5), 587–593.

Faryna, K., Tessier, L., Retamero, J., et al, 2024. Evaluation of artificial intelligence-based gleason grading algorithms "in the wild". Modern Pathology: an Official Journal of the United States and Canadian Academy of Pathology, Inc 37 (11), 100563.

Faure, G., Soglio, D.D., Patey, N., Oligny, L., Girard, S., Séoud, L., 2025. Self-supervised deep metric learning for prototypical zero-shot lesion retrieval in placenta whole-slide images. Comput. Biol. Med. 196 (Pt A).

Fehr, J., Jaramillo-Gutierrez, G., Oala, L., et al, 2022. Piloting a survey-based assessment of transparency and trustworthiness with three medical AI tools. Healthcare (Basel) 10 (10).

Feng, C., Liu, F., 2023. Artificial intelligence in renal pathology: current status and future. Biomol. Biomed. 23 (2), 225–234.

Fernandez, G., Prastawa, M., Madduri, A.S., et al, 2022. Development and validation of an AI-enabled digital breast cancer assay to predict early-stage breast cancer recurrence within 6 years. Breast Cancer Research: BCR 24 (1), 93.

Fitzgibbons, P.L., Page, D.L., Weaver, D., et al, 2000. Prognostic factors in breast cancer. College of American Pathologists Consensus Statement 1999. Arch. Pathol. Lab. Med. 124 (7), 966–978.

Flach, R.N., van Dooijeweert, C., Nguyen, T.Q., et al. 2025. Prospective clinical implementation of paige prostate detect artificial intelligence assistance in the detection of prostate cancer in prostate biopsies: CONFIDENT P trial implementation of artificial intelligence assistance in prostate cancer detection. JCO Clin. Cancer Inform. 9, e2400193.

Forlano, R., Mullish, B.H., Giannakeas, N., et al, 2020. High-throughput, machine learning-based quantification of steatosis, inflammation, ballooning, and fibrosis in biopsies from patients with nonalcoholic fatty liver disease. Clin. Gastroenterol. Hepatol. 18 (9), 2081–2090e2089.

Fremond, S., Andani, S., Wolf, J.B., et al, 2023. Interpretable deep learning model to predict the molecular classification of endometrial cancer from haematoxylin and eosin-stained whole-slide images: a combined analysis of the PORTEC randomised trials and clinical cohorts. Lancet Digit. Health 5 (2), e71–e82.

Froń, A., Semianiuk, A., Lazuk, U., et al, 2023. Artificial intelligence in urooncology: what we have and what we expect. Cancers 15 (17).

Fu, X., Fu, M., Li, Q., et al, 2020. Morphogo: an Automatic Bone Marrow Cell Classification System on Digital Images Analyzed by Artificial Intelligence. Acta Cytol. 64 (6), 588–596.

Gancberg, D., Jarvinen, T., Di Leo, A., et al, 2002. Evaluation of HER-2/NEU protein expression in breast cancer by immunohistochemistry: an interlaboratory study assessing the reproducibility of HER-2/NEU testing. Breast Cancer Res. Treat. 74 (2), 113–120.

Gao, Y., Wen, P., Liu, Y., et al, 2025. Application of artificial intelligence in the diagnosis of malignant digestive tract tumors: focusing on opportunities and challenges in endoscopy and pathology. J. Transl. Med. 23 (1), 412.

Gedefaw, L., Liu, C.F., RKL, I., et al, 2023. Artificial intelligence-assisted diagnostic cytology and genomic testing for hematologic disorders. Cells 12 (13).

Gelber, R.D., Gelber, S., 2009. Facilitating consensus by examining patterns of treatment effects. Breast 18, S2–S8.

Gerlinger, M., Rowan, A.J., Horswell, S., et al, 2012. Intratumor heterogeneity and branched evolution revealed by multiregion sequencing. N. Engl. J. Med. 366 (10), 883–892.

Ghahremani, P., Li, Y., Kaufman, A., et al, 2022. Deep learning-inferred multiplex immunofluorescence for immunohistochemical image quantification. Nat. Mach. Intell. 4 (4), 401–412.

Ghosh, A., Sirinukunwattana, K., Khalid Alham, N., et al, 2021. The potential of artificial intelligence to detect lymphovascular invasion in testicular cancer. Cancers (Basel 13 (6).

Gokhale, S., Rosen, D., Sneige, N., et al, 2007. Assessment of two automated imaging systems in evaluating estrogen receptor status in breast carcinoma. Appl. Immunohistochem. Mol. Morphol. 15 (4), 451–455.

Goodell, P.P., Krasinskas, A.M., Davison, J.M., Hartman, D.J., 2012. Comparison of methods for proliferative index analysis for grading pancreatic well-differentiated neuroendocrine tumors. Am. J. Clin. Pathol. 137 (4), 576–582.

Grala, B., Markiewicz, T., Kozłowski, W., Osowski, S., Słodkowska, J., Papierz, W., 2009. New automated image analysis method for the assessment of Ki-67 labeling index in meningiomas. Folia Histochem. Cytobiol. 47 (4), 587–592.

Greenson, J.K., Bonner, J.D., Ben-Yzhak, O., et al, 2003. Phenotype of microsatellite unstable colorectal carcinomas: well-differentiated and focally mucinous tumors and the absence of dirty necrosis correlate with microsatellite instability. Am. J. Surg. Pathol. 27 (5), 563–570.

Gruver, A.M., Lu, H., Zhao, X., et al, 2023. Pathologist-trained machine learning classifiers developed to quantitate celiac disease features differentiate endoscopic biopsies according to modified marsh score and dietary intervention response. Diagn. Pathol. 18 (1), 122.

Guitton, T., Allaume, P., Rabilloud, N., et al, 2023. Artificial intelligence in predicting microsatellite instability and KRAS, BRAF mutations from whole-slide images in colorectal cancer: a systematic review. Diagnostics (Basel 14 (1).

Gupta, A., Yu, A., Peyrin-Biroulet, L., Ananthakrishnan, A.N., 2021. Treat to target: the role of histologic healing in inflammatory bowel diseases: a systematic review and meta-analysis. Clin. Gastroenterol. Hepatol. 19 (9), 1800–1813e1804.

Habtemariam, L.W., Zewde, E.T., Simegn, G.L., 2022. Cervix type and cervical cancer classification system using deep learning techniques. Med. Devices Evid. Res., 163–176.

Hacking, S., 2025. Foundation models in pathology: bridging AI innovation and clinical practice. J. Clin. Pathol. 78 (7), 433–435.

Hagland, M., 2024. At Ohio state, a breakthrough leveraging AI in pathology. Healthcare Innovation. https://www.hcinnovationgroup.com/analytics-ai/article/53096393/at-ohio-state-a-breakthrough-leveraging-ai-in-pathology.

Hahnel, R., Woodings, T., Vivian, A.B., 1979. Prognostic value of estrogen receptors in primary breast cancer. Cancer 44 (2), 671–675.

Hall, A.R., Dhillon, A.P., Green, A.C., et al, 2013. Hepatic steatosis estimated microscopically versus digital image analysis. Liver Int. 33 (6), 926–935.

Hammond, M.E., Hayes, D.F., Dowsett, M., et al, 2010. American society of clinical oncology/college of American pathologists guideline recommendations for immunohistochemical testing of estrogen and progesterone receptors in breast cancer. J. Clin. Oncol. 28 (16), 2784–2795.

Han, Z., Wei, B., Zheng, Y., Yin, Y., Li, K., Li, S., 2017. Breast cancer multi-classification from histopathological images with structured deep learning model. Sci. Rep. 7 (1), 4172.

Harbeck, N., Rastogi, P., Martin, M., et al, 2021. Adjuvant abemaciclib combined with endocrine therapy for high-risk early breast cancer: updated efficacy and Ki-67 analysis from the monarchE study. Ann. Oncol. 32 (12), 1571–1581.

Hartage, R., Li, A.C., Hammond, S., Parwani, A.V., 2020. A validation study of human epidermal growth factor receptor 2 immunohistochemistry digital imaging analysis and its correlation with human epidermal growth factor receptor 2 fluorescence in situ hybridization results in breast carcinoma. J. Pathol. Inform. 11, 2.

He, Q.H., Tan, H., Liao, F.T., et al, 2022. Stratification of malignant renal neoplasms from cystic renal lesions using deep learning and radiomics features based on a stacking ensemble CT machine learning algorithm. Front. Oncol. 12, 1028577.

Hede, K., 2008. Breast cancer testing scandal shines spotlight on black box of clinical laboratory testing. J. Natl. Cancer Inst. 100 (12), 836–837,844.

Heindl, A., Sestak, I., Naidoo, K., Cuzick, J., Dowsett, M., Yuan, Y., 2018. Relevance of Spatial Heterogeneity of Immune Infiltration for Predicting Risk of Recurrence After Endocrine Therapy of ER+ Breast Cancer. J. Natl. Cancer Inst. 110 (2).

Hekler, A., Utikal, J.S., Enk, A.H., et al. 2019. Pathologist-level classification of histopathological melanoma images with deep neural networks. Eur. J. Cancer 115, 79–83.

Helin, H.O., Tuominen, V.J., Ylinen, O., Helin, H.J., Isola, J., 2016. Free digital image analysis software helps to resolve equivocal scores in HER2 immunohistochemistry. Virchows Arch. 468 (2), 191–198.

Hendry, S., Salgado, R., Gevaert, T., et al, 2017. Assessing tumor-infiltrating lymphocytes in solid tumors: a practical review for pathologists and proposal for a standardized method from the international immunooncology biomarkers working group: part 1: assessing the host immune response, TILs in invasive breast carcinoma and ductal carcinoma in situ, metastatic tumor deposits and areas for further research. Adv. Anat. Pathol. 24 (5), 235–251.

Hermsen, M., de Bel, T., den Boer, M., et al, 2019. Deep learning-based histopathologic assessment of kidney tissue. J Am Soc Nephrol 30 (10), 1968–1979.

Hirabayashi, K., Zamboni, G., Nishi, T., Tanaka, A., Kajiwara, H., Nakamura, N., 2013. Histopathology of gastrointestinal neuroendocrine neoplasms. Front. Oncol. 3, 2.

Holten-Rossing, H., Moller Talman, M.L., Kristensson, M., Vainer, B., 2015. Optimizing HER2 assessment in breast cancer: application of automated image analysis. Breast Cancer Res. Treat. 152 (2), 367–375.

Hong, R., Liu, W., DeLair, D., Razavian, N., Fenyö, D., 2021. Predicting endometrial cancer subtypes and molecular features from histopathology images using multi-resolution deep learning models. Cell Rep. Med. 2(9), 100400.

Hu, C., Qiao, X., Huang, R., Hu, C., Bao, J., Wang, X., 2024. Development and Validation of a Multimodality Model Based on Whole-Slide Imaging and Biparametric MRI for Predicting Postoperative Biochemical Recurrence in Prostate Cancer. Radiol. Imaging Cancer 6 (3), e230143.

Huang, B., Tian, S., Zhan, N., et al, 2021. Accurate diagnosis and prognosis prediction of gastric cancer using deep learning on digital pathological images: a retrospective multicentre study. EBioMedicine 73, 103631.

Huang, Z., Shao, W., Han, Z., et al, 2023. Artificial intelligence reveals features associated with breast cancer neoadjuvant chemotherapy responses from multi-stain histopathologic images. NPJ Precis. Oncol. 7 (1), 14.

Huijbers, A., Tollenaar, R.A., V Pelt, G.W., et al, 2013. The proportion of tumor-stroma as a strong prognosticator for stage II and III colon cancer patients: validation in the VICTOR trial. Ann. Oncol. 24 (1), 179–185.

Humphries, M.P., Hynes, S., Bingham, V., et al, 2018. Automated tumour recognition and digital pathology scoring unravels new role for PD-L1 in predicting good outcome in ER-/HER2+ breast cancer. J. Oncol. 2018, 2937012.

Huo, Y., Deng, R., Liu, Q., Fogo, A.B., Yang, H., 2021. AI applications in renal pathology. Kidney Int. 99 (6), 1309–1320.

Ibrahim, A., Jahanifar, M., Wahab, N., et al, 2024. Artificial intelligence-based mitosis scoring in breast cancer: clinical application. Modern Pathology: an Official Journal of the United States and Canadian Academy of Pathology, Inc 37 (3), 100416.

Idos, G.E., Kwok, J., Bonthala, N., Kysh, L., Gruber, S.B., Qu, C., 2020. The prognostic implications of tumor infiltrating lymphocytes in colorectal cancer: a systematic review and meta-analysis. Sci. Rep. 10 (1), 3360.

Inc. A. n.d. Clinical Solutions. https://www.aiforia.com/aiforia-clinical-solutions.

International Consensus Group for Hepatocellular Neoplasia, 2009. Pathologic diagnosis of early hepatocellular carcinoma: a report of the international consensus group for hepatocellular neoplasia. Hepatology 49 (2), 658–664.

Ivanova, M., Pescia, C., Trapani, D., et al, 2024. Early breast cancer risk assessment: integrating histopathology with artificial intelligence. Cancers 16 (11).

Jacobs, T.W., Gown, A.M., Yaziji, H., Barnes, M.J., Schnitt, S.J., 1999. Comparison of fluorescence in situ hybridization and immunohistochemistry for the evaluation of HER-2/neu in breast cancer. J. Clin. Oncol. 17 (7), 1974–1982.

Jaroensri, R., Wulczyn, E., Hegde, N., et al, 2022. Deep learning models for histologic grading of breast cancer and association with disease prognosis. NPJ Breast Cancer 8 (1), 113.

Jiang, J., Tekin, B., Guo, R., Liu, H., Huang, Y., Wang, C., 2021. Digital pathology-based study of cell-and tissue-level morphologic features in serous borderline ovarian tumor and high-grade serous ovarian cancer. J. Pathol. Inform. 12 (1), 24.

Jiao, J., Tang, H., Sun, N., Zhang, X., 2024. Artificial intelligence-aided steatosis assessment in donor livers according to the Banff consensus recommendations. Am. J. Clin. Pathol. 162 (4), 401–407.

Jimenez, J.E., Abdelhafez, A., Mittendorf, E.A., et al, 2022. A model combining pretreatment MRI radiomic features and tumor-infiltrating lymphocytes to predict response to neoadjuvant systemic therapy in triple-negative breast cancer. Eur. J. Radiol. 149, 110220.

Johnston, S.R.D., Harbeck, N., Hegg, R., et al, 2020. Abemaciclib combined with endocrine therapy for the adjuvant treatment of HR+, HER2-, Node-Positive, high-risk, early breast cancer (monarchE). J. Clin. Oncol. 38 (34), 3987–3998.

Kai, K., Aishima, S., Aoki, S., et al, 2016. Cytokeratin immunohistochemistry improves interobserver variability between unskilled pathologists in the evaluation of tumor budding in T1 colorectal cancer. Pathol. Int. 66 (2), 75–82.

Kapil, A., Meier, A., Zuraw, A., et al, 2018. Deep semi supervised generative learning for automated tumor proportion scoring on NSCLC tissue needle biopsies. Sci. Rep. 8 (1), 17343.

Kawakami, H., Zaanan, A., Sinicrope, F.A., 2015. Microsatellite instability testing and its role in the management of colorectal cancer. Curr. Treat. Options Oncol. 16 (7), 30.

Khoraminia, F., Fuster, S., Kanwal, N., et al, 2023. Artificial intelligence in digital pathology for bladder cancer: hype or hope? a systematic review. Cancers (Basel 15 (18).

Kidron, D., Vainer, I., Fisher, Y., Sharony, R., 2017. Automated image analysis of placental villi and syncytial knots in histological sections. Placenta 53, 113–118.

Klauschen, F., Wienert, S., Schmitt, W.D., et al, 2015. Standardized Ki67 diagnostics using automated scoring - clinical validation in the GeparTrio breast cancer study. Clin. Cancer Res. 21 (16), 3651–3657.

Klein, M.E., Dabbs, D.J., Shuai, Y., et al, 2013. Prediction of the Oncotype DX recurrence score: use of pathology-generated equations derived by linear regression analysis. Modern Pathology: an Official Journal of the United States and Canadian Academy of Pathology, Inc 26 (5), 658–664.

Klein, S., Gildenblat, J., Ihle, M.A., et al, 2020. Deep learning for sensitive detection of Helicobacter Pylori in gastric biopsies. BMC Gastroenterol. 20 (1), 417.

Kleppe, A., Skrede, O.J., De Raedt, S., et al, 2022. A clinical decision support system optimising adjuvant chemotherapy for colorectal cancers by integrating deep learning and pathological staging markers: a development and validation study. Lancet Oncol. 23 (9), 1221–1232.

Koopman, T., Buikema, H.J., Hollema, H., de Bock, G.H., van der Vegt, B., 2019. What is the added value of digital image analysis of HER2 immunohistochemistry in breast cancer in clinical practice? A study with multiple platforms. Histopathology 74 (6), 917–924.

Koopman, T., de Bock, G.H., Buikema, H.J., et al, 2018a. Digital image analysis of HER2 immunohistochemistry in gastric- and oesophageal adenocarcinoma: a validation study on biopsies and surgical specimens. Histopathology 72 (2), 191–200.

Kratz, A., S-h, L., Zini, G., et al, 2019. Digital morphology analyzers in hematology: ICSH review and recommendations. Int. J. Lab. Hematol. 41 (4), 437–447.

Kriegsmann, M., Haag, C., Weis, C.A., et al, 2020. Deep learning for the classification of small-cell and non-small-cell lung cancer. Cancers (Basel 12 (6).

Kris Lami, Y.T., Grinwald, M., Ziv, R., Shenav, L., Sandbank, J., Vecsler, M., 2025. Triage and diagnosis of primary gastric carcinoma using an AI solution: a real-world pilot study in Japan. January 30, 2025.

Krishnamurthy, S., Jain, P., Tripathy, D., et al. 2023. Predicting response of triple-negative breast cancer to neoadjuvant chemotherapy using a deep convolutional neural network-based artificial intelligence tool. JCO Clin. Cancer Inform. 7, e2200181.

Kulkarni, P.M., Robinson, E.J., Sarin Pradhan, J., et al, 2020. Deep learning based on standard H&E images of primary melanoma tumors identifies patients at risk for visceral recurrence and death. Clin. Cancer Res. 26 (5), 1126–1134.

Lakoumentas, J., Drakos, J., Karakantza, M., Nikiforidis, G.C., Sakellaropoulos, G.C., 2009. Bayesian clustering of flow cytometry data for the diagnosis of B-chronic lymphocytic leukemia. J. Biomed. Inform. 42 (2), 251–261.

Lami, K., Ota, N., Yamaoka, S., et al, 2023. Standardized classification of lung adenocarcinoma subtypes and improvement of grading assessment through deep learning. Am. J. Pathol. 193 (12), 2066–2079.

Langner, C., Aust, D., Ensari, A., et al, 2015. Histology of microscopic colitis-review with a practical approach for pathologists. Histopathology 66 (5), 613–626.

Laurinaviciene, A., Dasevicius, D., Ostapenko, V., Jarmalaite, S., Lazutka, J., Laurinavicius, A., 2011. Membrane connectivity estimated by digital image analysis of HER2 immunohistochemistry is concordant with visual scoring and fluorescence in situ hybridization results: algorithm evaluation on breast cancer tissue microarrays. Diagn. Pathol. 6, 87.

Laury, A.R., Blom, S., Ropponen, T., Virtanen, A., Carpén, O.M., 2021. Artificial intelligence-based image analysis can predict outcome in high-grade serous carcinoma via histology alone. Sci. Rep. 11(1), 19165.

Le Page, A.L., Ballot, E., Truntzer, C., et al, 2021. Using a convolutional neural network for classification of squamous and non-squamous non-small cell lung cancer based on diagnostic histopathology HES images. Sci. Rep. 11 (1), 23912.

Le, H., Gupta, R., Hou, L., et al, 2020. Utilizing automated breast cancer detection to identify spatial distributions of tumor-infiltrating lymphocytes in invasive breast cancer. Am. J. Pathol. 190 (7), 1491–1504.

Lebret, T., Paoletti, X., Pignot, G., et al, 2023. Artificial intelligence to improve cytology performance in urothelial carcinoma diagnosis: results from validation phase of the French, multicenter, prospective VISIOCYT1 trial. World J. Urol. 41 (9), 2381–2388.

Lee, J., Lim, J., Byeon, K., Kwak, J.T., 2025. Benchmarking pathology foundation models: adaptation strategies and scenarios. Comput. Biol. Med. 190, 110031.

Leung, S.C.Y., Nielsen, T.O., Zabaglo, L.A., et al, 2019. Analytical validation of a standardised scoring protocol for Ki67 immunohistochemistry on breast cancer excision whole sections: an international multicentre collaboration. Histopathology 75 (2), 225–235.

Li, C., Wang, X., Liu, W., Latecki, L.J., 2018. DeepMitosis: mitosis detection via deep detection, verification and segmentation networks. Med. Image Anal. 45, 121–133.

Li, M.Y., Pan, Y., Lv, Y., Ma, H., Sun, P.L., Gao, H.W., 2025. Digital pathology and artificial intelligence in renal cell carcinoma focusing on feature extraction: a literature review. Front. Oncol. 15, 1516264.

Li, X., Chi, X., Huang, P., Liang, Q., Liu, J., 2024. Deep neural network for the prediction of KRAS, NRAS, and BRAF genotypes in left-sided colorectal cancer based on histopathologic images. Comput. Med. Imaging Graph. 115, 102384.

Liu, L.L., Jing, B.Z., Liu, X., et al, 2025. MMRNet: ensemble deep learning models for predicting mismatch repair deficiency in endometrial cancer from histopathological images. Cell Rep. Med. 6 (5), 102099.

Liu, Y., Han, D., Parwani, A.V., Li, Z., 2023. Applications of artificial intelligence in breast pathology. Arch. Pathol. Lab. Med. 147 (9), 1003–1013.

Liu, Y., Kohlberger, T., Norouzi, M., et al, 2019. Artificial intelligence-based breast cancer nodal metastasis detection: insights into the black box for pathologists. Arch. Pathol. Lab. Med. 143 (7), 859–868.

Loeffler, C.M.L., Ortiz Bruechle, N., Jung, M., et al, 2022. Artificial Intelligence-based Detection of FGFR3 Mutational Status Directly from Routine Histology in Bladder Cancer: a Possible Preselection for Molecular Testing? Eur. Urol. Focus 8 (2), 472–479.

Loi, S., Drubay, D., Adams, S., et al, 2019. Tumor-infiltrating lymphocytes and prognosis: a pooled individual patient analysis of early-stage triple-negative breast cancers. J. Clin. Oncol. 37 (7), 559–569.

Lotan, Y., Krishna, V., Abuzeid, W.M., et al, 2025. Predicting response to intravesical bacillus calmette-guerin in high-risk nonmuscle-invasive bladder cancer using an artificial intelligence-powered pathology assay: development and validation in an international 12-center cohort. J. Urol. 213 (2), 192–204.

Lotan, Y., Shah, J.B., Krishna, V., et al, 2024. Pd27-12 Development and validation of generalizable interpretable ai biomarkers to predict clinical outcomes in bcg-treated patients with non-muscle invasive bladder cancer. J. Urol. 211 (5S), e555.

Lu, C., Romo-Bucheli, D., Wang, X., et al, 2018. Nuclear shape and orientation features from H&E images predict survival in early-stage estrogen receptor-positive breast cancers. Lab. Invest. 98 (11), 1438–1448.

Lu, M.Y., Chen, B., Williamson, D.F.K., et al, 2024a. A visual-language foundation model for computational pathology. Nat. Med. 30 (3), 863–874.

Lu, M.Y., Chen, B., Williamson, D.F.K., et al, 2024b. A multimodal generative AI copilot for human pathology. Nature 634 (8033).

Lu, M.Y., Williamson, D.F.K., Chen, T.Y., Chen, R.J., Barbieri, M., Mahmood, F., 2021. Data-efficient and weakly supervised computational pathology on whole-slide images. Nat. Biomed. Eng. 5 (6), 555–570.

Lugli, A., Kirsch, R., Ajioka, Y., et al, 2017. Recommendations for reporting tumor budding in colorectal cancer based on the international tumor budding consensus conference (ITBCC) 2016. Mod. Pathol. 30 (9), 1299–1311.

Lujan, G.M., Savage, J., Shana'ah, A., et al, 2021. Digital pathology initiatives and experience of a large academic institution during the coronavirus disease 2019 (COVID-19) pandemic. Arch. Pathol. Lab. Med. 145 (9), 1051–1061.

Luo, S., Nguyen, K.T., Nguyen, B.T.T., et al, 2021. Deep learning-enabled imaging flow cytometry for high-speed Cryptosporidium and Giardia detection. Cytometry A 99 (11), 1123–1133.

Luo, X., Zang, X., Yang, L., et al, 2017. Comprehensive computational pathological image analysis predicts lung cancer prognosis. J. Thorac. Oncol. 12 (3), 501–509.

Ma, C., Jairath, V., Feagan, B.G., et al, 2022. Responsiveness of a histologic scoring system compared with peak eosinophil count in eosinophilic esophagitis. Am. J. Gastroenterol. 117 (2), 264–271.

Mahmood, T., Arsalan, M., Owais, M., Lee, M.B., Park, K.R., 2020. Artificial intelligence-based mitosis detection in breast cancer histopathology images using faster R-CNN and deep CNNs. J. Clin. Med. 9 (3).

Maley, C.C., Koelble, K., Natrajan, R., Aktipis, A., Yuan, Y., 2015. An ecological measure of immune-cancer colocalization as a prognostic factor for breast cancer. Breast Cancer Research: BCR 17 (1), 131.

Malon, C., Brachtel, E., Cosatto, E., et al, 2012. Mitotic figure recognition: agreement among pathologists and computerized detector. Anal. Cell Pathol. (Amst. 35 (2), 97–100.

Mann, G.B., Fahey, V.D., Feleppa, F., Buchanan, M.R., 2005. Reliance on hormone receptor assays of surgical specimens may compromise outcome in patients with breast cancer. J. Clin. Oncol. 23 (22), 5148–5154.

Mantrala, S., Ginter, P.S., Mitkari, A., et al, 2022. Concordance in breast cancer grading by artificial intelligence on whole slide images compares with a multi-institutional cohort of breast pathologists. Arch. Pathol. Lab. Med. 146 (11), 1369–1377.

Marchal Bressenot, A., Riddell, R.H., Boulagnon-Rombi, C., et al, 2015. Review article: the histological assessment of disease activity in ulcerative colitis. Aliment. Pharmacol. Ther. 42 (8), 957–967.

Marletta, S., Eccher, A., Martelli, F.M., et al, 2024. Artificial intelligence-based algorithms for the diagnosis of prostate cancer: a systematic review. Am. J. Clin. Pathol. 161 (6), 526–534.

Marmé, F., Krieghoff-Henning, E.I., Kiehl, L., et al, 2025. Predicting benefit from PARP inhibitors using deep learning on H&E-stained ovarian cancer slides. Eur. J. Cancer 216, 115199.

Mastrosimini, M.G., Eccher, A., Nottegar, A., et al, 2022. WSI validation studies in breast and gynecological pathology. Pathol. Res. Pract. 240, 154191.

McCaffrey, C., Jahangir, C., Murphy, C., Burke, C., Gallagher, W.M., Rahman, A., 2024. Artificial intelligence in digital histopathology for predicting patient prognosis and treatment efficacy in breast cancer. Expert Rev. Mol. Diagn. 24 (5), 363–377.

McCurdy, J.D., Enders, F.T., Jones, A., et al, 2015. Detection of cytomegalovirus in patients with inflammatory bowel disease: where to biopsy and how many biopsies? Inflamm. Bowel Dis. 21 (12), 2833–2838.

McHugh, K., Pai, R.K., 2023. Deep learning and colon cancer interpretation: rise of the machine. Surg. Pathol. Clin. 16 (4), 651–658.

Medvedev, K.E., Acosta, P.H., Jia, L., Grishin, N.V., 2024. Deep learning for subtypes identification of pure seminoma of the testis. Clin Pathol 17, 2632010X241232302.

Mercan, E., Mehta, S., Bartlett, J., Shapiro, L.G., Weaver, D.L., Elmore, J.G., 2019. Assessment of machine learning of breast pathology structures for automated differentiation of breast cancer and high-risk proliferative lesions. JAMA Netw. Open 2 (8), e198777.

Misawa, M., Kudo, S.E., 2025. Current status of artificial intelligence use in colonoscopy. Digestion 106 (2), 138–145.

Mittendorf, E.A., Zhang, H., Barrios, C.H., et al, 2020. Neoadjuvant atezolizumab in combination with sequential nab-paclitaxel and anthracycline-based chemotherapy versus placebo and chemotherapy in patients with early-stage triple-negative breast cancer (IMpassion031): a randomised, double-blind, phase 3 trial. Lancet (London, England) 396 (10257), 1090–1100.

Modi, S., Jacot, W., Yamashita, T., et al, 2022. Trastuzumab deruxtecan in previously treated HER2-low advanced breast cancer. N. Engl. J. Med. 387 (1), 9–20.

Modi, S., Park, H., Murthy, R.K., et al, 2020. Antitumor activity and safety of trastuzumab deruxtecan in patients with HER2-low-expressing advanced breast cancer: results from a phase Ib study. J. Clin. Oncol. 38 (17), 1887–1896.

Mohammadi, M., Cooper, J., Arandelović, O., et al, 2022. Weakly supervised learning and interpretability for endometrial whole slide image diagnosis. Exp. Biol. Med. 247 (22), 2025–2037.

Mohlman, J.S., Leventhal, S.D., Hansen, T., Kohan, J., Pascucci, V., Salama, M.E., 2020. Improving augmented human intelligence to distinguish burkitt lymphoma from diffuse large B-cell lymphoma cases. Am. J. Clin. Pathol. 153 (6), 743–759.

Muhsen, I.N., Shyr, D., Sung, A.D., Hashmi, S.K., 2021. Machine learning applications in the diagnosis of benign and malignant hematological diseases. Clin. Hematol. Int. 3 (1), 13–20.

Mulliqi, N., Blilie, A., Ji, X., et al, 2025. Development and retrospective validation of an artificial intelligence system for diagnostic assessment of prostate biopsies: study protocol. BMJ Open 15 (7), e097591.

Nadji, M., Gomez-Fernandez, C., Ganjei-Azar, P., Morales, A.R., 2005. Immunohistochemistry of estrogen and progesterone receptors reconsidered: experience with 5,993 breast cancers. Am. J. Clin. Pathol. 123 (1), 21–27.

Najdawi, F., Sucipto, K., Mistry, P., et al, 2023. Artificial intelligence enables quantitative assessment of ulcerative colitis histology. Mod. Pathol. 36 (6), 100124.

Nanaa, A., Akkus, Z., Lee, W.Y., Pantanowitz, L., Salama, M.E., 2021. Machine learning and augmented human intelligence use in histomorphology for haematolymphoid disorders. Pathology 53 (3), 400–407.

Nateghi, R., Danyali, H., Helfroush, M.S., 2021. A deep learning approach for mitosis detection: application in tumor proliferation prediction from whole slide images. Artif. Intell. Med. 114, 102048.

Neil, D.A.H., Minervini, M., Smith, M.L., Hubscher, S.G., Brunt, E.M., Demetris, A.J., 2022. Banff consensus recommendations for steatosis assessment in donor livers. Hepatology 75 (4), 1014–1025.

Nero, C., Boldrini, L., Lenkowicz, J., et al, 2022. Deep-learning to predict BRCA mutation and survival from digital H&E slides of epithelial ovarian cancer. Int. J. Mol. Sci. 23 (19), 11326.

Niu, P.H., Zhao, L.L., Wu, H.L., Zhao, D.B., Chen, Y.T., 2020. Artificial intelligence in gastric cancer: application and future perspectives. World J. Gastroenterol. 26 (36), 5408–5419.

Olsson, H., Kartasalo, K., Mulliqi, N., et al, 2022. Estimating diagnostic uncertainty in artificial intelligence assisted pathology using conformal prediction. Nat. Commun. 13 (1), 7761.

Pai, R.K., Banerjee, I., Shivji, S., et al, 2022. Quantitative pathologic analysis of digitized images of colorectal carcinoma improves prediction of recurrence-free survival. Gastroenterology 163 (6), 1531–1546e1538.

Pai, R.K., Hartman, D., Schaeffer, D.F., et al, 2021. Development and initial validation of a deep learning algorithm to quantify histological features in colorectal carcinoma including tumour budding/poorly differentiated clusters. Histopathology 79 (3), 391–405.

Paige, n.d. Introducing FDA-Approved Paige Prostate. https://info.paige.ai/prostate. Accessed May 25, 2025.

Paige.ai. n.d. Paige launches new system-wide AI application for the gastrointestinal tract. https://www.paige.ai/press-releases/paige-launches-new-system-wide-ai-application-for-the-gastrointestinal-tract.

Paik, I., Lee, G., Lee, J., Kwak, T.-Y., Ha, H.K. Artificial intelligence–driven digital pathology in urological cancers: current trends and future directions. *Prostate International.* 2025.

Pan, J., Hong, G., Zeng, H., et al, 2023. An artificial intelligence model for the pathological diagnosis of invasion depth and histologic grade in bladder cancer. J. Transl. Med. 21 (1), 42.

Pantanowitz, L., Sinard, J.H., Henricks, W.H., et al, 2013. Validating whole slide imaging for diagnostic purposes in pathology: guideline from the College of American pathologists pathology and laboratory quality center. Arch. Pathol. Lab. Med. 137 (12), 1710–1722.

Park, J.W., Neve, R.M., Szollosi, J., Benz, C.C., 2008. Unraveling the biologic and clinical complexities of HER2. Clin. Breast Cancer 8 (5), 392–401.

Park, S.H., Kressel, H.Y., 2018. Connecting technological innovation in artificial intelligence to real-world medical practice through rigorous clinical validation: what peer-reviewed medical journals could do. J. Korean Med. Sci. 33 (22), e152.

Parker, C.T.A., Mendes, L., Liu, V.Y.T., et al, 2025. External validation of a digital pathology-based multimodal artificial intelligence-derived prognostic model in patients with advanced prostate cancer starting long-term androgen deprivation therapy: a post-hoc ancillary biomarker study of four phase 3 randomised controlled trials of the STAMPEDE platform protocol. Lancet Digit. Health, 100885.

Parwani, A.V., Patel, A., Zhou, M., et al, 2023. An update on computational pathology tools for genitourinary pathology practice: a review paper from the genitourinary pathology society (GUPS). J. Pathol. Inform. 14, 100177.

Patel, A., Balis, U.G.J., Cheng, J., et al, 2021. Contemporary whole slide imaging devices and their applications within the modern pathology department: a selected hardware review. J. Pathol. Inform. 12 (1), 50.

Patel, A.U., Atiya, S., Song, Y., Chu, W., Parwani, A.V., n.d. Novel liquid immunocytochemistry with machine learning analysis for bladder cancer detection. J. Histotechnol., 1–9.

Patel, A.U., Gu, Q., Esper, R., Maeser, D., Maeser, N., 2024. The crucial role of interdisciplinary conferences in advancing explainable AI in healthcare. BioMedInformatics 4 (2), 1363–1383.

Patel, A.U., Mohanty, S.K., Parwani, A.V., 2022. Applications of digital and computational pathology and artificial intelligence in genitourinary pathology diagnostics. Surgical Pathology Clinics 15 (4), 759–785.

PathAI, I. PathAI announces collaboration with the Crohn's & Colitis foundation to advance novel IBD research and development. 2024; https://www.pathai.com/news/pathai-announces-collaboration-with-the-crohns-colitis-foundation-to-advance-novel-ibd-research-and-development.

Pathmanathan, N., Balleine, R.L., Jayasinghe, U.W., et al, 2014. The prognostic value of Ki67 in systemically untreated patients with node-negative breast cancer. J. Clin. Pathol. 67 (3), 222–228.

Patil, D.T., Hammer, S., Langer, R., Yantiss, R.K., 2018. Lymphocytic esophagitis: an update on histologic diagnosis, endoscopic findings, and natural history. Ann. N.Y. Acad. Sci. 1434 (1), 185–191.

Peng, Z., Ayad, M.A., Jing, Y., Chou, T., Cooper, L.A.D., Goldstein, J.A., 2025. Benchmarking pathology foundation models for non-neoplastic pathology in the placenta. medRxiv. doi:10.1101/2025.03.19.25324282.

Perou, C.M., Sørlie, T., Eisen, M.B., et al, 2000. Molecular portraits of human breast tumours. Nature 406 (6797).

Petrelli, F., Viale, G., Cabiddu, M., Barni, S., 2015a. Prognostic value of different cut-off levels of Ki-67 in breast cancer: a systematic review and meta-analysis of 64,196 patients. Breast Cancer Res. Treat. 153 (3), 477–491.

Plass, M., Olteanu, G.E., Dacic, S., et al, 2025. Comparative performance of PD-L1 scoring by pathologists and AI algorithms. Histopathology 87 (1), 90–100.

Pohlkamp, C., Jhalani, K., Nadarajah, N., et al, 2020. Machine Learning (ML) Can Successfully Support Microscopic Differential Counts of Peripheral Blood Smears in a High Throughput Hematology Laboratory. Blood 136 (Supplement 1).

Pollack, A., Desilvio, M., Khor, L.Y., et al, 2004. Ki-67 staining is a strong predictor of distant metastasis and mortality for men with prostate cancer treated with radiotherapy plus androgen deprivation: radiation Therapy Oncology Group trial 92-02. J. Clin. Oncol. 22 (11), 2133–2140.

Post, C.S., Cheng, J., Pantanowitz, L., Westerhoff, M., 2023. Utility of machine learning to detect cytomegalovirus in digital hematoxylin and eosin-stained slides. Lab. Invest. 103(10), 100225.

Press, M.F., Pike, M.C., Chazin, V.R., et al, 1993. Her-2/neu expression in node-negative breast cancer: direct tissue quantitation by computerized image analysis and association of overexpression with increased risk of recurrent disease. Cancer Res. 53 (20), 4960–4970.

Raciti, P., Sue, J., Retamero, J.A., et al, 2023. Clinical validation of artificial intelligence-augmented pathology diagnosis demonstrates significant gains in diagnostic accuracy in prostate cancer detection. Arch. Pathol. Lab. Med. 147 (10), 1178–1185.

Ramwala, O.A., Lowry, K.P., Hippe, D.S., et al., 2024. ClinValAI: a framework for developing Cloud-based infrastructures for the external clinical validation of AI in medical imaging. Paper presented at: Biocomputing 2025: Proceedings of the Pacific Symposium.

Reisenbichler, E.S., Lester, S.C., Richardson, A.L., Dillon, D.A., Ly, A., Brock, J.E., 2013. Interobserver concordance in implementing the 2010 ASCO/CAP recommendations for reporting ER in breast carcinomas: a demonstration of the difficulties of consistently reporting low levels of ER expression by manual quantification. Am. J. Clin. Pathol. 140 (4), 487–494.

Remes, S.M., Tuominen, V.J., Helin, H., Isola, J., Arola, J., 2012. Grading of neuroendocrine tumors with Ki-67 requires high-quality assessment practices. Am. J. Surg. Pathol. 36 (9), 1359–1363.

Reta, C., Altamirano, L., Gonzalez, J.A., et al, 2015. Segmentation and classification of bone marrow cells images using contextual information for medical diagnosis of acute leukemias. PloS One 10 (6), e0130805.

Rewcastle, E., Skaland, I., Gudlaugsson, E., Fykse, S.K., Baak, J.P.A., Janssen, E.A.M., 2024. The Ki67 dilemma: investigating prognostic cut-offs and reproducibility for automated Ki67 scoring in breast cancer. Breast Cancer Res. Treat. 207 (1), 1–12.

Rexhepaj, E., Brennan, D.J., Holloway, P., et al, 2008. Novel image analysis approach for quantifying expression of nuclear proteins assessed by immunohistochemistry: application to measurement of oestrogen and progesterone receptor levels in breast cancer. Breast Cancer Research: BCR 10 (5), R89.

Riaz, I.B., Harmon, S., Chen, Z., Naqvi, S.A.A., Cheng, L., 2024. Applications of artificial intelligence in prostate cancer care: a path to enhanced efficiency and outcomes. Am. Soc. Clin. Oncol. Educ. Book 44 (3), e438516.

Ricaurte Archila, L., Smith, L., Sihvo, H.K., et al, 2023. Performance of an artificial intelligence model for recognition and quantitation of histologic features of eosinophilic esophagitis on biopsy samples. Mod. Pathol. 36 (10), 100285.

Rienda, I., Vale, J., Pinto, J., Polonia, A., Eloy, C., 2024. Using artificial intelligence to prioritize pathology samples: report of a test drive. Virchows Arch.

Rivenson, Y., Wang, H., Wei, Z., et al, 2019a. Virtual histological staining of unlabelled tissue-autofluorescence images via deep learning. Nat. Biomed. Eng. 3 (6), 466–477.

Rizzardi, A.E., Johnson, A.T., Vogel, R.I., et al. 2012. Quantitative comparison of immunohistochemical staining measured by digital image analysis versus pathologist visual scoring. Diagn. Pathol. 7, 42.

Røge, R., Riber-Hansen, R., Nielsen, S., Vyberg, M., 2016. Proliferation assessment in breast carcinomas using digital image analysis based on virtual Ki67/cytokeratin double staining. Breast Cancer Res. Treat. 158 (1), 11–19.

Romo-Bucheli, D., Janowczyk, A., Gilmore, H., Romero, E., Madabhushi, A., 2016. Automated tubule nuclei quantification and correlation with oncotype DX risk categories in ER+ breast cancer whole slide images. Sci. Rep. 6, 32706.

Romo-Bucheli, D., Janowczyk, A., Gilmore, H., Romero, E., Madabhushi, A., 2017. A deep learning based strategy for identifying and associating mitotic activity with gene expression derived risk categories in estrogen receptor positive breast cancers. Cytom., : j. Int. Soc. Anal. Cytol. 91 (6), 566–573.

Rudin, C., 2019. Stop explaining black box machine learning models for high stakes decisions and use interpretable models instead. Nat. Mach. Intell. 1 (5), 206–215.

Rudin, C.M., Brambilla, E., Faivre-Finn, C., Sage, J., 2021. Small-cell lung cancer. Nat. Rev. Dis. Primers 7 (1), 3.

Russell, P.A., Wainer, Z., Wright, G.M., Daniels, M., Conron, M., Williams, R.A., 2011. Does lung adenocarcinoma subtype predict patient survival?: a clinicopathologic study based on the new International Association for the Study of Lung Cancer/American Thoracic Society/European Respiratory Society international multidisciplinary lung adenocarcinoma classification. J. Thorac. Oncol. 6 (9), 1496–1504.

Sajjad, U., Chen, W., Rezapour, M., et al, 2024. Enhancing colorectal cancer tumor bud detection using deep learning from routine H&E-stained slides. Proc. SPIE Int. Soc. Opt. Eng., 12933.

Sandbank, J., Bataillon, G., Nudelman, A., et al, 2022. Validation and real-world clinical application of an artificial intelligence algorithm for breast cancer detection in biopsies. NPJ Breast Cancer 8 (1), 129.

Schmid, P., Cortes, J., Pusztai, L., et al, 2020. Pembrolizumab for early triple-negative breast cancer. N. Engl. J. Med. 382 (9), 810–821.

Schnegg-Kaufmann, A.S., Bacher, U., Rovó, A., et al, 2025.Artificial intelligence in haematologic diagnostics: current applications and future perspectives. Acta Haematologica, 1–25.

Scientific, F. Techcyte cytobay bladder cancer analysis (product page screenshot). 2025; https://www.fishersci.com/shop/products/cytobay-bladder-cancer-analysis/23888012. Accessed 10 Aug, 2025.

Sebai, M., Wang, X., Wang, T., 2020. MaskMitosis: a deep learning framework for fully supervised, weakly supervised, and unsupervised mitosis detection in histopathology images. Med. Biol. Eng. Comput. 58 (7), 1603–1623.

Sengupta, D., Ali, S.N., Bhattacharya, A., Mustafi, J., Mukhopadhyay, A., Sengupta, K., 2022. A deep hybrid learning pipeline for accurate diagnosis of ovarian cancer based on nuclear morphology. PloS One 17 (1), e0261181.

Shafi, S., Kellough, D.A., Lujan, G., Satturwar, S., Parwani, A.V., Li, Z., 2022. Integrating and validating automated digital imaging analysis of estrogen receptor immunohistochemistry in a fully digital workflow for clinical use. J. Pathol. Inform. 13, 100122.

Shah, A.H., Gami, A.J., Desai, N.H., Gandhi, J.S., Trivedi, P.P., 2022. Tumor budding as a prognostic indicator in colorectal carcinoma: a retrospective study of primary colorectal carcinoma cases in a tertiary care center. Indian J. Surg. Oncol. 13 (3), 459–467.

Sheikh, T.S., Kim, J.-Y., Shim, J., Cho, M., 2022. Unsupervised learning based on multiple descriptors for WSIs diagnosis. Diagnostics 12 (6), 1480.

Shen, L., Dai, B., Dou, S., Yan, F., Yang, T., Wu, Y., 2025. Estimation of TP53 mutations for endometrial cancer based on diffusion-weighted imaging deep learning and radiomics features. BMC Cancer 25 (1), 45.

Shen, Z., Simard, M., Brand, D., et al, 2024. A deep learning framework deploying segment anything to detect pan-cancer mitotic figures from haematoxylin and eosin-stained slides. Commun. Biol. 7 (1), 1674.

Shi, Z., Zhu, C., Zhang, Y., et al, 2022. Deep learning for automatic diagnosis of gastric dysplasia using whole-slide histopathology images in endoscopic specimens. Gastric Cancer 25 (4), 751–760.

Shivji, S., Conner, J.R., Barresi, V., Kirsch, R., 2020. Poorly differentiated clusters in colorectal cancer: a current review and implications for future practice. Histopathology 77 (3), 351–368.

Shouval, R., Fein, J.A., Savani, B., Mohty, M., Nagler, A., 2021. Machine learning and artificial intelligence in haematology. Br. J. Haematol. 192 (2), 239–250.

Shui, R., Yu, B., Bi, R., Yang, F., Yang, W., 2015. An interobserver reproducibility analysis of Ki67 visual assessment in breast cancer. PloS One 10 (5), e0125131.

Skaland, I., Ovestad, I., Janssen, E.A., et al, 2008. Comparing subjective and digital image analysis HER2/neu expression scores with conventional and modified FISH scores in breast cancer. J. Clin. Pathol. 61 (1), 68–71.

Skrede, O.J., De Raedt, S., Kleppe, A., et al, 2020. Deep learning for prediction of colorectal cancer outcome: a discovery and validation study. Lancet 395 (10221), 350–360.

Slamon, D.J., Clark, G.M., Wong, S.G., Levin, W.J., Ullrich, A., McGuire, W.L., 1987. Human breast cancer: correlation of relapse and survival with amplification of the HER-2/neu oncogene. Science 235 (4785).

Smith, A., Belanger, E.C., 2024. The paige prostate suite: assistive artificial intelligence for prostate cancer diagnosis. Canadian Agency for Drugs and Technologies in Health.

Soliman, A., Li, Z., Parwani, A.V., 2024. Artificial intelligence's impact on breast cancer pathology: a literature review. Diagn. Pathol. 19 (1), 38.

Song, J., Im, S., Lee, S.H., Jang, H.-J., 2022. Deep learning-based classification of uterine cervical and endometrial cancer subtypes from whole-slide histopathology images. Diagnostics 12 (11), 2623.

Stalhammar, G., Fuentes Martinez, N., Lippert, M., et al, 2016. Digital image analysis outperforms manual biomarker assessment in breast cancer. Modern Pathology: an Official Journal of the United States and Canadian Academy of Pathology, Inc 29 (4), 318–329.

Stålhammar, G., Robertson, S., Wedlund, L., et al, 2018. Digital image analysis of Ki67 in hot spots is superior to both manual Ki67 and mitotic counts in breast cancer. Histopathology 72 (6), 974–989.

Steiner, D.F., MacDonald, R., Liu, Y., et al, 2018. Impact of deep learning assistance on the histopathologic review of lymph nodes for metastatic breast cancer. Am. J. Surg. Pathol. 42 (12), 1636–1646.

Steiner, D.F., Nagpal, K., Sayres, R., et al, 2020. Evaluation of the Use of Combined Artificial Intelligence and Pathologist Assessment to Review and Grade Prostate Biopsies. JAMA Netw. Open 3 (11), e2023267.

Straumann, A., Katzka, D.A., 2018. Diagnosis and treatment of eosinophilic esophagitis. Gastroenterology 154 (2), 346–359.

Ström, P., Kartasalo, K., Olsson, H., et al, 2020. Artificial intelligence for diagnosis and grading of prostate cancer in biopsies: a population-based, diagnostic study. Lancet Oncol. 21 (2), 222–232.

Su, Z., Guo, Y., Wesolowski, R., et al, 2024a. Computational pathology for accurate prediction of breast cancer recurrence: development and validation of a deep learning-based tool. ArXiv.

Su, Z., Niazi, M.K.K., Tavolara, T.E., et al, 2023. BCR-Net: a deep learning framework to predict breast cancer recurrence from histopathology images. PloS One 18 (4), e0283562.

Sung, H., Ferlay, J., Siegel, R.L., et al, 2021. Global cancer statistics 2020: GLOBOCAN estimates of incidence and mortality worldwide for 36 cancers in 185 countries. CA Cancer J. Clin. 71 (3), 209–249.

Suzuki, H., Hamada, K., Hamanishi, J., et al, 2025. Artificial intelligence-based spatial analysis of tertiary lymphoid structures and clinical significance for endometrial cancer. Cancer Immunol. Immunother. 74 (3), 84.

Syed-Abdul, S., Firdani, R.P., Chung, H.J., et al, 2020. Artificial intelligence based models for screening of hematologic malignancies using cell population data. Sci. Rep. 10 (1), 4583.

Syrykh, C., Abreu, A., Amara, N., et al. 2020. Accurate diagnosis of lymphoma on whole-slide histopathology images using deep learning. NPJ Digit. Med. 3, 63.

Tahir, M., Hu, Y., Kumar, H., et al, 2025. A comprehensive ai-based approach in classifying breast lesions: focusing on improving pathologists' accuracy and efficiency. Clin. Breast Cancer.

Tandon, A.K., Clark, G.M., Chamness, G.C., Ullrich, A., McGuire, W.L., 1989. HER-2/neu oncogene protein and prognosis in breast cancer. J. Clin. Oncol. 7 (8), 1120–1128.

Tang, G., Fu, X., Wang, Z., Chen, M., 2021. A machine learning tool using digital microscopy (morphogo) for the identification of abnormal lymphocytes in the bone marrow. Acta Cytol. 65 (4), 354–357.

Teng, K., Ren, L., Yan, X., et al, 2025. Artificial intelligence-based model for diagnosing Helicobacter pylori in whole-slide images. Front. Med. (Lausanne 12, 1594614.

Thomsen, C., Nielsen, S., Nielsen, B.S., Pedersen, S.H., Vyberg, M., 2019. Estrogen receptor-alpha quantification in breast cancer: concordance between immunohistochemical assays and mRNA-In situ hybridization for ESR1 Gene. Appl. Immunohistochem. Mol. Morphol.

Tian, K., Rubadue, C.A., Lin, D.I., et al, 2019. Automated clear cell renal carcinoma grade classification with prognostic significance. PLoS One 14 (10), e0222641.

Tolkach, Y., Ovtcharov, V., Pryalukhin, A., et al, 2023. An international multi-institutional validation study of the algorithm for prostate cancer detection and Gleason grading. NPJ Precis. Oncol. 7 (1), 77.

Tschuchnig, M.E., Oostingh, G.J., Gadermayr, M., 2020. Generative adversarial networks in digital pathology: a survey on trends and future potential. Patterns (New York, NY) 1(6), 100089.

Turbin, D.A., Leung, S., Cheang, M.C., et al, 2008. Automated quantitative analysis of estrogen receptor expression in breast carcinoma does not differ from expert pathologist scoring: a tissue microarray study of 3,484 cases. Breast Cancer Res. Treat. 110 (3), 417–426.

Tward, B.V., Johnson, S.B., Tward, J.D., 2025. Using AI to identify optimal clinical, genomic, and radiographic prognostic features and novel risk classifiers compared to routinely available risk classifiers. J. Clin. Oncol. 43 (5_suppl), 396.

U.S. Food and Drug Administration. DEN200080: paige prostate de novo classification order. 2021; https://www.accessdata.fda.gov/cdrh_docs/pdf20/DEN200080.pdf.

Umemoto, M., Mariya, T., Nambu, Y., et al, 2024. Prediction of mismatch repair status in endometrial cancer from histological slide images using various deep learning-based algorithms. Cancers (Basel 16 (10).

van den Ende, N.S., Nguyen, A.H., Jager, A., Kok, M., Debets, R., van Deurzen, C.H.M., 2023. Triple-negative breast cancer and predictive markers of response to neoadjuvant chemotherapy: a systematic review. Int. J. Mol. Sci. 24 (3).

Veeraraghavan, H., Friedman, C.F., DeLair, D.F., et al, 2020. Machine learning-based prediction of microsatellite instability and high tumor mutation burden from contrast-enhanced computed tomography in endometrial cancers. Sci. Rep. 10 (1), 17769.

Vennalaganti, P., Kanakadandi, V., Goldblum, J.R., et al, 2017. Discordance among pathologists in the United States and Europe in diagnosis of low-grade dysplasia for patients with barrett's esophagus. Gastroenterology 152 (3), 564–570e564.

Verghese, G., Lennerz, J.K., Ruta, D., et al, 2023. Computational pathology in cancer diagnosis, prognosis, and prediction - present day and prospects. J. Pathol. 260 (5), 551–563.

Veta, M., van Diest, P.J., Willems, S.M., et al, 2015. Assessment of algorithms for mitosis detection in breast cancer histopathology images. Med. Image Anal. 20 (1), 237–248.

Viale, G., Regan, M.M., Maiorano, E., et al, 2007. Prognostic and predictive value of centrally reviewed expression of estrogen and progesterone receptors in a randomized trial comparing letrozole and tamoxifen adjuvant therapy for postmenopausal early breast cancer: BIG 1-98. J. Clin. Oncol. 25 (25), 3846–3852.

Volinsky-Fremond, S., Horeweg, N., Andani, S., et al, 2024. Prediction of recurrence risk in endometrial cancer with multimodal deep learning. Nat. Med. 30 (7), 1962–1973.

Vorontsov, E., Bozkurt, A., Casson, A., et al, 2024. A foundation model for clinical-grade computational pathology and rare cancers detection. Nat. Med. 30 (10), 2924–2935.

Wan, W., Wang, L., Liu, Y., Hu, Y., 2023. Improving the detection of Helicobacter pylori in biopsies of chronic gastritis: a comparative analysis of H&E, methylene blue, Warthin-Starry, immunohistochemistry, and quantum dots immunohistochemistry. Front. Oncol. 13, 1229871.

Wang, J., Wang, T., Han, R., Shi, D., Chen, B., 2025. Artificial intelligence in cancer pathology: applications, challenges, and future directions. CytoJournal 22, 45.

Wang, X., Janowczyk, A., Zhou, Y., et al, 2017. Prediction of recurrence in early stage non-small cell lung cancer using computer extracted nuclear features from digital H&E images. Sci. Rep. 7 (1), 13543.

Wang, X., Zou, C., Zhang, Y., et al, 2021. Prediction of BRCA gene mutation in breast cancer based on deep learning and histopathology images. Front. Genet. 12, 661109.

Wang, Y., Acs, B., Robertson, S., et al, 2022. Improved breast cancer histological grading using deep learning. Ann. Oncol. 33 (1), 89–98.

Wei, J.W., Tafe, L.J., Linnik, Y.A., Vaickus, L.J., Tomita, N., Hassanpour, S., 2019. Pathologist-level classification of histologic patterns on resected lung adenocarcinoma slides with deep neural networks. Sci. Rep. 9 (1), 3358.

Williams, B., Hanby, A., Millican-Slater, R., et al, 2020. Digital pathology for primary diagnosis of screen-detected breast lesions - experimental data, validation and experience from four centres. Histopathology 76 (7), 968–975.

Williams, B.J., Hanby, A., Millican-Slater, R., Nijhawan, A., Verghese, E., Treanor, D., 2018. Digital pathology for the primary diagnosis of breast histopathological specimens: an innovative validation and concordance study on digital pathology validation and training. Histopathology 72 (4), 662–671.

Woerl, A.C., Eckstein, M., Geiger, J., et al, 2020. Deep learning predicts molecular subtype of muscle-invasive bladder cancer from conventional histopathological slides. Eur. Urol. 78 (2), 256–264.

Wolff, A.C., Hammond, M.E.H., Allison, K.H., et al, 2018. Human epidermal growth factor receptor 2 testing in breast cancer: american society of clinical oncology/college of american pathologists clinical practice guideline focused update. J. Clin. Oncol. 36 (20), 2105–2122.

Wolff, A.C., Somerfield, M.R., Dowsett, M., et al, 2023. Human epidermal growth factor receptor 2 testing in breast cancer: American society of clinical oncology–college of American pathologists guideline update. Arch. Pathol. Lab. Med. 147 (9), 993–1000.

Wu, J., Liu, C., Liu, X., et al, 2022. Artificial intelligence-assisted system for precision diagnosis of PD-L1 expression in non-small cell lung cancer. Modern Pathology: an Official Journal of the United States and Canadian Academy of Pathology, Inc 35 (3), 403–411.

Wu, M., Yan, C., Liu, H., Liu, Q., 2018. Automatic classification of ovarian cancer types from cytological images using deep convolutional neural networks. Biosci. Rep. 38 (3).

Xie, P., Zuo, K., Zhang, Y., Li, F., Yin, M., Lu, K. Interpretable classification from skin cancer histology slides using deep learning: a retrospective multicenter study. *arXiv preprint arXiv:190406156.* 2019.

Xiong, Y., Yao, L., Lin, J., et al, 2025. Artificial intelligence links CT images to pathologic features and survival outcomes of renal masses. Nat. Commun. 16 (1), 1425.

Xu, H., Usuyama, N., Bagga, J., et al, 2024. A whole-slide foundation model for digital pathology from real-world data. Nature 630 (8015), 181–188.

Xu, Z., Zhou, Z., Son, J.B., et al, 2025. Deep learning models based on pretreatment MRI and clinicopathological data to predict responses to neoadjuvant systemic therapy in triple-negative breast cancer. Cancers 17 (6).

Y-x, L., Chen, F., J-j, S., Y-l, H., Wang, M., 2023. Convolutional neural networks for classifying cervical cancer types using histological images. J. Digit. Imaging 36 (2), 441–449.

Yamashita, R., Long, J., Longacre, T., et al, 2021. Deep learning model for the prediction of microsatellite instability in colorectal cancer: a diagnostic study. Lancet Oncol. 22 (1), 132–141.

Yarahmadi, P., Ahmadpour, E., Moradi, P., Samadzadehaghdam, N., 2025. Automatic classification of Giardia infection from stool microscopic images using deep neural networks. Bioimpacts 15, 30272.

Yerushalmi, R., Woods, R., Ravdin, P.M., Hayes, M.M., Gelmon, K.A., 2010. Ki67 in breast cancer: prognostic and predictive potential. Lancet Oncol. 11 (2), 174–183.

Yeung, M., Watts, T., Tan, S.Y.W., et al, 2023. Stain consistency learning: handling stain variation for automatic digital pathology segmentation. arXiv. doi:10.48550/ARXIV.2311.06552.

Yoon, C., Park, E., Misra, S., et al, 2024. Deep learning-based virtual staining, segmentation, and classification in label-free photoacoustic histology of human specimens. Light Sci. Appl. 13 (1), 226.

Yoshida, H., Kiyuna, T., 2021. Requirements for implementation of artificial intelligence in the practice of gastrointestinal pathology. World J. Gastroenterol. 27 (21), 2818–2833.

Yu, K.H., Wang, F., Berry, G.J., et al, 2020. Classifying non-small cell lung cancer types and transcriptomic subtypes using convolutional neural networks. J. Am. Med. Inform. Assoc. 27 (5), 757–769.

Yue, M., Zhang, J., Wang, X., et al, 2021. Can AI-assisted microscope facilitate breast HER2 interpretation? A multi-institutional ring study. Virchows Arch. 479 (3), 443–449.

Zeng, H., Chen, L., Zhang, M., Luo, Y., Ma, X., 2021. Integration of histopathological images and multi-dimensional omics analyses predicts molecular features and prognosis in high-grade serous ovarian cancer. Gynecol. Oncol. 163 (1), 171–180.

Zeng, X., Starr, D., Li, J., et al, 2024. AICyte-alone capabilities as an independent screener for triaging cervical cytology using a 50% negative cutoff value. Cancer Cytopathol. 132 (11), 723–730.

Zhang, H., Moisini, I., Ajabnoor, R.M., Turner, B.M., Hicks, D.G., 2020a. Applying the new guidelines of HER2 testing in breast cancer. Curr. Oncol. Rep. 22 (5), 51.

Zhang, K., Qiu, Y., Feng, S., et al, 2025. Development of model for identifying homologous recombination deficiency (HRD) status of ovarian cancer with deep learning on whole slide images. J. Transl. Med. 23 (1), 267.

Zhang, X., Ba, W., Zhao, X., et al, 2022. Clinical-grade endometrial cancer detection system via whole-slide images using deep learning. Frontiers in Oncology 12, 1040238.

Zhang, Y., Yu, P., Chen, Z., et al, 2021. Flagging performance of two automated hematology analyzers in blast cell screening. Ann. Transl. Med. 9 (20), 1606.

Zhao, Y., Schaafsma, E., Cheng, C., 2020. Gene signature-based prediction of triple-negative breast cancer patient response to Neoadjuvant chemotherapy. Cancer Medicine 9 (17), 6281–6295.

Zheng, Q., Mei, H., Weng, X., et al, 2025. Artificial intelligence-based multimodal prediction for nuclear grading status and prognosis of clear cell renal cell carcinoma: a multicenter cohort study. Int. J. Surg. 111 (6), 3722–3730.

Zheng, Q., Wang, X., Yang, R., et al, 2024. Predicting tumor mutation burden and VHL mutation from renal cancer pathology slides with self-supervised deep learning. Cancer Medicine 13 (16), e70112.

Zheng, T., Zheng, S., Wang, K., et al. n.d. An Automatic CD30 Scoring Method for Few Shot Whole Slide Images of Primary Cutaneous CD30+ Lymphoproliferative Diseases. *Available at SSRN 4029432.*

Zheng, X., Yao, Z., Huang, Y., et al, 2020. Deep learning radiomics can predict axillary lymph node status in early-stage breast cancer. Nat. Commun. 11 (1), 1236.

Zhou, C., Zhang, Y.F., Guo, S., et al, 2024. Multimodal data integration for predicting progression risk in castration-resistant prostate cancer using deep learning: a multicenter retrospective study. Front. Oncol. 14, 1287995.

Zhu, M., Liang, C., Zhang, F., Zhu, L., Chen, D., 2021. A nomogram to predict disease-free survival following neoadjuvant chemotherapy for triple negative breast cancer. Front. Oncol. 11, 690336.

Zhu, Y., Wang, Q.C., Xu, M.D., et al, 2019. Application of convolutional neural network in the diagnosis of the invasion depth of gastric cancer based on conventional endoscopy. Gastrointest. Endosc. 89 (4), 806–815e801.

Zielinski, B., Sroka-Oleksiak, A., Rymarczyk, D., Piekarczyk, A., Brzychczy-Wloch, M., 2020. Deep learning approach to describe and classify fungi microscopic images. PLoS One 15 (6), e0234806.

CHAPTER

10

Artificial intelligence in cytology

David Kim[a], *Swati Satturwar*[b], *Liron Pantanowitz*[c], *and Zaibo Li*[b]

[a]Department of Pathology and Laboratory Medicine, Memorial Sloan Kettering Cancer Center, New York, NY, United States

[b]Department of Pathology, The Ohio State University, Columbus, OH, United States

[c]Department of Pathology, University of Pittsburgh Medical Center, Pittsburgh, PA, United States

Introduction

The rapid adoption of whole-slide imaging (WSI) and the growing digital healthcare infrastructure have created an optimal setting for artificial intelligence (AI) to transform pathology practice. Cytology, perhaps more than any other pathology subspecialty, has a history of embracing technological innovation. Computer-assisted cervical screening represents one of the earliest and most successful clinical applications of AI in pathology, underscoring cytology's longstanding receptiveness to innovative methods (Biscotti et al., 2005; Doornewaard et al., 1999; Abulafia et al., 2003; Chute et al., 2010). Today, advances in machine learning and deep learning (DL) promise to extend these capabilities by offering new opportunities to improve diagnostic accuracy, efficiency, and overall patient care in cytology.

However, effective clinical integration of AI in cytology is not without challenges. Cytology presents distinctive complexities that need to be considered, from the diversity of specimen preparation methods to specialized workflows such as screening and rapid on-site evaluation (ROSE) (Schmidt et al., 2013a, 2013b). Moreover, the very foundation of cytology practice, which involves the careful evaluation at the cellular level, introduces nuances that necessitate unique AI approaches. Understanding these unique features is critical to designing systems that are not only technically robust but also practical and clinically useful. This chapter will examine the potential of AI in cytology by outlining the benefits, challenges, and workflow considerations that shape its development. It will also discuss strategies for implementation within the cytology laboratory, highlighting how these technologies may enhance both current practice and future directions in the field.

Advances in Digital Pathology and Artificial Intelligence in Pathology
https://doi.org/10.1016/B978-0-443-40383-5.00012-1

Developing AI for cytopathology: Approaches and strengths

To begin developing AI for cytology, it is necessary to identify tasks that are both appropriate for AI and beneficial for cytology practice. Developing algorithms that exceed current AI models (e.g., predicting malignancy and clinical outcomes from a single cell) or those that are difficult to integrate into cytology workflows would waste resources and cause frustration. Careful understanding of the various algorithms, consideration of the practice needs, and resources at hand for each individual laboratory is key to discovering useful AI tools. While there are cytology-specific challenges to consider when deploying AI, there are many unique advantages. This section reviews some of those benefits that can be leveraged to enhance workflow efficiency and diagnostic accuracy.

Approaching AI for cytology

An ideal starting point for identifying cytology practice areas amenable to AI is to think of repetitive, detailed tasks such as pattern recognition or summarization of complex data. AI algorithms, in simple terms, make predictions from rules determined by a human user (e.g., computer vision) or learned by the algorithm (e.g., neural networks) from a training set of data (Pantanowitz et al., 2025). More complex AI models, such as DL algorithms, consist of multiple layers of interconnected nodes that progressively extract more abstract and meaningful representations from the input data. During model training, the algorithm compares its predictions to known outcomes and gradually adjusts its parameters, improving performance through a process called backpropagation. This learning process makes DL algorithms unique compared to other computational models. Once trained, DL algorithms can analyze new data, identify patterns, and make predictions based on what the algorithm has learned. This ability to learn directly from data, such as images, text, or other complex inputs, is a key advantage, especially in cytology, where cellular morphology and clinical context are highly variable and nuanced. These tasks require large datasets to develop AI in a clinical setting, much like how a cytology fellow requires exposure to a diverse array of cases. Relevant cytology datasets may include digitized WSIs, diagnostic reports, genomic data, clinical information, or even multimodal data, to name a few. The effectiveness of AI depends on the availability of well-structured, standardized data, which enables robust model training and validation while ensuring reproducibility and generalizability.

AI opportunities in cytology

Given how AI models operate, cytology offers unique characteristics that make it particularly favorable for this framework. Cytology is well suited for AI applications because many of its diagnostic challenges resemble a classic "needle in a haystack" problem. Unlike surgical pathology, cytology specimens require the examination of thousands of individually dispersed cells on a single slide to identify even a small cluster of malignant cells. A major strength of cytology is its ability to diagnose abnormal cells in limited tissue samples acquired by minimally invasive means. However, this process is labor-intensive and subject to significant intra- and interobserver variability (Stoler and Schiffman, 2001; Viswanathan et al., 2020; Słowińska-Klencka et al., 2021). Computational methods, such as DL, can help overcome these challenges by preliminarily screening slides and highlighting regions of interest for review by a cytologist and cytopathologist. Doing so would reduce mental fatigue and improve efficiency and consistency. Region-of-interest detection may offer even greater benefits in cytology than in larger tissue surgical pathology samples, such as core biopsies or resections, where diagnostically relevant material is confined to relatively discrete areas across multiple slides.

AI development considerations for cytology

Cytology also offers unique practical advantages for AI training and development. Because diagnostic decision-making in cytology primarily occurs at the cellular or cell cluster level from fragmented tissue and fluid samples, each cytology slide can provide hundreds of thousands or even millions of individual cell images for algorithm training (Kitchener et al., 2015; Layfield et al., 1997; Michael et al., 2007). This provides a relatively large amount of image data from relatively few samples for model training. Many liquid-based preparations (i.e., urine, thyroid, and body cavity fluids) comprise a single slide representing the entire sample, simplifying data curation and reducing variability across slides. In contrast, surgical specimens often require identifying which of the multiple slides contains diagnostically relevant tissue. This process may require manual selection, making the creation of the AI training set more complex. Another advantage is the widespread adoption of standardized reporting systems in cytology, such as the Bethesda System for cervical and thyroid cytology, the Milan System for salivary gland cytology, and the Paris System for urine cytology (Wojcik et al., 2022; Faquin et al., 2023; Ali and VanderLaan, 2023; Nayar and Wilbur, 2015). While histology reporting often uses narrative or semistructured text, the system-based cytology reporting format facilitates standardized categorization of samples, inherently providing annotations for supervised learning in AI development. Rather than categorizing each case based on the pathology report, the cytology classification system can be used to easily label data into clinically significant categories.

Cytology is a practice that requires careful microscopic review, with a need for consistent interpretation of subtle cellular findings. This task can lead to fatigue, variability, and diagnostic errors over time. Due to its role in screening for certain precancerous and cancerous lesions, such as squamous intraepithelial lesions in cervical samples and high-grade urothelial carcinoma in urine, hundreds of the same specimen type are evaluated every day (Moriarty et al., 2014; Renshaw et al., 2016). Therefore AI has great potential in helping screen cytology samples due to the large volume of similar samples. Other potential AI solutions to help alleviate this burden include detecting rare malignant cells in fluid samples, assessing sample adequacy during ROSE, and triaging large volumes of negative or low-risk cases. These tasks are highly repetitive, rule-based, and can be quantified into discrete outcomes such as "negative," "atypical," "suspicious," or "positive for malignancy." This makes cytology samples particularly suitable for analysis, as each slide provides robust cell-level data, enabling algorithms to process far more diagnostic elements, where only a few regions of tissue architecture may be relevant. In addition, the absence of large tissue structures allows AI to apply cell-based classification methods that are more challenging to implement in histology without extensive segmentation of complex tissue architecture.

Cytology applications beyond screening

Another area amenable to AI in cytology is the use of LLMs, which can help synthesize millions of pathology reports to identify cases for QA/QC, research, and education. When applied to clinical data, language models can also help efficiently identify and summarize clinical information when reviewing pathology samples. A large portion of cytology case review is devoted to identifying and summarizing the patient's history to establish the correct clinical context for sample interpretation. AI can support sample identification for research beyond routine diagnostics, including mining archival cytology slides for rare entities or constructing large, annotated datasets for image-based DL model training. Cytology includes several specific quality assurance and quality control (QA/QC) metrics, such as ROSE adequacy rates, cytology-histology correlation, atypical squamous cell to squamous intraepithelial lesion (ASCUS/SIL) ratios, and atypia rates (Torous et al., 2023). Many of these metrics are manually curated by each cytology laboratory with the aid of, at best, cumbersome LIS software tools, creating opportunities for AI-driven automation. Image-based AI models can streamline cell screening and adequacy assessment, while large language models (LLMs) can automate text-based quality assurance (QA/QC) workflows by searching and extracting structured data from unstructured reports.

The potential for AI integration with recent advances in DL models, such as CNN and LLM, is promising. Cytology, after all, is no stranger to AI integration and outside-the-box thinking, with the adoption of automated GYN screening systems. The Hologic ThinPrep Imaging System and the Becton Dickinson SurePath Imaging System (BD FocalPoint GS Imaging System) for cervical cytology are some of the first clinical examples of the use of AI in pathology (Biscotti et al., 2005; Wilbur et al., 2009; Colgan et al., 2013; Miller et al., 2007). These systems are currently used across cytology laboratories today to screen for cervical cancer. As advanced AI applications for pathology mature, prioritizing tasks that synthesize large volumes of data in a repetitive yet precise manner will be essential to achieve maximal clinical benefit.

AI methodologies in cytopathology

The recent growth of AI has led to the development of a diverse range of algorithms, each designed to handle specific types of data and provide a specific output. While AI is often referred to as a singular tool, it comprises a wide variety of algorithmic models, spanning from simple computer vision methods to complex generative AI models. The choice of an AI model in cytology fundamentally depends on the nature of the input data and the intended task to be solved. For instance, some applications may require interpreting image data, such as WSIs from smear slides, while others involve textual data from cytology reports. Similarly, the intended output must be considered, whether it is a binary prediction (benign vs. malignant), a multiclass categorization, or a quantitative measurement. These considerations determine whether an image classification algorithm, such as a convolutional neural network (CNN), an object detection algorithm, or multiple instance learning (MIL), is appropriate for slide-level image analysis, or whether a large language model (LLM) utilizing self-supervised learning is more suitable for extracting information from pathology reports. The following section explores the most common AI paradigms to date as they relate to cytopathology applications.

AI paradigms for cytology

A fundamental strength of recent deep learning (DL) AI models is their ability to learn representations directly from data, rather than relying on manually set features or parameter adjustments by the user. As these models learn from the data provided, how they learn or use the data for training can differ. The method of learning from the data is termed an AI paradigm, whereas the structural design of the algorithm is defined by AI architectures (Pantanowitz et al., 2025; Hanna et al., 2025b). Understanding both is critical when applying AI to cytology workflows. The three most common paradigms in pathology AI are supervised, unsupervised, and weakly supervised learning. In supervised learning, models learn from already labeled datasets, where the correct classification (e.g., benign or malignant diagnosis) is provided for each data point. This paradigm is well-suited for classification tasks, such as distinguishing malignant from benign cells in cytology samples, provided that high-quality annotations are available. The model can learn distinctive morphological features associated with each class in a guided manner using large-scale annotations by expert users. This is akin to having a large, robust study set of images to learn the morphologic features of a tumor during pathology training. The main limitation of this paradigm is the need for large, high-quality annotated datasets, which can be resource-intensive to generate. This challenge is even more pronounced in cytology, where cell-level annotations are required for morphologic predictions. For instance, training a fully supervised model to identify malignant cells in pleural fluid would require hundreds to thousands of individually annotated cells. A cytopathologist would have to categorize each cell and object, a task that is time-consuming.

Unsupervised learning uses unlabeled data to identify hidden patterns or groupings in large datasets. This is commonly used in settings of large, complex data that are difficult to readily identify associations, such as molecular pathology. Examples of unsupervised learning used in molecular research include clustering tumors based on gene expression or methylation profiles. This may support cytology tasks such as grouping similar cellular patterns or discovering novel morphologic subtypes without prior labels. This approach is particularly useful for exploratory analyses or for uncovering subtle morphologic differences that might escape human perception. While unsupervised learning eliminates the need for curated annotations, it requires large datasets for the algorithm to discern meaningful patterns. Moreover, evaluation of performance and clinical applicability can be challenging because there are no ground-truth labels, and a thorough review of the output is required. Therefore the lack of interpretability may be another barrier to clinical workflow adoption (Wadden, 2021; Salahuddin et al., 2022).

Finally, a middle-ground approach between these paradigms is weakly supervised approaches, including multiple instance learning (MIL). This paradigm uses a label at the bag or group level so that individual patches of images do not need to be labeled. This allows model training with slide-level or case-level labels rather than labor-intensive cell-level annotations. Therefore this makes weakly supervised approaches attractive for digital cytology applications due to the decreased need for annotations. The model uses limited annotations to guide learning and then extrapolates to the unlabeled datapoints. MIL is particularly relevant in cytopathology because WSIs are too large to process in a single training pass. The assumption is that the features between a positive and a negative bag are learned even if there is one positive example in a bag—for example, one malignant cell in a cytology sample. This reduces the burden of exhaustive labeling, allowing the application of existing labels, such as those derived from a pathology report, to all image patches for that case. Existing pathology reports linked to digital slides or images can be easily used as labels without the need for further rigorous annotations. As an example, a dataset of thousands of cytology WSIs can be labeled as positive or negative for malignancy based upon their clinical pathology report. Although positive malignant cases may contain benign elements, a weakly supervised algorithm will learn the differences between positive and negative slides without the need for labels on each individual cell or region.

Nongenerative models for cytology

While these paradigms define how algorithms learn from data, their performance and suitability for cytology tasks also depend on the underlying architectural design of the models. Different architectures determine how features are extracted, relationships are modeled, and predictions are made from the data. Understanding these architectures provides important context for how AI systems process cytology images and related data inputs. Several models stand out for their unique strengths and limitations when considering AI architectures for cytology applications.

In some regards, a simpler AI approach is classical computer vision. Classical computer vision relies on hand-crafted features and image processing techniques rather than learned representations, as is the case for DL methods. Using relatively simple but powerful operations such as background–foreground separation and blur detection, these methods enabled object detection, segmentation, and quantification long before DL. In medicine, they are applied to tasks such as identifying cells, segmenting tissue regions, and measuring structures, though their performance was often limited by variability in staining, imaging conditions, and patient morphology. In cytology, tasks such as cell-of-interest detection and IHC quantification are some methods amenable to classical computer vision.

For DL applications, Convolutional Neural Networks (CNNs) are currently the most widely used architecture for image-based tasks, including cytology, due to their ability to capture image features and local patterns within WSIs. Their strength lies in their ability to detect subtle morphological features, such as nuclear shapes or cell clustering, just as cytopathologists do for diagnostic evaluation. However, CNNs require large labeled datasets, which can be challenging to obtain in pathology, as annotated individual cells or even slides are not readily available. Pathology foundation models based on a CNN architecture and finetuned on millions of pathology WSIs have recently become available (Chen et al., 2024; Wang et al., 2024b; Vorontsov et al., 2024). These CNN-based foundation models have already been trained on millions of pathology images; thus they have already learned histologic features to an extent. Therefore only further fine-tuning of the models to the intended task with a smaller dataset is needed. However, no foundation model currently exists for cytopathology that would consider the various morphologic differences one sees on cytology preparations. Potentially, a cytology foundation model would account for the cellular features found across cytology preparations to easily apply image-based AI methods to cytology predictions, such as assessing malignant features in a cell or case.

Generative models for cytology

In contrast, transformer-based models, originally designed for natural language processing and text-based data, have been adapted for image tasks (vision transformers). Transformer-based models can account for the spatial relationships between objects on a slide. This can be advantageous for cytology, where the distribution of abnormal cells across the slide provides diagnostic clues. Therefore more accurate predictions can be made for image-based cytology tasks. However, transformers are computationally expensive, require extensive pretraining, and often require even larger datasets than CNNs to perform optimally. Since CNNs handle low-level features, such as cell shape and contours, while transformer models handle spatial relationships, hybrid models that use both are also emerging to maximize performance on image-based tasks.

Large Language Models (LLMs), while not image-based, play a significant role in pathology by processing and interpreting unstructured text, such as cytology reports, clinical notes, and integrated diagnostic narratives. While previous ML models are types of nongenerative AI, LLMs are generative AI in that new data is provided based on example patterns. LLMs are particularly useful for extracting meaningful insights from textual data, which can enhance clinical decision support. In cytology, LLMs can summarize clinical, radiologic, and laboratory results prior to reviewing a sample. Rather than spending time searching for key pieces of information to help contextualize, LLMs can help collect all this for a cytologist and cytopathologist. In addition, LLMs can automate the drafting of diagnostic reports and clinical notes by extracting relevant clinical details from electronic health records or linking free-text pathology reports with image-based findings for weakly supervised learning in cytology. These systems represent a step toward AI-assisted integrated diagnostics, where textual, visual, and molecular information are analyzed holistically to improve diagnostic accuracy and efficiency. However, LLMs require extensive pretraining to learn the relationships and associations between texts. More recently, with the widespread use of generative AI, there have been concerns about false information, termed hallucination. While the source of hallucinations is uncertain, there is a possibility that false data may be presented by the LLM model.

Lastly, multimodal architectures combine image-processing backbones (CNNs or ViTs) with natural language processing (NLP) models to integrate cytology images with pathology reports, clinical narratives, or molecular data. Combining a variety of data, such as pathology WSI, molecular data, radiologic images, and clinical reports tied to outcomes data to predict patient outcomes, has also been explored. Cytology, much like the rest of healthcare, generates millions of data points in the form of text, numbers, and images. Combining different AI architectures allows the integration of all types of data to be used for prediction.

Together, these architectures highlight that the choice of AI model must align with the nature of the data and the specific task. Whether it involves detecting rare malignant cells on a slide, summarizing diagnostic reports, or integrating clinical, morphologic, and molecular features for precision diagnostics, the AI method used depends on the problem at hand. These models make predictions from large sets of complex data but lack the ability to reason or interpret data at a deeper level. AI is a tool for those practicing cytology, and as with any tool, one should be cognizant of how the tool works as well as its strengths and weaknesses. This allows for the proper use of AI solutions in cytology practice and further clinical development.

Challenges in AI development for cytopathology

AI offers many potential benefits for cytology, but obstacles exist in translating these innovations into clinical practice. The road to clinical AI implementation is lined with many technical, practical, and ethical hurdles. Some of these challenges are shared with surgical pathology; however, others are unique to cytology. Understanding these challenges is essential for the development and use of AI in the cytology laboratory.

Challenges in digital cytology

The availability of high-quality digital slides is one of the most fundamental requirements for image-based AI. However, cytology slides are far more complex to scan than standard histology sections. Unlike the relatively uniform formalin-fixed paraffin-embedded (FFPE) tissue blocks stained with hematoxylin and eosin (H&E), cytology specimens come in a wide range of preparations. Common cytology preparations include Diff-Quik, Papanicolaou, or H&E smears, as well as liquid-based methods such as ThinPrep, SurePath, and cytospins, in addition to FFPE cell block slides. Each of these techniques produces slides with distinct morphologic appearances due to differences in fixation and preparation (Fig. 10.1). Furthermore, cytology slides often contain dispersed 3-dimensional cell clusters with large areas of empty background and/or debris. These characteristics pose major difficulties for WSI scanners, which are primarily optimized for cohesive tissue sections. Achieving an adequate digital image for cytology may require the use of multilayer Z-stacking to maintain focus on overlapping cell groups (Hanna et al., 2017; Tanaka et al., 2025; Kim et al., 2022). While effective, this approach significantly increases both scanning time and file size, creating storage and processing burdens that complicate large-scale implementation. In essence, the process of generating suitable digital images prior to AI development remains a major bottleneck for cytopathology.

Even after optimal high-quality digitization, the dispersed nature of cells introduces new challenges in image analysis. Most AI pipelines start by segmenting slides into regions of interest, which are then processed into uniform image patches for analysis. This task is relatively straightforward in surgical pathology because tissue areas are contiguous and cohesive. In contrast, cytology slides may contain thousands of cells, ranging from isolated lymphocytes to large tissue fragments scattered across a wide background. The variability in

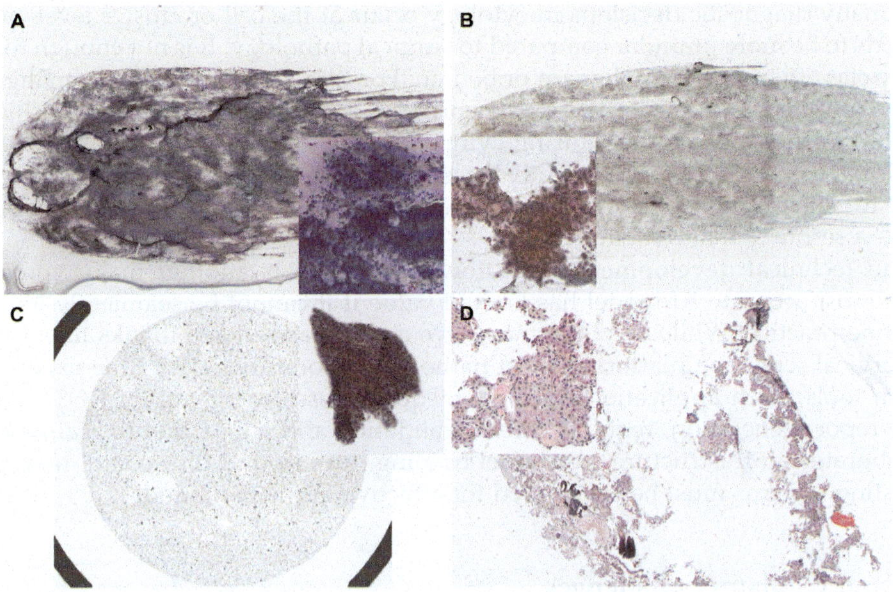

FIG. 10.1 Images of various cytology preparations displaying the morphology of the same tumor: (A) Diff-Quik smear, (B) Alcohol-fixed papanicolaou stain smear, (C) Liquid-based cytology ThinPrep, and (D) Formalin-fixed paraffin-embedded cell block.

size and distribution makes it difficult to define consistent regions for analysis. Since most neural networks rely on fixed patch sizes, this inconsistency complicates both training and inference, requiring additional preprocessing steps or novel algorithmic strategies.

Diversity of cytology preparations

Another source of complexity is the variability introduced by different preparation methods, as described above. Histology largely benefits from standardization, as most slides produced are FFPE and stained with H&E. This uniformity allows models trained on one set of images to transfer reasonably well to others, with only minimal fine-tuning. Cytology, however, lacks such uniformity. For example, a model developed on ThinPrep samples may perform poorly on Diff-Quik smears for the same tumor type and diagnostic task due to morphologic differences. Cytopathologists learn to integrate the various morphologic features across the various preparations, but AI models require explicit exposure to this diversity during training. This often means building separate models for each preparation type or amassing enormous multipreparation datasets, both of which present significant logistical and financial challenges.

Data curation and processing

Data requirements represent another major barrier for cytology. All machine learning models depend on large, well-annotated datasets, and cytology is no exception. However, because many diagnostic decisions in cytology occur at the cell or cluster level, annotation often needs to be more granular compared to surgical pathology. It is not enough for a model to simply classify a slide as malignant or benign. For true clinical utility, the malignant cells must be identified and localized to be incorporated into cytology practice. This level of annotation is labor-intensive, requiring cytopathologists to painstakingly mark thousands of individual cells. Although recent advances in weakly supervised learning have reduced the need for exhaustive labeling, robust model development still depends on large, curated datasets—a resource that remains scarce.

Beyond technical development, workflow integration is another major consideration. Even the most accurate AI model has limited value if it cannot be seamlessly incorporated into routine practice. While cytology may have a slight advantage thanks to existing automated cervical screening platforms, most pathology laboratories have little experience with AI-driven tools. Each application, such as adequacy assessment during ROSE, triaging of cases, or report generation, requires careful validation and adaptation to real-world conditions. Laboratory infrastructure, personnel training, turnaround times, and interoperability with existing systems must be considered for effective implementation.

Ethical and regulatory challenges

Ethical and regulatory considerations add yet another layer of complexity to implementation. Many DL algorithms function as "black boxes," making their decision-making process difficult to interpret. This raises questions about transparency, patient consent, and accountability when using patient data for model training and deployment. Furthermore, biases present in training datasets can propagate and even amplify through AI models, potentially affecting diagnostic equity (Drazen et al., 2023; Hanna et al., 2025a; Banerjee et al., 2023). While emerging guidelines aim to address these concerns, clear standards for ethical AI development and deployment remain looming questions. Similarly, regulatory frameworks for validation and quality assurance are still in their infancy. Unlike traditional laboratory assays, AI models pose unique challenges: How should performance be monitored over time? What constitutes acceptable QA/QC for an algorithm that may evolve with new data? These are open questions that pathology, as a field, must collectively address.

Despite these obstacles, the trajectory of AI innovation is promising, and cytopathology is poised to benefit substantially from these advances. Awareness of these barriers is the first step toward overcoming them. As with any laboratory technology, successful integration of AI requires not only technical sophistication but also rigorous validation, ethical vigilance, and adherence to the same high standards that govern all clinical testing. Ultimately, cytopathologists themselves must lead this transformation, ensuring that AI becomes a tool that enhances, rather than compromises, patient care.

Preparing the cytology lab for AI integration

When preparing the cytology laboratory for integrating AI algorithms into the routine clinical workflow, it is important to address a variety of laboratory aspects. This includes preimaging, imaging, and post-imaging factors. The evaluation of each aspect is no trivial task, as it runs the gamut from ensuring that compatible barcodes are placed on slides, image acquisition is designed to address focus issues of cellular material on the slides, workstations can accommodate multiple computer monitors, validation studies are undertaken and compliant to meet accreditation needs, cytology staff are adequately trained and remain proficient, there is a downtime procedure, and IT support is available to troubleshoot and maintain the digital system installed in the lab. This section will review some of the key cytology laboratory preparedness considerations for AI integration.

Digital infrastructure & imaging readiness

Most cytology labs still lack a full digital workflow, which is a prerequisite for AI deployment. To utilize AI tools, careful selection of a digital solution is therefore warranted, including a whole slide scanner capable of digitizing slides that are in focus. The reason this is important is that cytology slides often contain overlapping or out-of-focus cells and possibly obscuring material such as blood, mucus, or contaminating gel. Three-dimensional cell clusters and obscuring artifacts can lead to large out-of-focus areas on a digital slide, hindering AI evaluation. Solutions include Z-stacking or volumetric scanning to capture 3D cell groups and provide better in-focus images for AI evaluation. Dedicated digital cytology systems should not only overcome focus needs, but by coupling AI algorithms that assist with screening, they can also overcome workflow limitations. Such AI algorithms (i.e., killer applications) would potentially provide a large enough benefit to justify the upfront costs and workflow efforts. Ideally, the image management system (IMS) is integrated with both the AI and reporting system, viz., the laboratory information system (LIS). Making sure end users have ergonomic digital cockpits is important, as well as regular calibration for maintaining display performance (Marletta et al., 2025). If available, a QC algorithm should be employed to detect artifacts (e.g., air bubbles, out-of-focus regions) for prompt remediation.

Workflow adaptation

A digital image is only as good as the glass slide being scanned (i.e., garbage in = garbage out) (McAlpine et al., 2021). Therefore, for a successful digital transition, cytology labs must pay attention to pre-imaging steps in the workflow. This begins with sample preparation and slide standardization (e.g., staining, coverslips, labeling with barcodes). Case accession, assignment, and distribution to end-users will need to be handled in the IMS. It is unlikely that every cytology slide will be scanned, and labs are likely to only deploy AI algorithms for one use case at a time (e.g., Cervical Pap test screening, urine cytology, body cavity fluids). Hence, a laboratory initially should be prepared for a hybrid workflow (i.e., digital and manual slides) rather than a full AI workflow for all cytology samples. Until AI tools are cleared by regulatory bodies that do not require humans to oversee every analyzed case, Cytologist/Cytopathologist-in-the-Loop AI (CITL-AI) should be anticipated. CITL-AI combines automated triage with human review to ensure safety and accuracy. The ultimate objective of a CITL-AI is to aid the cytologist and cytopathologist in their workflow by automating repetitive, error-prone clinical tasks.

AI algorithm validation

Once an AI system has been approved, selected, and verified technically for clinical use, the cytology lab should undertake a "clinical" validation. Validation must emulate clinical workflow and include diverse sample types that reflect the complexity of cases encountered in routine practice. According to recommendations from the American Society of Cytology (ASC), these validation studies should include a case volume of 100–200 per AI category/ algorithm (Kim et al., 2024a). Performance metrics such as accuracy, sensitivity, specificity, and precision should be evaluated as part of the validation (Hanna et al., 2024). If any major changes to the AI workflow or a new component is introduced, such as the inclusion of different preparations or samples, a re-validation encompassing the changes should be undertaken.

Management responsibilities

AI output used for clinical reporting must be explainable and compatible with standardized and acceptable cytology terminologies. For now, final diagnostic responsibility remains with humans (i.e., cytologists and cytopathologists). Training is imperative and should include cytopathologists, cytologists, and trainees to build trust and proficiency (Cantley et al., 2024). Studies show that AI-assisted training improves interpretation scores and diagnostic consistency among trainees (Hassell et al., 2025). The cytology lab needs to adopt quality assurance (e.g., re-screening of cases) and proficiency testing (PT) methods appropriate for a digital workflow that relies on AI tools to analyze slides. Resources will also need to be dedicated to managing change resistance. Medicolegal considerations are important and include retaining analyzed images and AI outputs for review and documentation, as well as ensuring suitable malpractice coverage. Finally, it is important to educate staff on AI capabilities and limitations, and to foster a collaborative mindset between IT, pathology, and clinical provider teams.

Digital pap tests and AI-assisted screening

Historically, the first AI algorithms in pathology were developed for cervical cytology screening in the late 1990s (Halford et al., 1997). Since then, many of these AI algorithms for cervical cytology screening have been used in routine clinical practice, primarily by cytologists (Biscotti et al., 2005; Wilbur et al., 2009; Colgan et al., 2013; Miller et al., 2007; Dziura et al., 2006; Lozano, 2007; Zahniser et al., 1979; Tanaka et al., 1987; Lee et al., 1998; Levi et al., 2012; Dawson, 2004; Wilbur et al., 1998). Despite these advancements, laboratories face ongoing challenges such as increasing workloads, cytology workforce shortages, interobserver variability, and prolonged turnaround times, prompting AI vendors to discover novel ways to address these gaps. Emerging literature highlights the adoption of AI-assisted Pap screening techniques that offer automated review, detection, and classification of neoplastic lesions, thereby improving the diagnostic capabilities of cytologists as well as cytopathologists. ASC's guidelines for AI validation and implementation in cytology are still evolving (Kim et al., 2024a).

History of AI-assisted cervical pap tests

PAPNET is the first truly digital Pap test that uses DL neural network technology to identify abnormal cells in negative cases screened by cytologists and mails back a digital tape of 128 abnormal cells/cell clusters for review by a cytologist (Halford et al., 1997). This system received FDA approval in 1995 for screening negative cases, marking a new era for cervical cytology. Subsequently, several new devices (CytycTM, AutoPapTM, TripathTM) were developed despite initial setbacks and secured FDA approval. Of these systems, the Hologic ThinPrep and BD FocalPoint Slide Profiler System (BD FSPS) were introduced in the early 2000s. These two systems were designed for primary screening and soon became the prototypes of automated screening devices for cervical cytology in the United States. The BD FSPS system identifies abnormal cells by integrating optical density, N:C ratio, and nuclear contours, and assigns a score ranging from 0 to 1 for the likelihood of an abnormality. Slides below a cut-off require no further review (NFR), while those above the cut-off require full manual review (Wilbur et al., 2009; Colgan et al., 2013). Compared to manual screening, BD FSPS demonstrated the same or higher sensitivity for low-grade and high-grade squamous lesions (Wilbur et al., 2009; Colgan et al., 2013; Levi et al., 2012). The BD FSPS system underwent further development into the FocalPoint GS Imaging System (FIS). The FIS system provides 10 fields of view (FOV) of abnormal cells. If no abnormal cells are detected in the 10 FOVs, the case is reported as negative.

The ThinPrep Imaging System (TIS) was developed by Cytyc and received FDA approval in 2003. Hologic acquired Cytyc and the ThinPrep product in 2007. TIS scans the ThinPrep slides and provides 22 fields of view (FOV) that harbor abnormal cells on each slide by feature extraction based on optical density measurements, nuclear-to-cytoplasmic ratio, and hyperchromasia using a modified Papanicolaou staining protocol. The TIS image processor consists of an imaging station that scans the entire ThinPrep slide, an image processor controller which creates the images and analyzes the data to provide 22 FOVs, and a server and user interface allowing the operator to use the machine. Cytologists review these spatially oriented FOVs on the computer monitor. If no abnormal cells are identified, the case can be reported as negative. The presence of any abnormal cells on any FOVs requires review of the entire slide. The TIS system outperformed manual review and detected significantly more abnormal slides. Subsequent studies demonstrated improved productivity with doubled slide-reading efficiency and improved detection of LSIL and HSIL without affecting specificity after implementation of the TIS (Biscotti et al., 2005; Miller et al., 2007; Dziura et al., 2006; Lozano, 2007).

Advances in AI-assisted digital Pap tests

A new wave of WSI-based AI-assisted Pap systems emerged following the advancement of digital pathology and AI-assisted tools in surgical pathology (Cantley et al., 2024; OptraSCAN, n.d.; Crowell et al., 2019a; Rivera Rolon et al., 2025; Zhu et al., 2021; Wang et al., 2024a; Yang et al., 2024; Xue et al., 2023; Konfoong Biotech International Co., n.d.; Ikenberg et al., 2023; Elishaev et al., 2025; Bai et al., 2024; Zeng et al., 2024; Delga et al., 2014; Chantziantoniou, 2023; Crowell et al., 2019b). Emerging WSI-based AI systems (e.g., Hologic Genius Fig. 10.2, AICyte, BestCyte, CytoProcess, CytoSIA, KFBIO, LandingMED, and

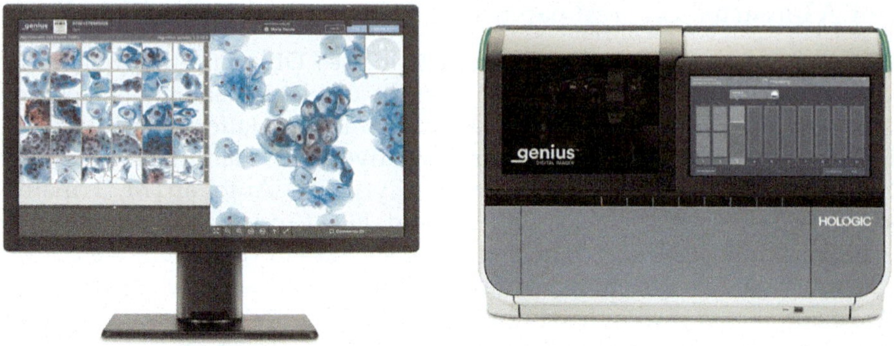

FIG. 10.2 Hologic genius digital diagnostic system slide scanner and review system with an image gallery of 30 FOVs. Image courtesy of Hologic, Inc.

Techcyte) for the screening of cervical cytology are summarized in Table 10.1, which includes details for slide preparation type, scanner type, z-stacking or volumetric scanning capabilities, magnification, and FDA approval status. All these systems have the potential to improve the efficiency of cervical cytology screening and allow remote review by cytologists and cytopathologists. All platforms provide an FOV/objects of interest (OOI) image gallery display of abnormal cells that parallels the job done by cytologists who manually mark cells and cell clusters of interest with markers. The image gallery is arranged in labeled rows. The number of images provided within the gallery varies by platform. For example, the Hologic Genius Digital Diagnostic System (HGDDS) displays 30 FOVs arranged in 5 labeled rows with 6 FOVs in each row and, if needed, provides 30 additional FOVs for diagnosis. AICyte and the LandingMED system highlight only 20 FOVs, while the BestCyte system displays tiles within 8 galleries. A subset of these systems allows further selection/tagging of image galleries by cytologists for cytopathologist's review, which provides AI-assisted interpretive Bethesda categories or binary negative/positive result categories. The American Society of Cytopathology's Digital Cytology Task Force recommends validation of AI-based tools in cytology practice based on performance matrices such as precision, accuracy, sensitivity, and specificity in a clinical setting intended for the use of AI algorithms (Kim et al., 2024a).

Table 10.2 summarizes recent studies using these emerging AI-assisted PAP systems. Different studies had different endpoints, reading time calculation methods, and numbers of participants, precluding optimal comparison. Of these novel systems, AIcyte has the fastest reported review speed, with an average review time of 22.23 s for screening (Delga et al., 2014).

Performance of AI-assisted digital pap test

Estimated calculations based on three HGDDS studies demonstrated an overall high sensitivity of 86.8% (Cantley et al., 2024; Ikenberg et al., 2023; Elishaev et al., 2025) with two studies highlighting time savings. Of these studies, Cantley et al. reported an average

TABLE 10.1 Emerging AI-assisted pap system details.

	Hologic genius	AIcyte	BestCyte	CytoProcessor	CytoSIA	KFBio	Landing Med	Techcyte	miLAB CER
Vendor	Hologic	AIcyte	CellSolutions	DATEXIM	OptraSCAN	KFBIO	Landing Med	Techcyte	NOUL
Scanner	Hologic genius	Ruiqian WSI-2400	3DHISTECH	3DHISTECH	OptraSCAN ULTRA	KFBIO	Ld Patho, ld Cyto2200	Nanozoomer, Grundium, Motic, 3DHISTECH	miLAB CER
Magnification	40x	40x	20x	40x	20x, 40x	20x	20x, 40x	40x	20x
z-stacking	Y, multiple planes	Y	Y	N	N	Y	Y	Y	Y
Slide preparations	ThinPrep	ThinPrep	BestPrep, LBC	ThinPrep, SurePath	ThinPrep, SurePath	LBC	LBC	ThinPrep, SurePath	LBC
FDA Approval	Y	N	N	N	N	N	N	N	N
Image Gallery	Y	Y	Y	Y	Y	Y	Y	Y	Y
FOVs	30 +additional 30, if needed	20	8 galleries	–	–	–	20	–	–
Automated Diagnostic categories	N	Y, 2 categories	N	N	N	Y	Y, 3	N	Y, 6 categories
Potential uses	Screening	Screening, reporting	Screening, reporting, quality control	Screening	Screening	Screening of gynecologic and nongynecologic cases	Screening	Screening, quality control	Screening, reporting, quality control
Availability	Y	N	Y	Y	Y	Y	Y	Y	Y
Time savings	Y	Y	Y	Y,1.6 times faster	Y	Y	Y	Y	Y

TABLE 10.2 Summary of studies evaluating novel AI-assisted pap systems.

References	AI system	Sample size	AI output/ analysis	Number of reviewers	Study end points
Ikenberg et al. (2023)	HGDDS	1994	Labeled image gallery of abnormal cells	4 CT, 2CP and 1 additional CP	Comparison between TIS and the new system; concordance was 81.22% and increased to 90.37% when compared to histology; 92.78% of CP matched the exact diagnosis; average screening time was 44.8 s versus 89.9 s (TIS)
Cantley et al. (2024)	HGDDS	319	Labeled image gallery of abnormal cells	6 CT, 3CP	Concordance between the original diagnosis and HGDD: 62.1% concordance; time-saving: 5.9 min for manual versus 3.2 min for digital
Elishaev et al. (2025)	HGDDS	890	Labeled image gallery of abnormal cells	3 CT, 3CP	100% sensitivity (ASCUS and above), NPV 92.4%–100%
Bai et al. (2024)	AICyte	32,451	Positive or negative binary result	2CP	Compared the performance of two CPs using AI to the original diagnosis; sensitivity was 97.89%, similar to manual review, and specificity improved from 6.77% to 16.19% for CIN2 and above lesions; overall average reading time was 22.23 s
Zeng et al. (2024)	AICyte	163,848	Positive or negative binary result	AI	AI standalone performance: Sensitivity 98.92% and NPV 98.80% for LSIL and above lesions; reduced workload using a 50% negative cut off value
Delga et al. (2014)	BestCyte	105	Image galleries of abnormal cells	6CT, 2CP	Compare BestPrep and BestCyte cell sorters with ThinPrep; 100% sensitivity for HSIL and higher specificity: 71.4% versus 54.8%
Chantziantoniou (2023)	BestCyte	500	Image galleries of abnormal cells	3CT	Interobserver comparison using BestCyte to the manual method. Sensitivity for ASCUS +, LSIL+, and ASCH+ thresholds: Manual (91.36%, 86.67%, 74.36%); Cytologist 1 (95.88%, 96.97%, 94.87%);

TABLE 10.2 Summary of studies evaluating novel AI-assisted pap systems—cont'd

References	AI system	Sample size	AI output/ analysis	Number of reviewers	Study end points
					Cytologist 2 (95.47%, 95.76%, 93.59%); Cytologist 3 (94.65%, 95.15%, 98.72%)
Crowell et al. (2019a)	CytoProcessor	1352	Image galleries of abnormal cells	NA	Compare CytoProcessor and TIS; 2.6 times fewer false negatives compared to TIS and 1.5 times time savings compared to TIS
Crowell et al. (2019a)	CytoProcessor	309	Image galleries of abnormal cells	NA	Compare CytoProcessor and NOVAPREP Processor System; improved sensitivity compared to NOVAPREP; 1.6 times faster compared to NOVAPREP
Rivera Rolon (2025)	TechCyte	1442	Image galleries of abnormal cells	1	Compare TechCyte with manual light microscopy; sensitivity: 82%, specificity: 99%, NPV: 98%, PPV: 86%; reduced review time compared to manual; integrated as a 100% QC tool
Zhu et al. (2021)	Artificial Intelligence-Assistive TBS (AIATBS) Diagnostic Solution	81,727 (training),34,403 (test)	Image galleries of abnormal cells	NA	AI development and validation; sensitivity >92% and specificity (82.14%); sensitivity increased to 94.7% for cases with follow-up biopsies; AI speed <180 s per case
Wang et al. (2024a)	Artificial Intelligence Cervical Cancer Screening (AICCS) System	11,468 (training) 4588 (validation)	Image galleries of abnormal cells	NA	Compare AI with cytopathologists and AI stand-alone; AUC 0.947; Sensitivity 94.6%, specificity 89%, and accuracy of 89.2%; Time reduced to 120 s compared to 180 s for manual
Yang et al. (2024)	Cell Image Analysis System (CIAS)	1231	Image galleries of abnormal cells	2CT	AI stands alone at LSIL and HSIL and LBC with HPV cotesting; AUC 0.621 versus 0.521 for HSIL and 0.637 versus 0.524 for LSIL
Xue et al. (2023)	Cytologist-In-The-Loop Artificial Intelligence (CITL-AI)	8000 (training), 3514 (validation)	TBS categories	NA	AI stand-alone sensitivity of 89.4% and specificity of 66.4% are superior to junior cytologists

time spent per case of 3.2 min, reduced from 5.9 min using the HGDDS. Ikenberg et al. reported a reduction in screening time by half using this system (Ikenberg et al., 2023). Currently, the HGDDS system is being validated at multiple academic institutions in the United States.

A study by Bai et al. (2024) compared the performance of pathologists using AICyte with the original diagnosis, which demonstrated similar sensitivity but improved specificity for CIN2 and above lesions. Given the initial results, Zeng et al. (2024) studied the standalone performance of AICyte with a 50% negative cut off value, achieving a sensitivity of 98.92% and an NPV of 98.80%. Two studies using the BestCyte AI system showed improved performance in sensitivity for the detection of ASCUS and above lesions compared to manual review by three cytologists without AI (Delga et al., 2014; Chantziantoniou, 2023). Time for review was not analyzed, and therefore, time savings for screening were not reported in these studies.

Studies using the CytoProcessor system demonstrated a slightly increased sensitivity for ASCUS and above lesions, with an overall improved diagnostic review time compared to manual review (Crowell et al., 2019a, 2019b). Another novel system by TechCyte has been implemented with a 100% QC in the cervical cytology workflow at a CorePlus laboratory in Puerto Rico (Rivera Rolon et al., 2025).

Although multiple AI-assisted systems exist for cervical cytology, researchers around the world continue to develop novel AI-assisted systems for cervical cytology. Of such efforts, three studies in China were conducted to develop and test the AI-assisted tests (Zhu et al., 2021; Wang et al., 2024a; Yang et al., 2024). Training cohorts ranged from 8000 to 81,727. These tests showed high sensitivity (up to 94.6%) and specificity (up to 89%) with reduced case review time. Similarly, a novel AI algorithm studied by Yang et al. (2024) on 1231 cases evaluated the performance of AI alone for cytology cases with discordant HPV results in LSIL and HSIL. This study described better performance for stand-alone-AI with an AUC of 0.637 vs. 0.524 for LSIL and 0.621 vs. 0.521 for HSIL cases. The reported sensitivity and specificity of another platform, KFBio, were 98.2% and 63.5%, respectively (Konfoong Biotech International Co., n.d.).

Main highlights of the novel emerging AI-assisted technologies include automation of slide preparation, staining, imaging, remote review capabilities for both cytologists as well as cytopathologists, advanced use of AI to identify abnormal cells and sub-categorize as per TBS for cervical cytology reporting, reduced review time, sharing cases digitally for second opinions, educational teaching, and multimodal research.

In summary, emerging AI-assisted digital Pap tests are set to transform cervical cytology practice around the world. However, many AI-based systems are relatively new, with limited real-world data on their validation and clinical utility. Based on preliminary studies, these tests provide opportunities to improve workload, reduce turnaround time, allow remote sign-out, and help standardize cytology practices across the globe. Large-scale clinical trials with the deployment of these systems are needed to showcase the real-world utility and long-term benefits of these systems. Proper guidance, education on these novel systems, recommendations for validation, and successful integration of these systems require a collaborative effort from the cytology community and are essential to embarking on this new AI-assisted cytology practice.

Review of current AI applications in cytopathology

Many recent image-based cytology AI applications have centered around GYN cytology samples; however, there has also been development in nongynecologic applications. While none of these algorithms are currently integrated into routine clinical use, recent advancements in AI technology have led to advancements in many different cytology specimens, including ROSE and QA/QC.

Automated screening of urine cytology

Urine cytology is currently widely used as a screening test for bladder cancer, specifically high-grade urothelial cancer. Given the similar specimen preparation to cervical cytology in liquid-based cytology (LBC), similar approaches were used for the development of AI for screening urine specimens. Preliminary studies demonstrate promising results (Table 10.3) (Pantazopoulos et al., 1998; Muralidaran et al., 2015; Wu et al., 2024; Liu et al., 2024; Hang et al., 2024; Sanghvi et al., 2019; Vaickus et al., 2019; Awan et al., 2021; Nojima et al., 2021; Lebret et al., 2022; Ou et al., 2022; Lebret et al., 2023; Tsuji et al., 2024). LBC was the most common specimen preparation type studied for urine cytology; however, conduit/diversion urine specimens were not included in any study design. A few of the urine cytology AI applications provided binary classifications of benign and malignant, while others were multiclass, encompassing the categories of benign, low-grade, and high-grade urothelial carcinoma. The reported specificity for most of these applications was 100% in detecting high-grade urothelial carcinoma. One recently developed DL AI algorithm used the nuclear-to-cytoplasmic ratio of the cells and thresholds analogous to the criteria used in The Paris System for Urine Cytology (TPSUC) (Liu et al., 2024). The AI algorithm, after analyzing each cell according to the TPSUC, displayed the cells in a gallery fashion, much like for HGDDS (Fig. 10.3). In this study, the sensitivity for detecting atypical urothelial cells increased to 63.9% from 25%–30.6%, and to 33.3% from 15.2%–27.3% for the category of suspicious for high-grade UC. Screening and review times were reduced as well with the system. The VISIOCYT1 pilot study by Lebret et al. (2023) developed and evaluated the VisioCyt test in France using DL. The first phase showed overall high improvement, with overall sensitivity increasing from 43% to 84.9%. Sensitivity for low-grade lesions improved from 26.3% to 77%, and from 61.1% to 92.6% for high-grade urothelial lesions. Vaickus et al. (2019) developed a hybrid DL and morphometric approach to detect atypical urothelial cells, squamous cells, and crystals in WSIs of urine cytology cases. This study reported an AUC of 0.92, sensitivity of 83%, and a false positive rate of 13%. Three false positive cases were rich with inflammatory cells in dense clusters. The only prospective study for urine cytology AI to date, conducted in China, demonstrated a sensitivity of 89.6% for the detection of atypical urothelial cells and 84% for detecting malignancy in atypical urothelial cells (Wu et al., 2024). Hang et al. (2024) explored the value of AI and heuristic scanning in 21 planes and demonstrated reduced scan time and reduced file size with noninferior detection of high-grade urothelial carcinoma cases (Hang et al., 2024). All these studies demonstrate that AI-assistance in urine cytology will eventually improve cytology workflow.

TABLE 10.3 Summary of AI-application-based studies for urine cytology.

References	AI algorithm	Cases	AI analysis	End points
Urine cytology				
Pantazopoulos et al. (1998)	Feature-based neural network	329 (test cases)	Detection of high-grade urothelial carcinoma	Sensitivity 94.5%; specificity 100%
Muralidaran et al. (2015)	ANN	115	Benign, malignant (low-grade v high-grade)	Distinguished all benign cases from malignant, and high-grade from low-grade (except for 1 case)
Sanghavi et al. (2019)	DL	2405	Detection of high-grade urothelial carcinoma	Sensitivity 79.5%, specificity 84.5%, AUC 0.88 for the detection of high-grade UC
Vaickus et al. (2019)	Hybrid DL model	217	Development of an AI algorithm	Automate the calculation of the N:C ratio, identify AUC, squamous cells, crystals, etc.
Awan et al. (2021)	DL	NA	Detect atypical and malignant cells	AUC 0.81 and 0.83 for atypical cells and malignant cells, respectively
Nojima et al. (2021)	DL	232	Development	AUC of 0.8661 for the detection of high-grade lesions
Lebret et al. (2022)	DL	598	14 centers	Sensitivity was 92.6% for high-grade UC with AI assistance compared to 61.1% without AI use, and 77% vs. 26.3% for low-grade tumors
Ou et al. (2022)	DL (AIxMed)	131 test cohort	1 Cytopathologist; 2 cytologists, with and without AI assistance, after 1 washout period Ground truth: Expert panel consensus	Sensitivity improved from 87.2% to 92.3% with AI assistance, and negative predictive value increased from 94.8% to 96.8%. $k = 0.944$
Lebret et al. (2023)	DL	391	14 centers	Sensitivity for high-grade tumors is 93.7%, and it is 66.7% for low-grade tumors
Tsuji et al. (2024)	DL	315 Test cohort	AI performance with respect to histologically proven high-grade UC, time savings	Sensitivity of 63% for HGUC vs. 46% by cytology; accuracy of 74% vs. 68%; time spent: 139 s with AI
Wu et al. (2024)	PUCAS (Precision Urine Cytology AI Solution)	2641: Training 2335: retrospective test validation 400: Prospective validation	Identify atypical urothelial cells	Sensitivity of 92%–100% for the retrospective study and 89.6% for the prospective validation study

TABLE 10.3 Summary of AI-application-based studies for urine cytology—cont'd

References	AI algorithm	Cases	AI analysis	End points
Urine cytology				
Hang et al. (2024)	Heuristic scan with 21 planes	NA	Multiple z-planes	Comparable suspicion for HGUC case numbers with reduced scan time and file size
Liu et al. (2024)	DL/AIxURO	116	Atypical urothelial cell, low-grade UC, high-grade UC	Sensitivity for AUC increased to 63.9% from 25%–30.6% and to 33.3% from 15.2%–27.3% for suspicion of high-grade UC; reduced screening time

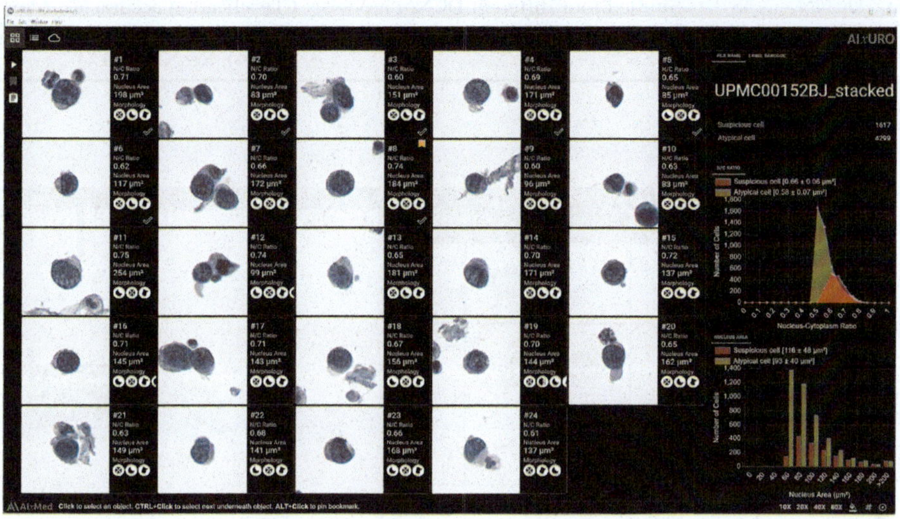

FIG. 10.3 AI-assisted screening of urine cytology using the AIxMed platform.
Image courtesy of AIxMed, Inc.

Lung fine needle aspiration cytology

For lung cytology, AI studies attempted the identification and subtyping (adenocarcinoma, squamous cell carcinoma, small cell carcinoma, etc.) of lung carcinoma using DCNN, achieving an overall accuracy of 71.1% for subtyping (Table 10.4) (Teramoto et al., 2017, 2020). Comparatively, one study focused on distinguishing small cell carcinoma (SCC) and large cell neuroendocrine carcinoma (LCNC) by using AI on Pap-stained, Diff-Quik-stained, and H&E slides. The algorithm correctly classified 6/7 (85.7%) SCC and 1/1 (100%) LCNC Pap-stained cases, as well as 7/8 (87.5%) SCLC and 1/1 (100%) LCNEC on a Diff-Quik-stained case (Gonzalez et al., 2020). Tsukamoto et al. (2022) tested the performance of four different DCNN models for lung carcinoma classification. They reported differing accuracies of 66% (GoogleNET), 74% (AlexNET and ResNET), and 76.8% (VGG16).

TABLE 10.4 Summary of AI-application-based studies for lung cytology.

Lung cytology				
References	**AI algorithm**	**Cases**	**AI analysis**	**End points**
Teramoto et al. (2017)	DCNN	76	Subclassification of cancer cases into adenocarcinoma, squamous cell carcinoma, and small cell carcinoma	Accuracy 71%
Gonzalez et al. (2020)	Open-source software	40	Small cell versus large cell neuroendocrine carcinoma on cytology smears as well as cell-block or biopsy	The algorithm correctly classified 6/7 (85.7%) SCLC and 1/1 (100%) LCNEC Pap-stained cases, and 7/8 (87.5%) SCLC and 1/1 (100%) LCNEC Diff-Quik -stained cases
Tsukamoto et al. (2022)	DCNN, different models	55	Studied four different DCNN models for lung cancer classification	Classification accuracies: AlexNET (74%), GoogleNET (66.8%), VGG16 (76.8%), ResNET (74%)

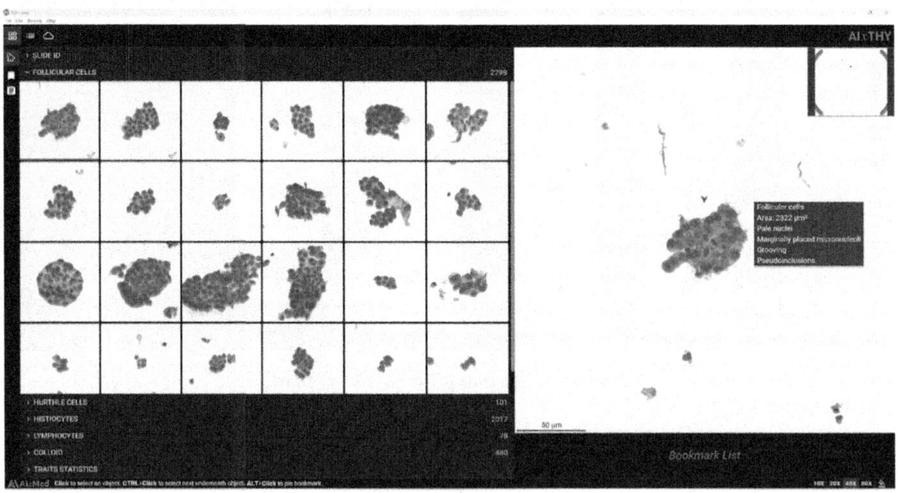

FIG. 10.4 AI-assisted screening of thyroid cytology using the AIxMed platform. Image courtesy of AIxMed, Inc.

Thyroid fine needle aspiration cytology

For thyroid FNAs, AI applications have been studied for identifying papillary thyroid carcinoma (Fig. 10.4). Some studies included virtual staining for thyroid smears and subtyping into benign and malignant categories (Table 10.5) (Jang et al., 2023; Ippolito et al., 2004; Lee et al., 2024; Cochand-Priollet et al., 2006; Varlatzidou et al., 2011; Collins and Collins, 2013; Sanyal et al., 2018; Lin et al., 2021; Maleki et al., 2019; Savala et al., 2018; Dov et al., 2023). Studies exploring AI applications for identifying papillary thyroid carcinoma (PTC) on FNA smears and ThinPrep slides demonstrate an accuracy range of 85.06% to 99%.

TABLE 10.5 Summary of AI-application-based studies for thyroid cytology.

Thyroid				
References	**AI algorithm**	**Cases**	**AI analysis**	**End points**
Ippoliti et al. (2004)	ANN	453	PTC identification	Sensitivity 86%, specificity 59%
Cochand-Priollet et al. (2006)	ANN	NA	Benign versus malignant	Accuracy of 75% to distinguish between benign and malignant
Varlatzidou et al. (2011)	ANN	335 FNA, LBC	Benign versus malignant	Sensitivity 91.5%; specificity 92.4%
Collins and Collins (2013)	DNN	44	Benign versus malignant	4.6 odds ratio for malignancy, N/C ratio of 0.5 or less, and 7.0 odds ratio for cases with an N/C ratio of 0.5 or less and greater than 20 cell groups
Sanyal et al. (2018)	CNN	174 test set	Distinguish PTC from other lesions	Sensitivity: 90.5%; Specificity: 83.3%; Accuracy: 85%
Maleki et al. (2019)	Text-based support vector matrix that uses 33 categories of synonymous keywords/phrases derived from microscopic descriptions of thyroid FNAs	59 cases	Detect PTC and distinguish PTC from NIFTP with surgical follow-up	Sensitivity of 88.9% and specificity of 96.0% in detecting PTC; distinguished PTC from NIFTP in 81.1% of cases
Savala et al. (2018)	ANN	57 (26 follicular adenomas, 31 carcinomas)	Distinguish follicular adenoma from carcinoma	100% accuracy to distinguish follicular adenoma from carcinoma
Lin et al. (2021)	DL	131	PTC detection on FNA smears and ThinPrep	99% accuracy for detection of PTC on FNA as well as ThinPrep slides
Dov et al. (2023)	DL	288	Classify benign and malignant, and assess the risk of malignancy	Classified 45.1% of cases as benign with a risk of malignancy of 2.7% and 94.7% for benign and malignant cases, and reduced the number of indeterminate cases ($n = 108$) by 21.3% by re-classifying them as benign
Lee et al. (2024)	Open AI	306 malignant, 228 benign FNAs	Comparison between AI and cytopathologists	Accuracy: 99.71% [95% confidence interval (CI): 99.38%–100.00%] vs. 88.91% [95% CI: 86.99%–90.83%], sensitivity: 99.81% [95% CI: 99.54%–100.00%] vs. 87.26% [95% CI: 85.22%–89.30%], and specificity: 99.61% [95% CI: 99.23%–99.99%] vs. 90.58% [95% CI: 88.80%–92.36%]

Not surprisingly, vague papillary formations in benign lesions were inaccurately classified as PTC in these studies. In a study by Jang et al. (2023), a DL/CNN model was developed to generate a virtual stain on unstained thyroid FNA smears to screen and detect nondiagnostic cases as per the adequacy criteria used in the Bethesda System for Reporting of Thyroid Cytology. This model showed an AUC of 0.84 for the detection of nondiagnostic cases. Bypassing slide staining with on-site virtual AI staining has the potential to reduce the nondiagnostic rate and eventually improve the quality of patient care. One of the first models (Cochand-Priollet et al., 2006) to distinguish benign thyroid lesions from malignant ones used four features: nuclear shape, size, chromatin texture, and distribution of chromatin. This model had an accuracy of 75.2% for this binary classification. A text-based model by Maleki et al. (2019) distinguished PTC from NIFTP with an accuracy of 81.1% compared to resection specimens. Savala et al. (2018) developed an ANN model to distinguish follicular adenoma from follicular carcinoma. In a recent study (Dov et al., 2023), the AI algorithm definitively classified 45.1% (130/288) of thyroid cases as benign or malignant, with a risk of malignancy of 2.7% and 94.7%, respectively. In addition, the number of indeterminate cases was reduced by reclassifying 21.3% (23/108) as benign. In a recent study by Lee et al. (2024), the standalone performance of AI was better than that of a cytopathologist, with an accuracy of 99.71% compared to 88.91% for subtyping benign and malignant lesions of the thyroid.

Other FNA cytology specimens

For pancreatic cytology, AI showed 100% accuracy in differentiating benign from malignant lesions using LBC (Momeni-Boroujeni et al., 2017) (Table 10.6). Image analysis algorithms also aided in the separation between atypical and suspicious cases, as well as improved sensitivity for differentiating benign and malignant lesions on biliary brush cytology (Collins and Weimholt, 2015; Monabbati et al., 2023).

Automated screening in effusion cytology

In effusion cytology, pleural fluid was the most targeted AI application (Barwad et al., 2012; Tosun et al., 2015; Xie et al., 2022; Park et al., 2023; Sanyal and Dey, 2023; Mavropoulos et al., 2023). Mavropoulos et al. (2023) used the Deepcell platform, which combined microfluidic sorting, brightfield imaging, and multidimensional morphology assessment by DL to enrich carcinoma cells in effusions without cell staining or labels. This study demonstrated a higher sensitivity for detecting tumor fractions using the Deepcell solution. Some studies developed AI applications for distinguishing benign from malignant effusions, while one study tested the performance of an AI algorithm to differentiate benign mesothelial cells from malignant mesothelioma.

Other AI applications in cytology

Other potential applications of AI include automated quantitation of biomarkers such as Ki-67 (Satturwar et al., 2020; Shaker et al., 2024), ER, PR, and HER2 on cytology cell blocks. These AI applications have the potential to improve the efficiency of screening, enhance

TABLE 10.6 Summary of AI-application-based studies for pancreaticobiliary and effusion cytology.

References	AI algorithm	Cases	AI analysis	End points
Pancreatico-biliary cytology				
Momeni-Boroujeni et al. (2017)	MNN	75	Differentiate benign and malignant on ThinPrep slides of the pancreas	100% accuracy to differentiate benign from malignant cases, 77% accuracy for atypical cases
Collins and Weimholt (2015)	Morphometry	28	Bile duct brushing with an indeterminate diagnosis (atypical or suspicious)	Quantitative image analysis showed that there was more nuclear pleomorphism and larger nuclei in suspicious cases compared to atypical cases
Monabbati et al. (2023)	Image Analysis	66 test cases	Bile duct brushing: benign vs. malignant	Sensitivity increased from 46% to 68% with AI
Effusion cytology				
Barwad et al. (2012)	ANN	114	Benign vs. malignant effusions	100% accuracy
Tosun et al. (2015)	ANN	34 pleural effusion cases	Distinguish benign mesothelial cells from mesothelioma	100% accuracy
Xie et al. (2022)	DL	60 test cases of pleural effusion	Detect metastatic lung cancer in pleural effusion compared to two cytopathologists	Accuracy (91.7%), sensitivity (87.5%), and specificity (94.4%)
Park et al. (2023)	DL	569 pleural effusions	Detect metastatic breast cancer in pleural effusion	Accuracy (81.1% vs. 68.7%); sensitivity (95% vs. 72.5%); and specificity (98.6% vs. 88.9%) by manual review
Sanyal and Dey (2023)	DCNN	112	Benign vs. malignant effusions	Sensitivity (88%), specificity (96%), AUC (0.92)

turnaround time, and optimize cytopathology workflow efficiency. Large-scale prospective studies are needed to explore the real-world utilization of these technologies. Currently, none of the AI algorithms have demonstrated performance superior to that of a pathologist.

Validation and implementation of AI in practice

Implementing AI in cytopathology demands structured planning, stakeholder engagement, validation, and governance guided by regulatory bodies. To ensure both technical success and cultural adoption in the laboratory, a systematic approach is needed for sustainable AI integration. AI validation guidelines are still under development. The following approach, outlined in this section, is recommended. It should be noted that every laboratory has its unique challenges and that no single implementation method exists for all laboratories.

Evaluating clinical relevance and identifying use cases

Align the capabilities of the AI being considered with clinical needs. Clearly identify/define the AI's intended use (assistance, primary diagnosis, quality control, etc.) so that the proper stakeholders are connected. Establish the appropriate success/outcome criteria upfront, such as the reduction in screening time, efficient resource utilization, and cost savings. Defining the scope of AI implementation upfront improves the whole process of validation and deployment. More time spent planning from beginning to end will ensure less time spent trouble shooting and addressing unforeseen obstacles.

Assess baseline performance

Understanding the state of the laboratory prior to AI is important not only for determining areas in the laboratory that can benefit but also for assessing AI performance. Metrics such as turn-around time, resource utilization (such as person-hours), laboratory costs, and error rates are just some of the baseline statistics that should be assessed in order to determine if the intended AI actually benefits the laboratory. These baseline benchmarks help in the assessment of the impact of AI after deployment.

Local validation of AI algorithms

Follow the CAP/ASC guidelines for local validation of AI. In addition, be familiar with any relevant laboratory regulatory and accreditation requirements so that the validation adequately addresses them. Validation cohorts could be prospective and/or retrospective. A recent survey by the Digital Cytology Taskforce of the ASC suggests testing 100–200 cases for AI validation (Kim et al., 2024b). However, due to limited literature, these values are subject to change. Aim for > 95% concordance between AI and pathologists for adequate validation. Metrics such as accuracy, sensitivity, specificity, and AUC should be assessed to ensure adequate AI performance. If any discrepancies occur, a thorough review should be conducted to re-assess and decide on deployment. If there is any deviation from the AI algorithm at the time of validation, whether it be intended use or even a new version of the AI algorithm, a revalidation should be considered.

User experience and personnel training

Arrange thorough training sessions from the vendors to educate the faculty/trainee/technician on using AI. The training should focus on interface usability to improve the pathologist's confidence and interpretability in the use of AI. Such training would allow for proper AI usage so that users are not misguided in its application.

Technical integration

Departmental and hospital-level IT teams should be engaged so that the AI is seamlessly integrated into the laboratory information systems (LIS), including the digital pathology image management systems. Data input and output compatibility, along with any hardware and software components, should be assessed to ensure it fits with the laboratory's workflow. Hospital IT should be continually engaged to ensure that proper support exists if any issues arise. All data involved in the AI implementation should follow clear data security and patient privacy policies. As with any hospital and patient data, all information, including WSI, should be stored and accessed in a secure manner.

Cost and return on investment (ROI) analysis

AI implementation in pathology involves both upfront costs (e.g., digital infrastructure, software, data preparation) and ongoing expenses for system integration, maintenance, updates, and staff training. Hidden costs, like workflow disruption during adoption, should also be anticipated.

To demonstrate value, performance indicators should align with institutional goals, including reduced turnaround time, improved diagnostic accuracy, greater workflow efficiency, and lower resource use. ROI goes beyond direct financial savings to include improved quality, enhanced patient safety, and strategic benefits such as institutional competitiveness and readiness for future innovations. Effective ROI assessment should combine financial analysis with broader measures of clinical impact and patient-centered outcomes.

Audit and quality assurance/quality control

Continuous quality assurance/quality control and periodic audits are essential to ensure AI tools in cytology remain accurate and reliable. Regular meetings should assess end-user satisfaction, address any technical issues, and record performance metrics such as diagnostic accuracy, turnaround time, and costs. Laboratories should ideally establish error-tracking systems and feedback loops so that users can flag questionable AI outputs. AI must also be monitored for data drift, which can occur with changes in specimen preparation, staining, or patient populations, and may, in turn, impact accuracy.

Future directions in AI for cytopathology

Most commercial AI algorithms for clinical use are currently marketed for assisting with Pap test screening for cervical cancer (Elishaev et al., 2025; Landau and Pantanowitz, 2019; Sanghvi et al., 2019; Harinath et al., 2025). However, AI applications are expanding into non-GYN cytology, including urine cytology, effusions, brushings, FNA interpretation, and ancillary test evaluation (Gonzalez et al., 2020; Marya et al., 2024). AI tools are also being developed for ROSE, enabling immediate feedback during procedures like FNAs. This will help reduce repeat biopsies and accelerate clinical decision-making, while lowering the demand to have busy cytologists present on-site. Moreover, computational staining (i.e., AI analysis without performing ancillary tests such as immunostaining or molecular testing) can predict biomarker expression or patient outcomes (e.g., prognosis, treatment response) directly from images of routine stains (e.g., H&E, Pap, DQ stains).

Future AI systems aim to combine cytology images with clinical data (e.g., radiology images, patient outcomes), genomic, proteomic, transcriptomic, and other "omics" data. This multimodal approach enhances diagnostic precision and supports personalized medicine. Generative AI and synthetic data are emerging tools that have already proven to be helpful in creating virtual teaching sets (Hanna et al., 2025b; McAlpine et al., 2022). By identifying cases the learner is interested in and automating case curation, generative AI, such as large language models, can help advance cytology education. Beyond the development of AI for cytology tasks, as the field becomes more familiar with AI integration, proper clinical guidance regarding validation, QA/QC, and regulations will become clearer. Lastly, AI models continue to advance, and multiple foundation models—models trained specifically for pathology image-based tasks—have been introduced. These foundation models are primarily intended for surgical FFPE slides rather than cytology slides. However, as cytology catches up in digitization, larger datasets of cytology images will become available, enabling the development of a cytology foundation model to improve image-based cytology AI applications.

Conclusion

The core mantra of pathology and cytopathology is to understand disease for the advancement of patient care. In the effort to better understand disease, pathology has always sought ways to integrate new technologies and methods to better detect, predict, and treat disease. Cytology is a field born out of that endeavor, aiming to do the same on individual cells acquired by minimally invasive means. As with molecular diagnostics, immunohistochemistry, electron microscopy, and even light microscopy, the integration of AI into cytology practice is just the next advancement. As the field continues to seek novel methods to improve patient care, core principles in integrating such technologies must remain steadfast. The next chapter of cytology practice must be written with close adherence to fundamental laboratory practices as patient care is enhanced in an increasingly digital world.

References

Abulafia, O., Pezzullo, J.C., Sherer, D.M., 2003. Performance of ThinPrep liquid-based cervical cytology in comparison with conventionally prepared Papanicolaou smears: a quantitative survey. Gynecol. Oncol. 0090-8258. 90 (1), 137–144. Available from: http://www.elsevier.com/inca/publications/store/6/2/2/8/4/0/index.htt. doi:10.1016/S0090-8258(03)00176-8.

Ali, S.Z., VanderLaan, P.A., (Eds.), 2023. The Bethesda System for Reporting Thyroid Cytopathology. Springer International Publishing. doi:10.1007/978-3-031-28046-7.

Awan, R., Benes, K., Azam, A., Song, T., Shaban, M., Verrill, C., Tsang, Y.W., Snead, D., Minhas, F., Rajpoot, N., 2021. Deep learning based digital cell profiles for risk stratification of urine cytology images. Cytometry A. 1552-4922. 99 (7), 732–742. doi:10.1002/cyto.a.24313.

Bai, X., Wei, J., Starr, D., Zhang, X., Wu, X., Guo, Y., Liu, Y., Ma, X., Wei, Y., Li, C., Zilla, M.L., Zhang, W., Zeng, X., Zhao, C., 2024. Assessment of efficacy and accuracy of cervical cytology screening with artificial intelligence assistive system. Mod. Pathol. 0893-3952. 37 (6), 100486. doi:10.1016/j.modpat.2024.100486.

Banerjee, I., Bhattacharjee, K., Burns, J.L., Trivedi, H., Purkayastha, S., Seyyed-Kalantari, L., Patel, B.N., Shiradkar, R., Gichoya, J., 2023. "Shortcuts" causing bias in radiology artificial intelligence: causes, evaluation, and mitigation. J. Am. Coll. Radiol. 1546-1440. 20 (9), 842–851. doi:10.1016/j.jacr.2023.06.025.

Barwad, A., Dey, P., Susheilia, S., 2012. Artificial neural network in diagnosis of metastatic carcinoma in effusion cytology. Cytometry B Clin. Cytom. 1552-4957. 82 (2), 107–111. doi:10.1002/cyto.b.20632.

Biscotti, C.V., Dawson, A.E., Dziura, B., et al, 2005. Assisted primary screening using the automated ThinPrep imaging system. Am. J. Clin. Pathol. 123 (2), 281–287. Available from: http://www.ncbi.nlm.nih.gov/pubmed/15842055.

Cantley, R.L., Jing, X., Smola, B., Hao, W., Harrington, S., Pantanowitz, L., 2024. Validation of AI-assisted ThinPrep pap test screening using the geniusTM digital diagnostics system. J. Pathol. Inform. 2153-3539. 15, 100391. doi:10.1016/j.jpi.2024.100391.

Chantziantoniou, N., 2023. BestCyte primary screening of 500 ThinPrep pap test thin-layers: 3 cytologists' interobserver diagnostic concordance with predicate manual microscopy relative to truth reference diagnoses defining NILM, ASCUS+, LSIL+, and ASCH+ thresholds for specificity. J. Pathol. Inform. 14, 100182. doi:10.1016/j.jpi.2022.100182.

Chen, R.J., Ding, T., Lu, M.Y., Williamson, D.F.K., Jaume, G., Song, A.H., Chen, B., Zhang, A., Shao, D., Shaban, M., Williams, M., Oldenburg, L., Weishaupt, L.L., Wang, J.J., Vaidya, A., Le, L.P., Gerber, G., Sahai, S., Williams, W., Mahmood, F., 2024. Towards a general-purpose foundation model for computational pathology. Nat. Med. 1546-170X. 30 (3), 850–862. Available from: https://www.nature.com/nm/. doi:10.1038/s41591-024-02857-3.

Chute, D.J., Lim, H., Kong, C.S., 2010. BD FocalPoint slide profiler performance with atypical glandular cells on SurePath papanicolaou smears. Cancer Cytopathol. 1934-6638. 118 (2), 68–74. doi:10.1002/cncy.20067.

Cochand-Priollet, B., Koutroumbas, K., Megalopoulou, T.M., Pouliakis, A., Sivolapenko, G., Karakitsos, P., 2006. Discriminating benign from malignant thyroid lesions using artificial intelligence and statistical selection of morphometric features. Oncol. Rep. 1791-2431. 15 (4), 1023–1026. Available from: http://www.spandidos-publications.com/serveFile/or_15_4_1023_PDF.pdf?type=article&article_id=or_15_4_1023&item=PDF. doi:10.3892/or.15.4.1023.

Colgan, T.J., Bon, N., Clipsham, S., Gardiner, G., Sumner, J., Walley, V., McLachlin, C.M., 2013. A validation study of the focalpoint gs imaging system for gynecologic cytology screening. Cancer Cytopathol. 1934-6638. 121 (4), 189–196. doi:10.1002/cncy.21271.

Collins, B.T., Collins, L.E., 2013. Assessment of malignancy for atypia of undetermined significance in thyroid fine-needle aspiration biopsy evaluated by whole-slide image analysis. Am. J. Clin. Pathol. 1943-7722. 139 (6), 736–745. Available from: http://ajcp.ascpjournals.org/content/139/6/736.full.pdf+html. doi:10.1309/AJCPQU29GHXYSZRR.

Collins, B.T., Weimholt, R.C., 2015. Whole slide image with image analysis of atypical bile duct brushing: quantitative features predictive of malignancy. J. Pathol. Inform. 2153-3539. 6 (1), 47. doi:10.4103/2153-3539.163986.

Crowell, E.F., Bazin, C., Saunier, F., Brixtel, R., Caillot, Y., Lesner, B., Toutain, M., Ferreri, C., Garcia, I., Mathieu, M. C., Vaussanvin, J., Depardon, J., Renouf, A., 2019b. CytoprocessorTM: a new cervical cancer screening system for remote diagnosis. Acta Cytol. 19382650. 63 (3), 215–223. Available from: http://content.karger.com/ProdukteDB/produkte.asp?Aktion=JournalHome&ProduktNr=254338&ContentOnly=false. doi:10.1159/000497111.

Crowell, E.F., Bazin, C., Thurotte, V., Elie, H., Jitaru, L., Olivier, G., Caillot, Y., Brixtel, R., Lesner, B., Toutain, M., Renouf, A., 2019a. Adaptation of CytoProcessor for cervical cancer screening of challenging slides. Diagn. Cytopathol. 1097-0339. 47 (9), 890–897. Available from: http://onlinelibrary.wiley.com/journal/10.1002/(ISSN)1097-0339. doi:10.1002/dc.24213.

Dawson, A.E., 2004. Can we change the way we screen?: the ThinPrep imaging system. Cancer. 0008-543X. 102 (6), 340–344. doi:10.1002/cncr.20721.

Delga, A., Goffin, F., Kridelka, F., Marée, R., Lambert, C., Delvenne, P., 2014. Evaluation of CellSolutions BestPrep automated thin-layer liquid-based cytology papanicolaou slide preparation and BestCyte cell sorter imaging system. Acta Cytol. 0001-5547. 58 (5), 469–477. doi:10.1159/000367837.

Doornewaard, H., Van Der Schouw, Y.T., Van Der Graaf, Y., Bos, A.B., Habbema, J.D.F., Van Den Tweel, J.G., 1999. The diagnostic value of computer-assisted primary cervical smear screening: a longitudinal cohort study. Mod. Pathol. 0893-3952. 12 (11), 995–1000.

Dov, D., Elliott Range, D., Cohen, J., Bell, J., Rocke, D.J., Kahmke, R.R., Weiss-Meilik, A., Lee, W.T., Henao, R., Carin, L., Kovalsky, S.Z., 2023. Deep-learning–based screening and ancillary testing for thyroid cytopathology. Am. J. Pathol. 0002-9440. 193 (9), 1185–1194. doi:10.1016/j.ajpath.2023.05.011.

Drazen, J.M., Ferryman, K., Mackintosh, M., Ghassemi, M., 2023. Considering biased data as informative artifacts in AI-assisted health care. N. Engl. J. Med. 0028-4793. 389 (9), 833–838. doi:10.1056/nejmra2214964.

Dziura, B., Quinn, S., Richard, K., 2006. Performance of an imaging system vs. manual screening in the detection of squamous intraepithelial lesions of the uterine cervix. Acta Cytol. 1938-2650. 50 (3), 309–311. doi:10.1159/000325959.

Elishaev, E., Harinath, L., Ye, Y., Matsko, J., Colaizzi, A., Wharton, S., Bhargava, R., Pantanowitz, L., Hanna, M.G., Harrington, S., Zhao, C., 2025. Assessment of the efficacy and accuracy of cervical cytology screening with the hologic genius digital diagnostics system. Cancer Cytopathol. 1934-662X. 133 (7). doi:10.1002/cncy.70022.

Faquin, W.C., Rossi, E.D., Baloch, Z., Barkan, G.A., Foschini, M.P., Kurtycz, D.F.I., Pusztaszeri, M.P., Vielh, P., 2023. The Milan System for Reporting Salivary Gland Cytopathology. Springer International Publishing. doi:10.1007/978-3-031-26662-1.

Gonzalez, D., Dietz, R.L., Pantanowitz, L., 2020. Feasibility of a deep learning algorithm to distinguish large cell neuroendocrine from small cell lung carcinoma in cytology specimens. Cytopathology. 1365-2303. 31 (5), 426–431. Available from: http://onlinelibrary.wiley.com/journal/10.1111/(ISSN)1365-2303. doi:10.1111/cyt.12829.

Halford, J.A., Wright, R.G., Ditchmen, E.J., 1997. Quality assurance in cervical cytology screening. Comparison of rapid rescreening and the PAPNET Testing System. Acta. Cytol. 41 (1), 79–81. doi:10.1159/000332309. PMID: 9022730.

Hang, J.F., Ou, Y.C., Yang, W.L., Tsao, T.Y., Yeh, C.H., Li, C.B., Hsu, E.Y., Hung, P.Y., Lin, M.Y., Hwang, Y.T., Liu, T. J., Tung, M.C., 2024. Evaluating urine cytology slide digitization efficiency: a comparative study using an artificial intelligence-based heuristic scanning simulation and multiple Z-plane scanning. Acta Cytol. 1938-2650. 68 (4), 342–350. Available from: http://content.karger.com/ProdukteDB/produkte.asp?Aktion=JournalHome&ProduktNr=254338&ContentOnly=false. doi:10.1159/000538985.

Hanna, M.G., Monaco, S.E., Cuda, J., Xing, J., Ahmed, I., Pantanowitz, L., 2017. Comparison of glass slides and various digital-slide modalities for cytopathology screening and interpretation. Cancer Cytopathol. 1934-6638. 125 (9), 701–709. Available from: http://onlinelibrary.wiley.com/journal/10.1002/(ISSN)1934-6638. doi:10.1002/cncy.21880.

Hanna, M.G., Olson, N.H., Zarella, M., Dash, R.C., Herrmann, M.D., Furtado, L.V., Stram, M.N., Raciti, P.M., Hassell, L., Mays, A., Pantanowitz, L., Sirintrapun, J.S., Krishnamurthy, S., Parwani, A., Lujan, G., Evans, A., Glassy, E.F., Bui, M.M., Singh, R., Souers, R.J., de Baca, M.E., Seheult, J.N., 2024. Recommendations for performance evaluation of machine learning in pathology: a concept paper from the College of American Pathologists. Arch. Pathol. Lab. Med. 1543-2165. 148 (10), e335. doi:10.5858/arpa.2023-0042-cp.

Hanna, M.G., Pantanowitz, L., Dash, R., Harrison, J.H., Deebajah, M., Pantanowitz, J., Rashidi, H.H., 2025b. Future of artificial intelligence—machine learning trends in pathology and medicine. Mod. Pathol. 0893-3952. 38 (4), 100705. doi:10.1016/j.modpat.2025.100705.

Hanna, M.G., Pantanowitz, L., Jackson, B., Palmer, O., Visweswaran, S., Pantanowitz, J., Deebajah, M., Rashidi, H.H., 2025a. Ethical and bias considerations in artificial intelligence/machine learning. Mod. Pathol. 0893-3952. 38 (3), 100686. doi:10.1016/j.modpat.2024.100686.

Harinath, L., Bai, X., Minkowitz, J., Zeng, X., Harrington, S., Zhao, C., Pantanowitz, L., 2025. Artificial intelligence in gynecologic cytology. Acta Cytol. 0001-5547. 1–20. doi:10.1159/000546182.

Hassell, L.A., Forsythe, M.L., Bhalodia, A., Lan, T., Rashid, T., Powers, A., Bui, M.M., Brickman, A., Gu, Q., Bychkov, A., Cree, I., Pantanowitz, L., 2025. Toward optimizing the impact of digital pathology and augmented intelligence on issues of diagnosis, grading, staging, and classification. Mod. Pathol. 0893-3952. 38 (7), 100765. doi:10.1016/j.modpat.2025.100765.

Ikenberg, H., Lieder, S., Ahr, A., Wilhelm, M., Schön, C., Xhaja, A., 2023. Comparison of the hologic genius digital diagnostics system with the thinprep imaging system—a retrospective assessment. Cancer Cytopathol. 1934-662X. 131 (7), 424–432. doi:10.1002/cncy.22695.

Ippolito, A.M., De Laurentiis, M., La Rosa, G.L., Eleuteri, A., Tagliaferri, R., De Placido, S., Vigneri, R., Belfiore, A., 2004. Neural network analysis for evaluating cancer risk in thyroid nodules with an indeterminate diagnosis at aspiration cytology: identification of a low-risk subgroup. Thyroid. 1050-7256. 14 (12), 1065–1071. doi:10.1089/thy.2004.14.1065.

Jang, J., Kim, Y.H., Westgate, B., Zong, Y., Hallinan, C., Akalin, A., Lee, K., 2023. Screening adequacy of unstained thyroid fine needle aspiration samples using a deep learning-based classifier. Sci. Rep. 2045-2322. 13 (1). doi:10.1038/s41598-023-40652-1.

Kim, D., Burkhardt, R., Alperstein, S.A., Gokozan, H.N., Goyal, A., Heymann, J.J., Patel, A., Siddiqui, M.T., 2022. Evaluating the role of Z-stack to improve the morphologic evaluation of urine cytology whole slide images for high-grade urothelial carcinoma: results and review of a pilot study. Cancer Cytopathol. 1934-6638. 130 (8), 630–639. Available from: http://onlinelibrary.wiley.com/journal/10.1002/(ISSN)1934-6638. doi:10.1002/cncy.22595.

Kim, D., Sundling, K.E., Virk, R., Thrall, M.J., Alperstein, S., Bui, M.M., Chen-Yost, H., Donnelly, A.D., Lin, O., Liu, X., Madrigal, E., Michelow, P., Schmitt, F.C., Vielh, P.R., Zakowski, M.F., Parwani, A.V., Jenkins, E., Siddiqui, M.T., Pantanowitz, L., Li, Z., 2024a. Digital cytology part 2: artificial intelligence in cytology: a concept paper with review and recommendations from the American Society of Cytopathology Digital Cytology Task Force. J. Am. Soc. Cytopathol. 2213-2945. 13 (2), 97–110. Available from: https://www.sciencedirect.com/science/journal/22132945. doi:10.1016/j.jasc.2023.11.005.

Kim, D., Thrall, M.J., Michelow, P., Schmitt, F.C., Vielh, P.R., Siddiqui, M.T., Sundling, K.E., Virk, R., Alperstein, S., Bui, M.M., Chen-Yost, H., Donnelly, A.D., Lin, O., Liu, X., Madrigal, E., Zakowski, M.F., Parwani, A.V., Jenkins, E., Pantanowitz, L., Li, Z., 2024b. The current state of digital cytology and artificial intelligence (AI): global survey results from the American Society of Cytopathology Digital Cytology Task Force. J. Am. Soc. Cytopathol. 2213-2945. 13 (5), 319–328. Available from: https://www.sciencedirect.com/science/journal/22132945. doi:10.1016/j.jasc.2024.04.003.

Kitchener, H.C., Gittins, M., Desai, M., Smith, J.H.F., Cook, G., Roberts, C., Turnbull, L., 2015. A study of cellular counting to determine minimum thresholds for adequacy for liquid-based cervical cytology using a survey and counting protocol. Health Technol. Assess. 1366-5278. 19 (22), 1–64. doi:10.3310/hta19220.

Konfoong Biotech International Co. n.d. Available from: https://www.kfbiopathology.com/pathology-ai/cervicalcancer-diagnosis/. (Accessed 1 September 2025).

Landau, M.S., Pantanowitz, L., 2019. Artificial intelligence in cytopathology: a review of the literature and overview of commercial landscape. J. Am. Soc. Cytopathol. 2213-2945. 8 (4), 230–241. doi:10.1016/j.jasc.2019.03.003.

Layfield, L.J., Mooney, E.E., Glasgow, B., Hirschowitz, S., Coogan, A., 1997. What constitutes an adequate smear in fine-needle aspiration cytology of the breast? Cancer. 81 (1), 16–21. doi:10.1002/(sici)1097-0142(19970225)81:1ESCAPED_CHARACTER_LESS_THAN16::aid-cncr5>3.0.co;2-e.

Lebret, T., Paoletti, X., Pignot, G., Roumiguié, M., Colombel, M., Savareux, L., Verhoest, G., Guy, L., Rigaud, J., De Vergie, S., Poinas, G., Droupy, S., Kleinclauss, F., Courtade-Saïdi, M., Piaton, E., Radulescu, C., Rioux-Leclercq, N., Kandel-Aznar, C., Renaudin, K., Cochand-Priollet, B., Allory, Y., Nivet, S., Rouprêt, M., 2023. Artificial intelligence to improve cytology performance in urothelial carcinoma diagnosis: results from validation phase of the French, multicenter, prospective VISIOCYT1 trial. World J. Urol. 1433-8726. 41 (9), 2381–2388. doi:10.1007/s00345-023-04519-4.

Lebret, T., Pignot, G., Colombel, M., Guy, L., Rebillard, X., Savareux, L., Roumigue, M., Nivet, S., Saidi, M.C., Piaton, E., Radulescu, C., 2022. Artificial intelligence to improve cytology performances in bladder carcinoma detection: results of the VisioCyt test. BJU Int. 1464-4096. 129 (3), 356–363. doi:10.1111/bju.15382.

Lee, J.S.J., Kuan, L., Oh, S., Patten, F.W., Wilbur, D.C., 1998. A feasibility study of the autopap system location-guided screening. Acta Cytol. 1938-2650. 42 (1), 221–226. doi:10.1159/000331550.

Lee, Y., Alam, M.R., Park, H., Yim, K., Seo, K.J., Hwang, G., Kim, D., Chung, Y., Gong, G., Cho, N.H., Yoo, C.W., Chong, Y., Choi, H.J., 2024. Improved diagnostic accuracy of thyroid fine-needle aspiration cytology with artificial intelligence technology. Thyroid. 1557-9077. 34 (6), 723–734. Available from: www.liebertonline.com/thy. doi:10.1089/thy.2023.0384.

Levi, A.W., Chhieng, D.C., Schofield, K., Kowalski, D., Harigopal, M., 2012. Implementation of focalpoint GS location-guided imaging system. Cancer Cytopathol. 1934-6638. 120 (2), 126–133. doi:10.1002/cncy.20187.

Lin, Y.J., Chao, T.K., Khalil, M.A., Lee, Y.C., Hong, D.Z., Wu, J.J., Wang, C.W., 2021. Article deep learning fast screening approach on cytological whole slides for thyroid cancer diagnosis. Cancers. 2072-6694. 13 (15). Available from: https://www.mdpi.com/2072-6694/13/15/3891/pdf. doi:10.3390/cancers13153891.

Liu, T.J., Yang, W.C., Huang, S.M., Yang, W.L., Wu, H.J., Ho, H.W., Hsu, S.W., Yeh, C.H., Lin, M.Y., Hwang, Y.T., Chu, P.Y., 2024. Evaluating artificial intelligence–enhanced digital urine cytology for bladder cancer diagnosis.

Cancer Cytopathol. 1934-6638. 132 (11), 686–695. Available from: http://onlinelibrary.wiley.com/journal/10.1002/(ISSN)1934-6638. doi:10.1002/cncy.22884.

Lozano, R., 2007. Comparison of computer-assisted and manual screening of cervical cytology. Gynecol. Oncol. 0090-8258. 104 (1), 134–138. doi:10.1016/j.ygyno.2006.07.025.

Maleki, S., Zandvakili, A., Gera, S., Khutti, S.D., Gersten, A., Khader, S.N., 2019. Differentiating noninvasive follicular thyroid neoplasm with papillary-like nuclear features from classic papillary thyroid carcinoma: analysis of cytomorphologic descriptions using a novel machine-learning approach. J. Pathol. Inform. 2153-3539. 10 (1). Available from: www.jpathinformatics.org/browse.asp?sabs=n. doi:10.4103/jpi.jpi_25_19.

Marletta, S., Caputo, A., Guidi, G., Pantanowitz, L., Pagni, F., Bavieri, I., L'Imperio, V., Brunelli, M., Dei Tos, A.P., Eccher, A., 2025. Digital pathology displays under pressure: benchmarking performance across market grades. J. Imaging Inform. Med. 2948-2933. 38 (6), 4090–4099. doi:10.1007/s10278-025-01452-3.

Marya, N.B., Powers, P.D., Bois, M.C., Hartley, C., Kerr, S.E., Thangaiah, J.J., Norton, D., Abu Dayyeh, B.K., Cantley, R., Chandrasekhara, V., Gores, G., Gleeson, F.C., Law, R.J., Maleki, Z., Martin, J.A., Pantanowitz, L., Petersen, B., Storm, A.C., Levy, M.J., Graham, R.P., 2024. Utilization of an artificial intelligence–enhanced, web-based application to review bile duct brushing cytologic specimens: a pilot study. Cancer Cytopathol. 1934-6638. 132 (12), 779–787. Available from: http://onlinelibrary.wiley.com/journal/10.1002/(ISSN)1934-6638. doi:10.1002/cncy.22898.

Mavropoulos, A., Johnson, C., Lu, V., Nieto, J., Schneider, E.C., Saini, K., Phelan, M.L., Hsie, L.X., Wang, M.J., Cruz, J., Mei, J., Kim, J.J., Lian, Z., Li, N., Boutet, S.C., Wong-Thai, A.Y., Yu, W., Lu, Q.-Y., Kim, T., Geng, Y., Masaeli, M. (Mahdokht), Lee, T.D., Rao, J., 2023. Artificial intelligence-driven morphology-based enrichment of malignant cells from body fluid. Mod. Pathol. 0893-3952. 36 (8), 100195. doi:10.1016/j.modpat.2023.100195.

McAlpine, E.D., Pantanowitz, L., Michelow, P.M., 2021. Challenges developing deep learning algorithms in cytology. Acta Cytol. 1938-2650. 65 (4), 301–309. Available from: http://content.karger.com/ProdukteDB/produkte.asp?Aktion=JournalHome&ProduktNr=254338&ContentOnly=false. doi:10.1159/000510991.

McAlpine, E., Michelow, P., Liebenberg, E., Celik, T., 2022. Is it real or not? Toward artificial intelligence-based realistic synthetic cytology image generation to augment teaching and quality assurance in pathology. J. Am. Soc. Cytopathol. 2213-2945. 11 (3), 123–132. doi:10.1016/j.jasc.2022.02.001.

Michael, C.W., Pang, Y., Pu, R.T., Hasteh, F., Griffith, K.A., 2007. Cellular adequacy for thyroid aspirates prepared by ThinPrep: how many cells are needed? Diagn. Cytopathol. 1097-0339. 35 (12), 792–797. doi:10.1002/dc.20768.

Miller, F.S., Nagel, L.E., Kenny-Moynihan, M.B., 2007. Implementation of the ThinPrep imaging system in a high-volume metropolitan laboratory. Diagn. Cytopathol. 8755-1039. 35 (4), 213–217. doi:10.1002/dc.20627.

Momeni-Boroujeni, A., Yousefi, E., Somma, J., 2017. Computer-assisted cytologic diagnosis in pancreatic FNA: an application of neural networks to image analysis. Cancer Cytopathol. 1934-662X. 125 (12), 926–933. doi:10.1002/cncy.21915.

Monabbati, S., Leo, P., Bera, K., Michael, C.W., Nezami, B.G., Harbhajanka, A., Madabhushi, A., 2023. Automated analysis of computerized morphological features of cell clusters associated with malignancy on bile duct brushing whole slide images. Cancer Med. 2045-7634. 12 (5), 6365–6378. doi:10.1002/cam4.5365.

Moriarty, A.T., Nayar, R., Auger, M., Kurtycz, D.F.I., Thomas, N., Laucirica, R., Padmanabhan, V., Souers, R.J., Schwartz, M.R., Fraig, M.M., 2014. Nongynecologic cytology practice patterns: a survey of participants in the College of American Pathologists interlaboratory comparison program in nongynecologic cytopathology. Arch. Pathol. Lab. Med. 0003-9985. 138 (7), 885–889. doi:10.5858/arpa.2012-0680-CP.

Muralidaran, C., Dey, P., Nijhawan, R., Kakkar, N., 2015. Artificial neural network in diagnosis of urothelial cell carcinoma in urine cytology. Diagn. Cytopathol. 8755-1039. 43 (6), 443–449. doi:10.1002/dc.23244.

Nayar, R., Wilbur, D.C., (Eds.), 2015. The Bethesda System for Reporting Cervical Cytology. Springer International Publishing. doi:10.1007/978-3-319-11074-5.

Nojima, S., Terayama, K., Shimoura, S., Hijiki, S., Nonomura, N., Morii, E., Okuno, Y., Fujita, K., 2021. A deep learning system to diagnose the malignant potential of urothelial carcinoma cells in cytology specimens. Cancer Cytopathol. 1934-662X. 129 (12), 984–995. doi:10.1002/cncy.22443.

OptraSCAN. n.d. Available from: https://optrascan.com/blogs/cytosia-toadvance-cytology-screening-throughunmatched-image-quality-and-artificial-intelligence-based-image-analysis-solution/. (Accessed 1 September 2025).

Ou, Y.C., Tsao, T.Y., Chang, M.C., Lin, Y.S., Yang, W.L., Hang, J.F., Li, C.B., Lee, C.M., Yeh, C.H., Liu, T.J., 2022. Evaluation of an artificial intelligence algorithm for assisting the Paris System in reporting urinary cytology:

a pilot study. Cancer Cytopathol. 1934-6638. 130 (11), 872–880. Available from: http://onlinelibrary.wiley.com/journal/10.1002/(ISSN)1934-6638. doi:10.1002/cncy.22615.

Pantanowitz, L., Pearce, T., Abukhiran, I., Hanna, M., Wheeler, S., Soong, T.R., Tafti, A.P., Pantanowitz, J., Lu, M.Y., Mahmood, F., Gu, Q., Rashidi, H.H., 2025. Nongenerative artificial intelligence in medicine: advancements and applications in supervised and unsupervised machine learning. Mod. Pathol. 0893-3952. 38 (3), 100680. doi:10.1016/j.modpat.2024.100680.

Pantazopoulos, D., Karakitsos, P., Iokim-Liossi, A., Pouliakis, A., Botsoli-Stergiou, E., Dimopoulos, C., 1998. Back propagation neural network in the discrimination of benign from malignant lower urinary tract lesions. J. Urol. 0022-5347. 159 (5), 1619–1623. doi:10.1097/00005392-199805000-00057.

Park, H.S., Chong, Y., Lee, Y., Yim, K., Seo, K.J., Hwang, G., Kim, D., Gong, G., Cho, N.H., Yoo, C.W., Choi, H.J., 2023. Deep learning-based computational cytopathologic diagnosis of metastatic breast carcinoma in pleural fluid. Cells. 2073-4409. 12 (14). Available from: https://www.mdpi.com/journal/cells. doi:10.3390/cells12141847.

Renshaw, A.A., Underwood, D., Aramoni, G., Cash, B., Croyle, M., Deeds, D., Dolar, S., Gmitro, S., Ray, N., Sabo, D., Shorie, J.A., Springer, B., Moffsinger, D.W., Elsheikh, T.M., 2016. Time consumed by microscopic and non-microscopic tasks in image-assisted gynecologic screening: implications for workload assessment. Cancer Cytopathol. 1934-6638. 124 (7), 501–507. Available from: http://onlinelibrary.wiley.com/journal/10.1002/(ISSN)1934-6638. doi:10.1002/cncy.21711.

Rivera Rolon, M.D.M., Gustafson, E., Cole, R., Matos, J., Hicken, K., Hicks, J., Cahoon, B., de Socarraz, M., Santa-Rosario, J.C., 2025. Implementing 100% quality control in a cervical cytology workflow using whole slide images and artificial intelligence provided by the Techcyte SureView system. Cancer Cytopathol. 1934-6638. 133 (6). Available from: http://onlinelibrary.wiley.com/journal/10.1002/(ISSN)1934-6638. doi:10.1002/cncy.70019.

Salahuddin, Z., Woodruff, H.C., Chatterjee, A., Lambin, P., 2022. Transparency of deep neural networks for medical image analysis: a review of interpretability methods. Comput. Biol. Med. 0010-4825. 140, 105111. doi:10.1016/j.compbiomed.2021.105111.

Sanghvi, A.B., Allen, E.Z., Callenberg, K.M., Pantanowitz, L., 2019. Performance of an artificial intelligence algorithm for reporting urine cytopathology. Cancer Cytopathol. 1934-6638. 127 (10), 658–666. Available from: http://onlinelibrary.wiley.com/journal/10.1002/(ISSN)1934-6638. doi:10.1002/cncy.22176.

Sanyal, P., Dey, P., 2023. Using a deep learning neural network for the identification of malignant cells in effusion cytology material. Cytopathology. 0956-5507. 34 (5), 466–471. doi:10.1111/cyt.13260.

Sanyal, P., Mukherjee, T., Barui, S., Das, A., Gangopadhyay, P., 2018. Artificial Intelligence in cytopathology: a neural network to identify papillary carcinoma on thyroid fine-needle aspiration cytology smears. J. Pathol. Inform. 2153-3539. 9 (1), 43. doi:10.4103/jpi.jpi_43_18.

Satturwar, S.P., Pantanowitz, J.L., Manko, C.D., Seigh, L., Monaco, S.E., Pantanowitz, L., 2020. Ki-67 proliferation index in neuroendocrine tumors: can augmented reality microscopy with image analysis improve scoring? Cancer Cytopathol. 1934-6638. 128 (8), 535–544. Available from: http://onlinelibrary.wiley.com/journal/10.1002/(ISSN)1934-6638. doi:10.1002/cncy.22272.

Savala, R., Dey, P., Gupta, N., 2018. Artificial neural network model to distinguish follicular adenoma from follicular carcinoma on fine needle aspiration of thyroid. Diagn. Cytopathol. 8755-1039. 46 (3), 244–249. doi:10.1002/dc.23880.

Schmidt, R.L., Walker, B.S., Howard, K., Layfield, L.J., Adler, D.G., 2013a. Rapid on-site evaluation reduces needle passes in endoscopic ultrasound-guided fine-needle aspiration for solid pancreatic lesions: a risk-benefit analysis. Dig. Dis. Sci. 1573-2568. 58 (11), 3280–3286. doi:10.1007/s10620-013-2750-6.

Schmidt, R.L., Witt, B.L., Lopez-Calderon, L.E., Layfield, L.J., 2013b. The influence of rapid onsite evaluation on the adequacy rate of fine-needle aspiration cytology: a systematic review and meta-analysis. Am. J. Clin. Pathol. 1943-7722. 139 (3), 300–308. Available from: http://ajcp.ascpjournals.org/content/139/3/300.full.pdf+html. doi:10.1309/AJCPEGZMJKC42VUP.

Shaker, N., Shen, R., Limbach, A.L., Satturwar, S., Kobalka, P., Ahmadian, S., Sun, S., Chen, W., Lujan, G., Esnakula, A., Parwani, A., Li, Z., 2024. Automated imaging analysis of Ki-67 immunohistochemistry on whole slide images of cell blocks from pancreatic neuroendocrine neoplasms. J. Am. Soc. Cytopathol. 2213-2945. 13 (3), 205–212. doi:10.1016/j.jasc.2024.02.001.

Słowińska-Klencka, D., Klencki, M., Duda-Szymańska, J., Szwalski, J., Popowicz, B., 2021. Low reproducibility of equivocal categories of the Bethesda system for reporting thyroid cytology makes the associated risk of malignancy specific to the diagnostic center. Endocrine. 1355-008X. 74 (2), 355–364. doi:10.1007/s12020-021-02781-3.

Stoler, M.H., Schiffman, M., 2001. Atypical squamous cells of undetermined significance-low-grade squamous intraepithelial lesion triage study (ALTS) group. Interobserver reproducibility of cervical cytologic and histologic interpretations: realistic estimates from the ASCUS-LSIL triage study. JAMA. 285 (11), 1500–1505. doi:10.1001/jama.285.11.1500.

Tanaka, K., Lami, K., Odate, T., Hori, T., Sato, T., Okoshi, E.N., Fukuoka, J., 2025. Cytology screening using Z-stack digital slides: a validation study. Diagn. Cytopathol. 8755-1039. 53 (11), 568–575. doi:10.1002/dc.70015.

Tanaka, N., Ikeda, H., Ueno, T., Mukawa, A., Watanabe, S., Okamoto, K., Hosoi, S., Tsunekawa, S., 1987. Automated cytologic screening system (CYBEST model 4): an integrated image cytometry system. Appl. Opt. 2155-3165. 26 (16), 3301–3307. doi:10.1364/AO.26.003301.

Teramoto, A., Tsukamoto, T., Kiriyama, Y., Fujita, H., 2017. Automated classification of lung cancer types from cytological images using deep convolutional neural networks. Biomed Res. Int. 2314-6133. 2017, 1–6. doi:10.1155/2017/4067832.

Teramoto, A., Tsukamoto, T., Yamada, A., Kiriyama, Y., Imaizumi, K., Saito, K., Fujita, H., Qiu, Y., 2020. Deep learning approach to classification of lung cytological images: two-step training using actual and synthesized images by progressive growing of generative adversarial networks. PLOS ONE. 1932-6203. 15 (3), e0229951. doi:10.1371/journal.pone.0229951.

Torous, V.F., VandenBussche, C.J., Randolph, M.L., Cuda, J.M., Manucha, V., 2023. Quality metrics in non-gynecologic cytology: results from the 2022 American Society of Cytopathology survey. J. Am. Soc. Cytopathol. 22132945. 12 (4), 258–266. Available from: http://www.elsevier.com/journals/journal-of-the-american-society-of-cytopathology/2213-2945. doi:10.1016/j.jasc.2023.03.003.

Tosun, A.B., Yergiyev, O., Kolouri, S., Silverman, J.F., Rohde, G.K., Rohde, G.K., 2015. Detection of malignant mesothelioma using nuclear structure of mesothelial cells in effusion cytology specimens. Cytometry A. 1552-4922. 87 (4), 326–333. doi:10.1002/cyto.a.22602.

Tsuji, K., Kaneko, M., Harada, Y., Fujihara, A., Ueno, K., Nakanishi, M., Konishi, E., Takamatsu, T., Horiguchi, G., Teramukai, S., Ito-Ihara, T., Ukimura, O., 2024. A fully automated artificial intelligence system to assist pathologists' diagnosis to predict histologically high-grade urothelial carcinoma from digitized urine cytology slides using deep learning. Eur. Urol. Oncol. 2588-9311. 7 (2), 258–265. doi:10.1016/j.euo.2023.11.009.

Tsukamoto, T., Teramoto, A., Yamada, A., Kiriyama, Y., Sakurai, E., Michiba, A., Imaizumi, K., Fujita, H., 2022. Comparison of fine-tuned deep convolutional neural networks for the automated classification of lung cancer cytology images with integration of additional classifiers. Asian Pac. J. Cancer Prev. 2476-762X. 23 (4), 1315–1324. doi:10.31557/apjcp.2022.23.4.1315.

Vaickus, L.J., Suriawinata, A.A., Wei, J.W., Liu, X., 2019. Automating the Paris system for urine cytopathology—a hybrid deep-learning and morphometric approach. Cancer Cytopathol. 1934-662X. 127 (2), 98–115. doi:10.1002/cncy.22099.

Varlatzidou, A., Pouliakis, A., Stamataki, M., Meristoudis, C., Margari, N., Peros, G., Panayiotides, J.G., Karakitsos, P., 2011. Cascaded learning vector quantizer neural networks for the discrimination of thyroid lesions. Anal. Quant. Cytol. Histol. 0884-6812. 33 (6), 323–334. Available from: http://www.aqch.com/toc/auto_article_process.php?year=2011&page=323&id=23159&sn=0.

Viswanathan, K., Patel, A., Abdelsayed, M., Rosado, L., Soong, L., Margolskee, E., Heymann, J.J., Goyal, A., Rao, R. A., 2020. Interobserver variability between cytopathologists and cytotechnologists upon application and characterization of the indeterminate category in the milan system for reporting salivary gland cytopathology. Cancer Cytopathol. 1934-662X. 128 (11), 828–839. doi:10.1002/cncy.22312.

Vorontsov, E., Bozkurt, A., Casson, A., Shaikovski, G., Zelechowski, M., Severson, K., Zimmermann, E., Hall, J., Tenenholtz, N., Fusi, N., Yang, E., Mathieu, P., van Eck, A., Lee, D., Viret, J., Robert, E., Wang, Y.K., Kunz, J.D., Lee, M.C.H., Bernhard, J.H., Godrich, R.A., Oakley, G., Millar, E., Hanna, M., Wen, H., Retamero, J.A., Moye, W. A., Yousfi, R., Kanan, C., Klimstra, D.S., Rothrock, B., Liu, S., Fuchs, T.J., 2024. A foundation model for clinical-grade computational pathology and rare cancers detection. Nat. Med. 1078-8956. 30 (10), 2924–2935. doi:10.1038/s41591-024-03141-0.

Wadden, J.J., 2021. Defining the undefinable: the black box problem in healthcare artificial intelligence. J. Med. Ethics 1473-4257. 48 (10), 764–768. Available from: http://jme.bmj.com/. doi:10.1136/medethics-2021-107529.

Wang, J., Yu, Y., Tan, Y., Wan, H., Zheng, N., He, Z., Mao, L., Ren, W., Chen, K., Lin, Z., He, G., Chen, Y., Chen, R., Xu, H., Liu, K., Yao, Q., Fu, S., Song, Y., Chen, Q., Zuo, L., Wei, L., Wang, J., Ouyang, N., Yao, H., 2024a. Artificial intelligence enables precision diagnosis of cervical cytology grades and cervical cancer. Nat. Commun. 2041-1723. 15 (1). doi:10.1038/s41467-024-48705-3.

Wang, X., Zhao, J., Marostica, E., Yuan, W., Jin, J., Zhang, J., Li, R., Tang, H., Wang, K., Li, Y., Wang, F., Peng, Y., Zhu, J., Zhang, J., Jackson, C.R., Zhang, J., Dillon, D., Lin, N.U., Sholl, L., Denize, T., Meredith, D., Ligon, K.L., Signoretti, S., Ogino, S., Golden, J.A., Nasrallah, M.P., Han, X., Yang, S., Kun-Hsing, Y., 2024b. A pathology foundation model for cancer diagnosis and prognosis prediction. Nature. 0028-0836. 634 (8035), 970–978. doi:10.1038/s41586-024-07894-z.

Wilbur, D.C., Black-Schaffer, W.S., Luff, R.D., Abraham, K.P., Kemper, C., Molina, J.T., Tench, W.D., 2009. The Becton Dickinson focalpoint GS imaging system: clinical trials demonstrate significantly improved sensitivity for the detection of important cervical lesions. Am. J. Clin. Pathol. 0002-9173. 132 (5), 767–775. Available from: http://ajcp.ascpjournals.org/content/132/5/767.full.pdf. doi:10.1309/AJCP8VE7AWBZCVQT.

Wilbur, D.C., Prey, M.U., Miller, W.M., Pawlick, G.F., Colgan, T.J., 1998. The AutoPap system for primary screening in cervical cytology: comparing the results of a prospective, intended-use study with routine manual practice. Acta Cytol. 1938-2650. 42 (1), 214–220. Available from: http://content.karger.com/ProdukteDB/produkte.asp?Aktion=JournalHome&ProduktNr=254338&ContentOnly=false. doi:10.1159/000331549.

Wojcik, E.M., Kurtycz, D.F.I., Rosenthal, D.L., 2022. The Paris System for Reporting Urinary Cytology. Springer International Publishing. doi:10.1007/978-3-030-88686-8.

Wu, S., Shen, R., Hong, G., Luo, Y., Wan, H., Feng, J., Chen, Z., Jiang, F., Wang, Y., Liao, C., Li, X., Liu, B., Huang, X., Liu, K., Qin, P., Wang, Y., Xie, Y., Ouyang, N., Huang, J., Lin, T., 2024. Development and validation of an artificial intelligence-based model for detecting urothelial carcinoma using urine cytology images: a multicentre, diagnostic study with prospective validation. eClinicalMedicine. 2589-5370. 71, 102566. doi:10.1016/j.eclinm.2024.102566.

Xie, X., Fu, C.C., Lv, L., Ye, Q., Yu, Y., Fang, Q., Zhang, L., Hou, L., Wu, C., 2022. Deep convolutional neural network-based classification of cancer cells on cytological pleural effusion images. Mod. Pathol. 1530-0285. 35 (5), 609–614. Available from: http://www.nature.com/modpathol/index.html. doi:10.1038/s41379-021-00987-4.

Xue, P., Hai-Miao, X., Tang, H.-P., Weng, H.-Y., Wei, H.-M., Wang, Z., Zhang, H.-Y., Weng, Y., Xu, L., Li, H.-X., Seery, S., Han, X., Ye, H., Qiao, Y.-L., Jiang, Y., 2023. Improving the accuracy and efficiency of abnormal cervical squamous cell detection with cytologist-in-the-loop artificial intelligence. Mod. Pathol. 0893-3952. 36 (8), 100186. doi:10.1016/j.modpat.2023.100186.

Yang, W., Jin, X., Huang, L., Jiang, S., Xu, J., Fu, Y., Song, Y., Wang, X., Wang, X., Yang, Z., Meng, Y., 2024. Clinical evaluation of an artificial intelligence-assisted cytological system among screening strategies for a cervical cancer high-risk population. BMC Cancer. 1471-2407. 24 (1). doi:10.1186/s12885-024-12532-y.

Zahniser, D.J., Oud, P.S., Raaijmakers, M.C.T., Vooys, G.P., Van de Walle, R.T., 1979. BioPEPR: a system for the automatic prescreening of cervical smears. J. Histochem. Cytochem. 0022-1554. 27 (1), 635–641. doi:10.1177/27.1.86581.

Zeng, X., Starr, D., Li, J., Bi, X., Wang, C., Bai, X., Yin, Y., Wu, X., Wei, J., Du, H., Dai, W., Li, C., Wu, X., Wu, R., Zhao, C., 2024. AICyte-alone capabilities as an independent screener for triaging cervical cytology using a 50% negative cutoff value. Cancer Cytopathol. 1934-662X. 132 (11), 723–730. doi:10.1002/cncy.22896.

Zhu, X., Li, X., Ong, K., Zhang, W., Li, W., Li, L., Young, D., Su, Y., Shang, B., Peng, L., Xiong, W., Liu, Y., Liao, W., Xu, J., Wang, F., Liao, Q., Li, S., Liao, M., Li, Y., Rao, L., Lin, J., Shi, J., You, Z., Zhong, W., Liang, X., Han, H., Zhang, Y., Tang, N., Hu, A., Gao, H., Cheng, Z., Liang, L., Yu, W., Ding, Y., 2021. Hybrid AI-assistive diagnostic model permits rapid TBS classification of cervical liquid-based thin-layer cell smears. Nat. Commun. 2041-1723. 12 (1). doi:10.1038/s41467-021-23913-3.

CHAPTER

11

Emerging advanced imaging technologies in digital pathology

Jeffrey L. Fine

Pathology, University of Pittsburgh, Pittsburgh, PA, United States

Introduction

Advanced imaging in pathology can be defined as the integration of multiple imaging modalities into pathology practice. In vivo and ex vivo microscopy (IVM and EVM) refer to novel imaging techniques that can produce diagnostic pathology images without traditional hematoxylin and eosin (H&E) microscopy (Wells et al., 2019; Krishnamurthy et al., 2019). This topic should also include the application of traditional pathology images, because the ability to use traditional images effectively may be a prerequisite for meaningful IVM/EVM adoption in pathology (Fine, 2023; Tosun et al., 2020). This could blur surgical pathology boundaries with other "turf" areas such as radiology, hyperspectral image analysis, and even point-of-care testing. Although IVM/EVM has great potential to disrupt pathology diagnosis, such changes have not yet occurred despite rapid developments in digital and computational pathology. However, advanced products are beginning to emerge that leverage IVM/EVM, such as a standalone breast resection margin tool that does not require pathologist involvement (Ha et al., 2018), or a portable probe-based device for rapid cervical cancer screening (Zhang et al., 2024). These developments may reflect realities where pathology practice has not been nimble enough to keep up with clinical needs, or where resources are so scarce that pathology assessment is unavailable or too slow to be effective. In addition, pathology has not made adequate use of existing images such as gross images or specimen radiographs, suggesting that pathologists are not yet ready to incorporate IVM/EVM into practice.

This chapter will briefly discuss major IVM/EVM techniques with the goal of providing starter references and superficial knowledge. Then we will explore applications for IVM and EVM, including those in other medical specialties. Finally, it is necessary to present how or where such images could fit into practice by highlighting new applications driven by better utilization of existing imaging modalities.

Advances in Digital Pathology and Artificial Intelligence in Pathology
https://doi.org/10.1016/B978-0-443-40383-5.00014-5

Example techniques

IVM and EVM techniques abound, and listing them comprehensively would be overwhelming. One trend worth noting is that for any given technique, engineering efforts frequently solve problems related to resolution, contrast, speed, and cost. Optical coherence tomography (OCT) is a great example, as it has improved markedly since its invention in 1991 in both acquisition speed and resolution (Leitgeb et al., 2021). OCT has been widely adopted in ophthalmology practice, and numerous variants of the technology exist. Full-field OCT (FF-OCT) can generate very high-quality histopathology images that approximate low-to-medium magnification whole slide images (WSIs) (Wells et al., 2019; Krishnamurthy et al., 2019). OCT has also been applied to upper gastrointestinal (GI) endomicroscopy (Dong et al., 2022) and is under development for cervical cancer screening (Zhang et al., 2024), coronary artery assessment (Yonetsu and Jang, 2024), and pulmonary bronchoscopy (Hariri et al., 2013).

OCT works from reflectance; different tissue structures have varying reflectivity, allowing them to be visualized. This is akin to ultrasound, only with light instead of sound, and with much higher resolution. Timing of the light "echoes" is accomplished using interferometry, a long-standing microscopy technique. OCT can penetrate tissue up to about 2–3 mm, enabling easy generation of 3D images (Fine, 2015; Wang et al., 2022). Although physicians often reduce these to 2D slices, there is likely additional value in using the third dimension, given adequate software tools. Many OCT images lack contrast relative to H&E-stained WSIs, but the patterns are present and represent a ripe target for machine learning for evaluation or contrast enhancement. For example, a breast margin assessment tool uses machine learning to help non-pathologists directly evaluate images (Ha et al., 2018). There are also methods for creating virtual H&E stains to facilitate image interpretation by pathologists (Abraham et al., 2023).

Despite its promise, OCT has not been adopted in pathology, even in breast pathology where frozen section limitations make intraoperative diagnosis difficult. At least one vendor uses OCT to generate histopathology images but does not target pathologists; rather, this effort utilizes machine learning to guide surgeon interpretation (Levy et al., 2023). This addresses a gap in breast pathology and does so in a limited fashion that is more akin to a laboratory test than a typical surgical pathology interpretation. Although it is unlikely that such a test could be politically controlled by the laboratory, it could represent a new "point-of-care" test. Despite not requiring pathologists, such a system could be modified to provide pathologists with basic histopathology analysis of gross specimens or to provide surgeons with expert pathology support in difficult cases.

Reflectance confocal microscopy (RCM) is another clinically important imaging modality, with adoption in dermatology (Ko et al., 2021). This is likely possible in part due to the wide availability of dermatopathology expertise within dermatology practices (Farabi et al., 2023). Like OCT, RCM images are derived from light reflectance. Unlike OCT, RCM uses a laser to focus on specific points within the tissue. Optics reject out-of-focus reflections, and the laser can illuminate a plane or slice of tissue, with some penetration into the skin. The focal depth can be adjusted to capture a 3D stack of 2D images that can be viewed by the dermatologist. For dermatologists, RCM is mostly helpful for rapid, non-invasive triage of skin lesions in real time, thereby preventing the need for tissue biopsy in many cases. Such a technique might also be helpful for rapid tissue imaging for specific purposes (e.g., Mohs surgery, pathology frozen section, or specimen grossing) (Levine and Markowitz, 2018).

Light sheet microscopy is a fluorescent method that uses a plane of UV light to illuminate a tissue specimen (Delgado-Rodriguez et al., 2022). Fluorescent dyes are usually employed, but it is also possible to stain tissue with labeled antibodies. The resulting images can be used to analyze histopathology or assess biomarkers. This is currently a research modality, but some work has proceeded with the goal of rapid biopsy assessment. An issue with LSM is that it requires some tissue processing to make the tissue transparent enough for good illumination, as illumination is done from the side. Another issue is that there are currently many different methods for biomarker analysis, including reliable hyperplexed methodologies (Uttam et al., 2020).

There are numerous other examples of imaging used in pathology research, including MUSE (microscopy with UV surface excitation), multiphoton microscopy (MPM), quantitative oblique back-illumination microscopy (qOBM), and others (Krishnamurthy et al., 2019; Wells et al., 2019; Abraham et al., 2023). There are also nontraditional methods for generating WSIs, often termed "computational photonics," such as ptychography (Eadie et al., 2025). This last method is fascinating because it could create WSIs of H&E-stained tissue using inexpensive and conventional optics, much more rapidly and without the usual technical constraints typical "brute force" WSIs face. There are many different ways to image tissue, each with unique strengths; in the setting of rapid engineering improvements, this is a landscape that will change quickly over time. The key step for pathologists is to figure out how or why these might be adopted.

Although not an IVM/EVM technique per se, image coregistration is an essential technique that is widespread in pathology diagnosis, albeit usually performed manually. This refers to the alignment and correlation of different images. With glass microscope slides or WSIs, this could involve aligning recuts or different stains from a tissue block, or arranging adjacent tissue sections to recreate larger tissue fragments (e.g., reconstructing tissue slices in breast or prostate resections). Taking this further, it could also involve aligning gross specimen images with WSIs, such as when tissue sections are documented on gross images (e.g., documenting sampling in soft tissue tumor resections). Coregistration is usually done manually, but nascent digital pathology tools are emerging to assist with this activity. This is also a potential place for IVM/EVM images to fit into existing pathology workflows (Zhang et al., 2023; Fine, 2023). Coregistration can also enable virtual multiplexing of biomarkers with pathological images (Furman et al., 2021).

Applications for IVM and EVM

For ophthalmologists and dermatologists, IVM is an immediate and non-invasive tool. Although more limited in usage, upper GI endomicroscopy IVM may represent an opportunity for surgical pathologists to provide diagnostic support in difficult cases or for clinicians who do not want the medicolegal responsibility of interpreting these images. This represents a large and potentially important area for expanding surgical pathology-type imaging, even if initial applications were developed for non-pathologists.

When tissue is excised, intraoperative frozen section (FS) pathology is already widely available and considered adequate in terms of accuracy, speed, and cost. Although FS pathology has sharp limitations, these are generally acceptable and do not prevent successful diagnosis, tissue banking, or even WSI scanning. However, FS pathology is unevenly available even within the United States for various reasons. EVM potentially addresses such gaps, especially if extensive gross specimen handling is not required. For example, an OCT-based margin tool is designed for use within the OR by surgical staff and does not substantially affect subsequent specimen handling (Ha et al., 2018). This provides the OR team with an agility that is not possible when consultation with other physicians is needed (e.g., radiology or pathology).

Specimen grossing is an area of great need for imaging. This includes not only the use of EVM to guide tissue sampling but also computer-assisted grossing, where an intelligent system monitors the process for automatic documentation of tissues, sampling, and measurement. Although gross specimen photography exists, it is severely underutilized by pathologists. Images may be of low quality if captured quickly during grossing without good control over lighting, background clutter, or without a clear software workflow for subsequent usage (Alan Rampy and Glassy, 2015). Gross specimen imaging may also present challenges related to the size of the field to be imaged; most EVM examples have been developed for small tissue pieces such as those analyzed in FS pathology. Larger fields may require more time to image. There have been developments in devices that combine radiology-type images with gross images; although not microscopic, these could serve as prototypical examples for follow-up development of IVM applications (Zhang et al., 2023).

As an example, many breast or gynecologic pathology specimens are highly sampled for histological assessment (Lester et al., 2009). This can lead to pathologist fatigue and strain laboratory resources due to the need to cut and stain numerous pathology slides. Imaging large cases as WSIs is also problematic from the standpoints of scanning personnel, image storage, and the burden of subsequent manual review of large cases with inadequate software tools. The ability to sharply reduce reliance on histologic oversampling would be transformative even without advanced digital pathology workflows.

Gross images could be co-registered or combined with WSIs and EVM to create hybrid image sets. For breast or prostate, such images might be useful for 3D mapping of tumors for staging or margin assessment. If developed in concert with computer-assisted diagnosis workflows, this could radically improve digital pathology workflow (Fine, 2023). In such a scenario, the addition of novel EVM images might become much more feasible than it has been in the past.

Tumor assessment is another area that should be considered. Existing biomarker assessment typically requires immunohistochemistry (IHC) and is largely a manual endeavor. There are novel imaging approaches for placing multiple biomarkers onto a single tissue section, referred to as multiplexed or hyperplexed depending on the number of simultaneous biomarkers. Coregistration also enables virtual multiplexing, as well as 3D assessments of both biomarkers and H&E. As traditional, isolated biomarker assessment gives way to holistic evaluation of the 3D tumor microenvironment, there is room for IVM/EVM techniques such as fluorescence-based methods like light sheet microscopy. These approaches are currently theoretical but are likely to begin appearing as proprietary tumor tests and, if successful, might also be applicable to routine pathology diagnosis (Khoshiwal et al., 2023).

Discussion

There are hundreds of different IVM/EVM technologies; some are already commercially available for patient care but the vast majority remain research projects. Given the history of engineering successes in addressing biomedical imaging shortfalls, it is likely that successful pathology imaging modalities will be developed in the proper setting of a compelling use case with adequate digital pathology imagination and infrastructure. While 3D histopathology is not an explicit endeavor in 2025, core needle biopsy pathology already does approximate 3D assessment whenever multiple H&E levels are examined. "Spatial" tissue analyses using hyperplexed biomarkers is also becoming an active area in research and testing (Furman et al., 2021; Khoshiwal et al., 2023).

This is a tall order, as digital pathology workflow is in its infancy and is, as of 2025, not capable of supporting advanced workflows such as "computer-assisted diagnosis for pathologists" (pCAD) (Fine, 2014). If realized, such applications could be compelling but could also begin the hard process of reimagining anatomic pathology workflow as a digital process that could include multiple imaging modalities, such as IVM/EVM but also including advanced gross imaging, specimen imaging, and WSIs, all supported by all-knowing information flows and machine learning.

This may seem overwhelming, but advanced imaging may represent a good opportunity to reimagine digital pathology workflow without dogmatically keeping cruft that currently encases pathology workflows. Starting this process with existing and long-standing modalities may help pave the way for more exotic technologies as they emerge. The challenge for pathology is to remain relevant in a world where nimbler specialties are willing to engage in EVM, without pathologist involvement and without regulatory oversight by the laboratory. Although the College of American Pathologists (CAP) has had an active IVM/EVM committee in the past, this effort appears to have stalled in recent years (College of American Pathologists, n.d.). In a world of automated surgical pathology, thanks to computational pathology advances, it will be imperative for surgical pathologists to understand and focus on how they add value to the multidisciplinary healthcare team.

References

Abraham, T.M., Costa, P.C., Filan, C., Guang, Z., Zhang, Z., Neill, S., Olson, J.J., Levenson, R., Robles, F.E., 2023. Label- and slide-free tissue histology using 3D epi-mode quantitative phase imaging and virtual hematoxylin and eosin staining. Optica 10 (12), 1605–1618. doi:10.1364/OPTICA.502859.

Alan Rampy, B., Glassy, E.F., 2015. Pathology gross photography: the beginning of digital pathology. Surg. Pathol. Clin. 8 (2), 195–211. doi:10.1016/j.path.2015.02.005.

College of American Pathologists n.d. In Vivo Microscopy Topic Center. https://www.cap.org/member-resources/councils-committees/in-vivo-microscopy-committee/in-vivo-microscopy-topic-center. (Accessed 31 August)

Delgado-Rodriguez, P., Brooks, C.J., José Vaquero, J., Muñoz-Barrutia, A., 2022. Innovations in ex vivo light sheet fluorescence microscopy. Prog. Biophys. Mol. Biol. 168, 37–51. doi:10.1016/j.pbiomolbio.2021.07.002.

Dong, J., Grant, C., Vuong, B., Nishioka, N., Gao, A.H., Beatty, M., Baldwin, G., Baillargeon, A., Bablouzian, A., Grahmann, P., Bhat, N., Ryan, E., Barrios, A., Giddings, S., Ford, T., Beaulieu-Ouellet, E., Hosseiny, S.H., Lerman, I., Trasischker, W., Reddy, R., Singh, K., Gora, M., Hyun, D., Quénéhervé, L., Wallace, M., Wolfsen, H., Sharma, P., Wang, K.K., Leggett, C.L., Poneros, J., Abrams, J.A., Lightdale, C., Leeds, S., Rosenberg, M., Tearney, G.J., 2022. Feasibility and safety of tethered capsule endomicroscopy in patients with Barrett's esophagus in a multi-center study. Clin. Gastroenterol. Hepatol. 20 (4). doi:10.1016/j.cgh.2021.02.008.

Eadie, F., Copeland, L., Di Caprio, G., McConnell, G., Kallepalli, A., 2025. Fourier ptychography microscopy for digital pathology. J. Microsc. doi:10.1111/jmi.70001.

Farabi, B., Khan, S., Jamgochian, M., Atak, M.F., Jain, M., Rao, B.K., 2023. The role of reflectance confocal microscopy in the diagnosis and management of pigmentary disorders: a review. J. Cosmet. Dermatol. 22 (12), 3213–3222. doi:10.1111/jocd.15827.

Fine, J.L., 2014. 21st century workflow: a proposal. J. Pathol. Inform. 5 (1). Available from: https://www.sciencedirect.com/science/article/pii/S2153353922003133?via%3Dihub. doi:10.4103/2153-3539.145733.

Fine, J.L., 2015. Advanced imaging techniques for the pathologist. Surg. Pathol. Clin. 8 (2), 213–221. doi:10.1016/j.path.2015.02.004.

Fine, J.L., 2023. Lining up the pieces: coregistration of digital pathology images is an underused technique with great promise in advancing computational pathology. Am. J. Clin. Pathol. 159 (3), 209–210. doi:10.1093/ajcp/aqac177.

Furman, S.A., Stern, A.M., Uttam, S., Taylor, D.L., Pullara, F., Chennubhotla, S.C., 2021. In situ functional cell phenotyping reveals microdomain networks in colorectal cancer recurrence. Cell Rep. Methods 1 (5). doi:10.1016/j.crmeth.2021.100072.

Ha, R., Friedlander, L.C., Hibshoosh, H., Hendon, C., Feldman, S., Ahn, S., Schmidt, H., Akens, M.K., Fitzmaurice, M., Wilson, B.C., Mango, V.L., 2018. Optical coherence tomography: a novel imaging method for post-lumpectomy breast margin assessment—a multi-reader study. Acad. Radiol. 25 (3), 279–287. doi:10.1016/j.acra.2017.09.018.

Hariri, L.P., Applegate, M.B., Mino-Kenudson, M., Mark, E.J., Medoff, B.D., Luster, A.D., Bouma, B.E., Tearney, G.J., Suter, M.J., 2013. Volumetric optical frequency domain imaging of pulmonary pathology with precise correlation to histopathology. Chest 143 (1), 64–74. doi:10.1378/chest.11-2797.

Khoshiwal, A.M., Frei, N.F., Pouw, R.E., Goldblum, J., Montgomery, E., Davison, J., Singh, J., Szymanski, J., Perry, A., Seldenrijk, K., ten Kate, F., Offerhaus, G.J.A., Drillenberg, P., Jansen, C., Leeuwis-Fedorovic, N., Chetty, R., Feakins, R., Jansen, M., Chinyama, C., Cooper, E., Vaziri, R., Baretton, G., Tannapfel, A., Vieth, M., Melcher, B., Mesteri, I., Müller, H., Wetzel, P., de Hertogh, G., Hoorens, A., Verschuere, S., Tamsin, A., Wetzels, K., van Caillie, M.-A., Smolko, C., Arora, M., Siegel, J.J., Duits, L.C., Critchley-Thorne, R.J., Bergman, J.J.G.H.M., 2023. The tissue systems pathology test outperforms pathology review in risk stratifying patients with low-grade dysplasia. Gastroenterology 165 (5). doi:10.1053/j.gastro.2023.07.029.

Ko, R.F., Smidt, A.C., Durkin, J.R., 2021. Reflectance confocal microscopy in pediatric dermatology: a state-of-the-art review. Pediatr. Dermatol. 38 (6), 1488–1499. doi:10.1111/pde.14837.

Krishnamurthy, S., Brown, J.Q., Iftimia, N., Levenson, R.M., Rajadhyaksha, M., 2019. Ex vivo microscopy a promising next-generation digital microscopy tool for surgical pathology practice. Arch. Pathol. Lab. Med. 143 (9), 1058–1068. doi:10.5858/arpa.2019-0058-RA.

Leitgeb, R., Placzek, F., Rank, E., Krainz, L., Haindl, R., Qian, L., Liu, M., Andreana, M., Unterhuber, A., Schmoll, T., Drexler, W., 2021. Enhanced medical diagnosis for doctors: a perspective of optical coherence tomography. J. Biomed. Opt. 26 (10). doi:10.1117/1.JBO.26.10.100601.

Lester, S.C., Bose, S., Chen, Y.-Y., Connolly, J.L., De Baca, M.E., Fitzgibbons, P.L., Hayes, D.F., Kleer, C., O'Malley, F. P., Page, D.L., Smith, B.L., Weaver, D.L., Winer, E., 2009. Protocol for the examination of specimens from patients with ductal carcinoma in situ of the breast. Arch. Pathol. Lab. Med. 133 (1), 15–25. doi:10.5858/133.1.15.

Levine, A., Markowitz, O., 2018. Introduction to reflectance confocal microscopy and its use in clinical practice. JAAD Case Rep. 4 (10), 1014–1023. doi:10.1016/j.jdcr.2018.09.019.

Levy, Y., Rempel, D., Nguyen, M., Yassine, A., Sanati-Burns, M., Salgia, P., Lim, B., Butler, S.L., Berkeley, A., Bayram, E., 2023. The fusion of wide field optical coherence tomography and AI: advancing breast cancer surgical margin visualization. Life 13 (12). doi:10.3390/life13122340.

Tosun, A.B., Pullara, F., Becich, M.J., Taylor, D.L., Fine, J.L., Chennubhotla, S.C., 2020. Explainable AI (xAI) for anatomic pathology. Adv. Anat. Pathol. 27 (4), 241–250. doi:10.1097/PAP.0000000000000264.

Uttam, S., Stern, A.M., Sevinsky, C.J., Furman, S., Pullara, F., Spagnolo, D., Nguyen, L., Gough, A., Ginty, F., Taylor, L., Chakra Chennubhotla, 2020. Spatial domain analysis predicts risk of colorectal cancer recurrence and infers associated tumor microenvironment networks. Nat. Commun. 11 (1). doi:10.1038/s41467-020-17083-x.

Wang, L., Fu, R., Xu, C., Xu, M., 2022. Methods and applications of full-field optical coherence tomography: a review. J. Biomed. Opt. 27 (5). doi:10.1117/1.JBO.27.5.050901.

Wells, W.A., Thrall, M., Sorokina, A., Fine, J., Krishnamurthy, S., Haroon, A., Rao, B., Shevchuk, M.M., Wolfsen, H. C., Tearney, G.J., Hariri, L.P., 2019. In vivo and ex vivo microscopy moving toward the integration of optical imaging technologies into pathology practice. Arch. Pathol. Lab. Med. 143 (3), 288–298. doi:10.5858/arpa.2018-0298-RA.

Yonetsu, T., Jang, I.-K., 2024. Cardiac optical coherence tomography: history, current status, and perspective. JACC: Asia 4 (2), 89–107. doi:10.1016/j.jacasi.2023.10.001.

Zhang, M., Liao, J., Jia, Z., Qin, C., Zhang, L., Wang, H., Liu, Y., Jiang, C., Han, M., Jinze, L., Wang, K., Wang, X., Bu, H., Yao, J., Liu, Y., 2023. High dynamic range dual-modal white light imaging improves the accuracy of tumor bed sampling after neoadjuvant therapy for breast cancer. Am. J. Clin. Pathol. 159 (3), 293–303. doi:10.1093/ajcp/aqac167.

Zhang, W., Fine, J.L., Pei, X., Cao, Y., Liu, Y., Yan, J., Ban, Z., Zhang, T., Wei, Y., Zhao, X., Wang, B., Zhao, C., Zeng, X., 2024. Characteristics and utility of high-resolution optical coherence microscopy images of endocervical canal lesions. Am. J. Clin. Pathol. 162 (4), 392–400. doi:10.1093/ajcp/aqae052.

CHAPTER

12

Generative artificial intelligence technologies in pathology

Ehsan Ullah
Health New Zealand–Counties District, Auckland, New Zealand
Department of Pathology, The Ohio State University, Columbus, OH, United States

The advent of generative artificial intelligence (GAI or GenAI) is a major breakthrough in the progress of artificial intelligence and the computational analysis of data in general. This advancement comes with the development of machine models showing human-level or superior abilities in learning, reasoning, and problem-solving in a variety of situations, thus enabling machine models to assist humans by performing tasks that would otherwise require human intelligence. GAI has accelerated the adoption of AI by making it accessible to organizations, businesses, and individuals alike. Since the release of the first GAI model, ChatGPT, there has been unprecedented interest in user uptake—followed by a flurry of other GAI models, both closed and open source. The scientific recognition of the potential of AI and GAI continues to grow, with the Nobel Prize in Physics 2024 being awarded to two experts and founders of modern AI methods. Meanwhile, real-world applications of GAI and the adoption of the technology are uncontrollably permeating all types of industries in some way or form.

Pathology in general, and anatomical pathology in particular, are behind radiology and other fields of medicine in the adoption of AI and GAI. However, the scope and potential are potentially even more far-reaching, as anatomical pathology is not only the backbone of diagnostic medicine but also serves as the foundation of biomedical research by providing the core material (tissue sections) for histomorphological, cellular, subcellular, molecular, and multiomic analyses. These analyses drive insights into health and disease mechanisms, disease progression and remission, response to treatment, and the development of novel biomarkers and therapies to improve diagnostic and therapeutic pathways, ultimately achieving better patient outcomes.

Advances in Digital Pathology and Artificial Intelligence in Pathology
https://doi.org/10.1016/B978-0-443-40383-5.00013-3

Introduction

The digital revolution in pathology: A key enabler of AI-driven transformation

The foundational shift from glass slides to digital scanning of whole slide images (WSI) in diagnostic pathology has bridged the chasm between the fields of pathology and artificial intelligence (AI) (Niazi et al., 2019). This technological advancement allows for the creation of vast, analyzable datasets of digitized pathology images, which are the raw materials required to train and deploy any AI models, from classical machine learning (ML) to sophisticated deep learning (DL) tools, including Generative AI (Oon et al., 2024; Parwani, 2019). The convergence of digital pathology and AI has continued to reshape the field from the 1990s to the 2020s (Pantanowitz et al., 2011; Parwani and Amin, 2020; Tizhoosh and Pantanowitz, 2018), and this is poised to completely deliver the precision pathology promise for holistic and personalized medicine (Bera et al., 2019).

For over a century, the core practice of pathology has relied on the microscopic examination of tissue on glass slides, a process that is fundamentally analog (Hussein and Raad, 2015; Titford, 2009). The advent of digital WSI has served as the critical catalyst, converting these physical specimens into gigapixel-scale digital files (Niazi et al., 2019; Oon et al., 2024; Parwani, 2019; Pantanowitz et al., 2011; Parwani and Amin, 2020; Tizhoosh and Pantanowitz, 2018). This digitization has not merely replicated the traditional workflow but has created a foundational "big data" substrate—a massive, structured repository of visual information that is accessible to computational analysis (Niazi et al., 2019; Hanna et al., 2019; Song et al., 2023). This digital ocean of data has enabled the development and refinement of DL models, which can now be trained on vast datasets of digitized histology images to identify patterns that are often imperceptible to the human eye (Gurcan et al., 2009; Madabhushi and Lee, 2016). While early DL models focused on discrete tasks such as cancer detection or cell counting, the emergence of generative AI represents a paradigm shift, introducing a new class of tools that can not only analyze but also create and synthesize novel data, text, and insights (Brodsky et al., 2025; Ullah et al., 2025; Rashidi et al., 2025).

This chapter will establish a conceptual framework for understanding the core generative AI models and their evolution. It will then demonstrate how these models are being applied to pathology, from enhancing image analysis and streamlining workflows to enabling unprecedented levels of diagnostic, prognostic, and predictive precision through the integration of multimodal data. Finally, it will highlight the significant ethical, regulatory, and practical challenges that must be navigated to ensure these technologies are integrated safely and equitably into clinical practice, where safety and medicolegal aspects are of paramount importance. The narrative in this chapter progresses from the foundational AI models to their advanced applications, culminating in a critical analysis of the hurdles and a forward-looking perspective on the future of human-machine collaboration in pathology within the broader context of personalized medicine.

Generative AI in context

Before the emergence of GAI, some of the most widely recognized applications of machine learning were concentrated on tasks such as image and data classification, object detection in visual fields, time series prediction, language translation, and speech recognition (Madabhushi and Lee, 2016; Rashidi et al., 2019; Kamath et al., 2019). These capabilities were embedded in everyday technologies outside diagnostic pathology: for instance, self-driving vehicles that relied on object detection to identify pedestrians and road signals, financial models predicting stock market fluctuations, Google Translate bridging linguistic barriers, and voice assistants such as Siri and Alexa interpreting and responding to spoken commands. While powerful, these systems were largely constrained to analysis, recognition, and prediction, without the capacity to generate novel data or content.

The development of generative adversarial networks (GANs) in 2014 (Goodfellow, 2016), followed by diffusion models in 2015 (Ratcliff et al., 2016), marked the beginning of generative AI. These models shifted the paradigm by enabling machines to generate realistic-looking new data rather than simply classify or predict. GANs rapidly found applications in generating synthetic images, art, and even pathology and molecular data for augmenting limited datasets, virtual staining, and other applications (Wölflein et al., 2023b; Ma et al., 2020; Mahmood et al., 2020; Vasiljević et al., 2022), while diffusion models advanced photorealistic image generation with high-fidelity pixel synthesis, almost making GANs nearly inferior for pathology use cases (Dhariwal and Nichol, 2021; Abraham and Levenson, 2023; Jeong et al., 2023; Müller-Franzes et al., 2023; Oh and Jeong, 2023; Harb et al., 2024; Kataria et al., 2024; Niehues et al., 2024). These innovations demonstrated that AI could operate as both a creative and an analytical tool, although training instability and scaling challenges limit their widespread deployment.

In parallel with the timeline where diffusion models were gaining momentum in pathology, another major breakthrough in AI came with the introduction of the transformer architecture in 2017 (Vaswani et al., 2017). By replacing recurrent and convolutional structures with self-attention mechanisms, transformers allowed models to capture long-range dependencies in sequences with unprecedented efficiency and scalability. This innovation catalyzed the rise of large-scale language models, such as the first generative pretrained transformer (GPT) in 2018 (Radford et al., 2018) by OpenAI, culminating in the release of OpenAI's ChatGPT in November 2022. For the first time, millions of users experienced an AI system capable of generating coherent and insightful conversational responses to free-text queries. The applications were immediate and diverse, spanning education (tutoring and summarization), healthcare (drafting clinical notes), business (customer service automation), and scientific research (literature reviews and code generation). ChatGPT gained millions of users within the first week post-launch and surpassed 100 million users by January 2023—that's within 2 months of its release (Wu et al., 2023) as shown in Fig. 12.1.

More recently, transformer-based algorithms have extended beyond language to encompass other modalities, establishing a new class of multimodal models (Zhan et al., 2023). These systems are capable of processing and generating images, video, audio, three-dimensional structures (Chen et al., 2024a), and even histologic and genomic data (Yellapragada et al., 2024; DaSilva et al., 2024). For example, models such as DALL·E and Stable Diffusion can generate high-quality images from text prompts, while video models synthesize realistic motion sequences for entertainment and simulation. Audio and speech models produce lifelike voices for accessibility applications. Notable examples of transformer-based Generative AI applications from anatomical pathology include the generation and evaluation of synthetic WSIs (Brodsky et al., 2025; Ullah et al., 2025; Rashidi et al., 2025; Carrillo-Perez et al., 2023) surpassing earlier generative and nongenerative models. Another powerful use case involves pathology co-pilots

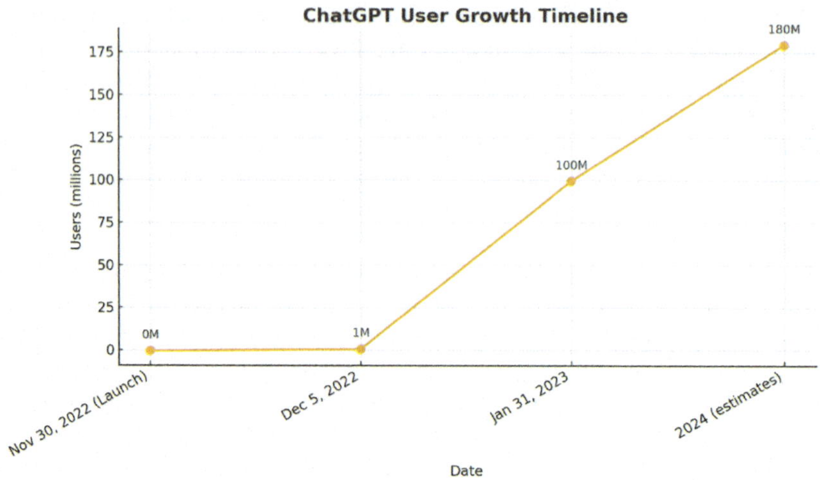

FIG. 12.1 Timeline of ChatGPT user growth.

(Lu et al., 2024a) combining image and text insights for a rich contextual understanding and downstream applications for AI-driven pathology workflows (Fei et al., 2022; Huang et al., 2023; Chen et al., 2024b). Protein-folding models like AlphaFold2 are another notable use case of these models, whereby the model predicts three-dimensional protein structures with remarkable accuracy (Guo et al., 2022). In genomics, multimodal architectures are beginning to integrate sequence data with phenotypic and imaging data using diffusion models, which promise advances in personalized medicine (Boehm et al., 2022; Steyaert et al., 2023).

Taken together, the progression from predictive machine learning to generative models, and now to multimodal transformers, represents a fundamental shift in AI's role. No longer limited to recognition and prediction, modern AI systems function as creative, integrative, and adaptive tools with wide-ranging applications across science, medicine, industry, and society.

The four most recognized iterative advancements in the field of AI over time, i.e., GANs, Diffusion Models, Large Language Models (LLMs), and Foundation Models, are summarized below for readers' interest.

Generative adversarial networks (GANs)

GANs were among the first models to demonstrate the ability to *generate* realistic-like synthetic data (Goodfellow, 2016). A GAN consists of two competing neural networks: a generator and a discriminator (Goodfellow, 2016). The generator creates synthetic images from random noise, attempting to make them as realistic as possible (Goodfellow et al., 2020). The discriminator, in turn, is tasked with distinguishing between real images from the training dataset and the fake images produced by the generator (Creswell et al., 2018) as shown in Fig. 12.2.

This adversarial process drives both networks to improve; the generator becomes more adept at creating convincing fakes, while the discriminator becomes more skilled at detecting them (Goodfellow et al., 2020; Creswell et al., 2018). The process continues

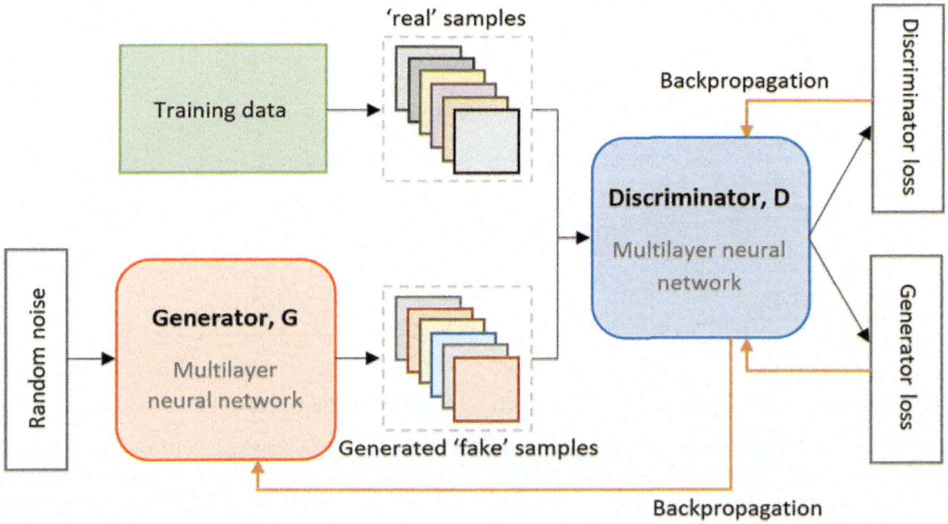

FIG. 12.2 A schematic representation of a generative adversarial network—an early form of generative AI model. *Reproduced from Little C, Elliot M, Allmendinger R, Samani S., 2021. Generative adversarial networks for synthetic data generation: a comparative study. under Creative Commons Attribution 4.0 International License.*

until the generated images are indistinguishable from real ones. GANs have been particularly valuable for data augmentation, a technique that addresses the common problem of data scarcity in medical imaging by artificially expanding a dataset, thereby improving the generalizability and performance of classifiers (Goodfellow et al., 2020; Creswell et al., 2018). It must be noted that, with further advancements in AI technology through diffusion models (described below), GANs have almost become a matter of historical mention with reference to Generative AI and have lost the interest they gained earlier, as diffusion models have virtually effaced them (Dhariwal and Nichol, 2021; Abraham and Levenson, 2023).

Diffusion models

Diffusion models have emerged as a powerful alternative and, in many pathology applications, a superior one to GANs (Ratcliff et al., 2016; Dhariwal and Nichol, 2021; Oh and Jeong, 2023; Harb et al., 2024). These models operate on a different principle, beginning with a training image and progressively adding random noise over multiple steps until the image is reduced to pure noise (Ratcliff et al., 2016). The model is then trained to learn the reverse process: to "denoise" the image by systematically removing the noise to reconstruct the original data (Ratcliff et al., 2016; Song et al., 2020). This iterative denoising process, as shown in Fig. 12.3, gives diffusion models a unique ability to model the complex distributions of image features with exceptional fidelity (Song et al., 2020; Yang et al., 2023; Moghadam et al., 2023; Pozzi et al., 2024).

Prior to diffusion, GANs dominated histopathology image synthesis, style transfer, and stain normalization (Zanjani et al., 2018; Hu et al., 2019). While GANs provided realistic outputs, they often suffered from artifacts, limited diversity, and difficulty in scaling to gigapixel WSIs (Tschuchnig et al., 2020). Diffusion models overcome these limitations by

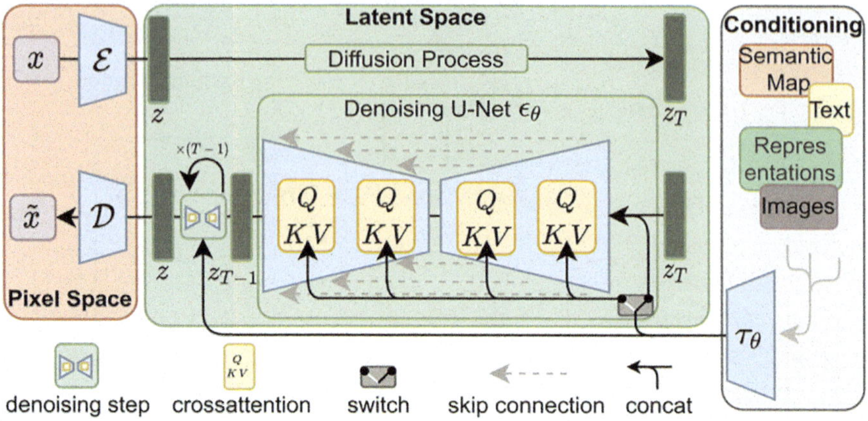

FIG. 12.3 A schematic representation of a diffusion model. *Reproduced from Rombach R, Blattmann A, Lorenz D, Esser P, Ommer B, editors., 2022. High-resolution image synthesis with latent diffusion models. In: Proceedings of the IEEE/CVF conference on computer vision and pattern recognition under Creative Commons Attribution 4.0 International License.*

combining the strengths of probabilistic modeling and iterative refinement, leading to sharper, more diverse, and controllable outputs. Moreover, their compatibility with conditional inputs (e.g., molecular labels, tissue type, or diagnostic categories) makes them particularly suited for precision pathology applications (Harb et al., 2024; Wang et al., 2022; Croitoru et al., 2023).

Diffusion models mark a notable advancement in digital pathology: from adversarial and reconstruction-based generative frameworks toward probabilistic, iterative denoising approaches that provide better training stability, higher fidelity, and greater flexibility for clinically relevant use cases (Harb et al., 2024; Niehues et al., 2024; Yellapragada et al., 2024). Compared to GANs, diffusion models are often more stable to train and have demonstrated superior performance in generating high-quality, diagnostically consistent images—a factor of critical importance in clinical settings (Dhariwal and Nichol, 2021; Abraham and Levenson, 2023; Harb et al., 2024; Niehues et al., 2024; Yellapragada et al., 2024; Yang et al., 2023; Wang et al., 2022).

Large language models (LLMs) and multimodal models

LLMs, such as ChatGPT and Gemini, represent a third major class of generative models that primarily operated on text initially rather than images or other modalities, but are ready to impact every field with a number of potential applications given their abilities and knowledge (Brodsky et al., 2025; Ullah et al., 2025; Rashidi et al., 2025; Semenova et al., 2024). A generative pretrained transformer (GPT) is a deep neural network architecture that uses stacked layers of self-attention and feed-forward modules, trained on massive text corpora in an autoregressive manner, to model long-range dependencies and generate coherent, contextually relevant sequences of natural language (Wu et al., 2023) as shown in Fig. 12.4.

Built on transformer architectures, these models are trained on vast and diverse corpora of biomedical literature, clinical notes, and general language data, enabling them to excel at

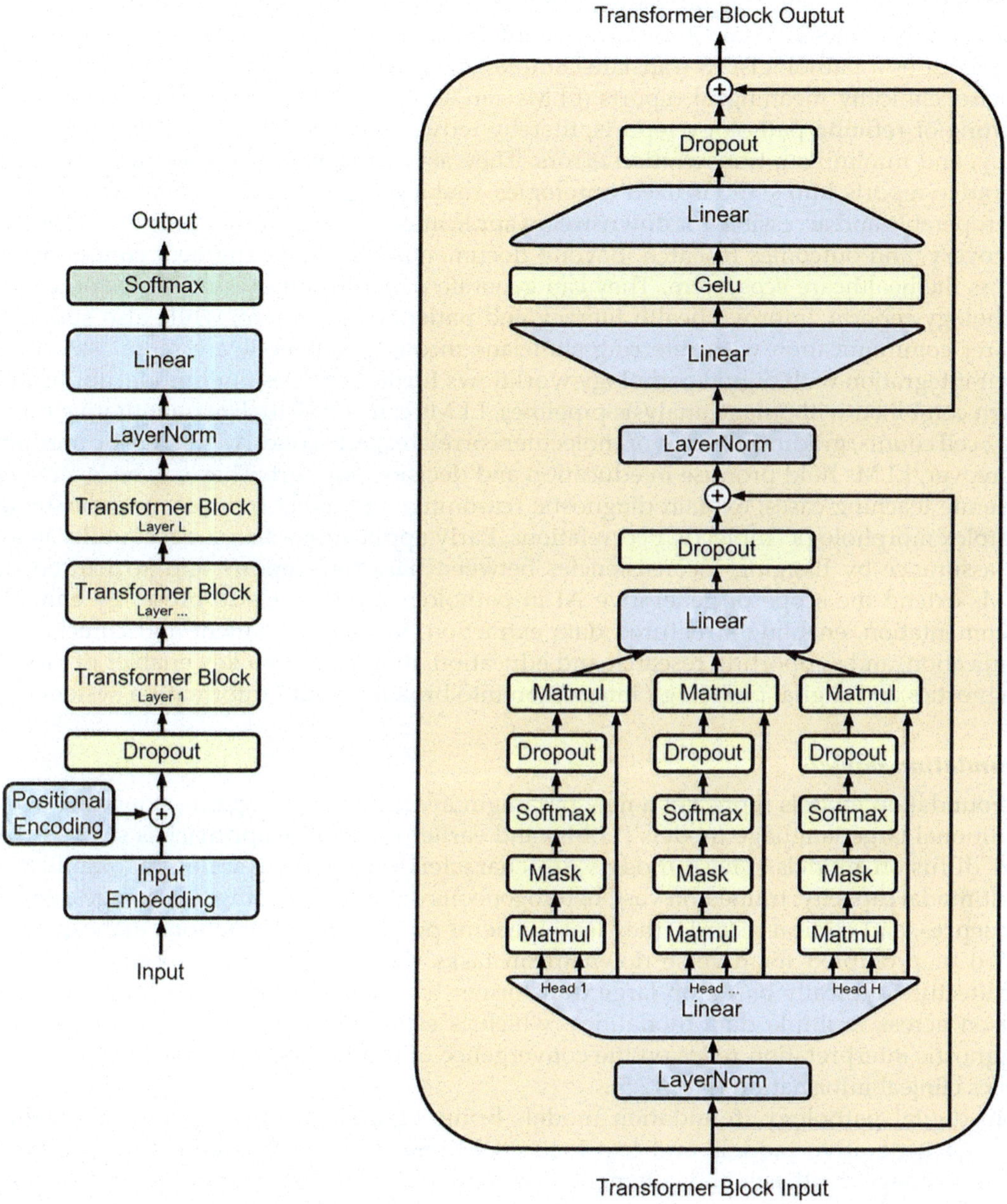

FIG. 12.4 A schematic representation of a GPT architecture, reproduced under Creative Commons CCO 1.0 Universal Public Domain Dedication.

a wide spectrum of natural language processing tasks. Their strengths include generating human-like narratives, summarizing complex documents, answering domain-specific queries, and extracting structured information from otherwise unstructured free-text sources.

In digital and diagnostic pathology, LLMs play a transformative role by bridging the gap between morphological data and narrative interpretation. Traditionally, the diagnostic process requires a pathologist to translate complex microscopic and macroscopic findings into concise, clinically meaningful reports. LLMs can assist in this translation by automatically drafting or refining pathology reports, thereby reducing clerical burden, enhancing consistency, and minimizing transcription errors. They are also capable of structuring data from narrative reports into standardized ontologies, making clinical and research datasets more interoperable and accessible for downstream applications such as cohort building, biomarker discovery, and outcomes research. Beyond documentation, LLMs augment communication across the healthcare ecosystem. They can generate patient-friendly summaries of technical pathology reports, improve health literacy and patient engagement, while also supporting clearer communication with referring clinicians through context-aware report structuring. Their integration with digital pathology workflows further enables multimodal applications: when combined with image analysis pipelines, LLMs can contextualize quantitative outputs (e.g., cell counts, grading metrics, or molecular correlates) into cohesive diagnostic narratives. Moreover, LLMs hold promise in education and decision support. They can be leveraged to generate teaching cases, explain diagnostic reasoning, and assist trainees in understanding complex morphologic–molecular correlations. Early applications also suggest utility in quality assurance by flagging inconsistencies between narrative reports and structured data. LLMs extend the scope of generative AI in pathology from pixels to prose. By enhancing documentation, enabling structured data extraction, facilitating patient and clinician communication, and supporting research and education, they serve as a key enabler of precision diagnostics and digital pathology integration into broader health information systems.

Foundation models

Foundation models represent a new paradigm in artificial intelligence, extending beyond traditional large language models (LLMs) and earlier generative approaches such as GANs and diffusion models. These models are characterized by their scale, adaptability, and multimodal capacity: trained on vast, heterogeneous datasets spanning text, images, genomic sequences, and clinical records, they learn general-purpose representations that can be fine-tuned or prompted for diverse downstream tasks with minimal task-specific data. Their architecture, typically based on large transformer backbones, allows them to integrate and reason across multiple data modalities, which is especially relevant in pathology, where diagnostic interpretation relies on the convergence of morphological, molecular, and multiomics clinical information (Fig. 12.5).

In digital pathology, foundation models bring significant advances over both image-focused generative models and text-focused LLMs. While GANs and diffusion models excel at image synthesis and augmentation, and LLMs at text generation and document understanding, foundation models unify these capabilities. They can simultaneously interpret histological slides, extract salient features, align them with molecular signatures or electronic health records, and generate coherent diagnostic narratives. This multimodal reasoning capacity enables more holistic decision support: for example, correlating immunohistochemistry results with morphological patterns, predicting genomic alterations from routine H&E slides, or suggesting differential diagnoses enriched by clinical context.

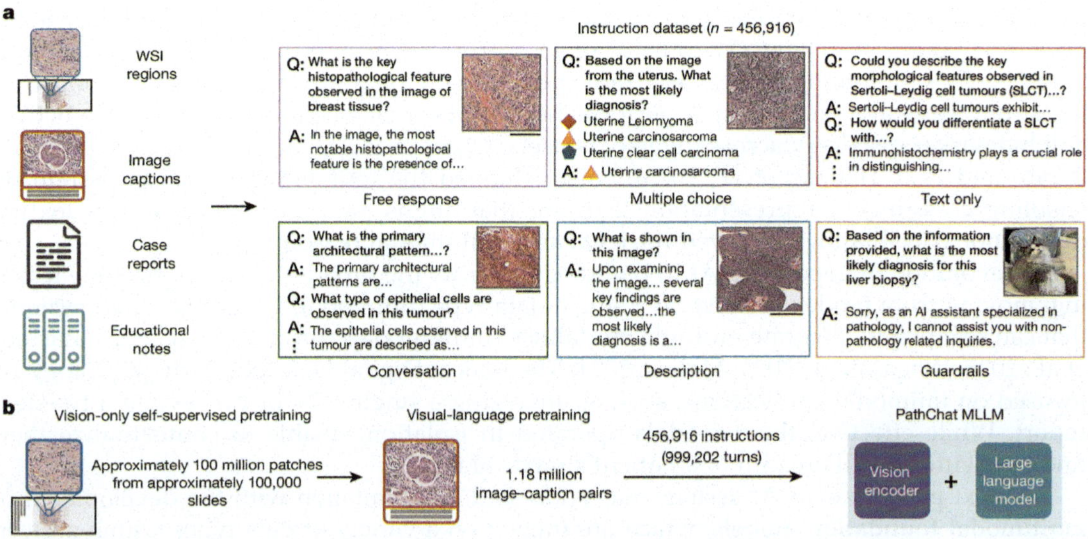

FIG. 12.5 A schematic representation of PathChat, a multimodal generative AI copilot for human pathology—an example of a Pathology Foundation Model. *Reproduced from Lu, M.Y., Chen, B., Williamson, D.F.K., Chen, R.J., Zhao, M., Chow, A.K., Ikemura, K., Kim, A., Pouli, D., Patel, A., Soliman, A., Chen, C., Ding, T., Wang, J.J., Gerber, G., Liang, I., Le, L. P., Parwani, A.V., Weishaupt, L.L., Mahmood, F., 2024a. A multimodal generative AI copilot for human pathology. Nature. 1476-4687. 634 (8033), 466–473. Available from: https://www.nature.com/nature/. doi:10.1038/s41586-024-07618-3. according to Creative Commons Attribution 4.0 International License.*

The superiority of foundation models lies in four domains. First, scalability and transferability: their pretraining on massive cross-domain datasets enables rapid adaptation to new tasks with minimal labeled pathology data, addressing the common bottleneck of annotation scarcity. Second, multimodal integration: unlike LLMs restricted to language or generative models confined to images, foundation models can fuse pathology images, laboratory results, and narrative reports into unified embeddings, supporting precision diagnostics. Third, robustness and generalization: their broad training base reduces bias toward narrow cohorts or institutions, improving reliability across diverse patient populations and histological variations. Finally, emergent reasoning abilities: foundation models demonstrate capacities beyond supervised training, such as in-context learning and cross-domain inference, which can be harnessed to derive novel biological insights and improve clinical decision-making.

In practice, foundation models are poised to underpin next-generation pathology workflows. They can serve as the "central intelligence" layer of digital pathology systems, orchestrating image analysis pipelines, molecular predictions, automated reporting, and communication with clinicians and patients. By combining the narrative fluency of LLMs, the image generation power of diffusion models, and the multimodal adaptability unique to their scale, foundation models hold the potential to shift pathology from siloed analyses toward integrated, patient-centric precision pathology and medicine—extracting data from pixels and other forms of clinical information into diagnostic, prognostic, and predictive insights. Several pioneering examples highlight their potential. PathChat (Lu et al., 2023a) applies LLM capabilities to WSIs, enabling interactive queries that link histopathological findings with natural language explanations, thereby supporting education and diagnostic decision-making. Virchow (Vorontsov et al., 2024) represents one of the first billion-parameter vision foundation models trained exclusively on pathology images, demonstrating strong zero-shot and few-shot performance in cancer classification, survival prediction, and biomarker discovery across diverse datasets. Other models, such as CONCH (Lu et al., 2024b) and PLIP (Huang et al., 2023), have extended the foundation model paradigm by enabling cross-modal representation learning that aligns pathology images with textual descriptions, facilitating improved retrieval and multimodal reasoning.

The evolution of generative AI in pathology mirrors the broader trend in machine learning, moving from models trained on a single data type (unimodal) to sophisticated systems that can seamlessly integrate multiple modalities (multimodal) (Brodsky et al., 2025; Rashidi et al., 2025; Waqas et al., 2023; Matias, 2024; The Lancet Digital H, 2024). Early pathology AI focused on unimodal applications, such as analyzing a single WSI or processing a free-text report. While effective, these models operated in isolation, unable to contextualize their findings with the full breadth of a patient's medical record.

The next generation of AI systems has addressed this limitation with the development of **multimodal foundation models**, which are trained on a variety of data types within a given domain. These models, exemplified by systems like CONCH and PathChat, can simultaneously ingest and analyze histology images, pathology reports, genomic data, and even clinical notes. This ability to link disparate data types is a crucial step toward enabling truly personalized and holistic diagnostics. For example, a multimodal model could be queried with a slide image and a patient's genetic profile to generate a tailored differential diagnosis, leveraging both morphological and molecular information. The progress in building these models is not a separate development but a direct consequence of the maturation of their unimodal predecessors. The success of multimodal systems like PathChat+ is predicated on the foundation of advanced visual encoders and language models that have been meticulously refined on large, pathology-specific datasets. This represents a developmental continuum where advances in one area directly enable progress in the next. The most advanced systems are now moving beyond simply integrating data to becoming autonomous agents that can mimic the entire human diagnostic reasoning process, a topic that will be explored in a later chapter. This demonstrates a clear cause-and-effect relationship where improvements in foundational AI capabilities directly enable the creation of more sophisticated and powerful diagnostic tools.

Together, these developments indicate that foundation models are not only improving the accuracy and efficiency of diagnostic pathology but are also enabling new workflows, such as natural language-based exploration of WSIs, automated biomarker identification, and integration of pathology with genomics and electronic health records (Fei et al., 2022; Lu et al., 2023a; Vorontsov et al., 2024; Lu et al., 2024b; Waqas et al., 2023; Truhn et al., 2024a; Xu et al., 2024b). These capabilities are setting the stage for scalable, generalizable, and clinically adaptable AI systems in precision pathology.

What makes generative AI so relevant in current times?

Several converging factors make the present moment uniquely appropriate for the integration of generative AI in digital pathology. Foundation models, pretrained on massive datasets comprising millions of images and documents, now display capabilities such as in-context learning and cross-modal reasoning, allowing them to be repurposed for pathology applications with relatively little domain-specific data. At the same time, the inherently multimodal nature of modern pathology, which integrates visual and textual information, aligns with the ability of generative AI systems to connect WSIs with clinical narratives and other multimodal insights, supporting workflows such as automated report generation, image retrieval through textual prompts, and interactive diagnostic querying. Advances in diffusion models have further improved image fidelity, enabling the synthesis of high-resolution, clinically realistic tissue images for augmentation, simulation, and quality control. Parallel developments in cloud infrastructure and the increasing availability of open, pretrained models have lowered the barriers to experimentation and deployment, extending access to smaller research groups and clinical centers. Finally, an urgent clinical demand, driven by rising diagnostic workloads, global workforce shortages, and the pressing need for scalable solutions in resource-limited settings—has created a strong impetus for the adoption of AI-driven augmentation in pathology workflows. Together, these technological advances, infrastructural shifts, and clinical pressures have transformed generative AI in pathology from a speculative possibility into a practical and actionable reality, positioning pathologists to play an active role in shaping its future use.

Image and data generation

Synthetic data augmentation

A persistent challenge in computational pathology is the scarcity of high-quality, annotated data for rare diseases and tumor subtypes (Tizhoosh and Pantanowitz, 2018; Brodsky et al., 2025; Rashidi et al., 2025). The process of obtaining and labeling these specimens is time-consuming and expensive. Generative AI offers a powerful solution through synthetic data augmentation (Ullah et al., 2025; Ye et al., 2025). By training models like GANs and diffusion models on limited real-world datasets, researchers can generate hundreds or even thousands of new, artificial slide images that are visually realistic and diagnostically consistent with the original data (Ye et al., 2025). This technique not only addresses the problem of data scarcity but also helps to balance class imbalances in datasets, exposing AI models to more variability and preventing them from overfitting to common pathologies (Brodsky et al., 2025; Rashidi et al., 2025).

Virtual staining

Beyond augmenting datasets, generative models are also revolutionizing the way pathologists view and interact with tissue samples. Virtual staining is a computational technique that uses AI to transform an image of a standard hematoxylin and eosin (H&E) stained slide into an image that mimics the appearance of a special histochemical or immunohistochemical (IHC) stain (Brodsky et al., 2025; Bai et al., 2023; Ma and Chen,

2024). This process offers significant clinical and operational advantages by eliminating the need for costly and time-consuming physical processing of specimens, while also preserving the original tissue for further analysis. By enabling rapid, on-demand visualization of different histochemical contrasts *in silico*, virtual staining allows pathologists to explore different diagnostic hypotheses without delay.

Recent advancements in this area have shown that diffusion models are often superior to traditional GANs for virtual staining (Brodsky et al., 2025). Diffusion models are able to model the plausible distributions of biomarker expression, which is a critical feature for diagnostic accuracy (Harb et al., 2024). To evaluate this, researchers have moved beyond simple image quality metrics to more clinically meaningful assessments. One such metric is the Semantic Fidelity Score (SFS), which quantifies diagnostic consistency by measuring a classifier's accuracy on a clinical task using the virtual stain (Liu et al., 2025a; Chugh et al., 2021). The StarDiff model (Liu et al., 2025a), a diffusion-based approach, demonstrated state-of-the-art performance, achieving the highest diagnostic accuracy of 0.68 ± 0.02 and the highest SFS of 0.74 ± 0.01 on a diagnostic evaluation, significantly outperforming GAN-based models. The shift in evaluation metrics from raw pixel quality to clinically grounded diagnostic utility demonstrates the field's increasing maturity and its prioritization of patient outcomes over technical performance.

AI-powered triage and intelligent quality control

A high volume of cases, often with long backlogs, places significant pressure on pathology labs, leading to potential delays in diagnosis (Smith et al., 2023; Bonert et al., 2021). Generative and deep learning tools are proving to be effective solutions by automating routine tasks and enabling intelligent workflow management. A key application is AI-driven case triage (Brodsky et al., 2025). By analyzing WSIs at high speed, AI algorithms can quickly identify high-priority findings, such as metastatic foci or other high-risk lesions, and automatically flag these cases for immediate review by a pathologist. This ensures that urgent cases are not delayed and allows pathologists to allocate their time more efficiently, focusing on the most diagnostically challenging cases rather than routine ones.

The real-world impact of such systems can be quantified by metrics like turnaround time (TAT). A retrospective study on the integration of an AI-driven triage tool for detecting intracranial hemorrhage in a teleradiology workflow provides a powerful, measurable example of this benefit (Bani-Sadr et al., 2025). The study found that the AI tool, with a high sensitivity of 97.3% and a specificity of 96.0%, significantly reduced key workflow metrics (Bani-Sadr et al., 2025). The mean PACS-to-assessment time (PTAT) was reduced by 13.72 min ($p = 0.002$), while the report turnaround time (RTAT) was significantly reduced by 26.62 min ($p = 0.001$) after AI integration (Bani-Sadr et al., 2025). This quantitative evidence, though from a radiology context, clearly demonstrates the potential for AI triage systems to enhance efficiency and accelerate critical diagnostic decisions in pathology, ultimately leading to improved patient outcomes (Bani-Sadr et al., 2025).

Integrated analysis of diagnostic enhancement

The applications detailed in this chapter, such as synthetic data augmentation, virtual staining, and AI-powered triage, illustrate a fundamental relationship between AI and pathology: the technology is being deployed to solve critical problems with measurable outcomes. The problem of data scarcity, a significant cause of performance plateaus in AI model development, is being directly addressed by generative models that can produce synthetic data, leading to a quantifiable increase in classifier accuracy (Rashidi et al., 2025; Pozzi et al., 2024). Similarly, the immense workload and time pressure on pathologists, a constant cause of operational strain, are being alleviated by AI triage systems, resulting in a demonstrable reduction in turnaround times and a reallocation of human expertise to more complex cases (Bani-Sadr et al., 2025).

A compelling trend is the evolution of AI evaluation in pathology. The shift from using simple, image-centric metrics like PSNR or SSIM to clinically grounded diagnostic metrics like the Semantic Fidelity Score (SFS) is a clear indication that the field is maturing (Liu et al., 2025a). This shift signals a movement away from the "black box" problem toward a focus on patient outcomes and the preservation of diagnostic consistency. In essence, the field is prioritizing trustworthy AI that can demonstrably assist clinicians in making better decisions, rather than simply producing visually appealing outputs.

Automating the clinical workflow and reporting

Streamlining case management with AI-assisted workflows

The applications of generative AI extend far beyond image analysis to encompass the entire pathology workflow, from the moment a specimen arrives at the lab to the final report (Brodsky et al., 2025; Ullah et al., 2025; Rashidi et al., 2025; Waqas et al., 2023). The integration of AI into Laboratory Information Systems (LIS) is revolutionizing how diagnostic centers operate by automating time-consuming administrative tasks (Mayall et al., 2023; Pati et al., 2023; Zhang et al., 2024). AI-driven LIS platforms can perform tasks such as automated data entry, case sorting, and quality control (QC) (Janowczyk et al., 2019), which are prone to human error and can significantly increase clerical burden. By autonomously verifying patient and specimen information and flagging discrepancies, these systems reduce the incidence of clerical mistakes and ensure a more consistent work-up (Brodsky et al., 2025). This automation is not simply about doing the same work faster; it is about fundamentally altering the pathologist's role. By offloading these "non-value-added" administrative tasks, AI enables pathologists to spend less time on routine, repetitive work and more time on high-level intellectual tasks, such as complex diagnostic cases, interdisciplinary collaboration, and research (NovoPath, 2025). The strategic application of AI-powered workflow tools is thus a mechanism for reducing burnout and improving overall efficiency within the laboratory, addressing the increasing pressure to deliver diagnoses with speed and accuracy amid rising case volumes (NovoPath, 2025) (Fig. 12.6).

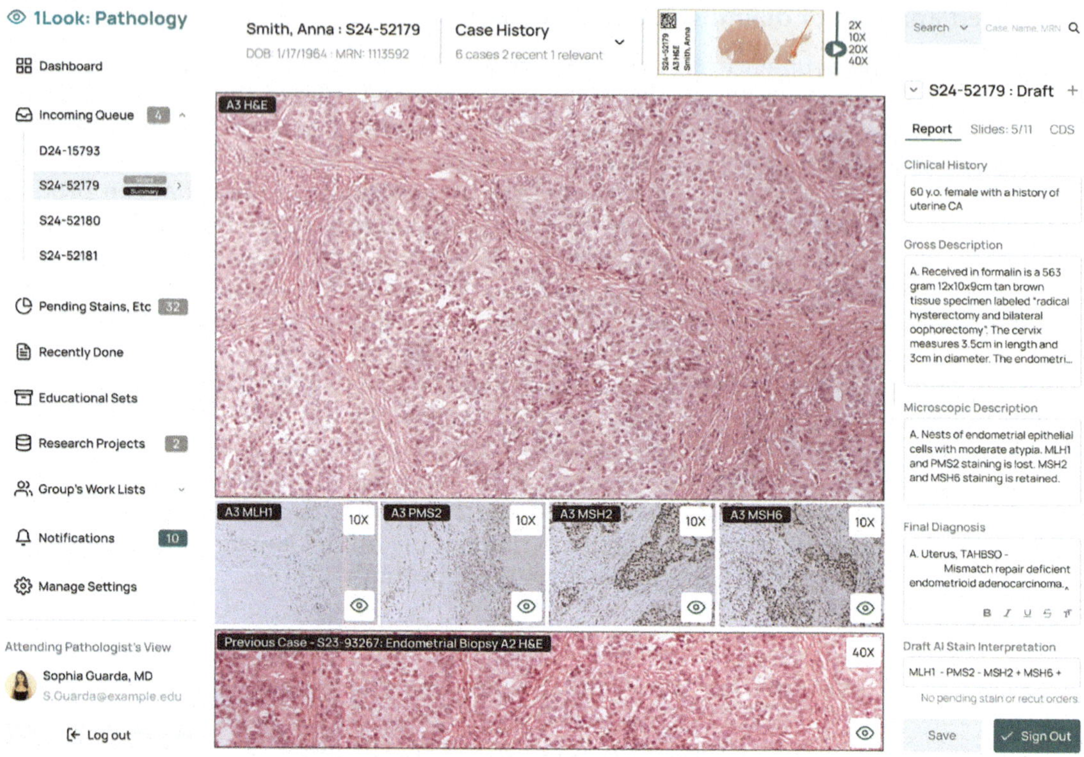

FIG. 12.6 A hypothetical interactive 'summary case view' presented by an AI-driven workflow to the reporting pathologist for review. *Reproduced from Brodsky, V., Ullah, E., Bychkov, A., Song, A.H., Walk, E.E., Louis, P., Rasool, G., Singh, R.S., Mahmood, F., Bui, M.M., Parwani, A.V., 2025. Generative Artificial Intelligence in Anatomic Pathology. Arch. Pathol. Lab. Med. 1543-2165. 149 (4), 298–318. Available from: https://meridian.allenpress.com/aplm/issue/149/4. doi:10.5858/arpa.2024-0215-RA.*

Automated report generation and structured data extraction

The narrative nature of pathology reports, while rich in detail, poses a significant challenge for large-scale data analysis and research (Balasubramanian et al., 2025; Tripathi et al., 2025). LLMs are proving to be a highly effective solution by enabling the automated extraction and structuring of key information from these free-text reports. By using natural language instructions, LLMs can be prompted to read an unstructured report and populate a synoptic template with specific fields such as tumor type, grade, dimensions, and margins (Brodsky et al., 2025; Tripathi et al., 2025).

The performance of these systems is remarkably high. A recent study evaluating LLMs for structured information extraction from breast cancer reports found that leading models, including GPT-4.0 and Llama 3.1, achieved an extraction accuracy of 96.1% and 94.7%, respectively (Balasubramanian et al., 2025). These results were statistically comparable to the accuracy of a human annotator, which stood at 95.4% (Balasubramanian et al., 2025). This capability is not limited to simple facts; the models can extract highly granular details (Truhn et al., 2024b), such as tumor laterality (right vs. left breast) (Tripathi et al., 2025), specimen dimensions, and even the presence of specific tissue compositions like skin or skeletal muscle (Shahid et al., 2025).

The ability of LLMs to convert unstructured data into a structured format has a profound cascading effect. Once reports are indexed and standardized, they can be used for a variety of purposes beyond their immediate clinical use. The structured data facilitates large-scale data mining for research, enables the identification of specific cohorts for clinical trials, and provides a cleaner, more usable data source for training the next generation of AI models (Tripathi et al., 2025; Truhn et al., 2024b; Shahid et al., 2025). The utility of this technology extends far beyond its initial application of saving a pathologist's time, creating a virtuous cycle of data generation and analysis.

Improving patient and clinician communication

In addition to streamlining internal workflows, generative AI is also enhancing communication between clinicians and patients (Reuter et al., 2025). Pathology reports are often filled with highly technical and domain-specific medical terminology that can be challenging for nonexpert physicians and patients to fully understand (Balasubramanian et al., 2025; Tripathi et al., 2025; Shahid et al., 2025). LLMs offer a promising solution by automatically rephrasing complex findings into clear, patient-friendly summaries. A recent study explored this application by providing hospitalized patients with AI-generated summaries of their discharge letters. The results were compelling. Among a cohort of 20 patients, a striking 90% reported a better understanding of their medical condition after reading the AI-generated summary, including those who felt they were already well informed. Furthermore, a significant majority (90%) found the summary more helpful than their discharge consultations with physicians. The study also noted that older patients, in particular, expressed a strong interest in receiving such summaries for future hospital stays (Reuter et al., 2025). This demonstration highlights that the same LLM technology that drafts a report template for a pathologist can simultaneously empower a patient with a clear explanation, demonstrating AI's capacity to bridge critical communication gaps in healthcare with a single source of truth.

Enabling equity and access

Generative AI also holds promise for reducing disparities in healthcare access and ensuring more equitable delivery of pathology services (Wachter and Brynjolfsson, 2024) Previous work has shown that LLMs can simplify complex medical language in radiology reports, and similar applications can extend to pathology (Jeblick et al., 2023). By translating technical jargon into plain language, potentially in real time, AI can improve patient understanding, reduce anxiety, and foster clearer communication between clinicians and patients. Enhancing health literacy in this way not only empowers patients but may also contribute to better clinical outcomes (Berkman et al., 2011). With appropriate validation and responsible implementation, generative AI could become an assistive tool that augments the capabilities of clinicians, patients, and even researchers, much like existing technologies designed to support people with disabilities.

Automating mundane administrative tasks

Generative AI has also demonstrated value in streamlining administrative responsibilities within medicine and pathology (Zhang and Kamel Boulos, 2023). Applications range from generating meeting minutes and improving dictation transcription to drafting insurance correspondence. Automating these repetitive tasks can reduce errors, improve workflow efficiency, and enhance communication within diverse, multilingual, and multigenerational teams (Tripathi et al., 2024; Melina et al., 2024; Wachter and Brynjolfsson, 2024). In doing so, AI not only saves time but also allows clinicians and pathologists to focus more on core clinical and research activities (Wachter and Brynjolfsson, 2024).

Transforming pathology training and education

Generative AI, particularly the foundation models, has significant potential to reshape training and education in medicine, especially within image-intensive specialties such as radiology and anatomical pathology (Waqas et al., 2023; Ravi et al., 2023; Dolezal et al., 2023; Cheng, 2024).

For example, models trained on pathology images and diagnostic algorithms could dynamically generate multiple-choice questions paired with illustrative microscopic images. Validated systems could also respond to trainee queries using a blend of text explanations and annotated images retrieved from pathology PACS (Benirschke et al., 2024; Daungsupawong and Wiwanitkit, 2024; Wölflein et al., 2023a). Furthermore, LLMs are capable of evaluating residency, fellowship, and faculty applications by generating preliminary rankings and offering transparent justifications for their recommendations, making program directors' jobs more efficient. A recent review (Cecchini et al., 2025) highlights the opportunities, use cases, and potential pitfalls of Generative AI tools for pathology and laboratory medicine. A few use cases are summarized below.

Several LLMs can generate diverse, high-quality practice questions in formats ranging from multiple-choice to essay-style, drawing on varied content sources such as articles, lecture notes, and online resources (Tan et al., 2023). They support both learners by providing feedback and clarifying concepts, and educators by expanding and diversifying question banks. LLMs also create realistic distractors, though output quality depends heavily on prompt design, model choice, and post-editing. Iterative prompting and manual refinement can further enhance the accuracy and educational value of generated questions. LLMs possess the capability to analyze both images captured via smartphones and those obtained from online sources, enabling the generation of customized questions and explanations that foster a more engaging and tailored educational experience (He, 2021; Ben Abacha et al., 2019). Recent refinements in LLMs, mainly during the early 2020s, have significantly improved their proficiency in processing visual data, allowing them to deconstruct intricate visuals into accessible elements for enhanced understanding and assessment.

Multimodal foundation models

Beyond the image: Multimodal integration in practice

The most advanced and transformative frontier of generative AI in pathology lies in the development of multimodal foundation models (Brodsky et al., 2025). These sophisticated systems are designed to go beyond the analysis of a single data type, combining visual information from histology slides with a patient's textual records, genomic data, and other clinical information. The goal is to create a unified analytical framework that can synthesize a patient's entire profile to enable a more holistic and personalized approach to medicine.

Leading examples of this technology include models like CONCH—CONtrastive learning from Captions for Histopathology (Lu et al., 2024b) and PathChat (Lu et al., 2024a). CONCH, for instance, has been trained on over 1.17 million image-caption pairs and has demonstrated state-of-the-art performance across 14 computational pathology tasks, including image classification and tissue segmentation. PathChat represents a significant leap forward, building upon its predecessor by training on over 1 million diverse visual language instruction samples, allowing it to provide a holistic, slide-level analysis that integrates essential contextual cues. The clinical utility of these models is vast. For example, a pathologist could use a multimodal model as a "copilot" by providing it with a slide image and the patient's history to receive a tailored differential diagnosis and a list of necessary tests (Brodsky et al., 2025; Ullah et al., 2025; Rashidi et al., 2025; Waqas et al., 2023).

Case studies in biomarker discovery and precision oncology

In the research domain, LLMs and generative AI can streamline multiple processes: drafting and reviewing grant proposals (Hosseini et al., 2023), assisting with institutional review board submissions (Chiu, 2023), identifying and refining patient cohorts (Liu et al., 2021), curating literature (Guo et al., 2024), evaluating data quality (Rasheed et al., 2025), and visualizing datasets. They also show potential in detecting biases and addressing class imbalances (Tasci et al., 2022), thereby improving research reliability. Beyond text processing, generative AI can support pathology research by enhancing slide quality control (Rodrigues et al., 2023), facilitating image annotation and segmentation (Mahmood et al., 2020; Liu et al., 2025b; Dhawan and Agrawal, 2023), and enabling disease detection in WSIs (Chen et al., 2024b; Lu et al., 2023a; Waqas et al., 2023).

These models can also be applied to data profiling, anonymization, synthesis, and even the generation of synthetic datasets for proof-of-concept studies in the absence of real-world data (Oon et al., 2024; Hou et al., 2019; Carrillo-Perez et al., 2024). Specifically, generative AI models can be used to produce training data for a proof-of-concept, even in the absence of previous data (Guo and Chen, 2024). Moreover, AI-driven molecular generation may accelerate the development of novel antibodies and stains relevant for both diagnosis and therapy (Chungyoun and Gray, 2023). Synthetic histology image datasets (Moghadam et al., 2023; Carrillo-Perez et al., 2024; Guo and Chen, 2024; Xue et al., 2021) can be used for algorithm training to discover antibodies for diagnosis and therapy (Zhao et al., 2023; Porebski et al., 2024). Emerging AI agents are already capable of proposing hypotheses, designing studies, and even forecasting potential outcomes, marking a shift toward AI-assisted scientific discovery.

Multimodal foundation models are also accelerating research and discovery by enabling powerful *in silico* experiments. By integrating vast datasets of genomics, transcriptomics, and proteomics, AI models can simulate drug-target interactions and predict tumor behavior with unprecedented accuracy (Chenthamarakshan et al., 2023). This approach significantly reduces the need for extensive in vitro and in vivo experiments, thereby accelerating drug discovery and screening (Chenthamarakshan et al., 2023; Pai et al., 2024).

A compelling example of this is the identification of a novel biomarker for lung cancer (Campanella et al., 2025). Through the integration of genomic and proteomic data, an AI-driven platform identified a new biomarker that was subsequently validated in clinical studies. This discovery is now being explored as a potential therapeutic target, demonstrating the power of these models to provide actionable insights for precision oncology (Campanella et al., 2025). Similar applications are being explored in the research of neurodegenerative diseases like Amyotrophic Lateral Sclerosis (ALS), where AI is used to integrate clinical data, genetic information, and physiological measurements to track disease progression and identify new therapeutic targets (Kitaoka et al., 2025).

The link between visual and molecular data is further solidified by models like PathCLIP. This system uses contrastive learning to connect gene names and relations with visual elements within biomedical pathway figures found in literature (He et al., 2024). This capability effectively creates a link between a visual diagram and molecular data, allowing researchers to uncover previously undiscovered pathways and enrich existing knowledge bases, further demonstrating the potential of AI to bridge traditionally siloed domains (He et al., 2024).

From copilot to agent: The emergence of autonomous AI systems

While multimodal models as "copilots" have been groundbreaking, the next paradigm shift is the move from a reactive assistant to a proactive, autonomous AI agent (Zou and Topol, 2025; Karunanayake, 2025; Huang, 2025). An agentic AI system is distinguished by its ability to take action, remember long-term context, and execute multistep plans to achieve a goal without continuous human prompting (Huang, 2025). In healthcare, this means an agent could not only answer a query but also autonomously perform a series of actions, such as scheduling appointments or ordering lab tests based on a single high-level instruction (Karunanayake, 2025; Huang, 2025).

Clinical decision-making in oncopathology demands the integration of diverse data and expertise. Ferber and coworkers have recently published an autonomous AI agent combining GPT-4 with precision oncology tools, including vision transformers for histopathology, MedSAM for radiological segmentation, and web-based search resources (OncoKB, PubMed, Google). In testing across 20 multimodal cases, their agentic system demonstrated 87.5% tool-use accuracy, 91.0% correct clinical conclusions, and 75.5% guideline citation accuracy. Compared to GPT-4 alone (30.3%), performance improved substantially (87.2%). This highlights the potential of integrated language model–driven systems to enhance personalized oncology decision support (Ferber et al., 2025).

The SlideSeek system is another example of this emerging paradigm in pathology (Chen et al., 2025a). It is a reasoning-enabled, multiagent AI system designed to autonomously analyze gigapixel WSIs, a task that is traditionally challenging for conventional AI models that require manual selection of regions of interest (ROIs). SlideSeek mimics the hierarchical diagnostic reasoning process of a human pathologist, employing a Supervisor Agent that formulates an initial hypothesis and a detailed exploration plan, and a team of Explorer Agents that use a powerful multimodal model, PathChat (Lu et al., 2024a) to analyze specific slide regions. The Supervisor Agent iteratively reviews the findings and refines the diagnostic plan until a conclusion is reached, demonstrating a metacognitive awareness of its diagnostic confidence. The performance of SlideSeek on the challenging DDxBench differential diagnosis benchmark was remarkable, achieving an 80.0% primary diagnosis accuracy, which is comparable to the performance of a human-assisted system that relied on manually preselected ROIs. These results indicate that the system's autonomous, multiagent reasoning is highly effective and represents a fundamental shift from a simple "copilot" that provides suggestions to an "autonomous partner" capable of performing an end-to-end diagnostic workup. The transition from a reactive assistant to a proactive agent fundamentally changes the pathologist's role from a primary diagnostician to a supervisor of AI systems, a change that will necessitate a reevaluation of training, workflow, and liability.

Integrated analysis of multimodal and agentic AI

The advancements in multimodal and agentic AI are part of a deeply interconnected and self-reinforcing cycle of innovation. The development of better unimodal vision models provides the foundational visual understanding necessary for more capable multimodal models (Lu et al., 2024b). These multimodal models, in turn, can be integrated as the core analytical engine within complex, multiagent systems like SlideSeek (Chen et al., 2025a). These agentic systems, by automating complex research and diagnostic tasks, can then accelerate biomarker discovery and other scientific endeavors (Alum, 2025). The new data and insights generated from these discoveries can then be used to train even more powerful and accurate models, creating a positive feedback loop that drives continuous progress.

The move toward an agentic paradigm is perhaps the most significant emerging trend. It signals a shift away from AI as a mere augmentation tool to a central component of the clinical decision-making process. The pathologist's role is transitioning from that of a primary diagnostician to a supervisor and validator of complex AI workflows. This evolution is not merely technological but also presents a new model for human and machine collaboration.

Summary of key multimodal foundation models for pathology

The following table provides a summary of key multimodal and vision-language foundation models that are shaping the future of computational pathology (Table 12.1).

TABLE 12.1 Summary of key multimodal foundation models for pathology.

Model	Key features	Uniqueness	Strengths	Applications
Prov-GigaPath (Xu et al., 2024a)	ViT architecture, hierarchical spatial tokenization, global self-attention; trained on 1.3B image tiles from 170k WSIs, 30k+patients	Unprecedented scale; open-source; benchmark standard	State-of-the-art on 25/26 benchmarks; generalizable across tissues and institutions; interpretable reasoning	WSI classification, morphologic feature extraction, slide captioning, mutation prediction; pretraining foundation
PathOrchestra (Yan et al., 2025)	Multistage ViT; trained on 300k slides, 112 tasks	Generates structured diagnostic reports mapped to EMRs	High throughput; >0.95 accuracy across many tasks	Tumor subtyping, grading, biomarker inference, margin assessment, automated EMR reporting
Atlas (Alber et al., 2025)	Efficient transformer with patch-wise token reduction, hierarchical attention; trained on 1.2 M WSIs	Focus on cross-institutional generalization and reproducibility	Robust across datasets, scanners, institutions; scalable and efficient	Classification, grading, metastasis detection; benchmarking
CPath-Omni (Sun et al., 2025; Lu et al., 2024a)	15B parameter VLM; hierarchical transformer +contrastive pretraining	Handles classification, captioning, VQA, and zero/few-shot inference	State-of-the-art on 39/42 benchmarks; flexible, instruction-aware	Tumor classification, biomarker detection, report generation, exploratory diagnostics
WSI-LLaVA (Liang et al., 2024)	VQA for gigapixel WSIs: morphology-aware alignment	Supports multi-turn dialogue, region-specific prompts	Interactive, explainable AI; clinical dialog grounded in pathology	Case triage, education, second-opinion support, region highlighting
PathChat (Lu et al., 2024b)	Instruction-tuned VLM; trained on ~1 M image–Q&A pairs	Domain-focused conversational AI	Handles nuanced queries; context-aware responses	Interactive Q&A, case review, clinical education, co-pilot for pathologists
CPath-CLIP (Radford et al., 2021)	Contrastive learning aligns images and text in a shared space	Foundation for multimodal systems	Zero-shot classification, retrieval, semantic search	Tumor classification, biomarker localization, WSI retrieval, and captioning
CONCH (Lu et al., 2024b)	Contrastive-pretrained on 1.17 M image–caption pairs	Uses clinical captions for supervision	Explainable embeddings, robust in low-supervision settings	Zero-shot classification, captioning, retrieval, education
PLIP (Huang et al., 2023)	Trained on open-source social media and web pathology image–text pairs	First model to leverage noisy real-world pathology discourse	Accessible, adaptable, generalizes phrasing and presentation styles	Semantic search, education, case exploration, retrieval

MI-Zero (Lu et al., 2023)	Contrastive MIL: weak supervision with slide-level labels	Zero-shot classification without patch annotations	Generalizes to rare/unseen diseases; annotation-efficient	Slide triage, label-efficient bootstrapping, and diagnostics in low-resource settings
SlideChat (Chen et al., 2025)	Dual-encoder multimodal architecture aligns WSIs with text	Interactive assistant with WSI navigation	Explainable outputs, zero/few-shot adaptability	VQA, WSI navigation, triage, education, descriptive explanations
PMPRG (Tan et al., 2024)	Multi-scale ViT + GPT-2 LM: integrates multiple WSIs per patient	Patient-level, organ-specific structured report generation	High METEOR score (0.68); attention maps for interpretability	Patient-level diagnostic reporting, organ-specific insights
Virchow2/2G (Zimmermann et al., 2024)	Trained on 3.1 M WSIs, 1.9B tiles; 225k patients, 200 tissue types	Universal histology model: distilled mini version (22 M params)	Outperforms SOTA in zero/few-shot; scalable generalist	Classification, retrieval, tissue-of-origin prediction, rare disease diagnosis
H-Optimus-0 (Saillard et al., 2024)	Self-supervised ViT G/14 + ABMIL; trained on 500K WSIs	Weakly supervised, pure vision model	Strong performance on TCGA; scalable inference	Tissue classification, segmentation, survival prediction
Phikon-V2 (Filiot et al., 2024)	ViT trained on 58 k public slides (~460 M tiles); ensembling strategy	Open access, trained only on public datasets	Competitive with billion-parameter models; strong biomarker prediction	MSI prediction, biomarker inference, external validation research

Ethical, regulatory, and practical challenges

The problem of bias

Despite the immense promise of GAI, its successful and ethical integration into clinical practice is contingent upon addressing significant challenges (Brodsky et al., 2025; Rashidi et al., 2025; Zhang and Kamel Boulos, 2023). A primary concern is the potential for algorithmic bias to perpetuate and even amplify existing healthcare disparities (Vaidya et al., 2024). Generative models learn from the data on which they are trained, and if that data is not representative of a diverse population, the resulting algorithms can perform unevenly across different demographic groups (Franklin et al., 2024). For example, one study found classification accuracy differences of 3%–17% between racial groups for certain cancers (Brodsky et al., 2025). This is not an abstract problem; it has direct implications for patient care.

A Cedars-Sinai study highlighted this issue in a different domain, finding that leading LLMs exhibited racial bias in their treatment recommendations for psychiatric patients (Bouguettaya et al., 2025). When presented with identical hypothetical cases, the models often proposed different treatments when the patient's identity was explicitly or implicitly stated as African American.

A separate, equally concerning example was found in a widely used commercial risk-prediction algorithm. The algorithm, which was applied to the care of some 200 million Americans annually, assigned Black patients the same health risk score as White patients, even though the Black patients were demonstrably sicker (The Lancet Digital H, 2019). This occurred because the algorithm was trained to use hospital costs as a proxy for health. Since hospitals spend less on care for sick Black patients, the AI inadvertently learned that lower cost meant better health, thus perpetuating systemic inequities on a massive scale. The solution lies not in abandoning the technology but in addressing the underlying data problem. This requires rigorous validation on diverse cohorts, continuous auditing of models for bias, and the adoption of techniques like federated learning and synthetic data generation to ensure a more equitable representation of data (Brodsky et al., 2025).

Data privacy, security, and accountability

The use of generative models, particularly those trained on vast amounts of patient data, raises concerns about data privacy and security (Chen and Esmaeilzadeh, 2024). Confidentiality must be paramount, and the sharing of sensitive information, such as pathology reports and clinical notes, with third-party servers for processing can be problematic (Rashidi et al., 2025). While proprietary models like GPT-4 offer impressive performance, their external hosting requires robust de-identification and data-handling protocols to mitigate risks (Balasubramanian et al., 2025). Open-source, self-hostable alternatives, such as the Llama model family and other open-source, privacy-preserving models (Niehues et al., 2024; Balasubramanian et al., 2025; Tripathi et al., 2025) provide a potential solution by allowing institutions to maintain full control over their data.

Beyond technological solutions, the safe deployment of AI requires a clear framework of accountability (Novelli et al., 2024). Brodsky and colleagues suggested a "shared accountability" model that involves vendors, labs, regulators, and pathologists/clinicians in ensuring an end-to-end chain of responsibility (Brodsky et al., 2025). If an AI model influences a diagnostic decision, there must be a clear chain of responsibility and established protocols for patient consent and incident review (Brodsky et al., 2025). The integration of AI into the pathology workflow is not just a technological challenge but a legal and ethical one, requiring a new social contract among all stakeholders (Brodsky et al., 2025).

Evolving regulatory pathways: A look at the US FDA guidance

The pace of AI development has historically outstripped the speed of regulatory frameworks and organizational preparedness (Hart et al., 2023; Warraich et al., 2025; Jackson et al., 2025). However, in early 2025, the U.S. Food and Drug Administration (FDA) made a significant move to address this gap with the release of new draft guidance documents for AI-enabled medical devices and the use of AI in regulatory decision-making (Food U Administration D, 2025). These documents provide recommendations for the contents of marketing submissions, emphasizing the need for comprehensive documentation that demonstrates the safety and effectiveness of these devices (Food U Administration D, 2025). A key concept introduced is the Total Product Lifecycle (TPLC) approach, which requires manufacturers to maintain a robust risk management framework throughout the entire life of the device, from design and development to post-market surveillance (Food U Administration D, 2025). This acknowledges that AI models are not static products but are dynamic and can change with new data and use. This new guidance suggests that regulators will treat generative AI features cautiously, requiring evidence of safety and effectiveness before they can be used for autonomous diagnosis (Brodsky et al., 2025; Jackson et al., 2025). For developers, this implies a strategy of incremental approval, where initial applications are validated as decision-support tools before seeking approval for more autonomous functions (Brodsky et al., 2025). This regulatory action is a direct and necessary response to the rapid proliferation of generative AI, compelling developers to adopt a more structured and safer pathway for clinical integration.

Transparency and explainability

The "black box" problem, where an AI model's decision-making process is opaque and unintelligible to humans, has understandably been a significant barrier to clinician trust and adoption. Generative AI outputs, whether a synthesized image or a drafted report, must be clearly labeled as AI-generated, and their provenance, including the input data and model version, must be tracked. To foster trust, developers are incorporating explainable AI (XAI) techniques. These methods can highlight the specific regions of an image that drove an AI's conclusion or provide a list of alternative hypotheses, allowing pathologists to understand and interpret the AI's suggestions. Ultimately, human oversight remains paramount. Pathologists must be able to review and correct AI outputs, and labs must retain final responsibility for all diagnoses (Brodsky et al., 2025). This reinforces the idea that AI is a tool to augment, not replace, human expertise (Goktas and Grzybowski, 2025).

The FUTURE-AI Consortium (Lekadir et al., 2025), established in 2021 with 117 experts from 50 countries, provides a consensus-driven framework to guide the development and regulation of trustworthy AI in healthcare. Through six core principles, i.e., fairness, universality, traceability, usability, robustness, and explainability, and 30 best practices, it addresses technical, clinical, socioethical, and legal dimensions across the AI lifecycle. By aligning with global regulatory priorities, the framework supports both developers in creating regulator-ready tools and policymakers in evaluating safety, equity, and compliance.

Integrated analysis of challenges

The challenges facing generative AI in pathology are not isolated technical issues but are deeply intertwined with existing systemic problems in healthcare (Weiner et al., 2025). The problem of AI bias, for instance, is not a flaw of the technology itself but a reflection of deep-seated societal inequities that exist within the data used for training (Brodsky et al., 2025; Reddy, 2024). This means that a purely technical solution is insufficient; addressing bias requires a multipronged approach that includes data diversity, continuous auditing, and a fundamental shift in how AI is designed and validated (Goktas and Grzybowski, 2025; Raji et al., 2020). The FDA's new guidance is a crucial first step in creating a structured, safe, and equitable pathway for AI integration. By moving toward a Total Product Lifecycle approach, regulators are compelling developers to engage in continuous monitoring, which creates a necessary feedback loop that will ultimately lead to safer and more reliable AI systems (Warraich et al., 2025).

Conclusion and future outlook

This chapter has outlined the potentially transformative impact of generative AI on diagnostic pathology, demonstrating how generative AI tools can partner with pathologists and technical laboratory staff to enhance diagnostic precision, streamline clinical workflows, and ensure quality control in laboratory practice. Beyond the diagnostic processes, these tools have the capability to complement human effort by accelerating research in drug design, discovering treatment targets, predicting molecular features and patient prognosis from WSIs, improving learners' experience through pathology education and training, enhancing clinician-patient interaction, and automating mundane administrative tasks.

From overcoming data scarcity with synthetic histology images to creating a new generation of multimodal foundation models that can unify a patient's entire clinical, genomic, and pathological profile, generative AI is truly reshaping the field at a pace that has never been seen with any technological advancement so far. Indeed, it seems to have come a little late to rescue the strained pathology workload, drowned in an ever-increasing workload, inconceivable (by humans alone) levels of complexity in the field, and the exceptional speed of pathology knowledge doubling time, mainly driven by molecular science and the duo of digitization and AI itself.

The quantitative data on reduced turnaround times and impressive extraction accuracies moves the conversation beyond theoretical potential to proven clinical utility. As the technology matures, it is shifting from a passive tool for image analysis to a proactive partner in diagnosis, prognostic/predictive work, and personalized care. The most significant emerging trend is the transition from a "copilot" to an "agentic" AI paradigm. The development of systems like SlideSeek, which can autonomously perform a multistep diagnostic workup, suggests a future where pathologists will increasingly serve as supervisors and validators of sophisticated AI agents rather than as primary diagnosticians. This shift will free up pathologists' time to focus on complex, multidisciplinary cases and research.

Furthermore, open-source initiatives, federated learning, and the increasing accessibility of powerful foundation models hold the potential to democratize pathology expertise. Even smaller laboratories in resource-limited settings may soon be able to access advanced AI tools, standardizing and improving the quality of care on a global scale. However, realizing this vision requires sustained, rigorous, and interdisciplinary collaboration.

Pathologists must work closely with computer scientists to ensure these models are clinically relevant and validated on diverse, representative data. Regulators must continue to develop flexible yet robust frameworks for oversight, and patients must be included in the conversation to ensure these technologies are deployed equitably and with their full consent.

The journey of generative AI in pathology is iterative, and its ultimate success depends on a shared commitment to addressing the ethical and practical challenges with the same rigor and innovation that have fueled the technology itself. The revolution is not just technological; it is a sociotechnical evolution that demands shared accountability and a new vision for holistic patient care.

AI disclosure

During the preparation of this work, the author(s) used ChatGPT to improve English and, in some paragraphs, to enhance flow and structure. After using this tool/service, the author(s) reviewed and edited the content as needed and take(s) full responsibility for the content of the publication.

References

Abraham, T.M., Levenson, R. A comparison of diffusion models and CycleGANs for virtual staining of slide-free microscopy images. 2023.

Alber, M., Tietz, S., Dippel, J., Milbich, T., Lesort, T., Korfiatis, P., Krügener, M., Cancer, B.P., Shah, N., Möllers, A., Seegerer, P., Carpen-Amarie, A., Standvoss, K., Dernbach, G., de Jong, E., Schallenberg, S., Kunft, A., von Ankershoffen, H.H., Schaeferle, G., Duffy, P., Redlon, M., Jurmeister, P., Horst, D., Ruff, L., Müller, K.R., Klauschen, F., Norgan, A., 2025. Atlas: a novel pathology foundation model by Mayo clinic, Charité, and Aignostics. arXiv. 2331-8422. Available from: https://arxiv.org. doi:10.48550/arXiv.2501.05409.

Alum, E.U., 2025. AI-driven biomarker discovery: enhancing precision in cancer diagnosis and prognosis. Discov. Oncol. 2730-6011. 16 (1). Available from: https://www.springer.com/journal/12672. doi:10.1007/s12672-025-02064-7.

Bai, B., Yang, X., Li, Y., Zhang, Y., Pillar, N., Ozcan, A., 2023. Deep learning-enabled virtual histological staining of biological samples. Light Sci. Appl. 2047-7538. 12 (1). doi:10.1038/s41377-023-01104-7.

Balasubramanian, J.B., Adams, D., Roxanis, I., de Gonzalez, A.B., Coulson, P., Almeida, J.S., García-Closas, M., 2025. Leveraging large language models for structured information extraction from pathology reports. J. Pathol. Inform. 2153-3539. 19. Available from: https://www.sciencedirect.com/science/journal/21533539. doi:10.1016/j.jpi.2025.100521.

Bani-Sadr, A., Cotena, M., Chaibi, Y., Chang, P., Chow, D., Cotton, F., Ayobi, A., 2025. Real-world impact of an AI-driven teleradiology workflow for intracranial hemorrhage detection: reducing diagnostic delays across a multicenter emergency network. Res. Sq. Available from: https://www.researchsquare.com/browse. doi:10.21203/rs.3.rs-6766169/v1.

Ben, A.A., Hasan, S.A., Datla, V.V., Demner-Fushman, D., Müller, H., editors. 2019. Vqa-med: overview of the medical visual question answering task at imageclef 2019. In: Proceedings of CLEF (Conference and Labs of the Evaluation Forum) 2019 Working Notes. 9-12 September 2019.

Benirschke, R.C., Wodskow, J., Prasai, K., Freeman, A., Lee, J.M., Groth, J., 2024. Assessment of a large language model's utility in helping pathology professionals answer general knowledge pathology questions. Am. J. Clin. Pathol. 1943-7722. 161 (1), 42–48. Available from: http://ajcp.oxfordjournals.org/content/by/year. doi:10.1093/ajcp/aqad106.

Bera, K., Schalper, K.A., Rimm, D.L., Velcheti, V., Madabhushi, A., 2019. Artificial intelligence in digital pathology—new tools for diagnosis and precision oncology. Nat. Rev. Clin. Oncol. 1759-4782. 16 (11), 703–715. Available from: http://www.nature.com/nrclinonc/archive/index.html. doi:10.1038/s41571-019-0252-y.

Berkman, N.D., Sheridan, S.L., Donahue, K.E., Halpern, D.J., Crotty, K., 2011. Low health literacy and health outcomes: an updated systematic review. Ann. Intern. Med. 1539-3704. 155 (2), 97–107. Available from: http://www.annals.org/content/155/2/97.full.pdf+html. doi:10.7326/0003-4819-155-2-201107190-00005.

Boehm, K.M., Khosravi, P., Vanguri, R., Gao, J., Shah, S.P., 2022. Harnessing multimodal data integration to advance precision oncology. Nat. Rev. Cancer. 1474-1768. 22 (2), 114–126. Available from: http://www.nature.com/cancer/. doi:10.1038/s41568-021-00408-3.

Bonert, M., Zafar, U., Maung, R., El-Shinnawy, I., Kak, I., Cutz, J.-C., Naqvi, A., Juergens, R.A., Finley, C., Salama, S., Major, P., Kapoor, A., Horton, S., 2021. Evolution of anatomic pathology workload from 2011 to 2019 assessed in a regional hospital laboratory via 574,093 pathology reports. PLoS ONE. 1932-6203. 16 (6), e0253876. doi:10.1371/journal.pone.0253876.

Bouguettaya, A., Stuart, E.M., Aboujaoude, E., 2025. Racial bias in AI-mediated psychiatric diagnosis and treatment: a qualitative comparison of four large language models. Npj Digit. Med. 2398-6352. 8 (1). doi:10.1038/s41746-025-01746-4.

Brodsky, V., Ullah, E., Bychkov, A., Song, A.H., Walk, E.E., Louis, P., Rasool, G., Singh, R.S., Mahmood, F., Bui, M.M., Parwani, A.V., 2025. Generative artificial intelligence in anatomic pathology. Arch. Pathol. Lab. Med. 1543-2165. 149 (4), 298–318. Available from: https://meridian.allenpress.com/aplm/issue/149/4. doi:10.5858/arpa.2024-0215-RA.

Campanella, G., Kumar, N., Nanda, S., Singi, S., Fluder, E., Kwan, R., et al, 2025.Real-world deployment of a fine-tuned pathology foundation model for lung cancer biomarker detection. Nat. Med. 31, 3002–3010.

Carrillo-Perez, F., Pizurica, M., Ozawa, M.G., Vogel, H., West, R.B., Kong, C.S., Herrera, L.J., Shen, J., Gevaert, O., 2023. Synthetic whole-slide image tile generation with gene expression profile-infused deep generative models. Cell Rep. Methods. 2667-2375. 3 (8). Available from: www.cell.com/cell-reports-methods. doi:10.1016/j.crmeth.2023.100534.

Carrillo-Perez, F., Pizurica, M., Zheng, Y., Nandi, T.N., Madduri, R., Shen, J., et al, 2024. Generation of synthetic whole-slide image tiles of tumours from RNA-sequencing data via cascaded diffusion models. Nat. Biomed. Eng. 9 (3), 320–332.

Cecchini, M.J., Borowitz, M.J., Glassy, E.F., Gullapalli, R.R., Hart, S.N., Hassell, L.A., Homer, R.J., Jackups, R., McNeal, J.L., Anderson, S.R., 2025. Harnessing the power of generative artificial intelligence in pathology education opportunities, challenges, and future directions. Arch. Pathol. Lab. Med. 1543-2165. 149 (2), 142–151. Available from: https://meridian.allenpress.com/aplm/article/doi/10.5858/arpa.2024-0187-RA/503226/Harnessing-the-Power-of-Generative-Artificial. doi:10.5858/arpa.2024-0187-RA.

Chen, C., Weishaupt, L.L., Williamson, D.F.K., Chen, R.J., Ding, T., Chen, B., Vaidya, A., Le, L.P., Jaume, G., Lu, M.Y., Mahmood, F., 2025a. Evidence-based diagnostic reasoning with multi-agent copilot for human pathology. arXiv. 2331-8422. Available from: https://arxiv.org. doi:10.48550/arXiv.2506.20964.

Chen, H., Wang, X., Zhou, Y., Huang, B., Zhang, Y., Feng, W., et al. 2024a. Multi-modal generative ai: multi-modal llm, diffusion and beyond. arXiv 2409, 14993.

Chen, R.J., Ding, T., Lu, M.Y., Williamson, D.F.K., Jaume, G., Song, A.H., Chen, B., Zhang, A., Shao, D., Shaban, M., Williams, M., Oldenburg, L., Weishaupt, L.L., Wang, J.J., Vaidya, A., Le, L.P., Gerber, G., Sahai, S., Williams, W., Mahmood, F., 2024b. Towards a general-purpose foundation model for computational pathology. Nat. Med. 1546-170X. 30 (3), 850–862. Available from: https://www.nature.com/nm/. doi:10.1038/s41591-024-02857-3.

Chen, Y., Esmaeilzadeh, P., 2024. Generative AI in medical practice: in-depth exploration of privacy and security challenges. J. Med. Internet Res. 1438-8871. 26 (1), e53008. doi:10.2196/53008.

Chen, Y., Wang, G., Ji, Y., Li, Y., Ye, J., Li, T., et al., editors. 2025b. Slidechat: a large vision-language assistant for whole-slide pathology image understanding. In: Proceedings of the Computer Vision and Pattern Recognition Conference.

Cheng, J., 2024. Applications of Large Language Models in Pathology. Bioengineering. 2306-5354. 11 (4), 342. doi:10.3390/bioengineering11040342.

Chenthamarakshan, V., Hoffman, S.C., Owen, C.D., Lukacik, P., Strain-Damerell, C., Fearon, D., Malla, T.R., Tumber, A., Schofield, C.J., Duyvesteyn, H.M.E., Dejnirattisai, W., Carrique, L., Walter, T.S., Screaton, G.R., Matviiuk, T., Mojsilovic, A., Crain, J., Walsh, M.A., Stuart, D.I., Das, P., 2023. Accelerating drug target inhibitor discovery with a deep generative foundation model. Sci. Adv. 2375-2548. 9 (25). Available from: https://www.science.org/doi/epdf/10.1126/sciadv.adg7865. doi:10.1126/sciadv.adg7865.

Chiu, T., 2023. The impact of generative AI (GenAI) on practices, policies and research direction in education: a case of ChatGPT and Midjourney. Interact. Learn. Environ. 32 (10), 6187–6203.

Chugh, G., Kumar, S., Singh, N., 2021. Survey on machine learning and deep learning applications in breast cancer diagnosis. Cogn. Comput. 1866-9956. 13 (6), 1451–1470. doi:10.1007/s12559-020-09813-6.

Chungyoun, M.F., Gray, J.J., 2023. AI models for protein design are driving antibody engineering. Curr. Opin. Biomed. Eng. 2468-4511. 28. Available from: https://www.journals.elsevier.com/current-opinion-in-biomedical-engineering. doi:10.1016/j.cobme.2023.100473.

Creswell, A., White, T., Dumoulin, V., Arulkumaran, K., Sengupta, B., Bharath, A.A., 2018. Generative adversarial networks: an overview. IEEE Signal Process. Mag. 1053-5888. 35 (1), 53–65. Available from: http://ieeexplore.ieee.org/xpl/RecentIssue.jsp?punumber=79&year=2008. doi:10.1109/MSP.2017.2765202.

Croitoru, F.A., Hondru, V., Ionescu, R.T., Shah, M., 2023. Diffusion models in vision: a survey. IEEE Trans. Pattern Anal. Mach. Intell. 1939-3539. 45 (9), 10850–10869. Available from: https://ieeexplore.ieee.org/servlet/opac?punumber=34. doi:10.1109/TPAMI.2023.3261988.

DaSilva, L.F., Senan, S., Patel, Z.M., Reddy, A.J., Gabbita, S., Nussbaum, Z., Córdova, C.M.V., Wenteler, A., Weber, N., Tunjic, T.M., Khan, T.A., Li, Z., Smith, C., Bejan, M., Louis, L.K., Cornejo, P., Connell, W., Wong, E. S., Meuleman, W., Pinello, L., 2024. DNA-diffusion: leveraging generative models for controlling chromatin accessibility and gene expression via synthetic regulatory elements. BioRxiv. 2692-8205. Available from: https://www.biorxiv.org. doi:10.1101/2024.02.01.578352.

Daungsupawong, H., Wiwanitkit, V., 2024. Large language model's utility in helping pathology professionals. Am. J. Clin. Pathol. 0002-9173. 161 (2), 210. doi:10.1093/ajcp/aqad130.

Dhariwal, P., Nichol, A. 2021 Diffusion models Beat GANs on image synthesis. In: Advances in Neural Information Processing Systems vol. 11. 9781713845393. Neural information processing systems foundation, United States, pp. 8780–8794. https://papers.nips.cc/.

Dhawan, K., Agrawal, U., 2023. Training and testing a CNN-based engine for brain MRI scan classification and segmentation. medRxiv Available from: https://www.medrxiv.org/. doi:10.1101/2023.07.03.23292122.

Dolezal, J.M., Wolk, R., Hieromnimon, H.M., Howard, F.M., Srisuwananukorn, A., Karpeyev, D., Ramesh, S., Kochanny, S., Kwon, J.W., Agni, M., Simon, R.C., Desai, C., Kherallah, R., Nguyen, T.D., Schulte, J.J., Cole, K., Khramtsova, G., Garassino, M.C., Husain, A.N., Li, H., Grossman, R., Cipriani, N.A., Pearson, A. T., 2023. Deep learning generates synthetic cancer histology for explainability and education. Npj Precis. Oncol. 2397-768X. 7 (1). Available from: https://www.nature.com/npjprecisiononcology/. doi:10.1038/s41698-023-00399-4.

Fei, N., Lu, Z., Gao, Y., Yang, G., Huo, Y., Wen, J., Lu, H., Song, R., Gao, X., Xiang, T., Sun, H., Wen, J.-R., 2022. Towards artificial general intelligence via a multimodal foundation model. Nat. Commun. 2041-1723. 13 (1). doi:10.1038/s41467-022-30761-2.

Ferber, D., Nahhas, O.S.M.E., Wölflein, G., Wiest, I.C., Clusmann, J., Leßmann, M.E., Foersch, S., Lammert, J., Tschochohei, M., Jäger, D., Salto-Tellez, M., Schultz, N., Truhn, D., Kather, J.N., 2025. Development and validation of an autonomous artificial intelligence agent for clinical decision-making in oncology. Nat. Cancer. 2662-1347. 6 (8), 1337–1349. Available from: https://www.nature.com/natcancer/. doi:10.1038/s43018-025-00991-6.

Filiot, A., Jacob, P., Mac Kain, A., Saillard, C., 2024. Phikon-v2, a large and public feature extractor for biomarker prediction. arXiv 240909173.

Food U, Administration D. Artificial Intelligence-Enabled Device Software Functions: lifecycle Management and Marketing Submission Recommendations. Draft Guidance January. 2025.

Franklin, G., Stephens, R., Piracha, M., Tiosano, S., Lehouillier, F., Koppel, R., Elkin, P., 2024. The sociodemographic biases in machine learning algorithms: a biomedical informatics perspective. Life. 2075-1729. 14 (6), 652. doi:10.3390/life14060652.

Goktas, P., Grzybowski, A., 2025. Shaping the future of healthcare: ethical clinical challenges and pathways to trustworthy AI. J. Clin. Med. 2077-0383. 14 (5), 1605. doi:10.3390/jcm14051605.

Goodfellow, I., 2016. NIPS 2016 tutorial: generative adversarial networks. arXiv. 2331-8422. Available from: https://arxiv.org. doi:10.48550/arxiv.1701.00160.

Goodfellow, I., Pouget-Abadie, J., Mirza, M., Bing, X., Warde-Farley, D., Ozair, S., Courville, A., Bengio, Y., 2020. Generative adversarial networks. Commun. ACM. 0001-0782. 63 (11), 139–144. doi:10.1145/3422622.

Guo, E., Gupta, M., Deng, J., Park, Y.-J., Paget, M., Naugler, C., 2024. Automated paper screening for clinical reviews using large language models: data analysis study. J. Med. Internet Res. 1438-8871. 26 (1), e48996. doi:10.2196/48996.

Guo, H.B., Perminov, A., Bekele, S., Kedziora, G., Farajollahi, S., Varaljay, V., Hinkle, K., Molinero, V., Meister, K., Hung, C., Dennis, P., Kelley-Loughnane, N., Berry, R., 2022. AlphaFold2 models indicate that protein sequence determines both structure and dynamics. Sci. Rep. 2045-2322. 12 (1). Available from: www.nature.com/srep/index.html. doi:10.1038/s41598-022-14382-9.

Guo, X., Chen, Y., 2024. Generative AI for synthetic data generation: methods, challenges and the future. arXiv. 2331-8422. Available from: https://arxiv.org. doi:10.48550/arXiv.2403.04190.

Gurcan, M.N., Boucheron, L.E., Can, A., Madabhushi, A., Rajpoot, N.M., Yener, B., 2009. Histopathological image analysis: a review. IEEE Rev. Biomed. Eng. 1937-3333. 2, 147–171. doi:10.1109/rbme.2009.2034865.

Hanna, M.G., Reuter, V.E., Samboy, J., England, C., Corsale, L., Fine, S.W., Agaram, N.P., Stamelos, E., Yagi, Y., Hameed, M., Klimstra, D.S., Sirintrapun, S.J., 2019. Implementation of digital pathology offers clinical and operational increase in efficiency and cost savings. Arch. Pathol. Lab. Med. 1543-2165. 143 (12), 1545–1555. Available from: https://www.archivesofpathology.org/doi/pdf/10.5858/arpa.2018-0514-OA. doi:10.5858/arpa.2018-0514-OA.

Harb, R., Pock, T., Müller, H., editors. 2024. Diffusion-based generation of histopathological whole slide images at a Gigapixel scale. In: Proceedings of the IEEE/CVF Winter Conference on Applications of Computer Vision.

Hart, S.N., Hoffman, N.G., Gershkovich, P., Christenson, C., McClintock, D.S., Miller, L.J., Jackups, R., Azimi, V., Spies, N., Brodsky, V., 2023. Organizational preparedness for the use of large language models in pathology informatics. J. Pathol. Inform. 2153-3539. 14. Available from: https://www.sciencedirect.com/journal/journal-of-pathology-informatics/about/aims-and-scope. doi:10.1016/j.jpi.2023.100338.

He, F., Liu, K., Yang, Z., Chen, Y., Hammer, R.D., Xu, D., Popescu, M., 2024. PathCLIP: detection of genes and gene relations from biological pathway figures through image-text contrastive learning. IEEE J. Biomed. Health Inform. 2168-2208. 28 (8), 5007–5019. Available from: http://ieeexplore.ieee.org/xpl/RecentIssue.jsp?punumber=6221020. doi:10.1109/JBHI.2024.3383610.

He, X., editor 2021. Towards visual question answering on pathology images. In: Proceedings of the 59th annual meeting of the association for computational linguistics and the 11th international joint conference on natural language processing.

Hosseini, M., Gao, C.A., Liebovitz, D.M., Carvalho, A.M., Ahmad, F.S., Luo, Y., MacDonald, N., Holmes, K.L., Kho, A., Clark, M.D., 2023. An exploratory survey about using ChatGPT in education, healthcare, and research. PLoS ONE. 1932-6203. 18 (10), e0292216. doi:10.1371/journal.pone.0292216.

Hou, L., Agarwal, A., Samaras, D., Kurc, T.M., Gupta, R.R., Saltz, J.H., editors. 2019. Robust histopathology image analysis: to label or to synthesize?. In: Proceedings of the IEEE/CVF conference on computer vision and pattern recognition.

Hu, B., Tang, Y., Chang, E.I.C., Fan, Y., Lai, M., Xu, Y., 2019. Unsupervised learning for cell-level visual representation in histopathology images with generative adversarial networks. IEEE J. Biomed. Health Inform. 2168-2208. 23 (3), 1316–1328. Available from: http://ieeexplore.ieee.org/xpl/RecentIssue.jsp?punumber=6221020. doi:10.1109/JBHI.2018.2852639.

Huang, K., 2025. AI agents in healthcare. In: Agentic AI: Theories and Practices. Springer. pp. 303–321.

Huang, Z., Bianchi, F., Yuksekgonul, M., Montine, T.J., Zou, J., 2023. A visual–language foundation model for pathology image analysis using medical Twitter. Nat. Med. 1078-8956. 29 (9), 2307–2316. doi:10.1038/s41591-023-02504-3.

Hussein, I.H., Raad, M., 2015. Once upon a microscopic slide: the story of histology. J. Cytol. Histol. 2157-7099. 06 (06). doi:10.4172/2157-7099.1000377.

Jackson, B.R., Rashidi, H.H., Lennerz, J.K., de Baca, M.E., 2025. Ethical and regulatory perspectives on generative artificial intelligence in pathology. Arch. Pathol. Lab. Med. 1543-2165. 149 (2), 123–129. Available from: https://meridian.allenpress.com/aplm/article/149/2/123/503126/Ethical-and-Regulatory-Perspectives-on-Generative. doi:10.5858/arpa.2024-0205-RA.

Janowczyk, A., Zuo, R., Gilmore, H., Feldman, M., Madabhushi, A., 2019. HistoQC: an open-source quality control tool for digital pathology slides. JCO Clin. Cancer Inform. 2473-4276. 3 (3), 1–7. doi:10.1200/cci.18.00157.

Jeblick, K., Schachtner, B., Dexl, J., Mittermeier, A., Stüber, A.T., Topalis, J., et al, 2023.ChatGPT makes medicine easy to swallow: an exploratory case study on simplified radiology reports. Eur. Radiol. 34 (5), 2817–2825.

Jeong, J., Kim, K.D., Nam, Y., Cho, C.E., Go, H., Kim, N., 2023. Stain normalization using score-based diffusion model through stain separation and overlapped moving window patch strategies. Comput. Biol. Med. 0010-4825. 152, 106335. doi:10.1016/j.compbiomed.2022.106335.

Kamath, U., Liu, J., Whitaker, J., 2019. Deep Learning for NLP and Speech Recognition. Springer International Publishing, United States, pp. 1–621. Available from https://link.springer.com/book/10.1007/978-3-030-14596-5. doi:10.1007/978-3-030-14596-5.

Karunanayake, N., 2025. Next-generation agentic AI for transforming healthcare. Inform. Health. 2949-9534. 2 (2), 73–83. doi:10.1016/j.infoh.2025.03.001.

Kataria, T., Knudsen, B., Elhabian, S.Y., 2024. StainDiffuser: multiTask dual diffusion model for virtual staining. arXiv 240311340.

Kitaoka, Y., Uchihashi, T., Kawata, S., Nishiura, A., Yamamoto, T., Hiraoka, S.-I., Yokota, Y., Isomura, E.T., Kogo, M., Tanaka, S., Spigelman, I., Seki, S., 2025. Role and potential of artificial intelligence in biomarker discovery and development of treatment strategies for amyotrophic lateral sclerosis. Int. J. Mol. Sci. 1422-0067. 26 (9), 4346. doi:10.3390/ijms26094346.

Lekadir, K., Frangi, A.F., Porras, A.R., Glocker, B., Cintas, C., Langlotz, C.P., Weicken, E., Asselbergs, F.W., Prior, F., Collins, G.S., Kaissis, G., Tsakou, G., Buvat, I., Kalpathy-Cramer, J., Mongan, J., Schnabel, J.A., Kushibar, K., Riklund, K., Marias, K., Amugongo, L.M., Fromont, L.A., Maier-Hein, L., Cerdá-Alberich, L., Martí-Bonmatí, L., Cardoso, M.J., Bobowicz, M., Shabani, M., Tsiknakis, M., Zuluaga, M.A., Fritzsche, M.C., Camacho, M., Linguraru, M.G., Wenzel, M., De Bruijne, M., Tolsgaard, M.G., Goisauf, M., Abadía, M.C., Papanikolaou, N., Lazrak, N., Pujol, O., Osuala, R., Napel, S., Colantonio, S., Joshi, S., Klein, S., Aussó, S., Rogers, W.A., Salahuddin, Z., Starmans, M.P.A., 2025. FUTURE-AI: international consensus guideline for trustworthy and deployable artificial intelligence in healthcare. BMJ. 1756-1833. Available from: http://www.bmj.com/. doi:10.1136/bmj-2024-081554.

Liang, Y., Lyu, X., Ding, M., Chen, W., Zhang, J., Ren, Y., et al. 2024. WSI-LLaVA: a multimodal large language model for whole slide image. arXiv 241202141. (preprint).

Liu, J., Deng, X., Li, H., Kazemi, A., Grashei, C., Wilkens, G., You, X., Groll, T., Navab, N., Mogler, C., Schüffler, P.J., 2025a. From Pixels to pathology: restoration diffusion for diagnostic-consistent virtual IHC. arXiv. 2331-8422. Available from: https://arxiv.org. doi:10.48550/arXiv.2508.02528.

Liu, X., Shi, C., Deore, U., Wang, Y., Tran, M., Khalil, I., et al., editors. 2021. A scalable AI approach for clinical trial cohort optimization. In: Joint European Conference on Machine Learning and Knowledge Discovery in Databases. Springer.

Liu, X., Wang, R., Lai, Y., Yongfei, W., Cheng, H., Yuanyue, L., Zhang, J., Hao, N., Ban, C., Wang, Y., Tang, S., Yang, Y., Ming, L., Zhou, X., Zheng, W., 2025b. MSMTSeg: multi-stained multi-tissue segmentation of kidney histology images via generative self-supervised meta-learning framework. IEEE J. Biomed. Health Inform. 2168-2194. 29 (6), 3906–3917. doi:10.1109/jbhi.2024.3381047.

Lu, M.Y., Chen, B., Williamson, D.F.K., Chen, R.J., Ikamura, K., Gerber, G., Liang, I., Le, L.P., Ding, T., Parwani, A.V., Mahmood, F., 2023a. A foundational multimodal vision language AI assistant for human pathology. arXiv. 2331-8422. Available from: https://arxiv.org. doi:10.48550/arXiv.2312.07814.

Lu, M.Y., Chen, B., Williamson, D.F.K., Chen, R.J., Liang, I., Ding, T., Jaume, G., Odintsov, I., Le, L.P., Gerber, G., Parwani, A.V., Zhang, A., Mahmood, F., 2024b. A visual-language foundation model for computational pathology. Nat. Med. 1546-170X. 30 (3), 863–874. Available from: https://www.nature.com/nm/. doi:10.1038/s41591-024-02856-4.

Lu, M.Y., Chen, B., Williamson, D.F.K., Chen, R.J., Zhao, M., Chow, A.K., Ikemura, K., Kim, A., Pouli, D., Patel, A., Soliman, A., Chen, C., Ding, T., Wang, J.J., Gerber, G., Liang, I., Le, L.P., Parwani, A.V., Weishaupt, L.L., Mahmood, F., 2024a. A multimodal generative AI copilot for human pathology. Nature. 1476-4687. 634 (8033), 466–473. Available from: https://www.nature.com/nature/. doi:10.1038/s41586-024-07618-3.

Lu, M.Y., Chen, B., Zhang, A., Williamson, D.F., Chen, R.J., Ding, T., et al., editors. 2023b. Visual language pretrained multiple instance zero-shot transfer for histopathology images. In: Proceedings of the IEEE/CVF conference on computer vision and pattern recognition.

Ma, J., Chen, H., 2024. Efficient supervised pretraining of swin-transformer for virtual staining of microscopy images. IEEE Trans. Med. Imaging. 0278-0062. 43 (4), 1388–1399. doi:10.1109/tmi.2023.3337253.

Ma, J., Yu, J., Liu, S., Chen, L., Li, X., Feng, J., Chen, Z., Zeng, S., Liu, X., Cheng, S., 2020. PathSRGAN: multi-supervised super-resolution for cytopathological images using generative adversarial network. IEEE Trans. Med. Imaging. 0278-0062. 39 (9), 2920–2930. doi:10.1109/tmi.2020.2980839.

Madabhushi, A., Lee, G., 2016. Image analysis and machine learning in digital pathology: challenges and opportunities. Med. Image Anal. 1361-8415. 33, 170–175. doi:10.1016/j.media.2016.06.037.

Mahmood, F., Borders, D., Chen, R.J., Mckay, G.N., Salimian, K.J., Baras, A., Durr, N.J., 2020. Deep adversarial training for multi-organ nuclei segmentation in histopathology images. IEEE Trans. Med. Imaging. 1558-254X. 39 (11), 3257–3267. Available from: http://ieeexplore.ieee.org/xpl/RecentIssue.jsp?punumber=42. doi:10.1109/TMI.2019.2927182.

Matias, J. 2024. Health [Internet]: Google. [cited 2024].

Mayall, F.G., Goodhead, M.D., de Mendonça, L., Brownlie, S.E., Anees, A., Perring, S., 2023. Artificial intelligence-based triage of large bowel biopsies can improve workflow. J. Pathol. Inform. 2153-3539. 14, 100181. doi:10.1016/j.jpi.2022.100181.

Melina, G., Panton, A.J., Pizzinelli, C., Rockall, E., Tavares, M.M. Gen-AI: artificial intelligence and the future of work. 2024.

Moghadam, P.A., Van Dalen, S., Martin, K.C., Lennerz, J., Yip, S., Farahani, H., et al., editors. 2023. A morphology focused diffusion probabilistic model for synthesis of histopathology images. In: Proceedings of the IEEE/CVF winter conference on applications of computer vision.

Müller-Franzes, G., Niehues, J.M., Khader, F., Arasteh, S.T., Haarburger, C., Kuhl, C., Wang, T., Han, T., Nolte, T., Nebelung, S., Kather, J.N., Truhn, D., 2023. A multimodal comparison of latent denoising diffusion probabilistic models and generative adversarial networks for medical image synthesis. Sci. Rep. 2045-2322. 13 (1). Available from: https://www.nature.com/srep/. doi:10.1038/s41598-023-39278-0.

Niazi, M.K.K., Parwani, A.V., Gurcan, M.N., 2019. Digital pathology and artificial intelligence. Lancet Oncol. 1474-5488. 20 (5), e253. Available from: http://www.journals.elsevier.com/the-lancet-oncology/. doi:10.1016/S1470-2045(19)30154-8.

Niehues, J.M., Müller-Franzes, G., Schirris, Y., Wagner, S.J., Jendrusch, M., Kloor, M., et al, 2024. Using histopathology latent diffusion models as privacy-preserving dataset augmenters improves downstream classification performance. Comput. Biol. Med. (175), 108410.

Novelli, C., Taddeo, M., Floridi, L., 2024. Accountability in artificial intelligence: what it is and how it works. AI Soc. 0951-5666. 39 (4), 1871–1882. doi:10.1007/s00146-023-01635-y.

NovoPath. Online: NovoPath. 2025. [2025]. Available from: https://www.novopath.com/blog/laboratory-information-systems/ai-in-pathology-labs/.

Oh, H.-J., Jeong, W.-K., editors. 2023. Diffmix: diffusion model-based data synthesis for nuclei segmentation and classification in imbalanced pathology image datasets. International Conference on Medical Image Computing and Computer-Assisted Intervention. Springer.

Oon, M.L., Syn, N.L., Tan, C.L., Tan, K.B., Ng, S.B., 2024. Bridging bytes and biopsies: a comparative analysis of ChatGPT and histopathologists in pathology diagnosis and collaborative potential. Histopathology. 1365-2559. 84 (4), 601–613. Available from: http://onlinelibrary.wiley.com/journal/10.1111/(ISSN)1365-2559. doi:10.1111/his.15100.

Pai, S., Bontempi, D., Hadzic, I., Prudente, V., Sokač, M., Chaunzwa, T.L., Bernatz, S., Hosny, A., Mak, R.H., Birkbak, N.J., Aerts, H.J.W.L., 2024. Foundation model for cancer imaging biomarkers. Nat. Mach. Intell. 2522-5839. 6 (3), 354–367. Available from: https://www.nature.com/natmachintell/. doi:10.1038/s42256-024-00807-9.

Pantanowitz, L., Valenstein, P.N., Evans, A.J., Kaplan, K.J., Pfeifer, J.D., Wilbur, D.C., Collins, L.C., Colgan, T.J., 2011. Review of the current state of whole slide imaging in pathology. J. Pathol. Inform. 2153-3539. 2 (1), 36. doi:10.4103/2153-3539.83746.

Parwani, A.V., 2019. Next generation diagnostic pathology: use of digital pathology and artificial intelligence tools to augment a pathological diagnosis. Diagn. Pathol. 1746-1596. 14 (1). Available from: http://www.diagnosticpathology.org/. doi:10.1186/s13000-019-0921-2.

Parwani, A.V., Amin, M.B., 2020. Convergence of digital pathology and artificial intelligence tools in anatomic pathology practice: current landscape and future directions. Adv. Anat. Pathol. 1533-4031. 27 (4), 221–226. Available from: http://www.anatomicpathology.com. doi:10.1097/PAP.0000000000000271.

Pati, P., Karkampouna, S., Bonollo, F., Compérat, E., Radic, M., Spahn, M., Martinelli, A., Wartenberg, M., Julio, M.K. D., Rapsomaniki, M.A., 2023. Multiplexed tumor profiling with generative AI accelerates histopathology workflows and improves clinical predictions. bioRxiv. 2692-8205. Available from: https://www.biorxiv.org. doi:10.1101/2023.11.29.568996.

Porebski, B.T., Balmforth, M., Browne, G., Riley, A., Jamali, K., Fürst, M.J.L.J., Velic, M., Buchanan, A., Minter, R., Vaughan, T., Holliger, P., 2024. Rapid discovery of high-affinity antibodies via massively parallel sequencing, ribosome display and affinity screening. Nat. Biomed. Eng. 2157-846X. 8 (3), 214–232. Available from: https://www.nature.com/natbiomedeng/. doi:10.1038/s41551-023-01093-3.

Pozzi, M., Noei, S., Robbi, E., Cima, L., Moroni, M., Munari, E., Torresani, E., Jurman, G., 2024. Generating and evaluating synthetic data in digital pathology through diffusion models. Sci. Rep. 2045-2322. 14 (1). doi:10.1038/s41598-024-79602-w.

Radford, A., Kim, J.W., Hallacy, C., Ramesh, A., Goh, G., Agarwal, S., Sastry, G., Askell, A., Mishkin, P., Clark, J., Krueger, G., Sutskever, I., 2021. Learning transferable visual models from natural language supervision. Proc. Mac. Learn. Rese. 2640-3498. vol. 139, pp. 8748–8763. https://proceedings.mlr.press/.

Radford, A., Narasimhan, K., Salimans, T., Sutskever, I. Improving language understanding by generative pre-training. 2018.

Raji, I.D., Smart, A., White, R.N., Mitchell, M., Gebru, T., Hutchinson, B., et al., editors. 2020. Closing the AI accountability gap: defining an end-to-end framework for internal algorithmic auditing. In: Proceedings of the 2020 conference on fairness, accountability, and transparency.

Rasheed, Z., Waseem, M., Ahmad, A., Kemell, K.K., Xiaofeng, W., Duc, A.N., Abrahamsson, P., 2025. Can large language models serve as data analysts? A multi-agent assisted approach for qualitative data analysis. arXiv. 2331-8422. Available from: https://arxiv.org. doi:10.48550/arXiv.2402.01386.

Rashidi, H.H., Pantanowitz, J., Chamanzar, A., Fennell, B., Wang, Y., Gullapalli, R.R., Tafti, A., Deebajah, M., Albahra, S., Glassy, E., Hanna, M.G., Pantanowitz, L., 2025. Generative artificial intelligence in pathology and medicine: a deeper dive. Mod. Pathol. 0893-3952. 38 (4), 100687. doi:10.1016/j.modpat.2024.100687.

Rashidi, H.H., Tran, N.K., Betts, E.V., Howell, L.P., Green, R., 2019. Artificial intelligence and machine learning in pathology: the present landscape of supervised methods. Acad. Pathol. 2374-2895. 6. Available from: https://www.sciencedirect.com/journal/academic-pathology. doi:10.1177/2374289519873088.

Ratcliff, R., Smith, P.L., Brown, S.D., McKoon, G., 2016. Diffusion decision model: current issues and history. Trends Cogn. Sci. 1879-307X. 20 (4), 260–281. Available from: www.elsevier.com/locate/tics. doi:10.1016/j.tics.2016.01.007.

Ravi, A., Neinstein, A., Murray, S.G., 2023. Large language models and medical education: preparing for a rapid transformation in how trainees will learn to be doctors. ATS Scholar. 2690-7097. 4 (3), 282–292. Available from: https://www.atsjournals.org/toc/ats-scholar/4/3. doi:10.34197/ats-scholar.2023-0036PS.

Reddy, S., 2024. Generative AI in healthcare: an implementation science informed translational path on application, integration and governance. Implement. Sci. 1748-5908. 19 (1). doi:10.1186/s13012-024-01357-9.

Reuter, N., von Lipinski, V.-N., Jeutner, J., Schlomm, T., Witzenrath, M., Sander, L.E., et al. 2025. AI-generated patient-friendly discharge summaries to empower patients. medRxiv 2025.07.04.25330804.

Rodrigues, D., Reinhard, S., Waldburger, T., Martin, D., Couto, S., Zlobec, I., et al, 2023. SlideQC: an AI-based tool for automated quality control of whole-slide digital pathology images. Cancer Res. 83 (7_Supplement), 5442.

Saillard, C., Jenatton, R., Llinares-López, F., Mariet, Z., Cahané, D., Durand, E., et al. H-optimus-0. Github; 2024.

Semenova, K., Reid, J., Szeto, W., Kakoti, L., Mannan, R., 2024. Comparative analysis of ChatGPT and Google Gemini in generating anatomic and clinical pathology board review questions. Arch. Pathol. Lab. Med. 148 (9), E305–E6.

Shahid, F., Hsu, M.H., Chang, Y.C., Jian, W.S., 2025. Using generative AI to extract structured information from free text pathology reports. J. Med. Syst. 1573-689X. 49 (1). Available from: https://www.springer.com/journal/10916. doi:10.1007/s10916-025-02167-2.

Smith, S.M., Liauw, D., Dupee, D., Barbieri, A.L., Olson, K., Parkash, V., 2023. Burnout and disengagement in pathology A prepandemic survey of pathologists and laboratory professionals. Arch. Pathol. Lab. Med. 1543-2165. 147 (7), 808–816. https://meridian.allenpress.com/aplm/article/147/7/808/487228/Burnout-and-Disengagement-in-Pathology-A. doi:10.5858/arpa.2022-0073-OA.

Song, A.H., Jaume, G., Williamson, D.F.K., Lu, M.Y., Vaidya, A., Miller, T.R., Mahmood, F., 2023. Artificial intelligence for digital and computational pathology. Nat. Rev. Bioeng. 2731-6092. 1 (12), 930–949. Available from: www.nature.com/natrevbioeng/. doi:10.1038/s44222-023-00096-8.

Song, J., Meng, C., Ermon, S., 2020. Denoising diffusion implicit models. arXiv. 2331-8422. Available from: https://arxiv.org. doi:10.48550/arXiv.2010.02502.

Steyaert, S., Pizurica, M., Nagaraj, D., Khandelwal, P., Hernandez-Boussard, T., Gentles, A.J., Gevaert, O., 2023. Multimodal data fusion for cancer biomarker discovery with deep learning. Nat. Mach. Intell. 2522-5839. 5 (4), 351–362. Available from: https://www.nature.com/natmachintell/. doi:10.1038/s42256-023-00633-5.

Sun, Y., Si, Y., Zhu, C., Gong, X., Zhang, K., Chen, P., et al., editors. 2025. Cpath-omni: a unified multimodal foundation model for patch and whole slide image analysis in computational pathology. In: Proceedings of the Computer Vision and Pattern Recognition Conference.

Tan, J.W., Kim, S., Kim, E., Lee, S.H., Ahn, S., Jeong, W.-K., editors. 2024. Clinical-grade multi-organ pathology report generation for multi-scale whole slide images via a semantically guided medical text foundation model. In: International Conference on Medical Image Computing and Computer-Assisted Intervention. Springer.

Tan, K., Pang, T., Fan, C., Yu, S., 2023. Towards applying powerful large AI models in classroom teaching: opportunities, challenges and prospects. arXiv. 2331-8422. Available from: https://arxiv.org. doi:10.48550/arxiv.2305.03433.

Tasci, E., Zhuge, Y., Camphausen, K., Krauze, A.V., 2022. Bias and class imbalance in oncologic data—towards inclusive and transferrable AI in large scale oncology data sets. Cancers. 2072-6694. 14 (12), 2897. doi:10.3390/cancers14122897.

The Lancet Digital H, 2019. There is no such thing as race in health-care algorithms. Lancet Digit. Health 1 (8), e375.

The Lancet Digital H, 2024. Pathology in the era of generative AI. Lancet Digit. Health 6 (8), e536.

Titford, M., 2009. Progress in the development of microscopical techniques for diagnostic pathology. J. Histotechnol. 0147-8885. 32 (1), 9–19. doi:10.1179/014788809794748024.

Tizhoosh, H.R., Pantanowitz, L., 2018. Artificial intelligence and digital pathology: challenges and opportunities. J. Pathol. Inform. 2153-3539. 9 (1), 38. doi:10.4103/jpi.jpi_53_18.

Tripathi, S., Sukumaran, R., Cook, T.S., 2024. Efficient healthcare with large language models: optimizing clinical workflow and enhancing patient care. J. Am. Med. Inform. Assoc. 1527-974X. 31 (6), 1436–1440. Available from: http://jamia.oxfordjournals.org/content/22/e1. doi:10.1093/jamia/ocad258.

Tripathi, W., Venkatesan, U., Khan, K., et al., Employing consensus-based reasoning with locally deployed LLMs for enabling structured data extraction from surgical pathology reports. medRxiv: the preprint server for health sciences. (2025),

Truhn, D., Eckardt, J.N., Ferber, D., Kather, J.N., 2024a. Large language models and multimodal foundation models for precision oncology. Npj Precis. Oncol. 2397-768X. 8 (1). Available from: https://www.nature.com/npjprecisiononcology/. doi:10.1038/s41698-024-00573-2.

Truhn, D., Loeffler, C.M.L., Müller-Franzes, G., Nebelung, S., Hewitt, K.J., Brandner, S., Bressem, K.K., Foersch, S., Kather, J.N., 2024b. Extracting structured information from unstructured histopathology reports using generative pre-trained transformer 4 (GPT -4). J. Pathol. 0022-3417. 262 (3), 310–319. doi:10.1002/path.6232.

Tschuchnig, M.E., Oostingh, G.J., Gadermayr, M., 2020. Generative adversarial networks in digital pathology: a survey on trends and future potential. Patterns. 2666-3899. 1 (6). Available from: www.cell.com/patterns. doi:10.1016/j.patter.2020.100089.

Ullah, E., Baig, M.M., Waqas, A., Rasool, G., Singh, R., Shandilya, A., GholamHossieni, H., Parwani, A.V., 2025. Multimodal generative AI for anatomic pathology-a review of current applications to envisage the future direction. Adv. Anat. Pathol. 1533-4031. Available from: http://www.anatomicpathology.com. doi:10.1097/PAP.0000000000000498.

Vaidya, A., Chen, R.J., Williamson, D.F.K., Song, A.H., Jaume, G., Yang, Y., Hartvigsen, T., Dyer, E.C., Lu, M.Y., Lipkova, J., Shaban, M., Chen, T.Y., Mahmood, F., 2024. Demographic bias in misdiagnosis by computational pathology models. Nat. Med. 1546-170X. 30 (4), 1174–1190. Available from: https://www.nature.com/nm/. doi:10.1038/s41591-024-02885-z.

Vasiljević, J., Nisar, Z., Feuerhake, F., Wemmert, C., Lampert, T., 2022. CycleGAN for virtual stain transfer: is seeing really believing? Artif. Intell. Med. 0933-3657. 133, 102420. doi:10.1016/j.artmed.2022.102420.

Vaswani, A., Shazeer, N., Parmar, N., Uszkoreit, J., Jones, L., Gomez, A.N., Kaiser, Ł., Polosukhin, I. 2017 Attention is all you need Advances in Neural Information Processing Systems 10495258 5999–6009 Neural information processing systems foundation United States

Vorontsov, E., Bozkurt, A., Casson, A., Shaikovski, G., Zelechowski, M., Severson, K., Zimmermann, E., Hall, J., Tenenholtz, N., Fusi, N., Yang, E., Mathieu, P., van Eck, A., Lee, D., Viret, J., Robert, E., Wang, Y.K., Kunz, J.D., Lee, M.C.H., Bernhard, J.H., Godrich, R.A., Oakley, G., Millar, E., Hanna, M., Wen, H., Retamero, J.A., Moye, W. A., Yousfi, R., Kanan, C., Klimstra, D.S., Rothrock, B., Liu, S., Fuchs, T.J., 2024. A foundation model for clinical-grade computational pathology and rare cancers detection. Nat. Med. 1078-8956. 30 (10), 2924–2935. doi:10.1038/s41591-024-03141-0.

Wachter, R.M., Brynjolfsson, E., 2024. Will generative artificial intelligence deliver on its promise in health care? JAMA. 1538-3598. 331 (1), 65–69. Available from: http://jama.jamanetwork.com/journal.aspx. doi:10.1001/jama.2023.25054.

Wang, Z., Zheng, H., He, P., Chen, W., Zhou, M., 2022. DIFFUSION-GAN: TRAINING GANS WITH DIFFUSION. arXiv. 23318422. Available from: https://arxiv.org. doi:10.48550/arXiv.2206.02262.

Waqas, A., Bui, M.M., Glassy, E.F., Naqa, I.E., Borkowski, P., Borkowski, A.A., Rasool, G., 2023. Revolutionizing digital pathology with the power of generative artificial intelligence and foundation models. Lab. Investig. 1530-0307. 103 (11). Available from: https://www.sciencedirect.com/journal/laboratory-investigation. doi:10.1016/j.labinv.2023.100255.

Warraich, H.J., Tazbaz, T., Califf, R.M., 2025. FDA perspective on the regulation of artificial intelligence in health care and biomedicine. JAMA. 1538-3598. 333 (3), 241–247. Available from: http://jama.jamanetwork.com/journal.aspx. doi:10.1001/jama.2024.21451.

Weiner, E.B., Dankwa-Mullan, I., Nelson, W.A., Hassanpour, S., Kuo, P.-C., 2025. Ethical challenges and evolving strategies in the integration of artificial intelligence into clinical practice. PLOS Digit. Health. 2767-3170. 4 (4), e0000810. doi:10.1371/journal.pdig.0000810.

Wölflein, G., Ferber, D., Meneghetti, A.R., El Nahhas, O.S., Truhn, D., Carrero, Z.I., et al. 2023a. A Good Feature Extractor Is All You Need for Weakly Supervised Learning in Histopathology. arXiv 231111772.

Wölflein, G., Um, I.H., Harrison, D.J., Arandjelović, O., editors. HoechstGAN: virtual Lymphocyte Staining Using Generative Adversarial Networks. Proceedings of the IEEE/CVF Winter Conference on Applications of Computer Vision; 2023b.

Wu, T., He, S., Liu, J., Sun, S., Liu, K., Han, Q.L., Tang, Y., 2023. A Brief Overview of ChatGPT: the History, Status Quo and Potential Future Development. IEEE/CAA J. Autom. Sin. 2329-9274. 10 (5), 1122–1136. Available from: https://www.ieee.org/membership-catalog/productdetail/showProductDetailPage.html?product=PER284-EPC. doi:10.1109/JAS.2023.123618.

Xu, H., Usuyama, N., Bagga, J., Zhang, S., Rao, R., Naumann, T., Wong, C., Gero, Z., González, J., Gu, Y., Xu, Y., Wei, M., Wang, W., Ma, S., Wei, F., Yang, J., Li, C., Gao, J., Rosemon, J., Bower, T., Lee, S., Weerasinghe, R., Wright, B.J., Robicsek, A., Piening, B., Bifulco, C., Wang, S., Poon, H., 2024a. A whole-slide foundation model for digital pathology from real-world data. Nature. 0028-0836. 630 (8015), 181–188. doi:10.1038/s41586-024-07441-w.

Xu, Y., Wang, Y., Zhou, F., Ma, J., Jin, C., Yang, S., et al. 2024b. A multimodal knowledge-enhanced whole-slide pathology foundation model. arXiv 240715362. (preprint).

Xue, Y., Ye, J., Qianying Zhou, L.R.L., Antani, S., Xue, Z., Cornwell, C., Zaino, R., Cheng, K.C., Huang, X., 2021. Selective synthetic augmentation with HistoGAN for improved histopathology image classification. Med. Image Anal. 1361-8415. 67, 101816. doi:10.1016/j.media.2020.101816.

Yan, F., Wu, J., Li, J., Wang, W., Lu, J., Chen, W., Gao, Z., Li, J., Li, H., Yan, H., Ma, J., Chen, M., Lu, Y., Chen, Q., Wang, Y., Ling, X., Wang, X., Wang, Z., Huang, Q., Hua, S., Liu, M., Ma, L., Shen, T., Zhang, X., He, Y., Chen, H., Zhang, S., Wang, Z., 2025. PathOrchestra: a Comprehensive Foundation Model for Computational Pathology with Over 100 Diverse Clinical-Grade Tasks. arXiv. 2331-8422. Available from: https://arxiv.org. doi:10.48550/arXiv.2503.24345.

Yang, L., Zhang, Z., Song, Y., Hong, S., Xu, R., Zhao, Y., et al, 2023. Diffusion models: a comprehensive survey of methods and applications. ACM Comput. Surv. 56 (4), 1–39.

Ye, J., Jin, P., Ni, H., Huang, S.X., Xue, Y., 2025. Histopathological synthetic augmentation with generative models. In: Generative Machine Learning Models in Medical Image Computing. Springer Nature, United States, pp. 183–207. Available from: https://link.springer.com/book/9783031809651. doi:10.1007/9783031809651_10.

Yellapragada, S., Graikos, A., Prasanna, P., Kurc, T., Saltz, J., Samaras, D., editors. Pathldm: text conditioned latent diffusion model for histopathology. In: Proceedings of the IEEE/CVF Winter Conference on Applications of Computer Vision; 2024.

Zanjani, F.G., Zinger, S., Bejnordi, B.E., van der Laak, J.A., de With, P.H., editors. 2018. Stain normalization of histopathology images using generative adversarial networks. In: 2018 IEEE 15th International symposium on biomedical imaging (ISBI 2018). IEEE.

Zhan, F., Yu, Y., Wu, R., Zhang, J., Lu, S., Liu, L., Kortylewski, A., Theobalt, C., Xing, E., 2023. Multimodal image synthesis and editing: the generative AI Era. IEEE Trans. Pattern Anal. Mach. Intell. 0162-8828. 45 (12), 15098–15119. doi:10.1109/tpami.2023.3305243.

Zhang, D.Y., Venkat, A., Khasawneh, H., Sali, R., Zhang, V., Pei, Z., 2024. Implementation of digital pathology and artificial intelligence in routine pathology practice. Lab. Investig. 0023-6837. 104 (9), 102111. doi:10.1016/j.labinv.2024.102111.

Zhang, P., Kamel Boulos, M.N., 2023. Generative AI in medicine and healthcare: promises, opportunities and challenges. Fut. Internet. 1999-5903. 15 (9), 286. doi:10.3390/fi15090286.

Zhao, W., Luo, X., Tong, F., Zheng, X., Li, J., Zhao, G., Zhao, D., 2023. Improving antibody optimization ability of generative adversarial network through large language model. Comput. Struct. Biotechnol. J. 2001-0370. 21, 5839–5850. doi:10.1016/j.csbj.2023.11.041.

Zimmermann, E., Vorontsov, E., Viret, J., Casson, A., Zelechowski, M., Shaikovski, G., et al. 2024. Virchow2: scaling self-supervised mixed magnification models in pathology. arXiv 240800738. (preprint).

Zou, J., Topol, E.J., 2025. The rise of agentic AI teammates in medicine. Lancet. 1474-547X. 405 (10477), 457. Available from: https://www.sciencedirect.com/science/journal/01406736. doi:10.1016/S0140-6736(25)00202-8.

CHAPTER

13

Pixelomics: The pixelome as a foundation of precision pathology

Ehsan Ullah[a,b], *Jingjing Hu*[c], *Hooman Rashidi*[d], *Liron Pantanowitz*[e], *Zaibo Li*[f], *and Anil V. Parwani*[f]

[a]Health New Zealand -Counties District, Auckland, New Zealand
[b]Department of Pathology, The Ohio State University, Columbus, OH, United States
[c]University of California, San Diego, San Diego, CA, United States
[d]University of Pittsburgh, Pittsburgh, PA, United States
[e]Department of Pathology, University of Pittsburgh Medical Center, Pittsburgh, PA, United States
[f]Department of Pathology, The Ohio State University, Columbus, OH, United States

Key highlights of the chapter

- Pixelomics defines an image-centric omics paradigm in which whole-slide images are treated as high-dimensional biological systems, enabling systematic, quantitative interrogation of tissue architecture, cellular ecosystems, and spatial organization.
- Advances in foundation models, particularly transformer-based architectures, shift computational pathology from task-specific predictions toward reusable, scalable representations that generalize across organs, diseases, and clinical contexts.
- Integration of pixelomic representations with molecular and clinical data positions digital pathology as a central, spatially-grounded layer within the emerging multiomics ecosystem, driving precision pathology and precision medicine.

Introduction

Whole slide imaging (WSI) refers to a digital pathology technology that uses a whole slide scanner to convert an entire glass microscope slide into a high-resolution digital image. This allows the whole slide image (digital slide) to be viewed or analyzed on a computer display

Advances in Digital Pathology and Artificial Intelligence in Pathology
https://doi.org/10.1016/B978-0-443-40383-5.00016-9

instead of through a traditional microscope. The multiresolution data used in WSI is efficiently stored in a hierarchical pyramid form of stacked images at progressively lower resolutions, often at gigapixel scale. Each level is divided into small tiles (e.g., 1024 × 1024 pixels) for independent compression and retrieval. To better understand the WSI infrastructure, let's borrow an analogy from modern geospatial mapping platforms such as Google Earth. A histopathology slide is similar to a vast, multilayered digital map; the scanner functions as a satellite, systematically capturing high-resolution images of every 0.25 square micrometer of tissue (that's roughly equivalent to the physical dimensions of tissue represented by a single pixel in a 40x digital slide), while network bandwidth and pyramidal file formats function as transport infrastructure and tiled map layers that make interactive exploration of otherwise unmanageable gigapixel datasets feasible. Yet this analogy extends beyond infrastructure to perception: classical training in diagnostic pathology emphasizes disease-entity recognition and targeted biomarker assessment, much as a mapmaker might focus on borders, land-use classes, or selected topographic features, leaving the majority of fine-grained spatial patterns underutilized. In routine practice, this focus on known entities and predefined features has served pathology well but inherently constrains interpretation to what the eye is taught to see, despite the fact that WSIs contain a far richer pixelome of spatial, morphological, and contextual information than is currently harvested. The pixelome of a whole slide image encompasses information about morphology, spatial organization, and subtle or latent features (e.g., molecular states) embedded within the pixels of these images. As such, every pixel in a digital slide carries diagnostic and prognostic signals that go beyond what can be assessed by a human. When WSI infrastructure is coupled with computational pathology and artificial intelligence (AI), it enables systematic interrogation of this latent information and shifts the discipline from a predominantly diagnostic discipline toward multiscale, data-driven precision pathology that is a fundamental requisite of precision medicine. Thus digital pathology represents more than slide digitization; it extends the pathologist's perceptual capacity from recognizing predefined disease entities to uncovering novel, emergent patterns embedded within tissue architecture.

Since the 1990s, digital pathology has revolutionized anatomical pathology by enabling high-resolution scanning of glass slides and the application of computational methods to derive quantitative data for correlative and prognostic studies (Cui and Zhang, 2021; Niazi et al., 2019). WSIs now form the substrate for deep learning (DL), foundation models (FMs), and multiomics integration, with the long-term goal of providing real-time clinical decision support (CDS), robust quality assurance and control (QA/QC), and integrative precision pathology workflows (Bera et al., 2019; Madabhushi and Lee, 2016; Vorontsov et al., 2024; Xu et al., 2024; Chen et al., 2024).

In parallel, advances in radiomics, genomics, transcriptomics, spatial transcriptomics, epigenomics, proteomics, and metabolomics have led to a multilayered view of disease biology, but a rigorous conceptualization of the image-based layer derived from WSIs remains incomplete. This chapter argues that the total complement of biologically meaningful information encoded in the pixels of WSIs should be referred to as the pixelome, and that the structured, high-throughput study of this space should be termed pixelomics, positioning it as a central pillar of precision pathology and, by extension, precision medicine.

From—ome to—omics: Conceptual foundations

The suffix "-ome" (from Greek -ōma, meaning "body" or "mass") in modern biology denotes the total complement of a given component, such as molecules, organelles, cells, or tissues, within a biological system (Ome suffix, 2025). Correspondingly, "-omics" names the field that systematically studies the entirety of that component, usually through high-throughput, quantitative methods. Core omics disciplines today include genomics, transcriptomics, proteomics, and metabolomics. Other emerging omics include epigenomics, lipidomics, microbiomics, and pixelomics.

Historically, the first widely recognized "-ome" was the **chondriome**, introduced by Friedrich Meves in the early 20th century to describe the total complement of mitochondria in a cell, following Carl Benda's description of mitochondria in 1898 (Logan, 2010). The term **genome**, coined by Hans Winkler in 1920, referred to the complete haploid set of chromosomes and their genes, giving rise to **genomics** as a discipline focused on mapping, sequencing, and functionally interpreting genomes (Winkler, 1920).

The term *genomics* emerged in the late 1980s, coinciding with the Human Genome Project, as a discipline devoted to mapping, sequencing, and analyzing genomes (Watson and Cook-Deegan, 1991). This model soon expanded, generating a family of *-omes* and *-omics* fields encompassing various molecular layers of biological organization. Commonly defined *-omes* and their corresponding *-omics* disciplines are discussed below.

Proteome and proteomics refer to the full complement of proteins, including their structure, abundance, and interactions, and their study, which was probably one of the first among the omics fields to make significant advancements in the 1990s (Wilkins et al., 1996).

Genome and genomics denote the total genetic content of an organism and its large-scale study, respectively (Church, 2006; Lander et al., 2001). Transcriptome and transcriptomics bear the same meaning for the complete set of RNA transcripts expressed under specific conditions and the field of study dedicated to it (Wang et al., 2009). Epigenome and epigenomics are the paired terms used to describe the total complement of heritable modifications that regulate gene activity without altering the DNA sequence and the corresponding field of study (Bird, 2007). Microbiome and microbiomics are the paired terms coined in 2007, which refer to the collective genomes of microbial communities within a given environment and the corresponding field of study (Turnbaugh et al., 2007).

A few additions to this list came in the 2010s, for example, "phenome," which refers to the complete set of observable phenotypic traits of an organism, and "phenomics," which refers to the corresponding field of study (Houle et al., 2010). "Metabolome" and "metabolomics," which denote the entire range of metabolites within a cell or tissue, reflecting metabolic state and their study, were also introduced in the early 2010s (Patti et al., 2012). Another addition in a similar timeframe consists of the "interactome" and "interactomics," the paired terms that refer to the total network of molecular interactions, especially protein–protein or protein–DNA interactions (Vidal et al., 2011).

Together, these layers form the conceptual and methodological scaffold of systems biology, in which multiomics data are integrated to understand health, disease, and environmental adaptation. The emerging notion of omni-omics emphasizes the convergent analysis of genomic, transcriptomic, proteomic, metabolomic, epigenomic, and other—omics data through advanced computational and AI-driven models to achieve a whole-system view of biology and personalized patient care (Hasin et al., 2017; Karczewski and Snyder, 2018; Sherman and Salzberg, 2020). Yet, one key layer, i.e., the total complement of insights residing in WSIs, remains ill-defined and inconsistently named (Nalisnik et al., 2017; Bülow et al., 2023; Gupta et al., 2019).

In practical terms, treating the pixelome as an—ome requires systematic feature extraction pipelines that convert raw pixel intensities into structured descriptors—for instance, nuclear morphometry, texture signatures, deep features, tissue graphs, and spatial patterns—analogous to how genomics converts reads to variants or expression profiles. Modern computational pathology platforms, for example, HistomicsTK (Pourakpour et al., 2025), TIAToolbox (Pocock et al., 2022), QuPath (Bankhead et al., 2017), and interactive AI frameworks like nuclei.io (Huang et al., 2025b) provide standardized ways to extract, visualize, and curate these pixelomic features at scale, making the "omics-style" study of the pixelome already feasible in routine research and, increasingly, in clinical workflows. A summary of the IT platforms with their role in Pixelomics is available in Table 13.1.

Pixelomics relies on a layered information technology ecosystem that spans image access, annotation, cohort management, and scalable analytics to interrogate pixel-level morphology and spatial organization in whole slide images. At the foundation are open, interoperable image readers and formats that enable reliable multiresolution access and reproducible preprocessing. Interactive WSI viewers support expert-driven region definition and ground-truth generation, while server-based platforms provide governance, metadata

TABLE 13.1 IT platform landscape for pixelomics.

Layer	Platform category	Representative examples	Primary role in pixelomics
Image access and formats	Open readers and interoperable formats	OpenSlide, Bio-Formats, OME-TIFF	Multiresolution WSI I/O, tile extraction, format standardization.
Interactive analysis	WSI viewers and annotation tools	QuPath, ASAP, CaseViewer, NDP.view	ROI definition, annotation, measurements, ground-truth creation.
Data management	Image management and collaboration servers	Digital Slide Archive, OMERO	Cohort curation, metadata control, reproducible pipelines.
Computational analytics	AI/ML toolkits for pathology	TIAToolbox, MONAI, CLAM, HoverNet	Tile-level feature learning, MIL, segmentation, representation modeling.
Cloud & standards	DICOM WSI and cloud services	DICOM WSI, Orthanc WSI, Cloud Healthcare APIs	Scalable storage, pixel-efficient retrieval, clinical interoperability
Enterprise deployment	Commercial digital pathology platforms	HALO, Visiopharm, Aiforia, Paige, Proscia, Sectra	Validated analytics, workflow integration, clinical-scale AI deployment.
Enabling technologies	Compression and pyramidal storage	JPEG/JPEG2000 pyramids	Efficient zoom, pan, and compute-aware pixel access

management, and collaborative workflows across large cohorts. On top of this substrate sit computational pathology toolkits that operationalize tile-based feature extraction, weakly supervised learning, and representation modeling to derive latent tissue signals. Increasingly, standards-compliant cloud and DICOM-based infrastructures enable performance-aware pixel retrieval and integration with clinical systems, while commercial digital pathology platforms translate these capabilities into validated, enterprise-scale deployments suitable for routine precision pathology.

Pixelome and pixelomics: Definitions and scope

Despite the increasing use of terms such as **pathome**, **pathomics**, **histome**, and **histomics**, there is no universal agreement on the terminology that captures the complete set of diagnostic, prognostic, predictive, and mechanistic insights derivable from WSIs (Bülow et al., 2023; Gupta et al., 2019; Nalisnik et al., 2017).

Existing terms, as highlighted in Table 13.2, often refer either to subsets of features (e.g., quantitative histomorphometry) or to particular analytic pipelines, rather than to a rigorously defined biological layer analogous to the genome or proteome.

This chapter proposes the following working definitions:

- **Pixelome**: The total complement of biologically and clinically relevant information embedded within the pixels constituting WSIs of tissue sections, including, but not limited to, morphology, spatial organization, and latent patterns that correlate with molecular states, clinical outcomes, and treatment responses.
- **Pixelomics**: The comprehensive, systematic, and typically high-throughput study of the pixelome, encompassing acquisition, harmonization, feature extraction, representation learning, and multimodal integration to enable precision pathology and downstream precision medicine.

In this framework, pixelomics is not merely synonymous with "computational pathology" or "pathomics," but rather represents an omics-oriented conceptualization and quantification of the entire information space encoded by the WSI, with explicit attention to completeness, standardization, and integration with other—omics layers, as depicted in Fig. 13.1.

TABLE 13.2 Key differences between omics terms relevant to pixelomics.

Term	Core idea	Data source	Computational emphasis
Pathome	Entire pathological landscape	Multiscale (molecular + tissue)	Conceptual, integrative
Pathomics	Quantitative image feature mining	WSIs (digital pathology)	Strong (AI/ML-driven)
Histome	Complete set of histological features	Microscopic tissue	Minimal (mostly descriptive)
Histomics	High-throughput histology analysis	WSIs + annotations	Moderate to strong

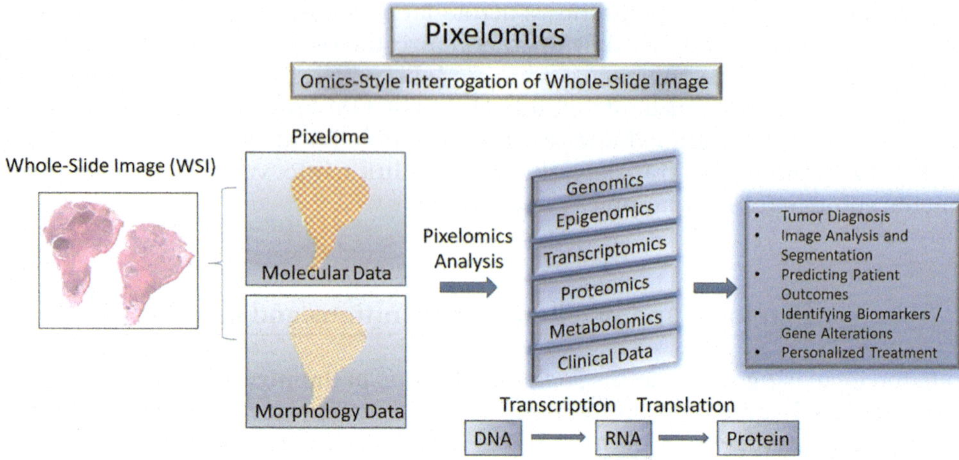

FIG. 13.1 Conceptual overview of pixelomics: omics-style interrogation of whole-slide images.

Pixels as fundamental units of the pixelome

The pixel pathway

In traditional histopathology, pathologists examine glass slides under a microscope, relying on visual interpretation constrained by optical hardware, human perception, and the logistics of slide handling. Digital pathology introduces a fundamental shift: the pathologist now interprets a digital representation of the tissue, mediated by a cascade of procedures, from tissue processing and staining through optical capture, sensor conversion, pixel creation, file assembly, storage, and display.

Professional bodies and regulators tend to refer to the notion of a "pixel pipeline," emphasizing that primary diagnostic validation must consider the entire end-to-end system rather than isolated components. Pixel pathway is a specific term used in digital pathology to describe the entire path that digital image data takes from initial scanning to final viewing and archiving. It is broadly used by a number of organizations all the way from vendors to the FDA, and professional bodies such as the College of American Pathologists, Digital Pathology Association, and American Telemedicine Association (Abels and Pantanowitz, 2017; Herrmann et al., 2018; Newitt, 2018; Pantanowitz et al., 2014, Pantanowitz et al., 2013; Sellaro et al., 2013; Snead et al., 2016; Zhao et al., 2015; Singh et al., 2011).

Pixels are the fundamental building blocks of digital pathology images. A pixel is the smallest unit of visual data in a digital image. Each pixel carries specific information regarding its intensity and color. Billions of these pixels combine to form a complete WSI, accurately replicating the microscopic field at high resolution.

In routine practice, pathologists examine and document a subset of tissue-level, cellular-level, and nuclei-level histologic features in diagnostic reports for every case, in which only key morphologic features are used to render a diagnosis. However, the images can be further analyzed to quantify the entirety of millions of features embedded within each pixel.

Beyond traditional histology, WSI analyze molecular data (biomarker expression), cellular features (cell counting, mitosis, nuclear size/shape analysis, and tumor cell segmentation), tumor size, vessel metrics, spatial relationships, and color variations, all enhanced by AI/deep learning for automated tasks like grading, segmentation, and identifying subtle patterns for a more accurate diagnosis, but also to identify prognostic and predictive "markers" through this process. WSI uncovers disease mechanisms and identifies therapeutic targets by enabling quantitative, AI-powered analysis of large volumes of high-resolution tissue data. Combined with image analysis software, quantitative tissue analysis allows for objective and precise measurement of tissue features, including cell counts, size, cellular patterns, and protein expression by immunohistochemistry studies. Researchers can observe and analyze molecular and subcellular changes within the tissue, which aids in understanding disease progression. Immunohistochemistry (IHC)-based biomarker screening through high-throughput, objective screening facilitates the development of potential biomarker heterogeneity. This would reveal nuanced insights into disease mechanisms, identify potential therapeutic targets, and ultimately enhance patient outcomes through more precise and personalized medical interventions (Ahuja and Zaheer, 2025; Bülow et al., 2023; Diao et al., 2021; Hölscher and Bülow, 2025). Encoding decisions at scan time, such as bit depth, dynamic range, and channel configuration for brightfield versus multiplex fluorescence, directly constrain which pixel-level variations can later be captured as morphologic or marker-intensity features.

Pixel creation and fidelity

Pixel creation in pathology involves complex interactions of physics, optics, and electronics (Ghaznavi et al., 2013; Park et al., 2012). Photons interacting with stained tissue undergo absorption, scattering, and transmission, encoding structural and molecular information into the light signal (Tuchin, 2015). Optical systems project this signal onto digital image sensor technologies such as Charge-Coupled Device (CCD) and Complementary Metal-Oxide -Semiconductor (CMOS), which convert it into electrical charges that are digitized into discrete pixel values across red, green, and blue channels, with fidelity shaped by parameters such as numerical aperture, bit depth, and signal-to-noise ratio (Hanna et al., 2020).

Any anomaly, whether due to staining variability, optical misalignment, sensor noise, or suboptimal compression, can introduce artifacts that either mimic disease or obscure genuine features, thereby compromising both human interpretation and downstream computational models (Kanwal et al., 2022; Shakhawat et al., 2023). Resolution choices (e.g., 10x, 20x, 40x) balance diagnostic detail against file size and storage requirements, while color calibration and monitor standardization are critical for reproducible interpretation (Romanchikova et al., 2022; Shrestha and Hulsken, 2014).

In this sense, pixels are not abstract data points but the most fundamental units of diagnostic truth, whose integrity underpins the reliability of the entire pixelome.

WSIs as gigapixel substrates

A single routine H&E slide scanned at 40x contains approximately ten billion pixels (Kumar et al., 2020) each potentially contributing to morphological patterns, subtle nuclear details, and spatial relationships that encode disease biology. Within this landscape, millions of objects, such as cells, subcellular structures, stromal elements, and histologic primitives, can be distinguished by computer vision methods, capturing a heterogeneous milieu of tumor and microenvironmental features. These computational methods treat these slides as sources of local measurements, e.g., nuclear size, shape, orientation, neighborhood density, and so on, which are computed after tissue and/or nuclear detection and segmentation, forming some of the first pixel-level morphometric descriptions of the pixelome.

In routine reporting, pathologists typically document a subset of tissue-level, cellular-level, and nuclear-level features essential for diagnosis, grading, and staging. Pixelomics, by contrast, aims to systematically quantify the entire spectrum of features embedded in the pixelome, moving from qualitative interpretation of "representative" fields to quantitative analysis of every pixel, object, and spatial configuration across the whole slide.

Digital pathology infrastructure for pixelomics

The core components of digital pathology infrastructure to support pixelomics include image acquisition and standardization (e.g., DICOM for interoperability), data management (e.g., scalable storage and annotation), computational infrastructure (e.g., high-performance computing and pipeline tools), secure workflow and system integration (e.g., image management system), as well as organizational and operational readiness (e.g., regulatory validation and successful deployment).

WSI acquisition and file formats

A whole slide scanner is a specialized imaging device used in digital pathology to convert an entire glass microscope slide into a high-resolution digital image. As noted, modern slide scanners produce WSIs as multiresolution image pyramids that capture entire tissue specimens at gigapixel scale, enabling pan-optic navigation from low-power overviews to high-power detail. However, widespread reliance on vendor-specific proprietary file formats (e.g., .svs, .ndpi, .mrxs, .iSyntax, .scn, .czi) has created data silos, limiting interoperability and multiinstitutional collaboration (Daniel et al., 2011; Dash et al., 2021; Herrmann et al., 2018; Singh et al., 2011).

The Digital Imaging and Communications in Medicine (DICOM) standard, which has been highly successful in radiology, is gradually being adopted for digital pathology, but uptake remains slow, delaying truly standardized data exchange and archiving (Herrmann et al., 2018; Singh et al., 2011; Zhang et al., 2024a). DICOM (extension .dcm) supports tiled multiframe images and metadata (e.g., patient and specimen information). Open-source libraries such as OpenSlide provide vendor-neutral APIs capable of reading many proprietary formats and thus mitigate some interoperability issues for research use, but they do not fully resolve regulatory, clinical deployment, or long-term archiving challenges (Goode et al., 2013).

Data management and annotation workflows

At gigapixel scale, effective pixelomics requires robust infrastructure for storage, versioning, access control, visualization, and collaborative annotation. Platforms such as the Digital Slide Archive (DSA; Gutman et al., 2013) and QuPath (Bankhead et al., 2017) integrate web servers, databases, visualization tools (e.g., HistomicsUI; (Gutman et al., 2017), and analysis toolkits (e.g., HistomicsTK), supporting large-cohort curation, human-in-the-loop (HIL) labeling, and integration with machine learning pipelines.

These infrastructures serve not only as repositories but also as active data ecosystems where pathologists and data scientists iteratively refine annotations, validate models, and deploy tools in research and clinical workflows. HIL systems allow experts to review AI-generated segmentations or predictions, correct errors, and thereby improve model performance while maintaining human oversight.

Harmonization and quality control

Preanalytic variability resulting from variations in tissue fixation, processing, section thickness, staining protocols, and scanner hardware introduces nonbiological variability in WSIs that can severely impair model generalizability and comparability across sites. Moreover, the introduction of artifacts (e.g., tissue folds, air bubbles, knife chatter) creates further inconsistencies that can affect image quality and, consequently, AI training and inference. Currently, there are no standard process guidelines or protocols for laboratories to follow to address these preimaging variables. In certain fields, such as cytopathology, attempts to standardize specimen and slide preparation (e.g., liquid-based cytology) have been successful. As a result, in the field of computational pathology, it has become essential to rely on alternate methods to handle preanalytical variability, including image normalization and preprocessing (e.g., color normalization, artifact detection), utilizing diverse and balanced training data to capture variability, as well as adapting AI workflows (e.g., implementing quality control steps for WSI, keeping a human-in-the-loop for safety).

Stain normalization methods (Anghel et al., 2019) ranging from classical histogram matching to modern deep learning–based generative models such as BottleGAN, aim to harmonize visual appearance across institutions while preserving tissue geometry and biologically relevant contrast (Wagner et al., 2022; Zanjani et al., 2018). This harmonization is particularly critical for federated learning, where models are trained on decentralized data without sharing raw patient images (Wagner et al., 2022). Local style normalization before global model aggregation can mitigate data drift and ensure that the aggregated model remains robust to inter-site variability, thereby addressing a core barrier to pixelomics at scale.

From handcrafted features to deep representation learning

Classical feature engineering

Early computational pathology relied on handcrafted features derived from established computer vision techniques and expert-defined histomorphometric descriptors (Mezei et al.,

2024). Examples include gray-level co-occurrence matrix (GLCM)–based texture statistics, local binary patterns (LBP) for rotation-invariant micro-texture, and measurements of nuclear size, shape, orientation, and spatial arrangement (Hao et al., 2022; Mezei et al., 2024; Niazi et al., 2019). These features provided the foundation for early diagnostic and prognostic models, enabling applications such as tumor grading, risk stratification, and survival prediction, often in tandem with traditional statistical or machine learning methods, such as support vector machines and random forests (Komura and Ishikawa, 2019; Madabhushi and Lee, 2016; Kumar et al., 2020). Before deep learning, early computational pathology relied on hand-crafted morphological and textural features focused on nuclei, cells, and tissue architecture: algorithms such as contour-tracing or hierarchical multilevel thresholding were used to segment nuclei, followed by measurements of nuclear color, size, roundness, irregularity, orientation, and local density. Texture descriptors such as Haralick gray-level co-occurrence matrix (GLCM) features (energy, contrast, homogeneity) and local binary patterns (LBP) captured subtle chromatin and tissue texture differences that distinguish, for example, breast from colorectal carcinoma or benign from malignant regions. These morpho-textural features were aggregated into patient-level signatures for tasks such as tumor grading, disease stratification, and survival prediction, including landmark work where hundreds of hand-crafted features from breast cancer H&E slides yielded the C-Path score, a stromal-driven risk index independently associated with patient survival.

While deep learning currently predominates, handcrafted features continue to play an important role in quality assurance, hypothesis formulation, and model auditing, as they offer interpretable and human-meaningful signals that are consistent with pathology expertise (Cui and Zhang, 2021; Diao et al., 2021; Huang et al., 2022).

Automated feature learning with deep learning

Deep learning has transformed whole-slide images (WSIs) from static digital replicas into rich sources of learned representations by enabling automated feature extraction directly from raw pixels (Janowczyk and Madabhushi, 2016; Madabhushi and Lee, 2016). Convolutional neural networks (CNNs) replace manual feature engineering by learning hierarchical filters, with early layers capturing edges and micro-textures, and deeper layers encoding nuclei, glands, and higher-order tissue architecture (Masci et al., 2011).

Given the gigapixel scale of WSIs, patch-based pipelines have become the norm: slides are decomposed into smaller tiles (e.g., 256 × 256 pixels), processed individually, and their features are subsequently aggregated at the slide or patient level (Hou et al., 2016). Because WSIs contain diagnostically relevant information across multiple spatial scales, ranging from cellular detail to global growth patterns, multiscale architectures are commonly employed (Xiang et al., 2023). These architectures may jointly encode high-magnification patches (e.g., 20x) for nuclear detail and lower-magnification patches (e.g., 5x) for broader tissue context, with feature fusion improving segmentation and classification performance in the multiple instance learning (MIL) and transformer models discussed in the sections below.

To further mitigate the annotation burden inherent to histopathology, weakly supervised learning paradigms train models using slide-level labels (such as the final diagnosis) rather than pixel-level annotations, substantially reducing labeling effort while enabling the effective use of large, partially annotated archival datasets (Xiang et al., 2023).

Multiple instance learning and transformer architectures

Multiple instance learning (MIL) conceptualizes each WSI as a "bag" of unlabeled patches (instances), with bag-level labels indicating the presence or absence of disease or other outcomes (Xiang et al., 2023). Attention-based MIL (ABMIL) introduced trainable attention mechanisms that assign weights to instances, allowing the model to highlight diagnostically salient regions via attention heatmaps that are interpretable by pathologists (Afonso et al., 2024).

Limitations of naïve attention mechanisms have motivated advanced frameworks such as CLAM (clustering-constrained attention MIL; Alamudun et al., 2022) and AttriMIL (attribute-driven MIL; Cai et al., 2025), which incorporate instance clustering and more robust scoring strategies to capture heterogeneous tumor phenotypes and mitigate attention misinterpretation. Parallel to these advances, vision transformers (ViTs), originally developed for natural language–inspired sequence modeling, have been adapted to WSIs (e.g., TransMIL), treating patches as tokens and modeling long-range dependencies that are crucial for capturing global tissue context (Chaurasia et al., 2025).

Vision Transformers (ViTs) further extend representational power by modeling long-range dependencies through multihead self-attention mechanisms (Chen et al., 2022) that operate across a sequence of patch tokens, unlike convolutional neural networks that rely on local receptive fields. ViTs divide an image or whole-slide image into a grid of fixed-size patches, project these into embedding vectors, and enrich them with positional encodings so spatial relationships are retained (Zhang et al., 2024b). The self-attention layers compute weighted interactions among all pairs of patch tokens, enabling each patch to incorporate information from distant regions of the slide and thereby capture global contextual patterns beyond local texture (Chen et al., 2022; Zhang et al., 2024b). This capacity to learn dependencies across entire slide representations is particularly valuable in computational pathology, where diagnostically salient structures such as immune cell clustering, stromal-tumor interfaces, or invasive fronts may span large spatial extents and require integrated context for accurate characterization.

Foundation models and generative AI in pixelomics

Pathology-specific foundation models

The transition from task-specific networks to large-scale, self-supervised foundation models has ushered in the next major transformation in computational pathology, enabling a single pretrained encoder to support a wide range of downstream diagnostic, prognostic, and biomarker discovery tasks with minimal task-specific fine-tuning (Lu et al., 2024; Chen et al., 2024). These pathology-specific foundation models are typically trained on tens of millions to billions of tiles derived from hundreds of thousands to millions of WSIs, using contrastive, masked modeling, or hybrid self-supervised objectives to learn generalizable, stain- and site-robust representations (Campanella et al., 2025; Ding et al., 2025; Lu et al., 2024; Wang et al., 2024; Chen et al., 2024a; Wang et al., 2022).

Early large-scale pathology models such as Virchow(Vorontsov et al., 2024), UNI (Chen et al., 2024), and GigaPath (Xu et al., 2024) illustrated that pan-cancer, multiinstitutional pretraining can match or surpass highly engineered tissue-specific models, especially for rare cancers and out-of-distribution data. Virchow, for example, was trained on 1.5 million WSIs and achieved specimen-level AUCs around 0.95 across common and rare cancers, while GigaPath used over 1.3 billion tiles from 171,000 WSIs and a LongNet-based ViT architecture to handle whole-slide context at scale.

A rapidly expanding ecosystem of pathology-specific foundation models has since emerged since 2022, but mainly over 2024–2025, as outlined below:

- **CTransPath**: CTransPath (Wang et al., 2022) is a hybrid convolutional–Transformer self-supervised model that combines a CNN backbone with a multiscale Swin Transformer to capture both local fine structure and global context in histopathology images. Pretrained on a large collection of unlabeled histopathology tiles using contrastive learning, CTransPath provides a stable and transferable feature extractor that has achieved state-of-the-art performance across five downstream tasks spanning nine public datasets, outperforming ImageNet-pretrained baselines in robustness and transferability.
- **UNI and UNI-2 (**Chen et al., 2024a**)**: UNI is a general-purpose self-supervised vision encoder trained on over 100 million image tiles from more than 100,000 WSIs, avoiding public datasets such as TCGA to prevent benchmark leakage and enabling resolution-agnostic classification across over 100 cancer types. Follow-on versions (often referred to as UNI-2) extend pretraining to larger and more diverse cohorts and incorporate architectural and training refinements, improving both biomarker prediction and throughput in independent benchmarks of histopathology foundation models.
- **Virchow-2 and Virchow-2G** (Vorontsov et al., 2024; Zimmermann et al., 2024): Building on the original Virchow model, Virchow-2 and Virchow-2G were trained on data from over 3.1 million WSIs (approximately 2.4 PB) covering more than 40 tissue types from 225,000 patients in 45 countries, representing the largest known digital pathology dataset to date. Virchow-2 maintains a parameter count of 632 million, whereas Virchow-2G scales to 1.85 billion parameters; together, they deliver state-of-the-art performance on pan-cancer detection and biomarker prediction tasks, with substantial gains in robustness across institutions and scanning protocols.
- **Phikon and Phikon-2**: Phikon (Filiot et al., 2024) was introduced as a ViT-based self-supervised model trained on approximately 43 million tiles from 6093 TCGA slides across 13 anatomic sites and was publicly released as a general histopathology backbone. Phikon-2 (Phikon-v2) scales this approach using a DINOv2-style objective on about 460 million tiles from over 50,000 slides across 30 cancer types, achieving significantly improved performance compared with Phikon-v1 and competitive results relative to larger models such as Virchow-2 and H-optimus-0 in multitask clinical benchmarks.
- **Hibou-B/Hibou-L:** The Hibou family of foundation models (Nechaev et al., 2024) comprises pathology-specific Vision Transformers pretrained using the DINOv2

self-supervised framework on a proprietary corpus of over one million whole-slide images spanning diverse tissue types and staining protocols. The two variants, i.e., Hibou-B (ViT-B/14) and Hibou-L (ViT-L/14), demonstrate strong and consistent performance across patch-level, slide-level, and segmentation benchmarks, with Hibou-L achieving state-of-the-art average accuracy on multiple public datasets and Hibou-B offering a favorable accuracy-efficiency trade-off. Notably, the models generalize well to unseen public cohorts despite being trained exclusively on private data, underscoring their robustness as transferable pathology foundation models.

- **H-optimus-0**: H-optimus-0 (Shvetsov et al., 2025) is a ViT-giant model trained on hundreds of millions of tiles derived from over 500,000 H&E WSIs, covering a broad range of tissue types. In recent large clinical benchmarks, H-optimus-0 consistently ranks among the best-performing models for both detection and biomarker prediction tasks, with particularly strong results for lung cancer biomarkers, and demonstrates favorable trade-offs between accuracy and throughput compared with other state-of-the-art foundation models.
- **Atlas**: Atlas (Alber et al., 2025) is a RudolfV-based vision foundation model trained on approximately 1.2 million WSIs from Mayo Clinic and Charité, representing one of the largest institution-spanning pathology datasets curated for foundation modeling. Despite not being the largest model in terms of parameters or dataset size, Atlas achieves state-of-the-art performance across 21 public benchmark datasets, underscoring that careful architectural design and data curation can rival brute-force scaling in pathology foundation models.
- **Lunit/CRISP and related clinical-grade models**: Commercial systems, such as those developed by Lunit and other industry groups, are increasingly built on foundation-model-like backbones trained on large, heterogeneous WSI cohorts to support CE-/FDA-grade applications, including breast cancer detection, MSI prediction, and PD-L1 scoring. In this context, models such as CRISP and other proprietary encoders demonstrate how foundation-model principles are being embedded into clinical product pipelines, although their training data and architectures are often less transparent than open academic models (Maleki et al., 2025).

Beyond purely vision-based encoders, multimodal foundation models explicitly link pathology images with text, reports, or other omics:

- **CONCH (CONtrastive learning from Captions for Histopathology)**: CONCH is a vision–language foundation model trained on approximately 1.17 million image–text pairs, one of the largest pathology-specific caption datasets assembled to date (Lu et al., 2024). Using contrastive learning between WSIs or patches and their free-text descriptions, CONCH achieves state-of-the-art performance across 14 tasks, ranging from zero-shot tissue classification and segmentation to text-to-image and image-to-text retrieval, and allows interactive querying of images with natural language prompts.
- **PLIP (Pathology Language–Image Pretraining)**: PLIP is a pathology language–image model trained on OpenPath, a large corpus of de-identified pathology images and associated text harvested from open social media and educational sources (Huang et al., 2023). When evaluated on four external datasets, PLIP substantially outperformed generic CLIP-like models for zero-shot pathology image classification, achieving F1 scores of

0.565–0.832 compared to 0.030–0.481 for prior contrastive language–image pretraining approaches.

- **TITAN, PathChat, and related multimodal assistants**: A new generation of models, including systems such as TITAN (Ding et al., 2025) and PathChat (Chen et al., 2024), extends pathology foundation models into interactive assistants capable of visual question answering, report drafting, and teaching use-cases by combining WSI encoders with large language models. These models illustrate how pixelomic representations can be embedded within conversational agents that ingest both images and text, enabling pathologists to query complex slide-level patterns, request region-of-interest explanations, and integrate pathology findings with clinical context in a unified interface.

To recap the section, collectively, these pathology-specific foundation models, ranging from uni-modal encoders such as Virchow-2, H-optimus-0, Phikon-2, CTransPath, and Atlas to multimodal systems like UNI-2, CONCH, PLIP, TITAN, and PathChat, demonstrate that large-scale self-supervised pretraining on pixel data produces broadly reusable representations across organs, cancer types, and tasks.

Despite their promise, foundation models for digital pathology introduce important limitations that warrant critical consideration. Many are trained on large, heterogeneous datasets that may embed latent biases related to tissue preparation, scanner optics, staining protocols, and population demographics, potentially limiting generalizability across institutions. Overparameterization, while beneficial for representation learning, increases computational cost, energy consumption, and inference latency, which may constrain deployment in routine clinical workflows. Furthermore, performance gains reported on curated benchmarks may not translate to real-world settings characterized by domain shift, rare subtypes, and imperfect labels. A balanced evaluation of foundation models must therefore consider not only peak benchmark performance but also robustness, efficiency, and operational feasibility.

To improve reproducibility and facilitate benchmarking, we recommend containerized model releases (e.g., Docker or Apptainer), explicit versioning of pretrained weights, and workflow orchestration using tools such as Nextflow or Snakemake. Public benchmark datasets (TCGA, TCIA, PANDA) should be paired with fixed train–validation–test splits and transparent preprocessing pipelines, including stain normalization and patch sampling strategies. Such practices are essential for translating foundation models from proof-of-concept to dependable research infrastructure.

Generative AI and multimodal foundation models

Generative AI extends foundation model capabilities to image synthesis, virtual staining, report generation, and cross-modal reasoning. Large multimodal models such as GPT-class language models, CLIP, and similar vision–language encoders (Jia et al., 2021), as well as visual instruction-following models (e.g., LLaVA; (Liu et al., 2023) can be adapted for pathology tasks including captioning, question answering, and standardized report generation without the need to train separate models for each task. In pathology, generative models are being used to generate realistic synthetic WSIs or regions to augment training datasets while preserving privacy; to perform virtual immunohistochemistry or multiplex immunofluorescence (e.g., ROSIE; (Wu et al., 2025) which can generate synthetic multiplex immunofluorescence (e.g., CODEX-like) images from routine histology; and to support interactive diagnostic reasoning, enabling pathologists to query both pixel patterns and textual clinical data within a unified interface. Such models reduce the barrier to deploying advanced pixelomics in clinical settings by leveraging pretrained generalist capabilities and focusing domain-specific fine-tuning on critical tasks. To read further about advanced generative and nongenerative AI methods in pathology, readers are encouraged to consult a few recent review articles referenced here (Brodsky et al., 2025; Pantanowitz et al., 2025; Rashidi et al., 2025a, 2025b; Ullah et al., 2025a).

Before moving further into this chapter, we would like the reader to pause here and focus on a few dots that are perfectly lining up by the end of the first quarter of the 21st century, as outlined in this paragraph. Ongoing advancements in ViTs technology, particularly in architectures, self-supervised and foundation model pretraining, multimodal integration, and scalable computation, are rapidly expanding the analytical capabilities underpinning whole-slide image (WSI) interpretation, placing computational pathology on the cusp of emerging as an omics-style discipline in its own right. Foundation models enable the encoding of complex spatial hierarchies and long-range dependencies across gigapixel WSIs, a key requirement for capturing biologically meaningful features such as tissue architecture and microenvironmental interactions, and recent work has demonstrated their effectiveness in slide-level classification, retrieval, and multimodal representation learning across hundreds of thousands of WSIs with minimal task-specific labels (Ding et al., 2025). The surge in publicly available and institutionally aggregated digital pathology datasets, coupled with high-capacity compute infrastructure (e.g., GPUs/TPUs, distributed frameworks, and cloud platforms), has enabled the training of large ViT-based foundation models such as TITAN and GigaPath that generalize across organs and disease contexts (Ding et al., 2025). Moreover, advances in self-supervised learning, efficient token compression, and hierarchical transformer designs address the unique challenges of WSI scale and heterogeneity, while integration with auxiliary data modalities (e.g., clinical text, molecular assays) further enriches the representational landscape (Chen et al., 2022; Li et al., 2023).

Taken together with improvements in scanner technology, data standards, and federated data ecosystems, these developments fuel a convergence of high-dimensional image phenotyping with systemic biological insight, justifying the need to "relabel" computational pathology as "pixelomics," a comprehensive, image-centric omics paradigm introduced earlier in this chapter. Within pixelomics, pathology-specific foundation models function as general-purpose encoders of the pixelome, pretrained on millions of tiles or whole slides in a self-supervised fashion and then adapted to downstream tasks such as cancer subtyping, biomarker prediction, and zero-shot recognition. The next sections will further emphasize this proposal.

Pathomics, histomorphometry, and precursors to pixelomics

The conceptual road to pixelomics has been paved by earlier work in pathomics and quantitative histomorphometry (Hölscher and Bülow, 2025). Early uses of "pathomics" appeared in the context of tissue microarrays and biosignature-based diagnostics for emerging infections in the early 2000s (Perkel, 2002; Turteltaub et al., 2006. In 2017, Nalisnik et al. (Nalisnik et al., 2017) used the suffix "-omics" to describe HistomicsML, a system capable of interactive phenotyping and feature mining across hundreds of millions of histologic objects in large-scale images. (Gupta et al., 2019) provided a clear roadmap for pathomics as the systematic extraction and analysis of quantitative features from WSIs using machine and deep learning. (Gupta et al., 2019; Diao et al., 2021) introduced human-interpretable image features (HIFs) derived from densely mapped cancer pathology slides to predict diverse molecular phenotypes, operationalizing an omics-like approach without explicitly naming it as such (Diao et al., 2021).

More recently, Bülow and colleagues (Bülow et al., 2023) coined "pathome" for the entirety of morphological histology features and "Next-Generation histoMorphometry (NGM)" for the associated technology, highlighting the conceptual parallels between pathomics and radiomics. They emphasize that radiomics focuses on macroscopic imaging modalities (e.g., CT, MRI), whereas pathomics pertains to microscopic histology, and they outline how pathomics fits within the broader omics ecosystem, arguing for standardized feature definitions and integration with other—omics layers.

These efforts collectively signal a deepening interest in omics-style approaches to computational pathology, yet they have not fully formalized the pixelome itself as an "ome" in the systems biology sense, nor have they systematically addressed completeness and integration at the scale envisioned for pixelomics.

Multiomics integration: From pixels to systems biology

Pixelomics and big data

Pixelomics (and its predecessors, such as pathomics and related image-omics approaches) applies advanced computational methods to digitized histopathology slides to quantify tissue architecture, cellular composition, and spatial organization at scale. Feature extraction pipelines can generate large data matrices containing thousands of quantitative descriptors for hundreds of thousands to millions of segmented structures, paralleling datasets from single-cell transcriptomics in structure and dimensionality.

These high-dimensional pixelomics datasets enable downstream analyses such as dimensionality reduction, clustering, trajectory inference, and multivariate modeling, facilitating detailed characterization of tumor heterogeneity, immune infiltration, and stromal remodeling. Standard machine learning methods such as support vector machines, random forests, gradient-boosted trees, and regularized regression, as well as unsupervised clustering algorithms (discussed elsewhere in this book), are already widely used to derive predictive models and discover novel phenotypic associations.

Spatial transcriptomics and visual–omics alignment

Spatial transcriptomics is an advanced molecular profiling approach that quantifies gene expression within intact tissue sections while preserving spatial context. In contrast to bulk or single-cell RNA sequencing methods that dissociate tissue and obscure positional information, spatial transcriptomics localizes mRNA transcripts within their native histological architecture. This spatially resolved molecular mapping enables systematic investigation of cellular heterogeneity, tissue microenvironments, and context-dependent molecular interactions directly in situ. Spatial transcriptomics (ST) technologies, such as STARmap (Wang et al., 2018) and barcoded NGS methods (Liu et al., 2020), quantify gene expression while preserving spatial coordinates, bridging morphology with molecular states at near-single-cell or single-cell resolution. Co-registration of H&E WSIs with ST data enables the identification of imaging biomarkers grounded in specific genes and pathways, transforming WSIs into mechanistically interpretable maps of tissue biology.

Graph-based approaches, such as SpaGCN, integrate gene expression profiles, spatial locations, and histologic context through graph convolutional networks to delineate tissue domains and identify spatially variable genes (Zhang et al., 2025). Visual–omics foundation models, such as OmiCLIP (Chen et al., 2025) and related methods, have already started to align latent representations of image patches and their corresponding transcriptomic profiles using contrastive learning, enabling queries such as "locate regions with high expression of immune checkpoint genes" directly from morphology.

Together, these approaches lay the groundwork for precision pathology, i.e., clinically actionable decision support for precision medicine, thereby enabling pathology workflows that connect routine morphology to molecularly informed diagnosis, prognostication, and therapy selection directly from whole-slide images.

Toward digital twins and multimodal platforms

Integration of spatial proteomics methods, such as t-CyCIF (Lin et al., 2018) and MELC (Pascual-Reguant et al., 2021) with histology and spatial transcriptomics moves the field toward the concept of a tissue digital twin—a unified coordinate space in which cellular morphology, molecular states, and microenvironmental interactions are jointly modeled. Platforms such as Loki, built on the Omni-CLIP backbone, have been constructed to support alignment, annotation, cell type decomposition, and gene expression prediction from histology, enabling holistic in situ modeling of disease (Chen et al., 2025). In parallel, generative models now enable synthetic multiplex imaging from H&E, such as ROSIE (Wu et al., 2025), suggesting that future pixelomics workflows will routinely infer or generate multilayered molecular views from standard histology, dramatically expanding the information content of routine slides without additional staining.

Pixelomics, radiomics, and genomics: Multiscale fusion

Moving forward, Pixelomics could be integrated with radiomics and genomics to create multiscale, multiomics models that link microscopic tissue architecture, macroscopic imaging, and molecular signatures. Radiomics quantifies macro features from CT, MRI, and other radiological modalities, while genomics and transcriptomics characterize molecular alterations and expression programs; together with Pixelomics, these layers provide complementary

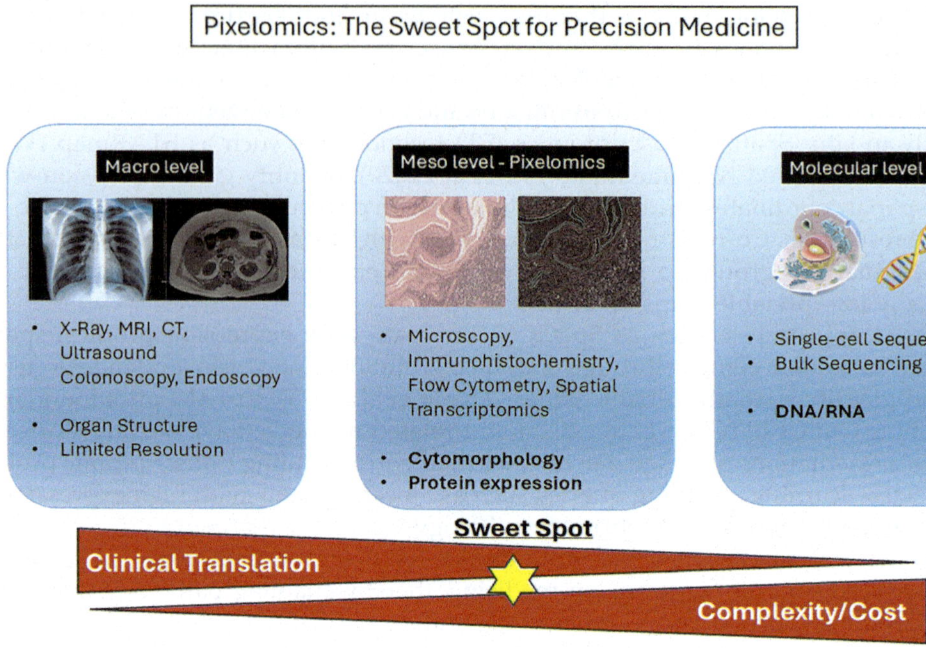

FIG. 13.2 Pixelomics represents the sweet spot between clinical translation and complexity/cost for precision medicine.

histological, anatomical, and molecular perspectives. It is also important to remember that Pixelomics is situated as a uniquely balanced sweet spot between clinical translation and complexity/cost when viewed in the context of multiomics fusion, thus poised to contribute more, but not less, than radiomics and genomics, as depicted in Fig. 13.2.

Correlating pixelomic features with genomic data uncovers mechanistic links between tissue morphology and gene expression or mutational status, using statistical methods such as Pearson correlation or sparse canonical correlation analysis. Integrated models combining histologic and genomic features via feature concatenation, attention-based fusion, or deep multimodal architectures have achieved superior prognostic performance in cancers such as glioma (Alleman et al., 2023), renal carcinoma (Zang et al., 2025), breast cancer (Akbari et al., 2026; Frascarelli et al., 2025), and prostate carcinoma (Varambally et al., 2005). Similarly, combining pixelomics with radiomics provides a pathway for radio-pixelomic (or radio-pathomic) classifiers that connect radiologic imaging signatures with their histologic underpinnings. Studies have shown that integrated models can better capture tumor heterogeneity and improve prediction of recurrence and survival compared with unimodal approaches (Captier et al., 2025; Wang et al., 2025; Yang et al., 2025; Huang et al., 2025a), although data scarcity and challenges in aligning multimodal datasets remain obstacles for all of us to collectively work on.

Challenges, standardization, and future directions

Methodological and infrastructural challenges

Despite substantial progress, current omics-style approaches to the pixelome have not yet quantified its total complement of insights, due in part to the lack of standardized protocols for image acquisition, preprocessing, feature definition, and analysis. Variability in these stages hampers cross-study comparability and undermines reproducibility, highlighting the need for widely accepted guidelines analogous to those developed for radiomics.

Interpretability remains another major barrier. When deep learning and foundation models function as "black boxes," providing limited transparency into how predictions are made, it complicates clinical deployment and thwarts regulatory approval. While explainability methods such as Grad-CAM, attention maps, and model-agnostic feature importance provide useful insights into model behavior, they often remain abstract and difficult to interpret in routine diagnostic practice (Ullah et al., 2025b). More intuitive explanatory frameworks that align with pathologists' cognitive workflows and established reasoning patterns are therefore needed to support trust, validation, and clinical adoption.

Data governance, sharing, and benchmarking

Data privacy regulations, such as the Health Insurance Portability and Accountability Act (HIPAA) in the United States and the General Data Protection Regulation (GDPR) in the European Union, together with security requirements including data encryption and access control, and legal constraints such as Institutional Review Board (IRB) approval, informed consent, and data ownership provisions, substantially impede large-scale sharing of WSIs and linked clinical data. These constraints limit the development of diverse, representative, and multiinstitutional datasets required for robust pixelomics research. Federated learning enables collaborative model training across institutions without centralizing patient data, while differential privacy and secure aggregation techniques mitigate the risk of information leakage. These approaches are particularly relevant for foundation models, whose scale and training regimes raise new governance challenges. Initiatives such as BigPicture (van der Laak et al., 2022) aim to create shared repositories for pathology WSIs, but broader strategies, including federated learning (Guan et al., 2024), de-identification standards, and privacy-preserving analytics, will be required.

Benchmark datasets such as The Cancer Genome Atlas (TCGA), which integrates genomics with histopathology, and The Cancer Imaging Archive (TCIA), which provides large-scale radiological imaging collections, have been foundational in advancing related computational domains. Their success underscores the need for analogous community-driven resources in pixelomics, supported by open-source software ecosystems and standardized feature ontologies to enable robust, comparable, and reproducible research. Independent validation cohorts, stain normalization protocols, and rigorous QC pipelines will be necessary to prevent overfitting and ensure generalizability.

Future of pixelomics in precision pathology

Looking ahead, the future of pixelomics will be shaped by advances in high-throughput imaging, computational power, and AI algorithms/models that will enable ever more detailed and integrated analyses of pathology images. The fusion of pixelomics with other omics, such as genomics, proteomics, and especially spatial transcriptomics, as discussed above, will deepen mechanistic understanding and yield more powerful biomarkers for screening, diagnosis, prognosis, and treatment selection.

Collaborative, interdisciplinary efforts among pathologists, bioinformaticians, data scientists, regulators, professional bodies, oncologists, and other clinicians will be crucial in defining standards, building infrastructure, and training the workforce required to leverage pixelomics. As these components mature, pixelomics will form a critical foundation of precision oncology and holistic precision medicine, ensuring that every pixel in a digital slide contributes to individualized, data-driven patient care.

Conclusion

Pixelomics reframes WSIs not merely as digital surrogates of glass slides but as exhaustive reservoirs of biologically and clinically meaningful information—the pixelome—that can be systematically interrogated using omics principles. By situating pixelomics within the historical evolution of—omes and—omics, this chapter has articulated a conceptual and technical framework spanning the full digital pathology pipeline, from pixel generation and quality control through representation learning, foundation models, and multimodal integration, in a comprehensive yet concise way. Crucially, pixelomics shifts computational pathology from task-specific prediction toward reusable, scalable, and biologically grounded representations that capture tissue architecture, cellular ecosystems, and spatial organization at scale. When integrated with genomic, transcriptomic, and clinical data, pixelomic representations provide a unifying spatial substrate through which molecular and phenotypic signals can be contextualized and interpreted. As the challenges and limitations described above are effectively managed, and as data standards mature, computational infrastructure expands, and explainable, regulation-ready methodologies emerge, pixelomics has the potential to embed quantitative image intelligence into routine diagnostic workflows rather than confining it to research settings. In doing so, it positions histopathology not as a passive endpoint but as an active, integrative engine for precision pathology and, by extension, precision medicine.

AI disclosure

During the preparation of this work, the author(s) used Microsoft Copilot to improve language and flow. After using this tool/service, the author(s) reviewed and edited the content as needed and take full responsibility for the content of the publication.

References

Abels, E., Pantanowitz, L., 2017. Current state of the regulatory trajectory for whole slide imaging devices in the USA. J. Pathol. Inform. 21533539. 8 (1), 23. doi:10.4103/jpi.jpi_11_17.

Afonso, M., Bhawsar, P.M.S., Saha, M., Almeida, J.S., Oliveira, A.L., 2024. Multiple instance learning for WSI: a comparative analysis of attention-based approaches. J. Pathol. Inform. 21533539. 15. Available from: https://www.sciencedirect.com/science/journal/21533539. doi:10.1016/j.jpi.2024.100403.

Ahuja, S., Zaheer, S., 2025. Advancements in pathology: digital transformation, precision medicine, and beyond. J. Pathol. Inform. 21533539. 16 (100408). doi:10.1016/j.jpi.2024.100408.

Akbari, Y., Abdullakutty, F., Al Maadeed, S., Bouridane, A., Hamoudi, R., 2026. Integrating histopathology and genomic data: a comparative study of fusion methods for breast cancer survival prediction. Complex Intell. Syst. 21986053. 12 (1). Available from: https://www.springer.com/journal/40747. doi:10.1007/s40747-025-02133-y.

Alamudun, F., Hinkle, J., Dash, S., Hernández, B., Tsaris, A., Yoon, H.J., 2022. Distilling knowledge from ensembles of cluster-constrained-attention multiple-instance learners for whole slide image classification. In: Paper presented at the 2022 IEEE International Conference on Big Data (Big Data), 17–20 December 2022.

Alber, M., Tietz, S., Dippel, J., Milbich, T., Lesort, T., Korfiatis, P., Krügener, M., Cancer, B.P., Shah, N., Möllers, A., Seegerer, P., Carpen-Amarie, A., Standvoss, K., Dernbach, G., de Jong, E., Schallenberg, S., Kunft, A., von Ankershoffen, H.H., Schaeferle, G., Duffy, P., Redlon, M., Jurmeister, P., Horst, D., Ruff, L., Müller, K.R., Klauschen, F., Norgan, A., 2025. Atlas: a novel pathology foundation model by Mayo Clinic, charité, and aignostics. arXiv. 23318422. Available from:: https://arxiv.org. doi:10.48550/arXiv.2501.05409.

Alleman, K., Knecht, E., Huang, J., Zhang, L., Lam, S., DeCuypere, M., 2023. Multimodal deep learning-based prognostication in glioma patients: a systematic review. Cancers. 2072-6694. 15 (2), 545. doi:10.3390/cancers15020545.

Anghel, A., Stanisavljevic, M., Andani, S., Papandreou, N., Rüschoff, J.H., Wild, P., Gabrani, M., Pozidis, H., 2019. A high-performance system for robust stain normalization of whole-slide images in histopathology. Front. Med. 2296858X. 6. Available from: journal.frontiersin.org/journal/medicine. doi:10.3389/fmed.2019.00193.

Bankhead, P., Loughrey, M.B., Fernández, J.A., Dombrowski, Y., McArt, D.G., Dunne, P.D., McQuaid, S., Gray, R.T., Murray, L.J., Coleman, H.G., James, J.A., Salto-Tellez, M., Hamilton, P.W., 2017. QuPath: open source software for digital pathology image analysis. Sci. Rep. 20452322. 7 (1). Available from: https://www.nature.com/srep/index.html. doi:10.1038/s41598-017-17204-5.

Bera, K., Schalper, K.A., Rimm, D.L., Velcheti, V., Madabhushi, A., 2019. Artificial intelligence in digital pathology — new tools for diagnosis and precision oncology. Nat. Rev. Clin. Oncol. 17594782. 16 (11), 703–715. Available from: http://www.nature.com/nrclinonc/archive/index.html. doi:10.1038/s41571-019-0252-y.

Bird, A., 2007. Perceptions of epigenetics. Nature. 14764687. 447 (7143), 396–398. Available from: http://www.nature.com/nature/index.html. doi:10.1038/nature05913.

Brodsky, V., Ullah, E., Bychkov, A., Song, A.H., Walk, E.E., Louis, P., Rasool, G., Singh, R.S., Mahmood, F., Bui, M.M., Parwani, A.V., 2025. Generative artificial intelligence in anatomic pathology. Arch. Pathol. Lab. Med. 15432165. 149 (4), 298–318. Available from: https://meridian.allenpress.com/aplm/issue/149/4. doi:10.5858/arpa.2024-0215-RA.

Bülow, R.D., Hölscher, D.L., Costa, I.G., Boor, P., 2023. Extending the landscape of omics technologies by pathomics. Npj Syst. Biol. Appl. 20567189. 9 (1). Available from: https://www.nature.com/npjsba/. doi:10.1038/s41540-023-00301-9.

Cai, L., Huang, S., Zhang, Y., Lu, J., Zhang, Y., 2025. AttriMIL: revisiting attention-based multiple instance learning for whole-slide pathological image classification from a perspective of instance attributes. Med. Image Anal. 103, 103631. doi:10.1016/j.media.2025.103631.

Campanella, G., Chen, S., Singh, M., Verma, R., Muehlstedt, S., Zeng, J., Stock, A., Croken, M., Veremis, B., Elmas, A., Shujski, I., Neittaanmäki, N., Huang, K.-L., Kwan, R., Houldsworth, J., Schoenfeld, A.J., Vanderbilt, C., 2025. A clinical benchmark of public self-supervised pathology foundation models. Nat. Commun. 2041-1723. 16 (1). doi:10.1038/s41467-025-58796-1.

Captier, N., Lerousseau, M., Orlhac, F., Hovhannisyan-Baghdasarian, N., Luporsi, M., Woff, E., Lagha, S., Feghali, P.S., Lonjou, C., Beaulaton, C., Zinovyev, A., Salmon, H., Walter, T., Buvat, I., Girard, N., Barillot, E., 2025. Integration of clinical, pathological, radiological, and transcriptomic data improves prediction for first-line immunotherapy outcome in metastatic non-small cell lung cancer. Nat. Commun. 2041-1723. 16 (1). doi:10.1038/s41467-025-55847-5.

Chaurasia, A.K., Toohey, P.W., Harris, H.C., Hewitt, A.W., 2025. Multi-resolution vision transformer model for histopathological skin cancer subtype classification using whole slide images. Comput. Biol. Med. 18790534. 196. Available from: https://www.sciencedirect.com/science/journal/00104825. doi:10.1016/j.compbiomed.2025.110724.

Chen, R.J., Chen, C., Li, Y., Chen, T.Y., Trister, A.D., Krishnan, R.G., Mahmood, F., 2022. Scaling vision transformers to gigapixel images via hierarchical self-supervised learning. In: Paper presented at the 2022 IEEE/CVF Conference on Computer Vision and Pattern Recognition (CVPR). https://doi.ieeecomputersociety.org/10.1109/CVPR52688.2022.01567.

Chen, R.J., Ding, T., Lu, M.Y., Williamson, D.F., Jaume, G., Song, A.H., et al., 2024. Towards a general-purpose foundation model for computational pathology. Nat. Med. 30 (3), 850–862. doi:10.1038/s41591-024-02857-3.

Chen, W., Zhang, P., Tran, T.N., Xiao, Y., Li, S., Shah, V.V., Cheng, H., Brannan, K.W., Youker, K., Lai, L., Fang, L., Yang, Y., Le, N.T., Abe, J.I., Chen, S.H., Ma, Q., Chen, K., Song, Q., Cooke, J.P., Wang, G., 2025. A visual–omics foundation model to bridge histopathology with spatial transcriptomics. Nat. Methods. 15487105. 22 (7), 1568–1582. Available from: https://www.nature.com/nmeth/. doi:10.1038/s41592-025-02707-1.

Church, G.M., 2006. Genomes for all. Sci. Am. 00368733. 294 (1), 47–54. Available from: http://www.nature.com/scientificamerican/archive/index.html. doi:10.1038/scientificamerican0106-46.

Cui, M., Zhang, D.Y., 2021. Artificial intelligence and computational pathology. Lab. Investig. 15300307. 101 (4), 412–422. Available from: http://www.nature.com/labinvest/index.html. doi:10.1038/s41374-020-00514-0.

D.C, L., 2010. Canada Mitochondrial fusion, division and positioning in plants. Biochem. Soc. Trans. 14708752. 38 (3), 789–795. http://www.biochemsoctrans.org/bst/038/0789/0380789.pdf. doi:10.1042/BST0380789.

Daniel, C., Macary, F., García Rojo, M., Klossa, J., Laurinavičius, A., Beckwith, B.A., Mea, V.D., 2011. Recent advances in standards for collaborative Digital Anatomic Pathology. Diagn. Pathol. 1746-1596. 6 (S1). doi:10.1186/1746-1596-6-s1-s17.

Dash, R.C., Jones, N., Merrick, R., Haroske, G., Harrison, J., Sayers, C., Haarselhorst, N., Wintell, M., Herrmann, M.D., Macary, F., 2021. Integrating the health-care enterprise pathology and laboratory medicine guideline for digital pathology interoperability. J. Pathol. Inform. 21533539. 12 (1). Available from: https://www.sciencedirect.com/journal/journal-of-pathology-informatics/about/aims-and-scope. doi:10.4103/jpi.jpi_98_20.

Diao, J.A., Wang, J.K., Chui, W.F., Mountain, V., Gullapally, S.C., Srinivasan, R., Mitchell, R.N., Glass, B., Hoffman, S., Rao, S.K., Maheshwari, C., Lahiri, A., Prakash, A., McLoughlin, R., Kerner, J.K., Resnick, M.B., Montalto, M.C., Khosla, A., Wapinski, I.N., Beck, A.H., Elliott, H.L., Taylor-Weiner, A., 2021. Human-interpretable image features derived from densely mapped cancer pathology slides predict diverse molecular phenotypes. Nat. Commun. 20411723. 12 (1). Available from: http://www.nature.com/ncomms/index.html. doi:10.1038/s41467-021-21896-9.

Ding, T., Wagner, S.J., Song, A.H., Chen, R.J., Lu, M.Y., Zhang, A., Vaidya, A.J., Jaume, G., Shaban, M., Kim, A., Williamson, D.F.K., Robertson, H., Chen, B., Almagro-Pérez, C., Doucet, P., Sahai, S., Chen, C., Chen, C.S., Komura, D., Kawabe, A., Ochi, M., Sato, S., Yokose, T., Miyagi, Y., Ishikawa, S., Gerber, G., Peng, T., Le, L.P., Mahmood, F., 2025. A multimodal whole-slide foundation model for pathology. Nat. Med. 1546170X. 31 (11), 3749–3761. Available from: https://www.nature.com/nm/. doi:10.1038/s41591-025-03982-3.

Filiot, A., Jacob, P., Kain, A.M., Saillard, C., 2024. Phikon-v2 A large and public feature extractor for biomarker prediction. arXiv. 23318422. Available from:: https://arxiv.org. doi:10.48550/arXiv.2409.09173.

Frascarelli, C., Venetis, K., Marra, A., Concardi, A., D'Ercole, M., Mangione, E., Negrelli, M., Porta, F.M., Keswani, S., Curigliano, G., Guerini-Rocco, E., Fusco, N., 2025. Computational pathology in breast cancer: optimizing molecular prediction through task-oriented AI models. Npj Breast Cancer. 2374-4677. 11 (1). doi:10.1038/s41523-025-00857-1.

Ghaznavi, F., Evans, A., Madabhushi, A., Feldman, M., 2013. Digital imaging in pathology: whole-slide imaging and beyond. Annu. Rev. Pathol. Mech. Dis. 1553-4006. 8 (1), 331–359. doi:10.1146/annurev-pathol-011811-120902.

Goode, A., Gilbert, B., Harkes, J., Jukic, D., Satyanarayanan, M., 2013. Open slide: a vendor-neutral software foundation for digital pathology. J. Pathol. Inform. 21533539. 4 (1), 27. doi:10.4103/2153-3539.119005.

Guan, H., Yap, P.-T., Bozoki, A., Liu, M., 2024. Federated learning for medical image analysis: a survey. Pattern Recognit. 151, 110424. doi:10.1016/j.patcog.2024.110424.

Gupta, R., Kurc, T., Sharma, A., Almeida, J.S., Saltz, J., 2019. The Emergence of Pathomics. Curr. Pathobiol. Rep. 2167485X. 7 (3), 73–84. Available from: https://www.springer.com/medicine/pathology/journal/40139. doi:10.1007/s40139-019-00200-x.

Gutman, D.A., Cobb, J., Somanna, D., Park, Y., Wang, F., Kurc, T., Saltz, J.H., Brat, D.J., Cooper, L.A.D., Kong, J., 2013. Cancer digital slide archive: an informatics resource to support integrated in silico analysis of TCGA pathology data. J. Am. Med. Inform. Assoc. 1067-5027. 20 (6), 1091–1098. doi:10.1136/amiajnl-2012-001469.

Gutman, D.A., Khalilia, M., Lee, S., Nalisnik, M., Mullen, Z., Beezley, J., Chittajallu, D.R., Manthey, D., Cooper, L.A. D., 2017. The digital slide archive: a software platform for management, integration, and analysis of histology for cancer research. Cancer Res. 15387445. 77 (21), e75. Available from: http://cancerres.aacrjournals.org//content/canres/77/21/e75full.pdf. doi:10.1158/0008-5472.CAN-17-0629.

Hanna, M.G., Parwani, A., Sirintrapun, S.J., 2020. Whole slide imaging: technology and applications. Adv. Anat. Pathol. 15334031. 27 (4), 251–259. Available from: http://www.anatomicpathology.com. doi:10.1097/PAP.0000000000000273.

Hao, Y., Zhang, L., Qiao, S., Bai, Y., Cheng, R., Xue, H., Hou, Y., Zhang, W., Zhang, G., Jianjun, H., 2022. Breast cancer histopathological images classification based on deep semantic features and gray level co-occurrence matrix. PLOS ONE. 1932-6203. 17 (5), e0267955. doi:10.1371/journal.pone.0267955.

Hasin, Y., Seldin, M., Lusis, A., 2017. Multi-omics approaches to disease. Genome Biol. 1474-760X. 18 (1). doi:10.1186/s13059-017-1215-1.

Herrmann, M.D., Clunie, D.A., Fedorov, A., Doyle, S.W., Pieper, S., Klepeis, V., Le, L.P., Mutter, G.L., Milstone, D.S., Schultz, T.J., Kikinis, R., Kotecha, G.K., Hwang, D.H., Andriole, K.P., Iafrate, A.J., Brink, J.A., Boland, G.W., Dreyer, K.J., Michalski, M., Golden, J.A., Louis, D.N., Lennerz, J.K., 2018. Implementing the DICOM standard for digital pathology. J. Pathol. Inform. 21533539. 9 (1). Available from: www.jpathinformatics.org/browse.asp?sabs=n. doi:10.4103/jpi.jpi_42_18.

Hölscher, D.L., Bülow, R.D., 2025. Decoding pathology: the role of computational pathology in research and diagnostics. Pflugers Arch. Eur. J. Physiol. 14322013. 477 (4), 555–570. Available from: https://www.springer.com/journal/424. doi:10.1007/s00424-024-03002-2.

Hou, L., Samaras, D., Kurc, T.M., Gao, Y., Davis, J.E., Saltz, J.H., 2016. Patch-based convolutional neural network for whole slide tissue image classification. In: Paper presented at the Proceedings of the IEEE conference on computer vision and pattern recognition.

Houle, D., Govindaraju, D.R., Omholt, S., 2010. Phenomics: the next challenge. Nat. Rev. Genet. 14710064. 11 (12), 855–866. doi:10.1038/nrg2897.

Huang, W., Wang, X., Zhong, R., Li, Z., Zhou, K., Lyu, Q., et al., 2025a. Multimodal radiopathomics signature for prediction of response to immunotherapy-based combination therapy in gastric cancer using interpretable machine learning. Cancer Lett. 631, 217930. doi:10.1016/j.canlet.2025.217930.

Huang, X., Li, Z., Zhang, M., Gao, S., 2022. Fusing hand-crafted and deep-learning features in a convolutional neural network model to identify prostate cancer in pathology images. Front. Oncol. 2234-943X. 12. doi:10.3389/fonc.2022.994950.

Huang, Z., Bianchi, F., Yuksekgonul, M., Montine, T.J., Zou, J., 2023. A visual–language foundation model for pathology image analysis using medical Twitter. Nat. Med. 1546170X. 29 (9), 2307–2316. Available from: https://www.nature.com/nm/. doi:10.1038/s41591-023-02504-3.

Huang, Z., Yang, E., Shen, J., Gratzinger, D., Eyerer, F., Liang, B., et al., 2025b. A pathologist–AI collaboration framework for enhancing diagnostic accuracies and efficiencies. Nat. Biomed. Eng. 9 (4), 455–470.

Janowczyk, A., Madabhushi, A., 2016. Deep learning for digital pathology image analysis: a comprehensive tutorial with selected use cases. J. Pathol. Inform. 21533539. 7 (1), 29. doi:10.4103/2153-3539.186902.

Jia, C., Yang, Y., Xia, Y., Chen, Y.T., Parekh, Z., Pham, H., Le, Q.V., Sung, Y., Li, Z., Duerig, T., 2021. ML research press undefined scaling up visual and vision-language representation learning with noisy text supervision. In: Proceedings of Machine Learning Research. 26403498. Pp. 4904–4916. https://proceedings.mlr.press/139

Kanwal, N., Perez-Bueno, F., Schmidt, A., Engan, K., Molina, R., 2022. The devil is in the details: whole slide image acquisition and processing for artifacts detection, color variation, and data augmentation: a review. IEEE Access. 21693536. 10, 58821–58844. Available from: http://ieeexplore.ieee.org/xpl/RecentIssue.jsp?punumber=6287639. doi:10.1109/ACCESS.2022.3176091.

Karczewski, K.J., Snyder, M.P., 2018. Integrative omics for health and disease. Nat. Rev. Genet. 14710064. 19 (5), 299–310. Available from: http://www.nature.com/reviews/genetics. doi:10.1038/nrg.2018.4.

Komura, D., Ishikawa, S., 2019. Machine learning approaches for pathologic diagnosis. Virchows Arch. 0945-6317. 475 (2), 131–138. doi:10.1007/s00428-019-02594-w.

Kumar, N., Gupta, R., Gupta, S., 2020. Whole slide imaging (WSI) in pathology: current perspectives and future directions. J. Digit. Imaging. 0897-1889. 33 (4), 1034–1040. doi:10.1007/s10278-020-00351-z.

Lander, E.S., Linton, L.M., Birren, B., Nusbaum, C., Zody, M.C., Baldwin, J., Devon, K., Dewar, K., Doyle, M., Fitzhugh, W., Funke, R., Gage, D., Harris, K., Heaford, A., Howland, J., Kann, L., Lehoczky, J., Levine, R., McEwan, P., McKernan, K., Meldrim, J., Mesirov, J.P., Miranda, C., Morris, W., Naylor, J., Raymond, C., Rosetti, M., Santos, R., Sheridan, A., Sougnez, C., Stange-Thomann, N., Stojanovic, N., Subramanian, A., Wyman, D., Rogers, J., Sulston, J., Ainscough, R., Beck, S., Bentley, D., Burton, J., Clee, C., Carter, N., Coulson, A., Deadman, R., Deloukas, P., Dunham, A., Dunham, I., Durbin, R., French, L., Grafham, D., Gregory, S., Hubbard, T., Humphray, S., Hunt, A., Jones, M., Lloyd, C., McMurray, A., Matthews, L., Mercer, S., Milne, S., Mullikin, J.C., Mungall, A., Plumb, R., Ross, M., Shownkeen, R., Sims, S., Waterston, R.H., Wilson, R.K., Hillier, L.W., McPherson, J.D., Marra, M.A., Mardis, E.R., Fulton, L.A., Chinwalla, A.T., Pepin, K.H., Gish, W.R., Chissoe, S.L., Wendl, M.C., Delehaunty, K.D., Miner, T.L., Delehaunty, A., Kramer, J.B., Cook, L.L., Fulton, R.S., Johnson, D.L., Minx, P.J., Clifton, S.W., Hawkins, T., Branscomb, E., Predki, P., Richardson, P., Wenning, S., Slezak, T., Doggett, N., Cheng, J.F., Olsen, A., Lucas, S., Elkin, C., Uberbacher, E., Frazier, M., Gibbs, R.A., Muzny, D.M., Scherer, S.E., Bouck, J.B., Sodergren, E.J., Worley, K.C., Rives, C.M., Gorrell, J.H., Metzker, M.L., Naylor, S.L., Kucherlapati, R.S., Nelson, D.L., Weinstock, G.M., Sakaki, Y., Fujiyama, A., Hattori, M., Yada, T., Toyoda, A., Itoh, T., Kawagoe, C., Watanabe, H., Totoki, Y., Taylor, T., Weissenbach, J., Heilig, R., Saurin, W., Artiguenave, F., Brottier, P., Bruls, T., Pelletier, E., Robert, C., Wincker, P., Rosenthal, A., Platzer, M., Nyakatura, G., Taudien, S., Rump, A., Smith, D.R., Doucette-Stamm, L., Rubenfield, M., Weinstock, K., Hong, M.L., Dubois, J., Yang, H., Yu, J., Wang, J., Huang, G., Gu, J., Hood, L., Rowen, L., Madan, A., Qin, S., Davis, R.W., Federspiel, N.A., Abola, A.P., Proctor, M.J., Roe, B.A., Chen, F., Pan, H., Ramser, J., Lehrach, H., Reinhardt, R., McCombie, W.R., Bastide, M.D.L., Dedhia, N., Blöcker, H., Hornischer, K., Nordsiek, G., Agarwala, R., Aravind, L., Bailey, J.A., Bateman, A., Batzoglou, S., Birney, E., Bork, P., Brown, D.G., Burge, C.B., Cerutti, L., Chen, H.C., Church, D., Clamp, M., Copley, R.R., Doerks, T., Eddy, S.R., Eichler, E.E., Furey, T.S., Galagan, J., Gilbert, J.G.R., Harmon, C., Hayashizaki, Y., Haussler, D., Hermjakob, H., Hokamp, K., Jang, W., Johnson, L.S., Jones, T.A., Kasif, S., Kaspryzk, A., Kennedy, S., Kent, W.J., Kitts, P., Koonin, E.V., Korf, I., Kulp, D., Lancet, D., Lowe, T.M., McLysaght, A., Mikkelsen, T., Moran, J.V., Mulder, N., Pollara, V.J., Ponting, C.P., Schuler, G., Schultz, J., Slater, G., Smit, A.F.A., Stupka, E., Szustakowki, J., Thierry-Mieg, D., Thierry-Mieg, J., Wagner, L., Wallis, J., Wheeler, R., Williams, A., Wolf, Y.I., Wolfe, K.H., Yang, S.P., Yeh, R.F., Collins, F., Guyer, M.S., Peterson, J., Felsenfeld, A., Wetterstrand, K.A., Myers, R.M., Schmutz, J., Dickson, M., Grimwood, J., Cox, D.R., Olson, M.V., Kaul, R., Raymond, C., Shimizu, N., Kawasaki, K., Minoshima, S., Evans, G.A., Athanasiou, M., Schultz, R., Patrinos, A., Morgan, M.J., 2001. Initial sequencing and analysis of the human genome. Nature 409 (6822), 860–921. doi:10.1038/35057062.

Li, Z., Cong, Y., Chen, X., Qi, J., Sun, J., Yan, T., Yang, H., Liu, J., Lu, E., Wang, L., Li, J., Hu, H., Zhang, C., Yang, Q., Yao, J., Yao, P., Jiang, Q., Liu, W., Song, J., Carin, L., Chen, Y., Zhao, S., Gao, X., 2023. Vision transformer-based weakly supervised histopathological image analysis of primary brain tumors. iScience. 25890042. 26 (1), 105872. doi:10.1016/j.isci.2022.105872.

Lin, J.R., Izar, B., Wang, S., Yapp, C., Mei, S., Shah, P.M., Santagata, S., Sorger, P.K., 2018. Highly multiplexed immunofluorescence imaging of human tissues and tumors using t-CyCIF and conventional optical microscopes. eLife. 2050084X. 7. Available from: https://elifesciences.org/magazine. doi:10.7554/eLife.31657.

Liu, H., Li, C., Wu, Q., Lee, Y.J., 2023. Visual Instruction Tuning. arXiv. 23318422. Available from:: https://arxiv.org. doi:10.48550/arXiv.2304.08485.

Liu, Y., Yang, M., Deng, Y., Su, G., Enninful, A., Guo, C.C., Tebaldi, T., Zhang, D., Kim, D., Bai, Z., Norris, E., Pan, A., Li, J., Xiao, Y., Halene, S., Fan, R., 2020. High-spatial-resolution multi-omics sequencing via deterministic barcoding in tissue. Cell. 10974172. 183 (6), 1665. Available from: https://www.sciencedirect.com/journal/cell. doi:10.1016/j.cell.2020.10.026.

Lu, M.Y., Chen, B., Williamson, D.F.K., Chen, R.J., Liang, I., Ding, T., Jaume, G., Odintsov, I., Le, L.P., Gerber, G., Parwani, A.V., Zhang, A., Mahmood, F., 2024. A visual-language foundation model for computational pathology.

Nat. Med. 1546170X. 30 (3), 863–874. Available from: https://www.nature.com/nm/. doi:10.1038/s41591-024-02856-4.

Madabhushi, A., Lee, G., 2016. Image analysis and machine learning in digital pathology: challenges and opportunities. Med. Image Anal. 13618415. 33, 170–175. doi:10.1016/j.media.2016.06.037.

Maleki, D., Shaikh, N., Li, X., Nie, Y., Venugopal, R., Kurkure, U., 2025. Understanding foundation models in digital pathology: performance, trade-offs, and model-selection recommendations. bioRxiv. 26928205. Available from: https://www.biorxiv.org. doi:10.1101/2025.09.12.675923.

Masci, J., Meier, U., Cireşan, D., Schmidhuber, J., 2011. Stacked convolutional auto-encoders for hierarchical feature extraction. In: Paper presented at the International conference on artificial neural networks, Berlin. pp. 52–59.

Mezei, T., Kolcsár, M., Joó, A., Gurzu, S., 2024. Image analysis in histopathology and cytopathology: from early days to current perspectives. J. Imaging. 2313-433X. 10 (10), 252. doi:10.3390/jimaging10100252.

Nalisnik, M., Amgad, M., Lee, S., Halani, S.H., Velazquez Vega, J.E., Brat, D.J., Gutman, D.A., Cooper, L.A.D., 2017. Interactive phenotyping of large-scale histology imaging data with HistomicsML. Sci. Rep. 20452322. 7 (1). Available from: https://www.nature.com/srep/index.html. doi:10.1038/s41598-017-15092-3.

Nechaev, D., Pchelnikov, A., Ivanova, E., 2024. HIBOU: a family of foundational vision transformers for pathology. arXiv. 23318422. Available from:: https://arxiv.org. doi:10.48550/arXiv.2406.05074.

Newitt, V., 2018. Digital pathology: a 1st anniversary report card. CAP today.

Niazi, M.K.K., Parwani, A.V., Gurcan, M.N., 2019. Digital pathology and artificial intelligence. Lancet Oncol. 14745488. 20 (5), e253. Available from: http://www.journals.elsevier.com/the-lancet-oncology/. doi:10.1016/S1470-2045(19)30154-8.

Ome suffix, 2025. Oxford University Press.

Pantanowitz, L., Dickinson, K., Evans, A.J., Hassell, L.A., Henricks, W.H., Lennerz, J.K., Lowe, A., Parwani, A.V., Riben, M., Daniel Smith, C.O.L., Mark Tuthill, J., Weinstein, R.S., Wilbur, D.C., Krupinski, E.A., Bernard, J., 2014. American Telemedicine Association clinical guidelines for telepathology. J. Pathol. Inform. 21533539. 5 (1), 39. doi:10.4103/2153-3539.143329.

Pantanowitz, L., Pearce, T., Abukhiran, I., Hanna, M., Wheeler, S., Soong, T.R., Tafti, A.P., Pantanowitz, J., Lu, M.Y., Mahmood, F., Gu, Q., Rashidi, H.H., 2025. Nongenerative artificial intelligence in medicine: advancements and applications in supervised and unsupervised machine learning. Mod. Pathol. 15300285. 38 (3). Available from: https://www.sciencedirect.com/journal/modern-pathology/about/insights. doi:10.1016/j.modpat.2024.100680.

Pantanowitz, L., Sinard, J.H., Henricks, W.H., Fatheree, L.A., Carter, A.B., Contis, L., Beckwith, B.A., Evans, A.J., Otis, C.N., Lal, A., Parwani, A.V., 2013. Validating whole slide imaging for diagnostic purposes in pathology: guideline from the College of American pathologists Pathology and Laboratory Quality Center. Arch. Pathol. Lab. Med. 137 (12), 1710–1722. Available from: http://www.archivesofpathology.org/doi/pdf/10.5858/arpa.2013-0093-CP. doi: 10.5858/arpa.2013-0093-CP. 15432165.

Park, S., Pantanowitz, L., Parwani, A.V., 2012. Digital imaging in pathology. Clin. Lab. Med. 15579832. 32 (4), 557–584. doi:10.1016/j.cll.2012.07.006.

Pascual-Reguant, A., Köhler, R., Mothes, R., Bauherr, S., Hernández, D.C., Uecker, R., Holzwarth, K., Kotsch, K., Seidl, M., Philipsen, L., Müller, W., Romagnani, C., Niesner, R., Hauser, A.E., 2021. Multiplexed histology analyses for the phenotypic and spatial characterization of human innate lymphoid cells. Nat. Commun. 20411723. 12 (1). Available from: http://www.nature.com/ncomms/index.html. doi:10.1038/s41467-021-21994-8.

Patti, G.J., Yanes, O., Siuzdak, G., 2012. Metabolomics: the apogee of the omics trilogy. Nat. Rev. Mol. Cell Biol. 1471-0072. 13 (4), 263–269. doi:10.1038/nrm3314.

Perkel, J.M., 2002. Tissue microarrays: advancing clinical genomics: inspired by DNA biochips, an NIH postdoc opens the door to "pathomics" (Lab Consumer). Sci. 16 (21), 39–43.

Pocock, J., Graham, S., Vu, Q.D., Jahanifar, M., Deshpande, S., Hadjigeorghiou, G., Shephard, A., Bashir, R.M.S., Bilal, M., Lu, W., Epstein, D., Minhas, F., Rajpoot, N.M., Raza, S.E.A., 2022. TIAToolbox as an end-to-end library for advanced tissue image analytics. Commun. Med. 2730664X. 2 (1). Available from: https://www.nature.com/commsmed/. doi:10.1038/s43856-022-00186-5.

Pourakpour, F., Szölgyén, Á., Nateghi, R., Gutman, D.A., Manthey, D., Cooper, L.A., 2025. HistomicsTK: a python toolkit for pathology image analysis algorithms. SoftwareX. 23527110. 31. Available from: https://www.sciencedirect.com/science/journal/23527110. doi:10.1016/j.softx.2025.102318.

Rashidi, H.H., Pantanowitz, J., Chamanzar, A., Fennell, B., Wang, Y., Gullapalli, R.R., Tafti, A., Deebajah, M., Albahra, S., Glassy, E., Hanna, M.G., Pantanowitz, L., 2025a. Generative artificial intelligence in pathology and medicine: a deeper dive. Mod. Pathol. 15300285. 38 (4). Available from: https://www.sciencedirect.com/journal/modern-pathology/about/insights. doi:10.1016/j.modpat.2024.100687.

Rashidi, H.H., Pantanowitz, J., Hanna, M.G., Tafti, A.P., Sanghani, P., Buchinsky, A., Fennell, B., Deebajah, M., Wheeler, S., Pearce, T., Abukhiran, I., Robertson, S., Palmer, O., Gur, M., Tran, N.K., Pantanowitz, L., 2025b. Introduction to artificial intelligence and machine learning in pathology and medicine: generative and non-generative artificial intelligence basics. Mod. Pathol. 15300285. 38 (4). Available from: https://www.sciencedirect.com/journal/modern-pathology/about/insights. doi:10.1016/j.modpat.2024.100688.

Romanchikova, M., Thomas, S.A., Dexter, A., Shaw, M., Partarrieau, I., Smith, N., Venton, J., Adeogun, M., Brettle, D., Turpin, R.J., 2022. The need for measurement science in digital pathology. J. Pathol. Inform. 21533539. 13. Available from: https://www.sciencedirect.com/journal/journal-of-pathology-informatics/about/aims-and-scope. doi:10.1016/j.jpi.2022.100157.

Sellaro, T.L., Filkins, R., Hoffman, C., Fine, J.L., Jon, H., Parwani, A.V., Pantanowitz, L., Montalto, M., 2013. Relationship between magnification and resolution in digital pathology systems. J. Pathol. Inform. 21533539. 4 (1), 21. doi:10.4103/2153-3539.116866.

Shakhawat, H., Hossain, S., Kabir, A., Mahmud, S.H., Islam, M.M., Tariq, F., 2023. Review of artifact detection methods for automated analysis and diagnosis in digital pathology. In: Mostefaoui, G.K., Islam, S.M., Tariq, F. (Eds), Artificial Intelligence for Disease Diagnosis and Prognosis in Smart Healthcare. CRC Press, pp. 177–202.

Sherman, R.M., Salzberg, S.L., 2020. Pan-genomics in the human genome era. Nat. Rev. Genet. 14710064. 21 (4), 243–254. Available from: http://www.nature.com/reviews/genetics. doi:10.1038/s41576-020-0210-7.

Shrestha, P., Hulsken, B., 2014. Color accuracy and reproducibility in whole slide imaging scanners. J. Med. Imaging. 2329-4302. 1 (2), 027501. doi:10.1117/1.jmi.1.2.027501.

Shvetsov, N., Kilvær, T.K., Tafavvoghi, M., Sildnes, A., Møllersen, K., Busund, L.T.R., Bongo, L.A., 2025. A lightweight and extensible cell segmentation and classification model for H&E-stained cancer whole slide images. Comput. Biol. Med. 18790534. 199. Available from: https://www.sciencedirect.com/science/journal/00104825. doi:10.1016/j.compbiomed.2025.111326.

Singh, R., Chubb, L., Pantanowitz, L., Parwani, A., 2011. Standardization in digital pathology: supplement 145 of the DICOM standards. J. Pathol. Inform. 21533539. 2 (1), 23. doi:10.4103/2153-3539.80719.

Snead, D.R.J., Tsang, Y.W., Meskiri, A., Kimani, P.K., Crossman, R., Rajpoot, N.M., Blessing, E., Chen, K., Gopalakrishnan, K., Matthews, P., Momtahan, N., Read-Jones, S., Sah, S., Simmons, E., Sinha, B., Suortamo, S., Yeo, Y., Daly, H.E., Cree, I.A., 2016. Validation of digital pathology imaging for primary histopathological diagnosis. Histopathology. 13652559. 68 (7), 1063–1072. Available from: http://www3.interscience.wiley.com/journal/118511588/home. doi:10.1111/his.12879.

Tuchin, V.V., 2015. Tissue optics and photonics: biological tissue structures. J. Biomed. Photonics Eng. 24112844. 1 (1), 3–21. doi:10.18287/JBPE-2015-1-1-3.

Turnbaugh, P.J., Ley, R.E., Hamady, M., Fraser-Liggett, C.M., Knight, R., Gordon, J.I., 2007. The human microbiome project. Nature. 14764687. 449 (7164), 804–810. Available from: http://www.nature.com/nature/index.html. doi:10.1038/nature06244.

Turteltaub, K., Ascher, M., Langlois, R., Fodor, I., Kercher, J., Laughlin, K., et al., 2006. Pathomics, Lawrence Livermore National Laboratory. https://www.osti.gov/servlets/purl/1036867.

Ullah, E., Baig, M.M., Waqas, A., Rasool, G., Singh, R., Shandilya, A., Gholam Hossieni, H., Parwani, A.V., 2025a. Multimodal generative AI for anatomic pathology-a review of current applications to envisage the future direction. Adv. Anat. Pathol., 15334031. Available from: http://www.anatomicpathology.com. doi:10.1097/PAP.0000000000000498.

Ullah, N., Guzmán-Aroca, F., Martínez-Álvarez, F., De Falco, I., Sannino, G., 2025b. A novel explainable AI framework for medical image classification integrating statistical, visual, and rule-based methods. Med. Image Anal. 13618415. 105, 103665. doi:10.1016/j.media.2025.103665.

van der Laak, J.A., Grünberg, K., Frisk, A.-L.A., Moulin, P., 2022. Building an EU scale digital pathology repository: the bigpicture initiative. J. Pathol. Inform. 13, 100026. doi:10.1016/j.jpi.2022.100026.

Varambally, S., Yu, J., Laxman, B., Rhodes, D.R., Mehra, R., Tomlins, S.A., Shah, R.B., Chandran, U., Monzon, F.A., Becich, M.J., Wei, J.T., Pienta, K.J., Ghosh, D., Rubin, M.A., Chinnaiyan, A.M., 2005. Integrative genomic and proteomic analysis of prostate cancer reveals signatures of metastatic progression. Cancer Cell. 15356108. 8 (5), 393–406. Available from: https://www.journals.elsevier.com/cancer-cell. doi:10.1016/j.ccr.2005.10.001.

Vidal, M., Cusick, M.E., Barabási, A.-L., 2011. Interactome networks and human disease. Cell 144 (6), 986–998.

Vorontsov, E., Bozkurt, A., Casson, A., Shaikovski, G., Zelechowski, M., Severson, K., Zimmermann, E., Hall, J., Tenenholtz, N., Fusi, N., Yang, E., Mathieu, P., van Eck, A., Lee, D., Viret, J., Robert, E., Wang, Y.K., Kunz, J.D., Lee, M.C.H., Bernhard, J.H., Godrich, R.A., Oakley, G., Millar, E., Hanna, M., Wen, H., Retamero, J.A., Moye, W. A., Yousfi, R., Kanan, C., Klimstra, D.S., Rothrock, B., Liu, S., Fuchs, T.J., 2024. A foundation model for clinical-grade computational pathology and rare cancers detection. Nat. Med. 1078-8956. 30 (10), 2924–2935. doi:10.1038/s41591-024-03141-0.

Wagner, N., Fuchs, M., Tolkach, Y., Mukhopadhyay, A., 2022. Federated stain normalization for computational pathology. In: Paper presented at the International Conference on Medical Image Computing and Computer-Assisted Intervention.

Wang, J., Dai, J., Cheng, Y., Wang, X., Deng, R., Zhu, H., 2025. Advances in the use of radiomics and pathomics for predicting the efficacy of neoadjuvant therapy in tumors. Transl. Oncol. 19365233. 58, 102435. doi:10.1016/j.tranon.2025.102435.

Wang, X., Allen, W.E., Wright, M.A., Sylwestrak, E.L., Samusik, N., Vesuna, S., Evans, K., Liu, C., Ramakrishnan, C., Liu, J., Nolan, G.P., Bava, F.A., Deisseroth, K., 2018. Three-dimensional intact-tissue sequencing of single-cell transcriptional states. Science. 10959203. 361 (6400). Available from: http://science.sciencemag.org/content/361/6400/eaat5691/tab-pdf. doi:10.1126/science.aat5691.

Wang, X., Yang, S., Zhang, J., Wang, M., Zhang, J., Yang, W., Huang, J., Han, X., 2022. Transformer-based unsupervised contrastive learning for histopathological image classification. Med. Image Anal. 13618415. 81, 102559. doi:10.1016/j.media.2022.102559.

Wang, X., Zhao, J., Marostica, E., Yuan, W., Jin, J., Zhang, J., Li, R., Tang, H., Wang, K., Li, Y., Wang, F., Peng, Y., Zhu, J., Zhang, J., Jackson, C.R., Zhang, J., Dillon, D., Lin, N.U., Sholl, L., Denize, T., Meredith, D., Ligon, K.L., Signoretti, S., Ogino, S., Golden, J.A., Nasrallah, M.P., Han, X., Yang, S., Yu, K.H., 2024. A pathology foundation model for cancer diagnosis and prognosis prediction. Nature. 14764687. 634 (8035), 970–978. Available from: https://www.nature.com/nature/. doi:10.1038/s41586-024-07894-z.

Wang, Z., Gerstein, M., Snyder, M., 2009. RNA-Seq: a revolutionary tool for transcriptomics. Nat. Rev. Genet. 1471-0056. 10 (1), 57–63. doi:10.1038/nrg2484.

Watson, J.D., Cook-Deegan, R.M., 1991. Origins of the human genome project. FASEB J. 08926638. 5 (1), 8–11. Available from https://onlinelibrary.wiley.com/journal/15306860.

Wilkins, M.R., Sanchez, J.C., Gooley, A.A., Appel, R.D., Humphery-Smith, I., Hochstrasser, D.F., Williams, K.L., 1996. Progress with proteome projects: why all proteins expressed by a genome should be identified and how to do it. Biotechnol. Genet. Eng. Rev. 20465556. 13 (1), 19–50. doi:10.1080/02648725.1996.10647923.

Winkler, H., 1920. Verbreitung Und Ursache der Parthenogenesis Im Pflanzen- Und Tierreiche. G. FIscher. doi:10.5962/bhl.title.1460.

Wu, E., Bieniosek, M., Wu, Z., Thakkar, N., Charville, G.W., Makky, A., Schürch, C.M., Huyghe, J.R., Peters, U., Li, C. I., Li, L., Giba, H., Behera, V., Raman, A., Trevino, A.E., Mayer, A.T., Zou, J., 2025. ROSIE: AI generation of multiplex immunofluorescence staining from histopathology images. Nat. Commun. 20411723. 16 (1). Available from: https://www.nature.com/ncomms/. doi:10.1038/s41467-025-62346-0.

Xiang, H., Shen, J., Yan, Q., Xu, M., Shi, X., Zhu, X., 2023. Multi-scale representation attention based deep multiple instance learning for gigapixel whole slide image analysis. Med. Image Anal. 13618415. 89, 102890. doi:10.1016/j.media.2023.102890.

Xu, H., Usuyama, N., Bagga, J., Zhang, S., Rao, R., Naumann, T., Wong, C., Gero, Z., González, J., Gu, Y., Yanbo, X., Wei, M., Wang, W., Shuming, M., Wei, F., Yang, J., Chunyuan, L., Gao, J., Rosemon, J., Bower, T., Lee, S., Weerasinghe, R., Wright, B.J., Robicsek, A., Piening, B., Bifulco, C., Wang, S., Poon, H., 2024. A whole-slide foundation model for digital pathology from real-world data. Nature. 0028-0836. 630 (8015), 181–188. doi:10.1038/s41586-024-07441-w.

Yang, H., Yang, M., Chen, J., Yao, G., Zou, Q., Jia, L., 2025. Multimodal deep learning approaches for precision oncology: a comprehensive review. Brief. Bioinform. 1467-5463. 26 (1). doi:10.1093/bib/bbae699.

Zang, X., Xia, Y., Xiao, H., Luo, H., Minggui, S., Hou, N., Haoni, A., Chen, T., Liu, Z., Xinyuan, P., Xiangyu, Z., Lijun, X., Zhu, J., Zhipeng, X., Wang, J., Wang, Z., Xia, J., Cao, D., Yin, Y., Wang, J., Xiaorong, W., Kong, W.,

Huang, J., Zhang, J., Chen, Y., Huang, Y., Leung, D.K.-W., Teoh, J.Y.-C., Wang, K., Liang, C., Zheng, J., Yu, Z., Zhai, W., 2025. A multimodal AI model for precision prognosis in clear cell renal cell carcinoma: a multicenter study. Npj Digit. Med. 2398-6352. 8 (1). doi:10.1038/s41746-025-02034-x.

Zanjani, F.G., Zinger, S., Bejnordi, B.E., Van Der Laak, J.A., de With, P.H., 2018. Stain normalization of histopathology images using generative adversarial networks. Paper Presented at the 2018 IEEE 15th Int. Symp. Biomed. Imaging (ISBI 2018.

Zhang, D.Y., Venkat, A., Khasawneh, H., Sali, R., Zhang, V., Pei, Z., 2024a. Implementation of digital pathology and artificial intelligence in routine pathology practice. Lab. Investig. 104(9), 102111.

Zhang, J., Fan, L., Zhang, X., Wang, H., Hei, X., 2024b. Automatic medical image segmentation with vision transformer. Appl. Sci. 2076-3417. 14 (7), 2741. doi:10.3390/app14072741.

Zhang, T., Zhang, H., Zhao, Z., Shao, S., Jiang, Y., Zhang, X., Wang, G., 2025. spaMGCN: a graph convolutional network with autoencoder for spatial domain identification using multi-scale adaptation. Genome Biol. 1474-760X. 26 (1). doi:10.1186/s13059-025-03637-z.

Zhao, C., Wu, T., Ding, X., Parwani, A.V., Chen, H., McHugh, J., Piccoli, A., Xie, Q., Lauro, G.R., Feng, X., Hartman, D.J., Seethala, R.R., Wu, S., Yousem, S., Liang, Y., Pantanowitz, L., 2015. International telepathology consultation: three years of experience between the University of Pittsburgh Medical Center and KingMed Diagnostics in China. J. Pathol. Inform. 21533539. 6 (1), 63. doi:10.4103/2153-3539.170650.

Zimmermann, E., Vorontsov, E., Viret, J., Casson, A., Zelechowski, M., Shaikovski, G., Tenenholtz, N., Hall, J., Klimstra, D., Yousfi, R., Fuchs, T., Fusi, N., Liu, S., Severson, K., 2024. VIRCHOW2: scaling self-supervised mixed magnification models in pathology. arXiv. 23318422. Available from:: https://arxiv.org. doi:10.48550/arXiv.2408.00738.

CHAPTER

14

Large language models utilizing digital pathology workflows: Impact on pathology practice

Zhi Huang[a], *Jialu Yao*[a], *and Tiancheng Bao*[a,b]

[a]Department of Pathology and Laboratory Medicine, Perelman School of Medicine at the University of Pennsylvania, Philadelphia, PA, United States

[b]Department of Bioengineering, University of Pennsylvania, Philadelphia, PA, United States

Introduction

Digital pathology is entering a new era with the emergence of large language models (LLMs) and multimodal AI systems. Since the public release of ChatGPT in late 2022, LLMs have gained broad attention and have been widely used across medical research and workflows, including pathology. For years, computational pathology focused on vision-only models for tasks such as tumor detection, grading, and survival prediction. However, in practice, pathologists interpret images using a combination of narrative reports, clinical notes, and accumulated expertise. LLMs open the door to unifying these modalities, transforming computational pathology into a more holistic and powerful framework.

Recent advances in vision-language models and multimodal LLMs now allow the integration of microscopic images with natural language, enabling AI "copilots" that can converse with pathologists, generate reports, retrieve reference cases, and assist in diagnostic reasoning. Early studies and prototypes have demonstrated that such systems can achieve impressive performance on pathology tasks, ranging from zero-shot image classification and image search to automated report generation and question answering (Huang, 2023; Lu et al., 2024a). In this chapter, we survey the latest developments in applying LLM-based technologies to digital pathology workflows, organized into four categories as shown in Fig. 14.1: (1) **vision-language foundation models without free-text generation** (e.g., PLIP, CONCH, and MUSK), (2) **multimodal chat models with text generation** (e.g., PathChat, HistoGPT),

Advances in Digital Pathology and Artificial Intelligence in Pathology
https://doi.org/10.1016/B978-0-443-40383-5.00005-4

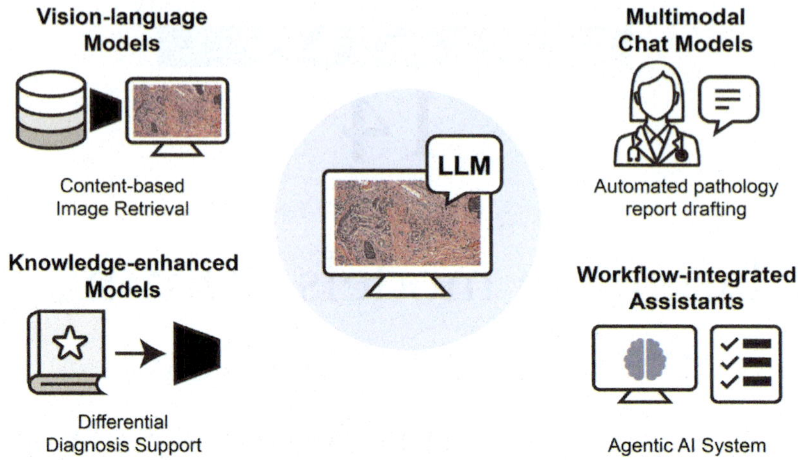

FIG. 14.1 Roles of large language models in digital pathology: LLMs support image retrieval, differential diagnosis, report drafting, and workflow integration through specialized model types.

(3) **LLM-driven knowledge-enhanced models that leverage domain-specific information**, and (4) **workflow-integrated LLM-driven assistants**.

We then discuss key applications such as content-based image retrieval, automated pathology report drafting, differential diagnosis support, and agentic AI systems—and consider their practical impact on pathology practice. We include technical details—for example, how image features are represented and fused with language models—and, where possible, preliminary findings from clinical and educational use. We also discuss constraints such as data scarcity, performance, and ethics, and conclude with a look at where LLM-driven digital pathology may be headed.

Vision-language foundation models for pathology

A major milestone in computational pathology has been the development of vision-language foundation models that learn joint representations of histology images and text. These models are typically trained using a contrastive approach to align image features with descriptive captions or labels, similar to the Contrastive Language-Image Pretraining (CLIP) model (Radford et al., 2021) in general computer vision. In pathology, researchers have assembled large image-text datasets to enable such pretraining. For example, Huang et al. curated **OpenPath**, a dataset of over 208,000 pathology images paired with text descriptions from publicly posted cases on social media, and used it to train **Pathology Language-Image Pretraining (PLIP;** Huang, 2023). PLIP achieved strong zero-shot classification performance on unseen pathology image datasets, far outperforming a generic CLIP model (zero-shot F1 scores 0.565–0.832 versus 0.03–0.481 for CLIP; Huang, 2023). In other words, without any task-specific fine-tuning, PLIP could recognize new image classes using only textual prompts, thanks to its learned image-text alignment. In addition, PLIP's embeddings support bidirectional image/text retrieval—users can search for cases by image or by natural-language description to find visually or diagnostically similar slides. This content-based retrieval capability promotes knowledge sharing among pathologists and can aid education and diagnosis by surfacing analogous examples.

Following PLIP's study, larger and more comprehensive foundation models have been developed and have demonstrated superior performance. Lu et al. introduced **CONtrastive learning from Captions for Histopathology (CONCH)**, which was trained on *over 1.17 million* image-caption pairs from diverse sources (Lu et al., 2024b). CONCH is a task-agnostic model that can be transferred to a wide range of pathology applications with little or no additional supervised training. Evaluated on 14 different benchmarks, CONCH attained state-of-the-art results in histology image *classification*, *segmentation*, *captioning*, and both text-to-image and image-to-text *retrieval*. This represents a substantial leap over earlier pathology VL models, indicating that scaling up multimodal pretraining can endow the model with a broad, flexible understanding of histopathology. Notably, these models operate mostly at the *patch* or image level and are not designed to produce long textual outputs—their strength lies in embedding images and text into a common space for recognition and search tasks (and, in the case of CONCH, also generating short captions).

Another breakthrough was integrating clinical context with images. Xiang et al. developed the **Multimodal Transformer with Unified Masked Modeling (MUSK)**, a vision–language foundation model for precision oncology that leverages both pathology images and accompanying textual data, such as pathology reports (Xiang et al., 2025). Uniquely, MUSK was pretrained in two stages: first, on extremely large unpaired datasets (50 million pathology image patches and 1 billion text tokens from reports and articles) using a masked modeling objective, and then on 1 million paired image-text examples for fine alignment. By training on unpaired image and text corpora, MUSK overcame the scarcity of perfectly annotated multimodal data and learned rich representations of each modality. The resulting model demonstrated superior performance across 23 different tasks, spanning patch-level and slide-level predictions. For example, MUSK set a new bar in cross-modal retrieval—matching images to text queries and vice versa—and in visual question answering, and it performed strongly on zero-shot classification of pathology images. Beyond diagnosis, MUSK performed strongly on prognosis. By combining histology with clinical narratives, it surpassed standard models in predicting melanoma relapse, pan-cancer survival, and immunotherapy response. This highlights the power of combining image features with textual (clinical) information—essentially infusing the model with domain knowledge and patient context. Like PLIP and CONCH, MUSK was not primarily a free-form text generator; instead, it outputs predictions or retrieved items and answers to specific prompts. Nevertheless, it demonstrates how foundation models can serve as versatile engines within pathology workflows (for example, retrieving similar cases, flagging high-risk features, or predicting biomarkers) by leveraging multimodal data, especially text in natural language.

Technical underpinnings: Vision-language foundation models in pathology generally consist of a deep image encoder (often a vision transformer (ViT) or convolutional neural network (ConvNet) pretrained on generic images, then adapted to pathology) and a text encoder (often a Transformer model pretrained on general text). During training, the model adjusts these encoders so that a pathology image and its corresponding description produce *matching* embeddings. For example, PLIP fine-tuned the OpenAI CLIP model on pathology data to align visual and text features specific to histology. A major advantage is zero-shot generalization: with broad grounding in visual patterns and terminology, the model can classify or search using plain-language labels it never saw during training. However, these models generally lack a generative language module; as their text encoders aren't decoders, they cannot produce long-form narratives or interactive responses. They are limited to identification, retrieval, or labeling. In summary, vision-language foundation models like PLIP, CONCH, and MUSK act as visual intelligence modules that embed pathology images and text in a common semantic space, enabling tasks like image diagnosis prediction and case retrieval without task-specific training. They set the stage for the next category of systems that do generate text: multimodal pathology chatbots and assistants.

Multimodal chat models with generative text output

Building on pathology foundation models, researchers have begun developing **interactive multimodal AI assistants** for pathology—systems that can accept image input and engage in text-based dialog or produce detailed textual outputs. These models typically combine a vision encoder (often initialized from a model like those mentioned above) with an LLM, enabling the system to "see" an image and then "speak" or answer questions about it. A prime example is **PathChat**, introduced by Lu et al. (2024a) as a "multimodal generative AI copilot for human pathology" (Lu et al., 2024a). PathChat was built by combining a strong pathology-specific vision encoder with a pretrained LLM and fine-tuning the unified model on *456,000* visual-language instruction examples (nearly one million question-answer turns) covering diverse pathology scenarios. The training data included synthetic Q&A generated to teach the model how to discuss histology images, similar to how general vision-language assistants (like LLaVA (Liu et al., 2023) or GPT5) are trained, but with domain-specific content. The result was a pathology chatbot capable of both multiple-choice and open-ended questioning formats. Impressively, PathChat achieved state-of-the-art accuracy on multiple-choice diagnostic questions drawn from real cases across various organs and diseases. In head-to-head comparisons of free-form responses, pathologists judged PathChat's answers to open-ended pathology queries as more accurate and preferable to those from a generic GPT-4 Vision system. This demonstrates that a domain-tuned multimodal LLM can outperform even powerful general models on specialized tasks, likely due to its exposure to pathology-specific images and terminology. PathChat is designed as an interactive assistant: it can take a whole slide image (or a region) plus a question in natural language and then generate a detailed answer or explanation. For instance, given an image and the prompt "Describe the diagnostic features", PathChat can output a paragraph describing the cell morphology and suggesting possible diagnoses, much like a pathologist's reasoning. The creators envision PathChat being useful for pathology **education**, allowing students or practitioners to query an AI about a case, as well as in **human-in-the-loop clinical decision-making** as a smart support tool.

PathChat is one of several emerging multimodal pathology assistants. Another early system, **PathAsst** (Sun et al., 2024), was introduced as a generative foundation AI assistant for pathology and fine-tuned on hundreds of thousands of image-caption and instruction examples (Sun et al., 2024). Similarly, **LLaVA-Med** (Li et al., 2023) and **PA-LLaVA** (Pathology-LLaVA;

Dai et al., 2024) adapted the open-source LLaVA vision-language model to medical images, including histopathology. These models use an architecture akin to LLaVA: a pretrained Vision Transformer (sometimes first adapted to pathology images) serves as a visual encoder, producing embeddings that are then fed into an LLM (like LLaMA or Vicuna) via a projection layer (i.e., a single-layer neural network). The language model is prompted in a way that incorporates the image features (often as special tokens) and then generates a conversational answer. Fine-tuning is done on Q&A pairs about pathology images, which are sometimes generated by GPT-4 or gathered from reports. A challenge specific to pathology is the enormous size of whole slide images (WSIs). Most models so far operate on image *patches* (e.g., 224×224 pixel regions) because processing an entire WSI is computationally infeasible for transformers. This means assistants like PathChat typically analyze one region at a time. Recent efforts like **ChatEXAONEPath** (2024; Kim et al., 2025) are tackling this by introducing specialized WSI-handling strategies—for example, using a patch-wise encoder and an aggregator so the model can take multiple patches or a summary of a WSI as input.

Another innovation to improve these chat models is enhancing their **spatial awareness**. **Quilt-LLaVA** (Seyfioglu et al., 2023; Seyfioglu et al., 2023) is a notable approach that leveraged annotated instructional videos: by tracking the narrator's mouse cursor on pathology slides, they captured which image region each comment referred to, thus providing spatial grounding for training. With this data, Quilt-LLaVA was trained to describe *localized* image features (e.g., "in this region, there are clusters of eosinophils") and to perform reasoning that considers the whole slide context. This led to significantly improved accuracy on pathology visual question answering, exceeding prior state-of-the-art by >10% on some benchmarks. In essence, Quilt-LLaVA taught the model to not just identify features but also indicate *where* in the image they are and reason beyond a single patch.

From a technical perspective, multimodal chat models integrate vision and language by *binding image features into the LLM's input sequence*. For example, an image encoder (often a ViT) produces a set of feature vectors for an input image or patch. A projection module then maps these vectors into the same space as the language model's embeddings, and they are fed as " pseudo-tokens" before the textual prompt. The language model (often an instruction-tuned GPT-like model) then continues generating text conditioned on both the visual and textual tokens. Many medical versions follow a similar idea. For example, **BLIP-2** (Li et al., 2023) and related architectures introduced a Q-former that queries image features and outputs a fixed set of tokens to feed into an LLM. In PathChat's case, a "foundational vision encoder" pretrained on pathology was combined with an LLM, and the whole system was fine-tuned end-to-end on QA data. This end-to-end training aligns the vision features with the LLM's knowledge and style, teaching it how to interpret pathology images and talk about them. Importantly, the language model brings a wealth of medical knowledge (from pretraining on text like medical literature), which complements the pattern-recognition ability of the image encoder. We see this synergy in PathChat's performance: it not only recognizes features in images but can relate them to disease entities and even suggest further investigations, implying it has "learned" pathology concepts during fine-tuning. As another example, an open model, **Med-PaLM** (Singhal et al., 2023) by Google, was reported as a multimodal LLM that can do VQA on pathology and radiology images, using the Med-PaLM (text-only) LLM augmented with vision encoders. All these systems remain in research or early-stage development, but their rapid improvement suggests that practical pathology chatbots are on the horizon.

Knowledge-enhanced and domain-specific LLMs

A recurring theme is that **domain-specific knowledge** significantly enhances LLM utility in pathology. General LLMs like ChatGPT, while fluent, often lack the detailed medical or histological knowledge required for accurate pathology outputs (Ullah et al., 2024). Thus a key direction is building **knowledge-enhanced models**—either by fine-tuning LLMs on pathology resources or by integrating external knowledge bases and retrieval mechanisms. One straightforward approach is curating a pathology-focused text corpus (e.g., textbooks, atlases, research papers) and using it to fine-tune or prompt LLMs so they produce content aligned with authoritative knowledge. Omar et al. (2024) describe an initiative where they created a "curated literature database" on pathology and built a user-interactive application for precise, referenced information retrieval. The idea is that a pathologist or researcher can query the system in natural language, and the LLM will return an answer backed by citations from the pathology literature, rather than an unverifiable black-box reply. They report that such domain-specific tools *substantially reduce* the risk of inaccurate or made-up answers, making the information extraction more reliable. This approach, often called **retrieval-augmented generation (RAG)**, combines the generative ability of LLMs with a search module that fetches relevant text (e.g., from PubMed or a pathology knowledge graph). The LLM then uses these retrieved facts to formulate its answer, significantly improving factual accuracy and trustworthiness.

Another complementary strategy is integrating **structured medical knowledge** into LLM-based reasoning. For example, researchers have proposed systems like **DR.KNOWS** (Diagnostic Reasoning Knowledge Graph System) that marry LLMs with ontology-based knowledge graphs (Gao et al., 2025). DR.KNOWS uses the Unified Medical Language System (UMLS) to retrieve a network of disease-symptom-pathology relationships relevant to a case and feeds this structured context into the LLM to guide its diagnostic predictions. In initial studies on clinical data, this approach improved diagnostic accuracy and produced reasoning more aligned with correct clinical logic. In pathology, one can imagine using knowledge graphs of histologic features, immunohistochemical markers, and diagnoses to help an LLM propose a differential diagnosis grounded in known disease patterns. By anchoring the model's suggestions to established knowledge (e.g., "keratin-positive, EMA-negative suggests X tumor"), such systems could reduce the chance of AI hallucinations and increase interpretability. While this is still experimental, it illustrates how LLMs can be **knowledge-enabled** for pathology decision support.

Notably, the foundation models themselves have implicitly injected knowledge by training on text from pathology reports and publications. MUSK, for instance, included pathology reports and medical journal text in its pretraining, meaning its visual representations were learned alongside rich cancer and pathology semantics. This likely contributed to MUSK's success in tasks such as biomarker prediction and disease prognosis. The model has effectively learned how certain histologic patterns correlate with molecular findings or outcomes, as it encountered these associations in text form. In a sense, MUSK and similar models are already knowledge-enhanced LLMs: they leverage large text corpora to enrich image analysis with clinical context. Going forward, we anticipate more explicit inclusion of external knowledge. For example, a future system might allow a pathology LLM to query a database of prior cases or WHO classification criteria when formulating a report or differential diagnosis. Indeed, one of the **applications** we discuss in **Section** Workflow Integration in Digital Pathology - Postanalytical reporting, regarding differential diagnosis assistance, would benefit from an LLM that can draw on a comprehensive knowledge of diseases, either from its training or via on-the-fly retrieval.

In summary, a crucial aspect of LLMs in pathology is ensuring they have access to reliable domain knowledge, whether through careful fine-tuning on trusted data, embedding them in a pipeline with literature retrieval, or linking to structured medical knowledge. Such enhancements can help overcome the "black box" and hallucination issues, increase factual accuracy, and build user trust in AI suggestions (Ullah et al., 2024). As noted in this recent scoping review, the lack of contextual medical understanding and potential biases are major challenges with deploying raw LLMs in diagnostic pathology. Tailoring these models with pathology-specific information and oversight is therefore not just beneficial but likely necessary for safe and effective use.

Workflow integration in digital pathology

This section presents an end-to-end digital pathology workflow comprising three phases—preanalytical preparation, analytical computation, and postanalytical reporting—and explains how AI systems, such as segmentation models and large language/vision-language models (LLMs/VLMs), assist pathologists. We emphasize data acquisition and processing, common segmentation strategies and result formats, the translation of computational outputs into pathologist-readable reports, and how AI supports diagnostic judgment and hypothesis generation for novel disease patterns. The chapter concludes with ethical, legal, and regulatory considerations related to data provenance, privacy, and equitable deployment.

Preanalytical preparation

A robust workflow starts upstream of modeling. Fig. 14.2 illustrates the preanalytical preparation workflow. Slides are created and scanned to preserve the diagnostic signal and capture context. Record specimen handling (anatomic source, fixation, thickness, stain) and scanner settings (magnification, numerical aperture, illumination, compression), along with microns-per-pixel and pyramid levels.

Upon ingestion, deidentify and validate: Is the metadata complete? Are the values plausible? Is tissue present? Automated quality control (QC) procedures can detect common artifacts such as defocus, folds, debris, pen marks, and scant tissue, enabling problematic slides to be re-scanned rather than introducing bias. Before measurement, harmonize the scale across devices (e.g., calibrate microns per pixel [MPP]), normalize the stain without erasing diagnostic contrast, and tile slides into patches while retaining coordinates so every metric maps back to the glass. Where clinical or molecular data exist, link them via stable identifiers. The result is a clean, auditable dataset with clear provenance and a documented chain of custody.

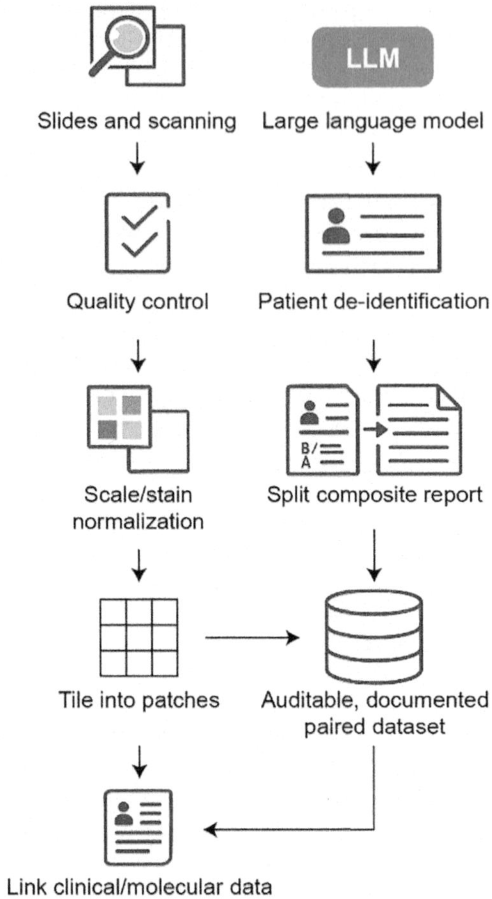

FIG. 14.2 Preanalytical preparation workflow: Steps from slide scanning, quality control, and stain harmonization to deidentification, report splitting, and dataset assembly for downstream analysis.

Critical preanalytical tasks include ensuring patient privacy and organizing textual data for computational analysis. For example, **deidentification using local LLMs** provides an automated solution for removing patient-identifiable information while maintaining data utility. Traditional anonymization relies on manual work or rigid keyword-based systems that are time-consuming and often miss contextual information. Local LLM-based systems operate on-site without cloud processing. For example, the LLM-Anonymizer (Wiest et al., 2025) pipeline, using Llama-3 70B, achieved 99.24% accuracy in removing personally identifiable information (PII), missing only 0.76% of sensitive data while incorrectly redacting 2.43% of characters. The system runs on local hardware with a web interface requiring no programming skills. The process targets key categories: patient names, birth dates, addresses, and patient IDs. LLMs use grammar-based sampling to ensure consistent output structure when extracting these elements from pathology reports. Current results show Llama-3 variants achieving the highest sensitivity for correctly identifying personal identifiers, particularly patient names and dates of birth. As larger and more capable models become available for local deployment, we can expect further improvements in deidentification accuracy and reliability.

Another essential task is the structured separation of pathology reports. Complex surgical cases often generate composite pathology reports describing multiple tissue specimens within a single document. Each specimen requires separate analysis and correlation with its corresponding histological slide. LLMs bring the potential to automatically identify and separate these individual specimen descriptions, distinguishing between different anatomical sites, tumor locations, and procedural findings. This separation enables precise matching between each slide and its diagnostic text, which is essential for multimodal AI training and ensures accurate image-text pairing. The structured approach also facilitates workflow management by organizing multiple specimens from complex cases while maintaining the overall case context.

Analytical phase

During the analytical phase, LLMs can read across the EHR (e.g., clinical notes, pathology requisitions, radiology reports, medication lists) and condense the information into a brief summary for pathology. The model highlights items that shift interpretation: prior malignancy, site/laterality, procedures and timing, morphology-altering therapies, infections, implants, and comorbidities. It understands negation and temporality, and it standardizes units and abbreviations, providing a concise, source-linked summary with the case. It handles negation and temporality ("no dysphagia since May", "started prednisone 10 mg 2 weeks ago") and normalizes units and abbreviations, so the pathologists can get a clean, source-linked synopsis attached to the case.

In addition to summarization, the LLM can compute a triage score that prioritizes cases for diagnosis. It considers clinical urgency (STAT flags, suspected sepsis, active bleeding), time-sensitive rules (frozen sections, transplant margins), specimen risk (positive margin likelihood, high-grade lesion on imaging), and operational factors (aging in queue, required ancillary tests). The score populates a dynamic worklist that prioritizes high-impact or time-critical cases while respecting service lines and sign-out assignments. As outcomes accumulate, the system learns which features best predicted "add-on work" or discordant revisions and updates the prioritization policy. In practice, each digital case opens with a one-screen "context card" that includes the distilled clinical story, a timeline of key events, relevant laboratory and imaging snippets, and LLM-generated prompts such as "consider deeper levels," "IHC likely required," or "prior biopsy at the same site (compare slide X)." The model can prefetch historical slides, suggest stain panels based on differential hints, and route specimens to subspecialists when the inferred organ/site or tumor pattern meets service thresholds.

While LLMs can be highly useful in the analytical phase, AI governance is equally essential. All summaries and triage scores should be considered recommendations with clear confidence levels and citations, and the pathologist should remain the final authority. Regarding data, PHI should be deidentified upfront when appropriate, access should be role-controlled, and every model action should be auditable. Calibration and bias monitoring prevent systematic over- or under-prioritization of certain patient groups or specimen types, while feedback loops (overrides, turnaround outcomes) continuously retrain or adjust the system to keep safety and quality at the center of faster, more focused sign-out.

Postanalytical reporting

LLMs play an important role in postanalytical reporting. In digital pathology, LLMs can act both as a "reporting assistant" and a "teaching mentor" (Fig. 14.3). For reporting, an LLM can extract key elements from the LIS and image-analysis outputs (e.g., specimen and site, tumor size and grade, margin distances, LVI/PNI, and quantitative IHC). It can then generate a structured, guideline-aligned summary (e.g., CAP/WHO style), standardize terminology, flag omissions or inconsistencies, and link each statement to corresponding tables, coordinates, and figures to ensure auditability. It can harmonize messages across cases, generate evidence-based discussion snippets, and perform pre-sign-out checks for completeness and internal consistency. For training next-generation pathologists, the same LLM can transform real or synthetic cases into tiered learning modules, provide stepwise hints tied to a trainee's annotations and reasoning, compare answers with the reference standard to deliver granular feedback, and automatically compile "error books", flashcards, or oral-exam prompts at varying difficulty. With multimodal interfaces, it can couple what the trainee marks on the slide to what they write, reinforcing a "see—think—justify" loop. All of this must operate within a safe governance layer: deidentification and access controls, clear attribution of model provenance, and a human-in-the-loop sign-out where every draft remains a suggestion and the pathologist has final authority.

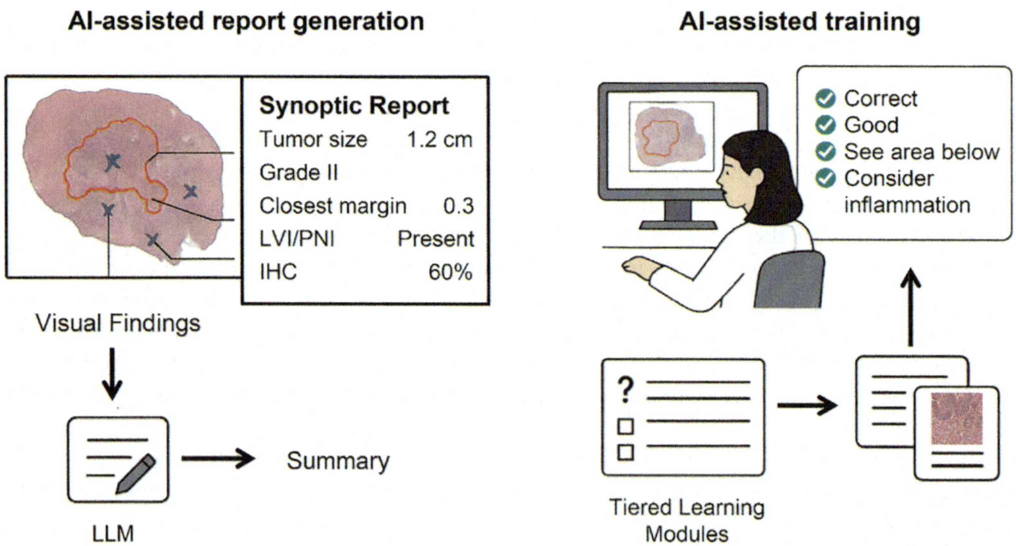

FIG. 14.3 LLM used in postanalytical reporting. AI can help with report generation by leveraging vision models to capture important components and using LLMs to generate reports. LLMs can also create tiered training modules with feedback and study tools.

Ethics and legal compliance

Lastly, guardrails are nonnegotiable. Every slide must have a lawful basis for use, be properly deidentified, and be accessed only as needed. We should routinely audit performance across sites, scanners, stains, and patient subgroups; display uncertainty; and let clinicians override easily. Keep complete provenance so an external reviewer can see what data and model produced which decision. If local policy states that data can't leave the hospital or must sit in a restricted cloud, deploy it that way. The aim is straightforward: deliver real clinical benefit with transparency and respect for patients' data.

Applications in pathology practice

LLMs integrated into digital pathology can impact many parts of the pathologist's workflow (Fig. 14.4). In this section, we will highlight several key applications and use cases that have been explored.

Case retrieval and image search

LLMs are widely adopted in pathology practice. One practical use of vision-language pathology models is **content-based retrieval** of similar cases or images. A pathologist encountering an unfamiliar finding could benefit from instantly retrieving reference cases with matching morphology or diagnosis. Models like PLIP and CONCH enable exactly this: they embed images and text in the same space, so one can search through a large database of images using a text query (or vice versa). For example, using PLIP, one could type *"glomerulus with crescent formation"* and retrieve kidney pathology images showing crescents, or submit an image of an unusual tumor and retrieve the most similar images along with their diagnoses (Huang, 2023). In PLIP's demonstration, users could *"retrieve similar cases by either image or natural language search"*, greatly aiding knowledge sharing and reference lookup. Such tools

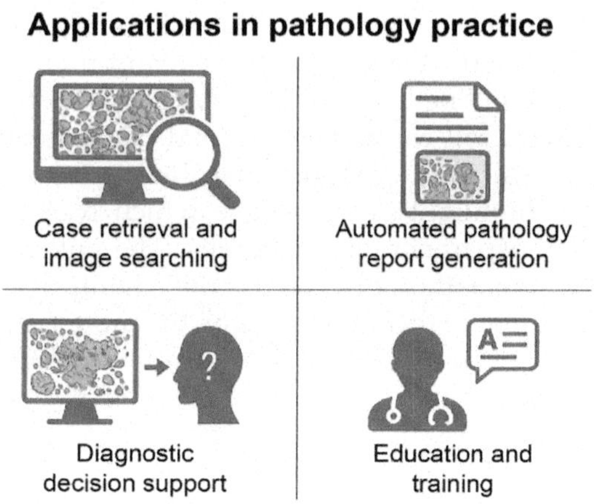

FIG. 14.4 LLM applications in pathology practice include case retrieval and image searching, automated pathology report generation, diagnostic decision support, and pathology education and training.

essentially act as a Google for pathology images—a powerful aid for differential diagnosis, teaching, and even quality assurance. CONCH further showed that zero-shot image retrieval works well: without training on a specific dataset, it could find relevant images and captions based purely on the multimodal embedding, outperforming previous systems (Lu et al., 2024b). In practice, this means a pathology department that has digitized archives could, in the future, allow a pathologist to quickly query that archive (or an external reference database) for similar cases when they face a challenging slide. This is akin to getting a second opinion from thousands of prior cases in seconds. Early examples of such a system in use are mostly at the research stage, but the concept has clear value in clinical decision support. It can help ensure rare patterns are recognized by finding precedents and help pathologists include or exclude entities in their differential diagnosis by comparing visuals.

Another retrieval use case is in tumor boards or multidisciplinary discussions, where relevant prior cases or images need to be pulled up quickly. A vision-language model could, for instance, retrieve all instances of "ALK-positive lung adenocarcinoma with signet-ring cells" from a hospital's database when preparing for a lung cancer tumor board. Moreover, the same technology can facilitate **knowledge sharing and education**: pathologists on online forums (or trainees in a teaching session) could use it to find example images to discuss without laborious manual searching (Huang, 2023). The OpenPath dataset itself was built on the idea that pathologists were already sharing and describing cases on Twitter; a retrieval model just leverages that collective knowledge more efficiently.

Automated pathology report generation

Generating a complete pathology report from digitized slides has long been a goal, recently made feasible by multimodal LLMs. **HistoGPT** (2025) is a landmark in this area—a vision-language model specifically designed to write full **dermatopathology reports** given a patient's whole slide images (Tran et al., 2025). HistoGPT was trained on over 15,000 WSIs with their accompanying reports, learning to translate microscopic findings into the kind of narrative a pathologist would dictate. The model's performance is remarkable: for common skin cancers, AI-generated reports were judged to *match the quality of human-written reports* by both NLP metrics and expert reviewers. In a multicenter evaluation, HistoGPT's reports accurately identified tumor type, subtype, invasion depth (Breslow thickness), and margin status, even in a zero-shot manner for external cases it had never seen. This suggests that for routine diagnoses (like typical basal cell or squamous cell carcinomas), an AI can indeed produce a reliable first draft of the report. The HistoGPT study shows that the model can describe the tissue, provide a diagnosis, and even highlight in the image which regions correspond to each descriptive phrase, providing transparency via heatmaps. The authors emphasize this as a *"second opinion"* or drafting tool—the pathologist can use the AI output as a general overview and then adjust or verify it for the final report. Importantly, the model can also populate structured templates by extracting key findings (e.g., tumor size, margin) as needed.

Beyond HistoGPT, the general capability of LLMs to assist in report generation is supported by other studies. Cheng (2024) notes that LLMs can "populate a template using unstructured text, extract data from different sources, and combine [them] into a single report," thereby automating much of the clerical aspect of report writing (Cheng, 2024). For instance, if a pathologist has free-text notes or a preliminary diagnosis, an LLM could structure it into the standard report format or even translate it into lay language for patient communication (Siepmann et al., 2025). LLMs have been used experimentally to simplify pathology reports into patient-friendly versions, improving doctor-patient communication (e.g., explaining a cancer pathology report in plain language; Yang et al., 2025).

In a routine workflow, an AI that generates draft reports could save significant time for pathologists, allowing them to focus on slide interpretation and difficult cases. As the HistoGPT introduction points out, experienced pathologists often diagnose quickly by sight, but *documenting* those findings takes longer—dictating a full report is time-consuming and not always standardized (Tran et al., 2025). Automating that documentation can improve efficiency and consistency. It may also reduce errors of omission (the AI will always fill in required fields, whereas a human might occasionally miss something when hurried). Of course, these systems must be thoroughly validated. HistoGPT's authors stress that it works best on "common and homogeneous" cases—rare or complex cases still pose challenges. But even handling the bulk of straightforward cases could be a game-changer in busy labs.

Looking ahead, similar models might be developed for other specialties (e.g., a HistoGPT for GI biopsies or cytology). Some early work in radiology report generation provides a template, and pathology—with its detailed textual reports—is a natural next frontier. It is conceivable that within a few years, digital pathology systems will come with an "AI scribe" feature: as soon as the slides are scanned and analyzed, the system presents a draft report including diagnosis, key histologic findings, tumor staging elements, and even ancillary study results for the pathologist to review and sign out. This could speed up turnaround time, especially for straightforward cases, and help address workload issues in pathology.

Diagnostic decision support and differential diagnosis

One of the most exciting applications of LLMs in pathology is as a **diagnostic assistant**—essentially a smart companion that can help the pathologist consider possibilities, double-check findings, and even suggest next steps. Because LLMs—especially when combined with vision models—can interpret both images and text, they are uniquely suited to support the cognitive aspects of diagnosis. A near-future scenario described by Lu et al. (PathChat's team) is illustrative: "An AI copilot would be able to ingest a histopathology image, provide an initial assessment of the morphology, and identify potential features of malignancy." Then, the pathologist can provide clinical context or ask the model to "suggest a differential diagnosis." The model, using its learned pathology knowledge, would list likely diagnoses based on the image and context. The pathologist might follow up by asking for suggested ancillary tests or immunohistochemical stains to distinguish those possibilities—and the AI can *"provide helpful suggestions for ancillary testing"*. After the additional results are provided, the model could *"make a final deduction and arrive at a diagnosis."* This conversation mimics how a human consultant might assist: the AI highlights key features, weighs the differential, and recommends confirmatory steps, but the human makes the ultimate judgment. Such a system could be invaluable in complex cases or in training settings.

Even in simpler usage, an LLM could serve as a **safety net** by suggesting diagnoses that should be considered. For example, if a pathologist describes a case to the LLM (or if the LLM reads the report draft), it could suggest, "Have you considered X?" for a rare entity that matches certain findings, thus reducing the chance of missed diagnoses. In a study of multiple-choice pathology questions, PathChat demonstrated expert-level skill, indicating it can reliably identify the correct diagnosis from an image-plus-question context. This suggests these models have learned features of many diseases. They can also incorporate patient data: since models like MUSK and PathChat were trained with text inputs, patient history or lab results can be included in the prompt along with image descriptions. The integrated analysis (clinical-pathologic correlation) is something pathologists do mentally; an AI might highlight, for instance, that a certain morphologic pattern together with a specific clinical background is highly indicative of a particular diagnosis.

For differential diagnosis specifically, LLM-based tools could generate a list of possible entities given a set of findings, along with a brief rationale for each. This could be especially useful for trainees as a learning aid ("digital tutor") or for general pathologists encountering a case outside their usual specialty. Some early work outside pathology has shown that LLMs can assist with diagnostic reasoning when grounded in medical knowledge (Gao et al., 2025). In pathology, a small study in *Diagnostic Pathology* (2024) evaluated ChatGPT on pathology tasks and noted that it can draft plausible differentials but with caveats (Ullah et al., 2024). The main concerns are ensuring the AI's suggestions are accurate and not biased by its training data—hence the value of domain-specific fine-tuning and knowledge integration, as discussed. With those improvements, an AI differential generator could become a standard feature in digital viewers: imagine hovering over an image region and the system listing "likely entities: 1. X (>90% match), 2. Y (~10%), consider special stain Z to confirm".

In terms of **clinical impact**, such decision support could reduce diagnostic errors and provide confidence in challenging cases. It is well known that even expert pathologists benefit from second opinions in difficult cases—an AI that is always available to consult could catch things a tired human might miss. On the flip side, pathologists must be cautious to avoid overreliance. The AI might occasionally suggest an incorrect diagnosis; thus, as with any clinical decision support, the final decision must rest with the human who interprets the AI input judiciously. Overreliance on AI can risk "de-skilling and automation bias" in health professionals, so the goal is to **assist/augment** rather than replace the pathologist's expertise.

Pathology education and training

LLM-based tools also hold great promise for medical education, particularly in pathology, which requires mastering both visual pattern recognition and extensive textual knowledge (terminology, criteria, clinical implications). An interactive pathology chatbot can serve as a **tutor or study partner** for trainees. For example, a resident can show an image to the AI and ask, "What features do you see, and what diagnoses should I consider?"—essentially simulating a discussion with an attending pathologist. The AI, like PathChat, can respond with a description of the histology and a differential diagnosis, possibly explaining why each differential is considered based on the image. This immediate feedback and explanation are invaluable for learning. It also allows students to learn at their own pace: the AI is infinitely patient and can answer any number of "simple" questions without judgment.

Furthermore, LLMs can generate **educational content**: practice quiz questions, flashcards, and explanatory summaries. In fact, ChatGPT and similar models have been used informally by students to quiz themselves on pathology topics. With domain-specific fine-tuning, the questions can be made more high-yield and accurate. LLMs can also summarize complex journal articles or textbook sections into concise outlines for review. For instance, one could ask an LLM, "Summarize the WHO criteria for diagnosing myelodysplastic syndrome," and get a quick answer. In pathology training, gross pathology and microscopic findings must be correlated—a multimodal model could even take an image of a gross specimen or radiology scan and explain the pathologic correlation.

Another educational use is **assessment**: providing an AI with a pathology report or description and having it point out errors or ask for clarification could train residents in report writing. Some have suggested using LLMs to simulate pathology board exams or tumor boards for practice, where the AI takes on the role of a colleague asking questions or a clinician providing new information, testing the trainee's knowledge in real-time. The interactive and on-demand nature of LLM assistants aligns well with the needs of modern education, which values active learning and accessibility.

Initial experiences with PathChat in research settings indicate that pathologists found its explanations and answers generally helpful and aligned with expert reasoning. Of course, educators will need to vet these tools for accuracy and ensure they supplement, rather than replace, critical thinking. Given the complexity of pathology, having a knowledgeable AI that a trainee can consult anytime (even outside of rotations) could **democratize access to expertise**, especially in regions or institutions with limited subspecialty mentors (Omar et al., 2024). It also offers a way to standardize training by providing all learners with a baseline of high-quality explanations and references.

Challenges and future directions

Despite the rapid progress and promising potential of LLMs in digital pathology, significant challenges remain before these tools can be widely adopted in routine practice. One major concern is **accuracy and reliability**. LLMs, especially when not properly fine-tuned, are prone to *hallucinations*—producing incorrect statements with confident phrasing. In a pathology context, a hallucination could mean a falsely identified feature or a made-up diagnostic criterion, which could be dangerous. For example, if asked to provide references for a rare diagnosis, a generic LLM might fabricate a citation (Cheng, 2024). In the study by Bhattacharyya et al. (Bhattacharyya et al., 2023), nearly half (47%) of the references provided by ChatGPT-3.5 in generated medical text were fabricated, another ~46% were authentic but inaccurate, and only 7% were error-free. This is unacceptable for clinical or scientific use. Domain-specific tuning and knowledge integration (as discussed) are crucial to mitigate this, and any AI output should ideally be traceable to evidence—hence the interest in models that can cite literature. In addition, *interpretability* remains a challenge, as these models are black boxes in how they weigh visual features and textual knowledge. Some solutions, such as

HistoGPT's attention maps and DR.KNOWS's knowledge paths, provide glimpses into why the model suggests certain outputs, which is reassuring. Ongoing research into explainable AI for pathology will likely augment LLM-based systems with heatmaps, highlighted text, and case-based reasoning to justify their outputs.

Data limitations pose another challenge. Training these advanced models requires extremely large and diverse datasets of images and paired text. Unlike natural images, pathology data is harder to obtain due to privacy concerns and the need for expert annotation. Projects like OpenPath, CONCH, and HistoGPT have assembled some of the largest pathology datasets to date, but these remain orders of magnitude smaller than the datasets used for general-purpose models. This raises the risk that models may not generalize to rare diseases or unusual histology patterns. It can also introduce **bias**—if a model is trained predominantly on data from a certain institution, it may reflect that lab's diagnostic criteria or patient population, potentially reducing performance on underrepresented groups. Ensuring training data diversity (across geographies, patient demographics, and lab techniques) is important to prevent AI that works great in one setting but poorly in another. The community is moving toward sharing larger datasets (some are public, like TCGA, or involve multicenter collaborations) to improve this.

There are also **regulatory and ethical considerations**. Using LLMs in patient care raises concerns about data privacy (slides and reports are personal health information). If an LLM is cloud-hosted (as many currently are), uploading images or text could violate privacy regulations unless proper agreements are in place. This is driving interest in on-premises or open-source models that can be deployed within a hospital's secure network. In addition, in many jurisdictions, introducing AI for diagnosis triggers regulatory oversight (e.g., FDA approval in the United States). Currently, no LLM-based pathology tool is FDA-approved; any clinical use would be experimental. Validation through clinical trials will be needed to demonstrate that these tools improve outcomes or efficiency without compromising quality. Some recent works have started prospective validation—e.g., HistoGPT's multicenter trial with pathologists reviewing AI reports is a step in that direction.

From an ethical standpoint, there is concern about the impact on the pathology workforce and training pipeline. Will reliance on AI for easy cases erode the skills of junior pathologists ("de-skilling")? On the other hand, many labs are overloaded and understaffed; AI assistance could alleviate drudgery and free up experts for more critical tasks, which is a positive. It will be important to maintain a balance: using AI as a *tool* rather than a crutch, and continuing to cultivate robust training for new pathologists with AI as an adjunct. Professional guidelines and consensus will be needed on how to integrate AI into pathology practice safely. Encouragingly, pathology as a field has adapted to new technologies in the past (immunohistochemistry, molecular testing, digital imaging) by setting standards and ensuring quality control—AI will likely follow a similar path with validation studies, proficiency testing, and clear definitions of its role.

Future directions for LLMs in pathology are broad. We can expect even more capable models that handle *multimodality* beyond images and text—for example, integrating genomics, radiology, and pathology data together for a comprehensive diagnostic AI (some work in multimodal oncology is already underway). Foundation models might expand to include **cytology** images, which pose unique challenges like cell focus and thickness (some specialized scanners and z-stacking techniques are being explored for digital cytology, which an AI could take advantage of—as per **Chapter 6**). Another promising area is using generative models to create synthetic pathology images for data augmentation, though it carries risks if not carefully marked as synthetic. **Continual learning** is another aspect: LLMs could be updated with the latest medical knowledge or a hospital's new cases to remain current, much like a pathologist's knowledge evolves.

In summary, LLMs and VLMs have the potential to transform pathology workflows by automating routine tasks, providing decision support, and enhancing education. The progress between 2023 and 2025 has been rapid—from basic CLIP adaptations to sophisticated chat assistants and report generators. The impact on pathology practice could be significant: faster turnaround times, more consistent reports, better access to expertise, and improved diagnostic confidence. Yet, careful implementation is required. By combining peer-reviewed research advances with thoughtful clinical validation and ethical guardrails, the pathology community can harness these AI tools to ultimately improve patient care while elevating the role of the pathologist in the diagnostic process.

References

Bhattacharyya, M., Miller, V.M., Bhattacharyya, D., Miller, L.E., 2023. High rates of fabricated and inaccurate references in ChatGPT-generated medical content. Cureus. 2168-8184. 15 (5). doi:10.7759/cureus.39238.

Cheng, J., 2024. Applications of large language models in pathology. Bioengineering. 2306-5354. 11 (4), 342. doi:10.3390/bioengineering11040342.

Dai, D., Zhang, Y., Xu, L., Yang, Q., Shen, X., Xia, S., Wang, G., 2024. PA-LLaVA: a large language-vision assistant for human pathology image understanding. In: Proceedings – 2024 IEEE International Conference on Bioinformatics and Biomedicine, BIBM 2024. 9798350386226. Institute of Electrical and Electronics Engineers Inc., China, pp. 3138–3143. http://ieeexplore.ieee.org/xpl/mostRecentIssue.jsp?punumber=10821710. 10.1109/BIBM62325.2024.10821785

Gao, Y., Li, R., Croxford, E., Caskey, J., Patterson, B.W., Churpek, M., Miller, T., Dligach, D., Afshar, M., 2025. Leveraging medical knowledge graphs into large language models for diagnosis prediction: design and application study. JMIR AI. 2817-1705. 4 (1), e58670. doi:10.2196/58670.

Huang, Z., et al, 2023. A visual-language foundation model for pathology image analysis using medical Twitter. Nat. Med. 29 (9), 2307–2316. doi:10.1038/s41591-023-02504-3. 37592105.

Kim, S., Lee, S., Jang, J., 2025. ChatEXAONEPath: an expert-level multimodal large language model for histopathology using whole slide images. arXiv. 23318422. Available from: https://arxiv.org. doi:10.48550/arXiv.2504.13023.

Li, C., Wong, C., Zhang, S., Usuyama, N., Liu, H., Yang, J., Naumann, T., Poon, H., Gao, J., 2023. LLaVA-Med: training a large language-and-vision assistant for biomedicine in one day. arXiv. 23318422. Available from: https://arxiv.org.

Li, J., et al., 2023. BLIP-2: bootstrapping language-image pre-training with frozen image encoders and large language models. In: Proceedings of the 40th International Conference on Machine Learning, Art. No. 814, 13. JMLR.org.

Liu, H. et al., 2023. Visual instruction tuning (Oh, A., et al., Eds.). Adv. Neural Inf. Process. Syst. 36, 34892–34916.

Lu, M.Y., Chen, B., Williamson, D.F.K., Chen, R.J., Liang, I., Ding, T., Jaume, G., Odintsov, I., Le, L.P., Gerber, G., Parwani, A.V., Zhang, A., Mahmood, F., 2024b. A visual-language foundation model for computational pathology. Nat. Med. 1546170X. 30 (3), 863–874. Available from: https://www.nature.com/nm/. doi:10.1038/s41591-024-02856-4.

Lu, M.Y., Chen, B., Williamson, D.F.K., Chen, R.J., Zhao, M., Chow, A.K., Ikemura, K., Kim, A., Pouli, D., Patel, A., Soliman, A., Chen, C., Ding, T., Wang, J.J., Gerber, G., Liang, I., Le, L.P., Parwani, A.V., Weishaupt, L.L., Mahmood, F., 2024a. A multimodal generative AI copilot for human pathology. Nature. 14764687. 634 (8033), 466–473. Available from: https://www.nature.com/nature/. doi:10.1038/s41586-024-07618-3.

Omar, M., Ullanat, V., Loda, M., Marchionni, L., Umeton, R., 2024. ChatGPT for digital pathology research. Lancet Digit. Health. 25897500. 6 (8), e595. doi:10.1016/s2589-7500(24)00114-6.

Radford, A., et al., 2021. Learning transferable visual models from natural language supervision. In: Meila, M., Zhang, T., (Eds.), Proceedings of the 38th International Conference on Machine Learning, PMLR, pp. 8748–8763.

Seyfioglu, M.S., Ikezogwo, W.O., Krishna, R., Shapiro, L., Ghezloo, F., 2023. Quilt-LLaVA: visual instruction tuning by extracting localized narratives from open-source histopathology videos. arXiv. 23318422. Available from: https://arxiv.org. doi:10.48550/arXiv.2312.04746.

Siepmann, R.M., Baldini, G., Schmidt, C.S., Truhn, D., Müller-Franzes, G.A., Dada, A., Kleesiek, J., Nensa, F., Hosch, R., 2025. An automated information extraction model for unstructured discharge letters using large language models and GPT-4. Healthc. Anal. 27724425. 7, 100378. doi:10.1016/j.health.2024.100378.

Singhal, K., Azizi, S., Tu, T., Mahdavi, S.S., Wei, J., Chung, H.W., Scales, N., Tanwani, A., Cole-Lewis, H., Pfohl, S., Payne, P., Seneviratne, M., Gamble, P., Kelly, C., Babiker, A., Schärli, N., Chowdhery, A., Mansfield, P., Demner-Fushman, D., Agüera Arcas, B., Webster, D., Corrado, G.S., Matias, Y., Chou, K., Gottweis, J., Tomasev, N., Liu, Y., Rajkomar, A., Barral, J., Semturs, C., Karthikesalingam, A., Natarajan, V., 2023. Large language models encode clinical knowledge. Nature. 0028-0836. 620 (7972), 172–180. doi:10.1038/s41586-023-06291-2.

Sun, Y., et al., 2024. PathAsst: a generative foundation AI assistant towards artificial general intelligence of pathology. In: Proceedings of the Thirty-Eighth AAAI Conference on Artificial Intelligence and Thirty-Sixth Conference on Innovative Applications of Artificial Intelligence and Fourteenth Symposium on Educational Advances in Artificial Intelligence, Art. No. 560, AAAI Press, p. 9.

Tran, M., Schmidle, P., Guo, R.R., Wagner, S.J., Koch, V., Lupperger, V., Novotny, B., Murphree, D.H., Hardway, H. D., D'Amato, M., Lefkes, J., Geijs, D.J., Feuchtinger, A., Böhner, A., Kaczmarczyk, R., Biedermann, T., Amir, A.L., Mooyaart, A.L., Ciompi, F., Litjens, G., Wang, C., Comfere, N.I., Eyerich, K., Braun, S.A., Marr, C., Peng, T., 2025. Generating dermatopathology reports from gigapixel whole slide images with HistoGPT. Nat. Commun. 2041-1723. 16 (1). doi:10.1038/s41467-025-60014-x.

Ullah, E., Parwani, A., Baig, M.M., Singh, R., 2024. Challenges and barriers of using large language models (LLM) such as ChatGPT for diagnostic medicine with a focus on digital pathology – a recent scoping review. Diagn. Pathol. 1746-1596. 19 (1). doi:10.1186/s13000-024-01464-7.

Wiest, I.C., Leßmann, M.-E., Wolf, F., Ferber, D., Van Treeck, M., Zhu, J., Ebert, M.P., Westphalen, C.B., Wermke, M., Kather, J.N., 2025. Deidentifying medical documents with local, privacy-preserving large language models: the LLM-anonymizer. NEJM AI. 2836-9386. 2 (4). doi:10.1056/aidbp2400537.

Xiang, J., Wang, X., Zhang, X., Xi, Y., Eweje, F., Chen, Y., Li, Y., Bergstrom, C., Gopaulchan, M., Kim, T., Yu, K.-H., Willens, S., Olguin, F.M., Nirschl, J.J., Neal, J., Diehn, M., Yang, S., Li, R., 2025. A vision–language foundation model for precision oncology. Nature. 0028-0836. 638 (8051), 769–778. doi:10.1038/s41586-024-08378-w.

Yang, X., Xiao, Y., Liu, D., Zhang, Y., Deng, H., Huang, J., Shi, H., Liu, D., Liang, M., Jin, X., Sun, Y., Yao, J., Zhou, X., Guo, W., Yang, H., Tang, W., Xu, C., 2025. Enhancing doctor-patient communication using large language models for pathology report interpretation. BMC Med. Inform. Decis. Mak. 1472-6947. 25 (1). doi:10.1186/s12911-024-02838-z.

CHAPTER

15

Usability of artificial intelligence tools in pathology practice

Andrey Bychkov[a] *and Yosep Chong*[b]

[a]**Department of Pathology, Kameda Medical Center, Kamogawa, Chiba, Japan**
[b]**Department of Hospital Pathology, The Catholic University of Korea College of Medicine, Seoul, South Korea**

Introduction

The integration of artificial intelligence (AI) into pathology practice has progressed rapidly, with a growing number of AI-based tools being developed and evaluated for diagnostic support, screening, triage, and quality assurance (Matthews et al., 2024; McGenity et al., 2024; Chong and Bychkov, 2025; Chong et al., 2025). Despite promising performance in research settings, the successful deployment of these tools in real-world clinical workflows remains uneven (Berbís et al., 2023; Pinto et al., 2023). One of the key determinants of implementation success is usability—a multifaceted concept that encompasses the effectiveness, efficiency, and satisfaction with which users can achieve their intended goals in a particular context of use (Riihiaho, 2018).

A persistent challenge lies in bridging the gap between academic research and clinical AI deployment. While research outputs often prioritize novelty, benchmark performance, and academic recognition, clinical adoption depends on consistency, safety, trustworthiness, and usability.

Theoretical advances and elegant codebases may excel in controlled environments, yet they can falter when confronted with the realities of clinical data that are incomplete, noisy, or heterogeneous. Edge cases in the lab frequently become routine in practice. Similarly, what counts as success in academia (e.g., developing innovative algorithms) differs markedly from success in healthcare, where impact on patient outcomes matters most. Addressing this translational gap requires more than technical refinement; it necessitates collaboration between developers and clinicians, alignment of incentives, and a design philosophy that integrates human factors, operational constraints, and user experience from the outset.

Advances in Digital Pathology and Artificial Intelligence in Pathology
https://doi.org/10.1016/B978-0-443-40383-5.00017-0

In the context of AI-assisted pathology, usability plays a central role in bridging the gap between technological innovation and clinical utility. Tools that fail to align with the needs, expectations, and workflows of pathologists and laboratory personnel are unlikely to gain sustained adoption, regardless of their algorithmic accuracy. Conversely, systems designed with a deep understanding of clinical realities can facilitate smoother integration, increase user trust, and ultimately contribute to improved diagnostic performance and patient safety.

Central to this discussion is the concept of human factors engineering (HFE), which emphasizes the importance of designing systems that accommodate human cognitive and physical capabilities (Daniels et al., 2021). In AI tool development for pathology, this includes considerations such as interface design, feedback mechanisms, error tolerance, transparency in decision-making, and adaptability to diverse practice settings. By incorporating human factors principles early and systematically into the development lifecycle, developers can create tools that not only perform well technically but also align with user behavior, expectations, and constraints in clinical environments.

This chapter aims to elucidate the importance of usability and human factors in the design, validation, and implementation of AI tools in pathology. It further explores strategies to foster meaningful collaboration between developers and end-users, thereby enhancing the clinical readiness of AI-powered solutions.

Key terminology

Building upon the foundational importance of usability, it is essential to establish a clear understanding of the key terminologies (Table 15.1) and the regulatory frameworks that govern the design and evaluation of software-based medical tools, particularly Software as a Medical Device (SaMD).

Usability is defined by international standards (e.g., IEC 62366-1:2015, FDA guidance, and MFDS guidelines) as the extent to which specified users can use a product to achieve specific goals with effectiveness, efficiency, and satisfaction in a particular context of use. Within pathology, this implies that AI tools must support pathologists and laboratory staff in performing diagnostic tasks without undue complexity or error-prone processes. Usability engineering, often referred to interchangeably with HFE in regulatory contexts, denotes a structured process of design and evaluation that aims to ensure this alignment between system functionality and user capabilities.

A critical distinction exists between usability engineering as an iterative, formative design process and human factors validation testing as a summative, regulatory-mandated demonstration that the final product can be used safely and effectively. While formative evaluations occur early in development to refine user interfaces and interaction flows, human factors validation focuses on confirming that users can reliably perform critical tasks—those interactions where failure could result in serious harm to the patient or user. For AI tools in pathology, such tasks may include confirming diagnostic outputs, navigating images overlaid with heatmaps, or managing ambiguous algorithmic recommendations. Note that the 2022 FDA document remains in draft form (Docket FDA-2015-D-4599) as of this writing; the binding U.S. instrument is the 2016 final guidance on Applying Human Factors and Usability Engineering to Medical Devices.

TABLE 15.1 Key terminologies on usability testing.

Term	Definition	Source/context
Core usability concepts		
Usability	The extent to which a product can be used by specific users to achieve specific goals with effectiveness, efficiency, and satisfaction in a specified context of use.	IEC 62366-1:2015, FDA 2016, MFDS
Usability engineering/testing	A structured evaluation process to assess the ease of use, learnability, and user satisfaction of a product or software in typical usage scenarios focuses on improving the user interface and experience.	GMP Usability Guideline; MFDS SaMD Guideline
Human factors engineering (HFE)/validation testing	A structured testing method to confirm that a device can be used safely and effectively by the intended users, focusing especially on critical tasks that, if performed incorrectly, could result in harm.	GMP Usability Guideline; FDA Guidance on Human Factors
Formative evaluation	Early-stage usability testing is conducted to explore the design and identify potential use errors. It is iterative and intended to inform improvements.	FDA 2016, FDA 2022 Draft
Scenario-based testing	Usability test method using clinical scenarios to mimic real-world conditions in which the AI system will be used.	Planning phase in both guidelines
User interaction and tasks		
Task	One or more user interactions with a device are intended to achieve a specific goal. They may consist of multiple steps or inputs.	FDA 2022 Draft
Critical task	A user task which, if performed incorrectly or not performed at all, could cause serious harm to the patient or the user.	GMP Usability Guideline
Task analysis	The process of examining how a user interacts with a system in order to identify steps, cognitive loads, or potential failure points.	Foundational step in usability and HF validation.
Use error	User action or inaction that differs from the intended action and leads to, or could lead to, harm—provided the cause is not solely due to device failure.	FDA 2016, FDA 2022 Draft
Normal use	Use of a medical device as intended by the manufacturer, including setup, operation, cleaning, and maintenance under expected conditions.	FDA 2022 Draft
Abnormal use	Deliberate or reckless behavior beyond the scope of reasonably anticipated use may include sabotage.	FDA 2022 Draft
Intended use	The objective intent of the manufacturer regarding the use of a product is reflected in the labeling, instructions, and promotional materials.	FDA 2022 Draft, MFDS software guideline—required for risk classification and documentation.
User	Any individual interacting with the medical device, including clinicians, technicians, caregivers, or patients.	FDA 2022 Draft, MFDS
Use environment	The actual conditions and settings in which a device is intended to be used (e.g., hospital, home, transport vehicle).	FDA 2022 Draft
User Interface (UI)	The means by which the user interacts with the device, including physical, visual, auditory, and software-based elements (e.g., displays, labels, alarms, touchscreens), are crucial.	IEC 62366-1, FDA 2022 Draft

Continued

TABLE 15.1 Key terminologies on usability testing—cont'd

Term	Definition	Source/context
Risk and safety terminology		
Use-related Risk	Risk that arises from the interaction between the user and the device, potentially leading to harm, is evaluated through task analysis and hazard identification.	FDA 2022 Draft, ISO 14971
Use-related risk analysis (URRA)	Structured method to identify hazards related to use, assess their severity and likelihood, and propose mitigations. Often tabulated.	FDA 2022 Draft
Residual risk	Risk remaining after all mitigation measures have been applied and requires justification if not eliminated.	ISO 14971, FDA 2022 Draft
Use safety	The extent to which user interaction with a device has minimized use errors to an acceptable level or eliminated them.	FDA 2022 Draft
Safety classification	The categorization of software functions is based on the potential severity of harm resulting from malfunction or misuse (Class A: no harm, Class B: minor injury, Class C: serious injury or death).	MFDS software guideline; also linked to IEC 62304.
Risk management	Systematic application of policies, procedures, and practices for analyzing, evaluating, controlling, and monitoring risk.	Required in software technical documentation (MFDS and IEC 14971).
Traceability	The ability to link usability test results and design decisions to specific user needs, regulatory requirements, and risk mitigation strategies.	Cited in both documents as part of the documentation and audit trail.
Software and system categories		
Embedded software	Software that is embedded, installed, or connected wired/wirelessly to medical devices for the purpose of controlling or operating the medical device, or for storing, transmitting, signal processing, and image processing of data generated from the medical device.	MFDS SaMD Guideline—e.g., firmware in X-ray equipment.
Standalone software (Software as a Medical Device—SaMD)	Software that is not embedded in a hardware device and performs medical functions on general computing platforms (e.g., AI diagnosis tools running on PCs).	MFDS software guideline aligns with the IMDRF SaMD framework.
Medical device software	Software developed and manufactured for use in purposes corresponding to Article 2 of the Medical Device Act, including embedded software, standalone software, mobile medical apps, etc. This includes software that transmits data generated from medical devices to servers for storage, analysis, signal processing, image processing, etc.	Korean Medical Device Act, Article 2
Mobile medical app (Mobile Medical Application)	Software application or web-based application software that runs on mobile devices and is used for the functions of embedded software, standalone software, or other similar functions.	Mobile Medical Device Regulation
Software item	A software item that cannot be further divided into other items.	Software Engineering Standards
Software system	An integrated collection of software items configured to perform specific functions or a series of functions.	Software Engineering Standards
SOUP (Software of Unknown Provenance)	Commercial off-the-shelf software or third-party libraries are integrated into medical software without full source or validation control.	MFDS SaMD Guideline—special caution is advised for validation and risk control.

In this context, understanding the causes and implications of use errors is pivotal. A use error refers to any user action or inaction that deviates from intended use and may result in harm, provided that the cause is not solely due to device malfunction. Regulatory authorities emphasize that these errors arise from the design of the human-system interaction, and not from isolated user mistakes. Task analysis, a foundational step in usability assessment, involves mapping out the steps users must take to achieve their goals and identifying where breakdowns may occur—particularly in high-stakes settings like diagnostic laboratories.

Several other concepts enrich this regulatory landscape. For example, normal use encompasses all interactions intended by the manufacturer, including setup, operation, and maintenance, whereas abnormal use refers to deliberate or reckless behaviors that fall outside foreseeable contexts. Use-related risk captures the potential for harm arising specifically from user-device interactions and is analyzed through structured frameworks such as use-related risk analysis. Residual risks—those that remain after all design mitigations—must be carefully justified and documented. Collectively, these concepts inform the usability engineering file, a core component of regulatory submissions, which must trace user requirements, risk assessments, design decisions, and validation results with rigorous traceability.

Regulatory bodies

Across international regulatory bodies (Table 15.2), there is a convergence in recognizing the centrality of usability and human factors in medical device development. The US Food and Drug Administration (FDA), the European Union Medical Device Regulation (EU MDR), Korea's Ministry of Food and Drug Safety (MFDS), Japan's Ministry of Health, Labor and Welfare (MHLW), together with the Pharmaceuticals and Medical Devices Agency (PMDA), the International Medical Device Regulators Forum (IMDRF), and other national regulatory bodies all mandate human-centered evaluation frameworks. Common elements across these jurisdictions include the requirement to test with representative end-users under realistic use conditions and to focus particularly on the performance of critical tasks. The internationally harmonized standard IEC 62366-1 serves as the backbone of these requirements, offering a systematic process for integrating usability engineering into the broader quality management system.

National-level implementations reflect this harmonization while adapting to local regulatory environments. The FDA, for example, categorizes submissions based on risk and requires detailed documentation for devices with the potential for harm through use error. The EU MDR mandates usability as part of general safety and performance requirements and emphasizes post-market surveillance. The MFDS in Korea complements these international norms with specific expectations for Korean-language interfaces, risk-based software classification (aligned with IEC 62304), and a cybersecurity checklist. Japan's MHLW and PMDA similarly embed usability principles within a broader regulatory framework that is tailored to the characteristics of SaMD. Their 2023 guidance on program medical devices formalizes a two-step approval process that enables early market entry for products with sufficient performance data but incomplete clinical validation, provided that post-market evidence generation is planned. It also emphasizes thorough clinical positioning, bias-controlled evaluation of AI and machine learning algorithms, and consideration of local clinical practice, cultural context, and Japanese-language interfaces.

TABLE 15.2 International guidelines and recommendations on usability testing of software as a medical device (SaMD).

Body	Key guideline/standard	Main focus areas	Documentation requirements	SaMD-Specific considerations
FDA (USA)	• Applying Human Factors and Usability Engineering to Medical Devices (2016) • Content of Human Factors Information in Medical Device Marketing Submissions (2022, Draft) • Content of Premarket Submissions for Device Software Functions (2023)	• Human Factors Validation Testing • Critical Task Identification • Representative Users & Realistic Use • Risk-Based Approach	• Human Factors Submission Category (Matthews et al., 2024; McGenity et al., 2024; Chong and Bychkov, 2025) • Risk-based documentation depth	• Differentiation of Basic vs. Enhanced Documentation • Mobile apps, clinical decision support tools, digital UI layout impacts
EU (European Union)	• EU MDR 2017/745 Annex I • Regulation (EU) 2024/1689 (AI Act) • IEC 62366-1:2015 (+AMD1:2020)	• Usability Engineering Process • Use Error Reduction • Ergonomics, Labeling, Lay Users • Integration with QMS	• Complete Usability Engineering File • Both formative and summative evaluation records	• IEC 62366-1 directly applies to SaMD • Emphasis on post-market surveillance and user feedback
IMDRF (Global)	• SaMD Risk Categorization Framework • QMS for SaMD (2015)	• Risk Categorization (I–IV) • Lifecycle-based Quality Management • Global Regulatory Harmonization	• Risk-stratified documentation • Adaptable to local regulatory frameworks	• Emphasis on software clinical evaluation, UI reliability, and post-market updates
MHLW/PMDA (Japan)	• Guidance on Appropriate and Prompt Approval and Development Based on the Characteristics of Program Medical Devices (MHLW, 2023) • Guidance on Review of Medical Device Programs (2016) • Related notifications (Rebalance Notification, 0929 Notification, IDATEN system, etc.)	• Use of a two-step approval system tailored for SaMD • Continuous safety and effectiveness assurance from development to post-market • Clarification of clinical positioning, organization of conceptual requirements, and risk management • Evaluation of AI/machine learning-based programs (bias elimination, appropriate dataset coverage) • Utilization of existing noninvasive data (0929 Notification) for expedited approval	• Detailed description of intended use, performance, principles, and safety in the approval application • Documentation of compliance with JIS T 14971 (Risk Management), JIS T 2304 (Software Life Cycle), and JIS T 81001-5-1 (Cybersecurity). • Separate clinical evaluation plans/results for first- and second-step approvals • Reliability assurance and ethical consent when using existing data.	• Two-step approval enabling early market entry prior to full clinical validation. • Clearly defined eligibility and exclusion criteria for diagnostic, therapeutic, and preventive programs • Obligation to describe AI model training data characteristics and algorithm transparency • Post-market change management for OS updates or algorithm modifications (IDATEN) • Consideration of Japanese-language UI, cultural context, and local clinical practice
MFDS (Korea)	• Guideline on the Application of Usability in Medical Device GMP Standards (2022) • Guideline on the Approval and Review of Medical Device Software (2023)	• Differentiation of Usability Test vs. HF Validation • Critical Task Identification & Residual Risk Justification • Grading system based on clinical consequences	• Usability Engineering File (if applicable) • Risk-based classification of software (A/B/C) • Submission of cybersecurity self-checklist	• Required labeling, SOUP management, safety class (IEC 62304-aligned) • Localized guidelines for Korean-language UIs—emphasis on task complexity, training, and clinical context

Notably, all of these bodies stress that usability is not an ancillary feature but a primary determinant of safety and effectiveness. For pathologists and academic researchers engaged in AI tool development or evaluation, understanding these terminologies and regulatory principles is important. It not only facilitates regulatory compliance but also ensures that the tools under development genuinely serve the clinical end-user. By foregrounding usability and human factors from the outset, the field of digital pathology can move toward more meaningful, safe, and sustainable integration of AI technologies in daily practice. Most recently, in January 2025, the International Medical Device Regulators Forum (IMDRF) issued the final document Good Machine Learning Practice for Medical Device Development: Guiding Principles (IMDRF/AIML WG/N88), which sets out ten cross-jurisdictional principles designed to further harmonize international expectations for AI/ML-enabled medical devices, including those used in pathology (IMDRF, 2025).

Usability engineering in digital and computational pathology

As established in the introduction, usability and human factors are critical components for ensuring that AI-powered tools in pathology can be safely and effectively integrated into clinical workflows. To operationalize these concepts, developers must follow a structured *usability engineering process,* which is now embedded in regulatory expectations worldwide. Figure 15.1 outlines the process adapted from the Korean MFDS usability guideline—based on the U.S. FDA's human factors guidance and aligned with IEC 62366-1:2015—which provides a systematic framework for evaluating and optimizing the user interface (UI) of medical devices, including SaMD.

Overview of the usability engineering process

The usability engineering process begins with the definition of the *intended use, user characteristics,* and *use environment*—collectively captured in the user specification document. In digital pathology, this includes clarifying whether the tool is designed for primary diagnosis, triage, or quality assurance, who the primary users are (e.g., board-certified pathologists vs. residents), and where the tool will be used (e.g., high-volume labs vs. remote settings). Next, developers identify safety-related UI characteristics and potential use errors that could lead to patient harm. For example, a misaligned heatmap or an ambiguous alert from an AI system could cause a user to misinterpret findings. A preliminary hazard analysis is conducted to explore foreseeable risks associated with user interactions, often referencing past incidents, similar products, or published literature. Based on this, developers construct use scenarios that reflect how the device could fail during typical or high-risk interactions. These scenarios are assessed for risk severity and likelihood, and a subset is selected for summative usability validation. Simultaneously, these scenarios inform design requirements, which may include interface layout, font size, alert positioning, or system feedback mechanisms to mitigate identified risks.

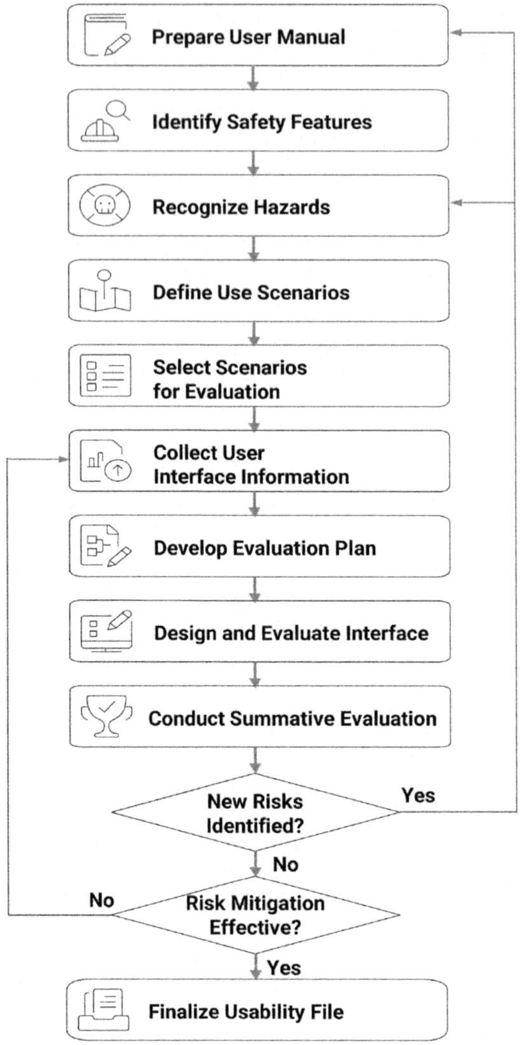

FIG. 15.1 Usability engineering process. Stepwise usability engineering flowchart.

Following design implementation, formative evaluations are conducted iteratively with representative users to refine the UI and correct usability issues. Finally, a summative evaluation—also called human factors validation testing—is performed under simulated clinical conditions to ensure that all critical tasks can be completed safely and effectively. This rigorous assessment verifies that the product meets both usability and safety requirements before regulatory submission. Throughout the process, each stage must be documented in the usability engineering file, ensuring traceability between user needs, risk mitigations, and test results. For high-risk applications such as cancer diagnosis, even a single critical use error during testing may necessitate design revisions and retesting.

Embedded versus standalone software: Usability considerations

In digital pathology, AI tools may be deployed either as embedded software (integrated into a medical device, such as a slide scanner or microscope) or as standalone SaMD (e.g., browser- or app-based diagnostic tools). Each deployment model has distinct usability implications.

Embedded software requires seamless interaction between the hardware and the software UI. Users often interact with embedded AI systems through limited display panels or physical controls, necessitating streamlined interfaces that are easy to interpret. For example, if an AI algorithm is built into a scanner, the system should present results clearly on the device's screen without requiring users to navigate complex menus. In addition, any alerts or warnings must be designed not to interfere with or duplicate the device's native alarm system.

Standalone SaMD, by contrast, typically runs on general-purpose computing platforms in varied environments (e.g., hospital desktops, personal laptops, remote workstations). These tools must account for differences in screen resolution, lighting conditions, network reliability, and user workflows. Responsive UI design, customizable settings, and compatibility with hospital IT infrastructure are essential. Moreover, because these tools are updated more frequently, clear communication of interface changes and adequate user retraining are necessary to avoid introducing new usability risks.

The regulatory approach also differs slightly: embedded systems are evaluated as part of the device hardware approval, while standalone SaMD often follows a distinct regulatory pathway with dedicated documentation on usability testing, cybersecurity, and clinical effectiveness. In both cases, however, the principles of usability engineering must be rigorously applied to minimize use-related risks.

Risk classification and safety grading

The depth and rigor of usability evaluation depend significantly on the device's risk classification, which reflects the clinical consequences of potential failure or misuse. Tools whose use errors could delay or misguide critical diagnostic decisions—such as cancer detection—are generally considered high-risk and require comprehensive usability testing. Conversely, tools with limited clinical impact may follow a simplified process, although basic usability validation is still expected.

Regulatory bodies such as the FDA, EU, and MFDS all emphasize the identification of critical tasks—those which, if performed incorrectly or not at all, could result in serious harm. These tasks must be included in both formative and summative testing. For example, if an AI tool flags suspicious regions on a digital slide, the user must reliably recognize and act upon that flag in every test scenario. If even one evaluator fails to do so, the software may be deemed not yet suitable for release. As the risk level increases, so too does the expectation for larger sample sizes in validation studies, more complex and realistic use scenarios, more robust training materials and user manuals, and stronger justifications for any residual risks.

Ultimately, usability engineering in high-risk diagnostic digital pathology tools is not merely a design best practice—it is a regulatory and ethical imperative. By systematically addressing user interaction risks, developers can produce AI systems that are not only innovative and accurate but also safe, intuitive, and clinically acceptable.

Planning usability evaluation for AI tools

Successful usability evaluation of pathology AI tools depends on several core elements. It begins with defining the intended use and user environment, followed by participant selection, task analysis, and workflow mapping. Scenario development and contextual inquiry are then addressed to ensure authenticity, with particular attention given to identifying critical tasks that reflect safety and efficiency concerns (Table 15.3). When these factors are appropriately addressed, usability findings become both representative of real-world practice and actionable for clinical integration, as well as regulatory submission (Russ and Saleem, 2018; Shin and Lee, 2023; Bitkina et al., 2020).

Define intended use and user environment

Planning should begin by clearly defining the AI tool's intended use, user groups, and use environment. The end-users (e.g., pathologists, lab technicians, clinicians) and their experience levels (experts, trainees) should be identified, along with where and how the tool will be used (pathology lab, hospital, remote vs. onsite viewing, etc.). This forms the foundation for planning a realistic evaluation. In a pathology AI context, for example, the primary users are pathologists who will interact with a digital pathology interface and AI decision support within a clinical workflow. It is critical to understand user needs and constraints from the outset—guidelines emphasize formative user research as the first step to capture these requirements (Chen et al., 2022). Techniques like interviews and contextual inquiry (observing users in their actual work environment) can uncover workflow patterns, usability needs, and pain points. Such early user-centered research ensures the AI tool's design addresses real-world conditions and reduces the risk of design-induced use errors. For instance, observing a pathologist reading slides in practice may reveal the need for the AI's interface to fit into the dual-monitor setup and not obstruct the slide image—insights that inform both design and testing scenarios.

Selection of target users

Participant recruitment should focus on representative end-users as defined in the device's intended use profile. For most computational pathology applications, the primary users will be practicing pathologists with varying years of experience, subspecialty backgrounds, and degrees of familiarity with digital workflows. Inclusion of pathology residents and fellows can be valuable to assess usability across the training spectrum, capturing potential learning curve effects. In certain use cases—such as slide scanning, image QC, or report export—laboratory technicians or administrative staff may also be relevant.

TABLE 15.3 Critical task design template.

Critical task ID	Task description	User group	Intended use context	Potential use errors	Associated hazards	Risk level (severity/ probability)	Design controls/ mitigations	Verification/ validation methods
CT-001	Upload whole slide image (WSI) for analysis	Pathologists, Lab Technicians	Clinical pathology lab with WSI scanners	Wrong case uploaded; low-quality/ partial scan	Misdiagnosis due to incomplete or incorrect input	High (Severity: Critical harm; Probability: Occasional)	Automated QC checks (focus, completeness); confirmation prompts	Usability testing with real-case scenarios; Image QC validation
CT-002	Interpret AI-generated diagnostic output (e.g., tumor classification, prognosis prediction)	Pathologists	Diagnostic workflow with a digital pathology viewer	Over-reliance on AI output; misinterpretation of probability scores.	Incorrect diagnosis, inappropriate treatment planning	High (Severity: Serious harm; Probability: Remote)	Confidence scores, explainable AI visualizations, training modules	Summative evaluation with pathologists: Diagnostic concordance studies
CT-003	Integrate AI results into the pathology report	Pathologists	EHR/LIS integration environment	Copy-paste errors; incorrect mapping to patient records	Erroneous clinical report issued	Medium (Severe: Moderate harm; Probability: Occasional)	Auto-population with patient ID verification; audit trail	System integration validation; Workflow simulation
CT-004	Input clinical metadata for combined analysis (e.g., age, tumor site)	Clinicians, Pathologists	Hospital environment	Wrong or missing metadata	AI model outputs irrelevant or misleading results	Medium (Severe: Moderate harm; Probability: Frequent)	Input validation; mandatory field checks	Human factors testing; review of usability engineering file
CT-005	Monitor AI-assisted triage results (e.g., flagging suspicious cases)	Pathologists, Cytotechnologists	Screening/triage workflow	Failure to review flagged cases; overlooking unflagged positives	Missed cancer detection	High (Severity: Critical harm; Probability: Remote)	Alert dashboards, double-read workflow design	Summative usability test; clinical validation study
CT-006	Perform periodic recalibration of the AI model or system updates	IT Staff, Administrators	Post-market maintenance	Model version mismatch; update failure	Inconsistent diagnostic performance	Medium (Severity: Serious harm; Probability: Remote)	Version control, rollback protection, validation dataset checks.	Software verification; compliance with ISO 13485/14971
CT-007	Export and share AI analysis results with the multidisciplinary team (MDT)	Pathologists, Oncologists	Integrated with PACS/EHR systems	Data export errors; unauthorized data sharing	Patient privacy breach, data loss	Medium (Severe: Privacy violation; Probability: Improbable)	Encryption, access control, export confirmation	Cybersecurity & usability testing for data transfer
CT-008	Handle error recovery when the AI system fails during a case review	Pathologists, Lab Staff	Any diagnostic session	Inability to reaccess AI output; workflow interruption	Delay in diagnosis, incomplete report	High (Severe: Serious harm; Probability: Occasional)	Fail-safe operation, clear recovery instructions	Simulated failure mode testing; hazard analysis

It is important to ensure that all user groups who perform critical tasks are represented. For instance, if a tool requires a lab technician to calibrate a scanner or manage WSI uploads before a pathologist's review, those roles must be tested as well. Regulatory guidance (FDA, MFDS, PMDA) stresses that omitting a user group from validation may render results incomplete if that group is involved in hazard-related use scenarios.

While IEC 62366-1 does not prescribe an exact participant number, most HFE validation studies adopt 8–15 participants per distinct user group to achieve saturation of usability findings (Borsci et al., 2014). FDA guidance suggests that this range balances statistical confidence with feasibility. Equally important is diversity in participant characteristics, including experience level (novice versus expert in digital workflows), institution type (academic centers, community hospitals, or private laboratories), geographic and language background (especially relevant for multilingual user interfaces), and comfort with technology (users with high versus low familiarity with digital tools). Such heterogeneity in participant experience can help uncover subtle usability issues that might only surface in certain subgroups. For example, less tech-savvy users may overlook small on-screen cues, while high-volume pathologists may encounter fatigue-related errors under time pressure.

Task analysis and workflow mapping

With the intended context defined, a thorough task analysis of how users will interact with the AI tool should be performed. The pathology workflow can be broken down into discrete user tasks and sequences of steps, from opening a case and loading a whole slide image (WSI), through using the AI's diagnostic suggestions, to making a final diagnosis and report. Each user interaction (e.g., selecting a region of interest, adjusting an AI-recommended diagnosis, saving results) should be mapped.

Task analysis is a cornerstone of HFE—it helps reveal potential use errors or confusing steps by examining each action's requirements (both physical and cognitive). In digital pathology, this may involve mapping how a pathologist navigates high-resolution WSIs at different magnifications, how they access AI-generated heatmaps or annotations, and how they confirm or override AI findings. It is important to detail the workflow and UI sequence in which these tasks occur, as this mapping will guide scenario design. Ensuring the tasks are described with appropriate granularity is key—tasks defined too broadly might mask critical subtasks where errors could occur. By contrast, a detailed task analysis tied to use-related risk analysis allows the team to pinpoint *hazardous use scenarios* and plan to mitigate them. Each task should be assessed for what could go wrong (e.g., clicking the wrong case or slide file, misinterpreting an AI overlay, ignoring an alert) and what *design controls* could prevent errors (such as confirmation dialogs or clearer labeling). This preparatory analysis not only feeds into risk management (per ISO 14971) but also sets the stage for usability test design. In fact, international standards like IEC 62366 (usability engineering) explicitly require manufacturers to identify use-related hazards and verify that critical tasks can be performed safely.

Scenario development and contextual inquiry

Using the task analysis, develop realistic use scenarios that will be enacted during usability testing. A use scenario is a representative storyline or situation that combines a sequence of tasks a user would perform with the device to achieve a goal (for example, "A pathologist uses the AI tool to evaluate a lymph node biopsy for metastatic cancer during a busy workday"). Each scenario should reflect real-world conditions as closely as possible, including the clinical context and any relevant stressors or environmental factors. Insights from contextual inquiry and user feedback should be leveraged to ensure scenarios are authentic. For instance, if pathologists often handle multiple cases under time pressure, a test scenario might involve reviewing several WSIs with AI assistance within a limited timeframe, amidst potential interruptions. The intended use environment should also be simulated—e.g., using a similar display setup, lighting conditions, and interface hardware as would be present in an actual pathology lab. This includes the physical setting (e.g., a quiet office vs. a busy lab, whichever matches reality), the equipment (using the actual digital pathology workstation setup, input devices, display resolution, etc.), and even time-of-day or workload context if relevant (to incorporate typical fatigue or multitasking levels). The digital test environment also extends to network and system performance, where latency, image loading speed, and AI inference times should match those of the production system. Finally, the data context should reflect a realistic case mix and workload pattern, incorporating both straightforward and challenging slides.

Early formative evaluations (heuristic reviews or pilot tests) can be conducted to refine scenarios and the UI before formal validation. The INTRPRT guidelines for medical imaging AI emphasize that adhering to human-centered design principles and involving end-users in prototype evaluations greatly increases the likelihood that the AI will be usable and trustworthy in practice (Chen et al., 2022). Therefore scenario design is iterative: initial user research informs scenario creation, and dry-run tests with surrogate users can further refine the tasks to ensure they address all important aspects of device use. Recent comparative usability assessments of computer input devices for WSI navigation provide concrete examples of how such factors can be evaluated systematically in digital pathology (Rogers et al., 2025).

Identifying critical tasks

A core part of planning the validation is determining which user tasks are *critical* from a safety perspective. According to the FDA, a critical task is "a user task which, if performed incorrectly or not performed at all, could cause serious harm to the patient or user". In other words, these are the steps where a mistake or omission could lead to significant adverse consequences (harm could mean a misdiagnosis, delayed treatment, or injury). Not every task in using an AI will carry equal risk—many interactions are benign or recoverable—so the focus is on those that directly impact patient safety or clinical decisions.

In digital and computational pathology, examples of potential critical tasks include: (1) confirming a case or patient identity before analysis (mix-ups could lead to a wrong patient report); (2) interpreting the AI's diagnostic output correctly (e.g., understanding an AI-proposed Gleason grade in a prostate biopsy—misinterpretation could alter therapy); (3) ensuring no important region is overlooked (if the AI fails to highlight a cancerous region, the pathologist must still catch it, or vice versa, not be falsely reassured by an AI false negative); and (4) making the final diagnosis decision (the pathologist must integrate AI results with pathology expertise without over-relying on incorrect suggestions). Each of these, if done incorrectly, could result in compromised patient care—hence, they are *critical tasks* requiring rigorous evaluation. These critical tasks should be prioritized in the test planning: the FDA's human factors guidance and draft submissions guidance both insist that all critical tasks be included in the summative usability test scenarios. Likewise, any scenario that contains one or more critical tasks is considered a *hazard-related use scenario* and must be evaluated in the validation testing. Identifying critical tasks upfront ensures that the test will specifically probe those high-risk interactions and provide evidence that users can perform them safely.

Simulation-based validation under realistic conditions

The human factors validation test (summative usability test) should be conducted as a simulation of actual use, using the final device interface and encompassing all critical tasks (Vincent and Blandford, 2015). Test participants must be representative of the intended users—for a pathology AI, this means recruiting practicing pathologists (with a range of experience levels reflective of the user population) and any other user group that interacts with the system (such as lab technologists for image scanning or IT personnel for certain maintenance tasks, if those are in scope). The test scenarios defined earlier are then executed by these users in an environment that mimics real-use conditions as described above. Realism is crucial: users should be asked to perform the tasks exactly as they would in practice, and they should be given any training or instructions that real users would have. If training is provided in actual use, the test should include a training session and then a realistic "washout period" between training and testing (e.g., an hour or more) to simulate retention of knowledge.

During the test, observers should record any difficulties, use errors, hesitations, or subjective feedback as the participant works through each scenario. Particular attention should be paid to how users navigate the critical tasks—e.g., does the pathologist correctly interpret the AI's highlighted areas and warnings? Do they safely override or double-check AI outputs when needed? Each critical task must be completed without *unacceptable* errors or problems for the design to pass validation. The FDA's guidance essentially requires evidence that the device "can be used by the intended users without serious errors or problems, for the intended uses, and under the expected use conditions". This means that if any serious use errors occur during testing of critical tasks, the design may need refinement and retesting. Therefore the test is both a verification of the design's safety and an ultimate validation of the UI's adequacy. Generally, a summative test is run in a controlled manner (often one-on-one sessions in a usability lab or simulated workplace) but aims to recreate the cognitive and ergonomic context of actual pathology workflows. By the end of testing, evaluators should have concrete data on task success rates, error frequencies, root causes of any use difficulties, and user feedback on the AI tool's usability.

In summary, planning a usability evaluation for a digital pathology AI tool involves a blend of user-centered design preparation and rigorous risk-focused testing. One must define how and by whom the tool will be used, analyze and prioritize the tasks (especially critical tasks) that must be performed correctly, and craft realistic test scenarios to observe actual users in action. The evaluation should simulate real clinical conditions and stressors, ensuring that the AI's interface and outputs can be interpreted and used correctly by pathologists. By following international HFE guidelines and integrating risk management at each step, the development team can validate that the AI tool's user interface supports error-free operation in its intended context. This human factors validation process is now an essential part of regulatory submissions for AI medical devices, and when done well, it not only satisfies compliance but also greatly improves the tool's adoption and effectiveness in everyday pathology practice.

Usability testing and risk management

Usability evaluation is not only a design verification activity; it is an integral part of risk management for digital pathology AI tools. Because these systems directly influence diagnostic decisions, the usability process must be tightly coupled with the device's use-related risk analysis.

Application of risk analysis to software workflows

A risk management framework, such as that outlined in ISO 14971 (Flood et al., 2015), should be applied across the entire software interaction workflow. In digital pathology, this involves identifying hazards not only from hardware interactions, such as navigation controls or scanner operation, but also from AI-specific failure modes, including misleading heatmaps, overly confident probability scores, or ambiguous textual outputs. Each user task, particularly critical ones, must be mapped to potential use errors, their clinical consequences, and the design controls intended to prevent or mitigate them. This mapping provides the foundation for prioritizing which workflows must be explicitly validated during usability testing.

Managing residual risks from AI output interpretation

Even with interface refinements and user training, residual risks often remain, especially in how AI outputs are interpreted. These risks may include overreliance on AI suggestions without verification, disregarding alerts due to fatigue or low trust, or misunderstanding confidence scores and uncertainty indicators. Mitigation strategies can involve presenting redundant information through multiple modalities, providing training that helps calibrate user interpretation of AI confidence, or developing adaptive alert mechanisms that emphasize high-severity findings while reducing unnecessary interruptions. Post-market surveillance data should also feed into this process, enabling continuous reassessment of whether residual risks remain acceptable and whether additional mitigations are needed.

Traceability and documentation requirements

Regulatory expectations require full traceability between identified risks, design mitigations, usability test scenarios, and observed outcomes. Documentation must demonstrate how formative evaluations inform design changes and how summative validation provides evidence that risk controls are effective (Table 15.4). Essential records include the use-related risk analysis report, a list of critical tasks with associated hazard analysis, scenario protocols and participant demographics, raw usability test data with error analyses, and justifications for residual risks together with acceptance criteria. Agencies such as the FDA and MFDS require this documentation within the marketing submission package for SaMD, and equivalent requirements exist under PMDA and EU MDR technical file frameworks.

Failure mode analysis and mitigation

Failure mode analysis strengthens this process both before and after usability testing. Prior to testing, the team should hypothesize potential failure modes for each task, such as the risk of a user overlooking an AI alert if the notification is too subtle, or confusing similar icons and selecting the wrong function. Many of these issues can be proactively mitigated through design interventions, including distinct color-coding and symbols for critical alerts, confirmation dialogs for irreversible actions, or simplified terminology that matches users' mental models. Best practice is to incorporate such design controls before validation testing so that the evaluated interface is already optimized for safety. During task analysis and risk assessment, each critical step should be explicitly linked to hazard mitigations, ensuring that potential errors, their hazards, and corresponding controls are documented in a traceable manner.

Following usability testing, any observed use errors or near misses should be analyzed through a structured root cause assessment. This involves determining whether the error arose from a confusing interface layout, an ambiguous AI output, insufficient training, or an extreme scenario outside the intended use context. The resulting failure mode analysis should guide additional design refinements or updated training materials to further reduce residual risks. Regulators expect such findings to be transparently documented, with validation reports linking observed outcomes to previously identified hazards and demonstrating that all critical use-related risks have been reduced to an acceptable level.

In essence, the usability evaluation plan must embed a closed-loop process of risk management: identify risks, mitigate them through design, validate through testing, and refine where issues remain. By integrating usability testing within this cycle, AI developers ensure that products entering clinical practice not only satisfy regulatory requirements but also minimize the likelihood of harmful use errors. This approach protects patients from misdiagnoses or delays while fostering clinicians' trust that AI tools will reliably support, rather than hinder, pathology workflows.

TABLE 15.4 Point-by-point comparison of usability testing vs. human factors validation.

Category	General usability testing (formative)	Formal human factors validation testing (summative)
Primary objective	• Improve the design by identifying usability issues, pain points, and user feedback early. • Focus is on iterative design refinement and user satisfaction. • Not a regulatory requirement, but a best practice to ensure a user-friendly UI.	• Demonstrate the safe and effective use of the finalized design under expected conditions. • Focus is on showing that use-related risks are mitigated—i.e., no use errors leading to harm. • This is a regulatory validation step required for the approval of high-risk medical AI devices.
Timing in lifecycle	• Conducted early and throughout development (multiple iterative tests). • Performed on prototypes or mid-development versions to inform ongoing design changes. May occur at several points (e.g., after major UI updates).	• Conducted near the end of development, on a production-equivalent device (final hardware/software) once the design is frozen. • Occurs during the design validation phase, prior to regulatory submission and after formative improvements are complete.
User group & participants	• May use a convenience sample of users; often representative of target users, but with flexibility. • Small numbers of end-users, sometimes including internal staff or subject-matter experts, are acceptable for getting quick feedback. • Participants are encouraged to freely discuss issues (think-aloud).	• Uses strictly representative end-users from each distinct intended user group (e.g., practicing pathologists, lab technicians). • No company insiders or biased experts as participants. • Typically requires ≥15 users per group to satisfy regulatory expectations, covering diversity in age, experience, etc., to reflect the actual user population.
Scenario realism & fidelity	• Can involve partial or simplified scenarios. • Often uses low- to mid-fidelity prototypes or software simulations. • Testing may occur in a lab or office setting, not necessarily mimicking the exact clinical environment. • Training and instructions might be informal. • The goal is to observe general usability, so some realism can be sacrificed for convenience during early design tests.	• Involves high-fidelity, realistic use scenarios in a simulated use environment that closely mimics real clinical settings. • The device interface is the final version, with all features, labels, and packaging as in production. • Scenarios cover actual clinical workflows (e.g., pathologist diagnosing cases with the AI tool) with realistic timing, distractions, lighting, and environmental conditions simulated to the extent practical. • Any training provided to participants mirrors real-world training and is documented.
Critical task coverage	• Not guaranteed to cover all critical tasks; the focus is on general usability and areas where developers suspect issues. • Tests may target new features or known trouble-spots rather than every possible use scenario. • If a serious use-error is observed, it leads to a redesign in formative iterations. • Overall risk analysis is informal at this stage.	• All critical tasks (using steps that could cause harm if done incorrectly) must be identified via use-related risk analysis and tested. • The test protocol ensures that each critical task is performed at least once by participants through realistic clinical scenarios. • No critical functionality is omitted—from logging in, case selection, viewing slides with AI overlays, making diagnoses, to report generation—everything safety-related is evaluated. • This ensures the final design supports users in executing every high-risk task without error.

Continued

TABLE 15.4 Point-by-point comparison of usability testing vs. human factors validation—cont'd

Category	General usability testing (formative)	Formal human factors validation testing (summative)
Sample size & diversity	• Small sample sizes per round (often 5–8 users) are typical. • The emphasis is on quickly finding major usability problems, not statistical measures. • Participants might be relatively homogeneous (e.g., all from one hospital) in early tests; diversity is nice-to-have but not strictly required. Multiple small tests are run until usability issues are resolved.	• A larger sample with a minimum of ~15 users per distinct user group is required to achieve adequate coverage of user variability and to satisfy FDA/IEC guidelines. • Participants should represent the full range of end-users (e.g., junior and senior pathologists, and those with varying experience with digital tools). • Demographic and background diversity are important to ensure the device is universally safe and effective. The sample size is justified to regulators as being sufficient to detect use errors or use difficulties across user populations.
Data collection methods	• Qualitative and flexible: observers often use the think-aloud protocol, ask participants to describe their thought processes, and conduct informal debrief interviews. • Video/audio recording may be used, but the analysis is typically observational (note-taking on usability issues, user comments). • Metrics like task time or error counts are secondary; more emphasis is placed on insights and design feedback. • Documentation can be a simple list of findings and recommendations.	• Rigorous data collection: testers observe silently (no coaching), logging every use error, close call, and difficulty systematically. • Sessions are often video-recorded. • After tasks, a structured interview is conducted to probe the root causes of any observed issues (without leading the participant). • Both objective performance data (errors, task success/failure, time on task, if relevant) and subjective feedback (user's explanations, Likert-scale ratings of ease, etc.) are collected. • All data is recorded in detail (often transcripts, annotated videos) to support a thorough analysis.
Documentation & traceability	• Informal documentation: findings are primarily for internal use. • Typically includes a summary of usability issues found, along with design team notes on planned improvements. • There is no formal requirement to trace each finding to risk controls or to retain extensive records. • The focus is on actionable design changes rather than compliance. (Formative test reports may be kept as part of the design history, but they are not usually submitted to regulators.)	• Extensive documentation: a formal Human Factors Validation Report is produced and included in regulatory submissions. • This report details the test protocol, participant demographics, task scenarios, and results for each critical task. • All use errors and difficulties are documented, along with their root cause analysis, and whether design mitigations are adequate. • Traceability is maintained, linking critical tasks and identified use errors back to the risk assessment (use-related risk analysis) and design specifications. • Regulators (FDA, MFDS, PMDA, etc.) expect complete documentation to verify that the summative test confirms that risk controls are effective and user needs are met. • Any residual use-related risks must be justified.

Co-design approach and knowledge translation

A co-design approach, where pathologists and other clinician scientists collaborate with developers from the outset, is crucial for creating effective and clinically relevant AI tools. If clinicians are only involved at the final testing or deployment stages, it fails to leverage their domain expertise and can even foster mistrust of the AI within the clinical community (Banerji et al., 2025). By contrast, engaging an interdisciplinary team (including clinicians, engineers, and even ethicists) early in development leads to AI systems better attuned to real-world clinical workflows and needs. Studies have found that co-design methods are *paramount* for improving the adoption of AI in healthcare settings (Zicari et al., 2021). In practice, involving pathologists throughout the AI lifecycle (from data collection and labeling to model training and validation) helps ensure the tool is well-matched to pathology workflows and addresses relevant clinical problems. This collaborative design process not only improves usability but also builds a sense of ownership among end-users, laying a foundation for smoother integration into routine practice.

Successful knowledge translation between the technical and clinical domains requires bridging the "AI literacy" gap among healthcare professionals. Many clinicians currently lack formal training in machine learning or data science, which can hinder their ability to understand or trust AI-driven systems. In one survey, the majority of medical students agreed that AI should be included in medical training curricula, reflecting broad recognition of this need (Pinto Dos Santos et al., 2019). To address this, task-specific education programs can be implemented in pathology departments, providing foundational knowledge about what AI is, how it works, its possibilities and limitations, and how it uses data (Drogt et al., 2022). Such training ensures a shared understanding of AI's role and fosters a support base for new tools, aligning them with clinicians' expectations. Notably, a recent review identified the literacy gap as a significant hurdle for AI usability: poor usability often leads to increased workload, especially when users are not properly trained to use the AI system (Asan and Choudhury, 2021). Thus improving AI literacy through continuous professional development is key. This is a two-way street—developers of pathology AI also benefit from training in clinical workflows and terminology, enabling effective communication with clinicians. Ultimately, AI solutions should be co-developed by teams of pathologists, laboratory scientists, and computer scientists, combining diverse expertise and values. Such an inclusive approach facilitates knowledge exchange and ensures the resulting tool embodies both technical robustness and clinical relevance.

For AI to be embraced in a clinical environment, pathologists must trust its outputs and understand its reasoning. A black-box model that produces a result with no explanation will understandably face skepticism. Transparency is widely regarded as critical for fostering clinician trust in AI. By making an AI system's workings more transparent and interpretable, we enable users to see *why* a recommendation was made, allowing them to make informed decisions rather than blindly accepting the output (Tucci et al., 2022). In practical terms, this means integrating user-friendly explanations or visualizations of the algorithm's decision criteria (for example, highlighting which cell features or image regions influenced a diagnosis). Research indicates that explainable AI can indeed bolster clinicians' trust in AI tools, particularly when the explanations are clear, concise, and relevant to clinical decision-making. Transparency and explainability also contribute to safety: an explainable and transparent AI system helps physicians detect potential biases or errors in the model's reasoning. Clinicians thus feel less "overruled" by the technology and more confident in taking responsibility for AI-assisted diagnoses. Importantly, building trust is not about blind faith in AI, but about achieving *calibrated* trust—clinicians should feel comfortable relying on the AI when it performs well, yet remain vigilant and override it if it conflicts with their expert judgment. Providing meaningful explanations and involving end-users in design are both instrumental in reaching this balance. In summary, a co-design approach coupled with strong knowledge translation efforts (through education and transparent, explainable tools) is vital to build user trust and ensure that pathology AI systems are adopted and used responsibly in clinical practice.

Cybersecurity considerations for AI-powered pathology tools

Pathology laboratories are increasingly adopting digital workflows and AI-powered diagnostic tools, making cybersecurity a critical concern. This section examines the importance of securing AI pathology systems, the regulatory obligations for cybersecurity (with reference to MFDS, FDA, and EU requirements), the balance between security and usability, real-world case examples, and a checklist of best practices (Table 15.5).

TABLE 15.5 Cybersecurity checklist for AI-powered pathology tools.

Security domain	Key requirements	Relevant standards/guidance
Authentication and authorization	Unique user IDs; strong passwords or MFA; role-based access control (RBAC); automatic session timeout/logoff	EU MDR Annex I; FDA, MFDS SaMD guidance; IEC 62304; ISO 14971
Data confidentiality and encryption	Encrypt sensitive data at rest (AES) and in transit (HTTPS/TLS); encrypt portable media; use secure protocols (VPN, SFTP)	ISO/IEC 27001; HIPAA; IEC 81001-5-1:2021
Secure software update mechanisms	Digitally signed and verified updates; rollback protection; validate in test environment; maintain SBOM for vulnerability tracking	FDA Cybersecurity Guidance; IEC 81001-5-1:2021; NIST SP 800-53
Logging and audit trail	Detailed logs of user access, data modifications, AI model updates; tamper-resistant logs; timestamps and user IDs; regular review	ISO/IEC 27001; EU MDR Annex I; IEC 62304
Emergency access protocol	Break-glass procedure for emergencies with strict control, alerts, and post-incident review; ensure patient care continuity	EU MDR Annex I; ISO/IEC 27001
Compliance with cybersecurity standards	Follow ISO/IEC 27001, IEC 81001-5-1:2021, IEC 62304, and ISO 14971; adopt a structured approach to risk, security measures, and incident response	ISO/IEC 27001; IEC 81001-5-1:2021; IEC 62304; ISO 14971

Importance of cybersecurity in digital and computational pathology

Modern pathology relies heavily on networked digital systems and AI, which introduce new cyber risks (Patel et al., 2023). WSIs and AI analysis results are often stored or processed in cloud-based platforms, making them accessible "anywhere" and integrated into hospital networks. This connectivity demands rigorous security to protect sensitive patient data and ensure diagnostic integrity. Legacy imaging standards like DICOM, originally designed decades ago, have known vulnerabilities; for instance, experts have linked DICOM-based PACS systems to large-scale data breaches, even demonstrating exploits at cybersecurity conferences (Gershkovich, 2025). These examples underscore that AI pathology tools can be targets for cyberattacks.

The risks range from AI model tampering to data breaches. A dramatic illustration comes from radiology: researchers showed that attackers could maliciously alter medical images (e.g., inserting or removing cancerous nodules in CT scans) in ways that misled both radiologists and AI diagnostic algorithms (Mirsky et al., 2019). Such tampering can cause dangerous misdiagnoses or fraudulent outcomes, highlighting the clinical consequences if an AI model or its input data are compromised. Likewise, unauthorized access to pathology databases can lead to data leakage, exposing confidential patient information. Healthcare data are highly valuable to attackers—hacking and IT incidents are the leading causes of healthcare data breaches, and stolen medical records fetch hundreds of dollars each on the dark web (Seh et al., 2020). A successful breach could not only violate patient privacy but also disrupt clinical operations. For example, ransomware attacks on hospital systems (including lab and pathology information systems) can force delays in diagnoses and even divert patients, directly endangering patient care (Stowman et al., 2022). Given these stakes, robust cybersecurity in digital pathology is as much about patient safety as it is about data protection.

Regulatory requirements and reporting obligations

Regulators around the world now mandate that manufacturers of SaMD, including AI pathology tools, address cybersecurity as part of product approval. In South Korea, the MFDS has updated its regulations (via the Digital Medical Products Act, enacted in 2024 and in force from 24 January 2025) to require cybersecurity documentation aligned with international standards. Starting January 2025, any AI-based medical device in Korea must comply with revised cybersecurity guidelines, including an "AI security" verification step during approval. This means companies must submit a dedicated cybersecurity report detailing their device's protections.

Similarly, the U.S. FDA now enforces Section 524B of the FD&C Act and has issued final guidance (June 2025) on the cybersecurity content of premarket submissions, requiring submissions for "cyber devices" to include comprehensive cybersecurity information. Manufacturers must provide a plan for post-market monitoring of vulnerabilities, a commitment to timely security updates and patches, and a Software Bill of Materials (SBOM) listing all software components. The FDA's guidance also recommends adopting a Secure Product Development Framework and integrating cybersecurity risk management throughout the device lifecycle. In the EU, the Medical Device Regulation (MDR) and its guidance (MDCG 2019-16) similarly insist on "secure by design" development. Regulators expect technical dossiers to include a dedicated cybersecurity section covering key safeguards—for example, user authentication and access control measures, data encryption, audit trails, and secure update mechanisms. Annex I of the MDR explicitly lists security capabilities like authentication, authorization, encryption, and auditing that devices must implement. Manufacturers need to document how they enforce access controls and user authentication (e.g., unique logins, role-based permissions, multifactor login) and how they protect data integrity and confidentiality (e.g., encryption of sensitive data at rest and in transit). They must also describe software update mechanisms—updates should be digitally signed, authenticated, and include rollback protection to prevent introducing new vulnerabilities—and demonstrate logging and audit trails for security-relevant events.

Importantly, regulators impose reporting obligations for cybersecurity. In the EU, if a serious cyber incident (e.g., a breach affecting diagnoses or safety) occurs post-market, the manufacturer is obliged to report it to authorities within a tight timeframe (often 15 days or less). Compliance submissions should outline an incident response plan and audit trail functions that facilitate detecting, responding to, and reporting breaches. In summary, today's regulatory dossiers for an AI pathology tool must include a dedicated cybersecurity report addressing: user access control, data encryption/integrity, secure update and patching processes, system monitoring and audit logs, and plans for incident response and post-market surveillance of security. Failing to meet these cybersecurity submission requirements can result in approval delays or rejections, as global regulators consider cybersecurity integral to device safety and effectiveness.

Interaction between usability and cybersecurity

Effective cybersecurity must be balanced with usability, especially in fast-paced clinical environments. Overly stringent security controls can inadvertently induce user errors or unsafe workarounds—the classic usability-security trade-off. In general, more secure systems tend to be less convenient, and vice versa (Di Nocera and Tempestini, 2022). If security measures make an AI pathology tool cumbersome to use, healthcare staff might circumvent them, undermining security entirely. For example, requiring extremely complex 16-character passwords that expire frequently may improve theoretical security, but in practice, clinicians often respond by writing passwords on sticky notes or reusing credentials—behaviors that reduce overall security. Frequent logins and multifactor authentication prompts can consume substantial time and frustrate users. This "security fatigue" leads to burnout and risky shortcuts, like shared generic accounts or leaving systems unlocked.

In pathology AI systems, designers must avoid security measures that hinder the diagnostic workflow. For instance, if an AI-powered digital slide viewer automatically logs out a pathologist every few minutes (for security), it could lead to mistakes or delays in diagnosis. A complex multistep authentication process to access an AI analysis result might prompt time-pressed staff to share a single logged-in session. HFE should be applied to cybersecurity features: interfaces should clearly indicate when a session is active or has timed out, and security steps should align with clinical workflows. Examples of poor design include overly complicated authentication apps or tokens that are difficult to use in a lab setting, or alarm prompts that are so frequent that users begin to ignore them.

To prevent these issues, it's recommended to implement user-friendly security measures—for example, single sign-on (SSO) integration so that pathologists authenticate once to access all needed tools, or biometric logins (fingerprint, badge tap) to replace typing long passwords. Such approaches can actually strengthen security (by removing the incentive to bypass it) while improving user satisfaction. Design principles for secure yet intuitive UIs in pathology AI include: keep authentication as simple as possible (but no simpler), provide rapid access for trusted users (e.g., cached credentials or SSO during a shift), give informative feedback for security events (so users understand, say, why a login failed), and involve clinicians in usability testing of security features. By incorporating usable security design, pathology tools can maintain robust protection without impeding the diagnostic process.

Case examples in pathology AI

Example 1: Secure cloud-based pathology viewer

A large hospital deploys a cloud platform for digital pathology images and AI analysis. To protect patient data, the system uses HTTPS for all web access and a virtual private network (VPN) for external connections, ensuring encrypted channels. Data in transit is secured with TLS 1.2 or 1.3, and images stored in the cloud use strong encryption algorithms such as AES-128 or AES-256. Slide data are sent over TLS/SSL and stored encrypted on the server disks. The system does not cache images on user workstations, so no local copies exist that could be compromised. These measures, including HTTPS, VPN, and encryption, maintain data confidentiality for pathology images both in transit and at rest. Even if network traffic is intercepted, the content remains unintelligible. Fine-grained access rules on the cloud server ensure that only authorized users from the pathology department can access specific cases. This example shows how standard IT security tools can integrate into pathology AI workflows without affecting performance. Cloud-based WSI viewers demonstrate the ability to meet stringent standards such as HIPAA using these protections. Pathologists can review slides and AI-generated results on any device with confidence that patient data remains secure.

Example 2: Role-based access control in a computational pathology platform

Consider an AI-assisted platform for prostate cancer diagnosis used by pathologists, urologists, and lab technicians. It is crucial to ensure that each user accesses only the functions appropriate to their role. The system implements role-based authorization to manage access. In one digital pathology system, this is achieved through role-based access control with predefined user roles and permissions. For example, a pathologist's role allows viewing all AI-analyzed slides and the ability to approve or override AI findings, whereas a lab technician's role permits slide scanning and initial AI runs but not final diagnosis approval. The platform integrates with the hospital's directory using SSO protocols such as SAML, so that user accounts and roles are centrally managed. Upon login, the system issues a secure token that encodes the user's role, and all actions are checked against assigned permissions. This RBAC scheme enforces access control: only an authorized pathologist can confirm an AI-generated diagnosis, and only IT administrators can adjust AI algorithm settings. Role-based controls prevent inadvertent or malicious misuse, such as a nonclinician attempting to alter an AI model. By logging all user actions, the system maintains a detailed audit trail to support accountability and traceability. This example demonstrates how authorization mechanisms can align with clinical workflows, reflecting hierarchical responsibilities in a pathology laboratory. By tailoring access control to user roles, the system upholds security without disrupting collaboration among team members.

Example 3: Secure mobile deployment for cytology AI in resource-limited settings

AI pathology tools are also deployed on mobile devices for point-of-care use. In Japan and Korea, pilot programs equip small clinics and screening centers with mobile cytology AI applications to assist in Pap smear analysis. A similar project in rural Kenya provided a blueprint: clinicians digitized microscope slides using a portable scanner and uploaded them via cellular networks to a cloud AI for cervical cancer screening (Holmström et al., 2021). In these deployments, special care secures data on the mobile device and during transmission, even when infrastructure is limited. The cytology application requires user authentication each time it is opened, and patient data on the phone are stored in encrypted form using the device's hardware encryption. Communication with the cloud AI service is encrypted over HTTPS or a VPN tunnel to protect privacy on mobile networks. In the Kenya trial, data entry occurred through a secure web application on a password-protected, encrypted local server at the clinic. This arrangement ensures that patient records and slide data remain protected and do not persist on unsecured devices. The app also provides an offline mode, allowing images and results to be cached locally in an encrypted container and synchronized with the cloud once a secure connection is available. The mobile AI tool includes a remote wipe capability so that administrators can erase data if a device is lost or stolen, preventing potential breaches. This example illustrates that even in resource-limited settings, robust cybersecurity practices, such as device encryption, secure networking, and user authentication, can be implemented effectively. This deployment demonstrates that strong security is feasible on low-cost hardware and in challenging environments through careful system design and use of built-in encryption.

Cybersecurity checklist for AI tools in pathology

Table 15.5 outlines essential cybersecurity safeguards that developers and pathology departments should verify in any AI-powered pathology system. Ensuring robust authentication and authorization is a fundamental step. Each user should have a unique identity and be authenticated, eliminating the use of shared accounts. Strong password policies or two-factor authentication should be implemented where appropriate, and role-based access control should ensure that users only have access to functions and data relevant to their responsibilities. Automatic session timeouts or logoff mechanisms further help prevent unauthorized access when devices are left unattended. Together, these measures provide a secure foundation for controlling access to sensitive systems and data.

Data confidentiality and encryption must be maintained throughout the entire data lifecycle. Patient information should be encrypted both at rest, for example, in databases storing pathology images or AI-generated results using AES encryption, and in transit, with network communications secured via HTTPS or TLS. This protects sensitive information against eavesdropping or data theft. Additional precautions include encrypting data stored on portable media or devices and using secure protocols such as VPNs or SFTP for remote access. By implementing comprehensive encryption and secure communication practices, pathology departments can safeguard patient information while maintaining compliance with regulatory and ethical requirements.

Secure software update mechanisms are another critical element of a strong cybersecurity framework. AI software and algorithms must be updated through trusted methods that use digital signatures to verify authenticity and prevent tampering. Updates should include rollback protection, allowing the system to revert to a known safe version if an update fails or is compromised. All updates should first be validated in a test environment before deployment, and a Software Bill of Materials should be maintained to track individual components and ensure that any new vulnerabilities can be promptly addressed. By integrating these practices, pathology departments can minimize the risk of introducing compromised software or vulnerabilities into clinical workflows.

Maintaining comprehensive logging and audit trails is essential for traceability and regulatory compliance. Detailed logs should record user access, data modifications, AI model updates, and other critical events in a tamper-resistant format that includes timestamps and user identifiers. Regular review of these logs is necessary to identify suspicious activity or potential security breaches. Audit trails not only provide a mechanism for forensic analysis after an incident but also demonstrate that only authorized personnel accessed specific diagnostic results, supporting compliance with relevant regulations and standards.

Emergency access protocols should be established to ensure that security measures do not impede patient care in critical situations. A controlled "break-glass" procedure allows supervisors to access essential information or override normal authentication procedures during life-threatening scenarios, while maintaining robust auditing and accountability. Such accounts should be tightly regulated, trigger alerts upon use, and require post-incident review to ensure transparency. EU guidance recommends combining automatic session logoff with emergency access procedures to maintain an appropriate balance between security and patient safety.

Finally, adherence to established cybersecurity standards is essential for both regulatory compliance and a security-conscious organizational culture. Healthcare information systems can follow ISO/IEC 27001, which provides a comprehensive framework for risk assessment, access control, and incident management. For medical device software, IEC 81001-5-1:2021 provides specific guidance for integrating security throughout the product lifecycle. Additional standards, including IEC 62304 for software lifecycle processes and ISO 14971 for risk management, ensure a structured approach to identifying potential risks, implementing protective measures, and preparing response plans (Puder et al., 2023). By following these standards, pathology departments can create a robust cybersecurity framework that protects patient data, supports safe AI deployment, and fosters confidence among users.

Recommendations and future directions

The future of usability in pathology AI lies in treating it as an evolving, pathologist-centered process that extends across the entire product lifecycle. Moving forward, three areas require particular emphasis. First, usability must be maintained as a continuous process from early design through post-deployment monitoring. Second, regulatory frameworks need to adapt and harmonize internationally to ensure usability and safety without hindering innovation. Third, AI tools must be developed and deployed with pathologists at the center, embracing an augmented intelligence model that enhances rather than replaces their expertise. Together, these directions chart a pathway for AI systems that are safe, trusted, and seamlessly integrated into clinical workflows.

Usability as a continuous lifecycle process

Usability in AI-driven pathology systems is not a one-off checkbox but a continuous lifecycle process spanning design, deployment, and maintenance. Modern usability engineering standards for medical devices (e.g., IEC 62366-1) prescribe an iterative approach to analyzing user needs, designing with human factors in mind, and evaluating usability at each development phase (Engler et al., 2024). In practice, this means conducting early formative usability tests (e.g., think-aloud sessions) to catch interface issues and workflow mismatches, followed by summative evaluations before clinical roll-out. A recent usability study of an AI pathology reporting tool illustrates this iterative refinement: pathologists identified several ease-of-use issues during think-aloud testing, and poor AI accuracy actually *created* extra work, undermining trust (Alvarado et al., 2025). This finding highlights that technical performance and usability are intertwined—an AI must be accurate and reliable enough to integrate seamlessly into the workflow if it is to improve efficiency and be accepted by clinicians.

Crucially, usability monitoring continues post-deployment. Data distributions, lab protocols, and user expectations can drift over time; consequently, an AI's clinical utility may change between the validation phase and real-world practice. Regulators and researchers recommend deploying tools to detect such changes (e.g., monitoring for input data shifts and performance drops) to support ongoing quality assurance (Reniewicz et al., 2024). In essence, achieving and maintaining high usability demands a continuous feedback loop: collecting user feedback and performance data, refining the system, and reevaluating—a cycle that persists throughout the AI's lifecycle. By treating usability as an evolving process rather than a final task, pathology AI systems can remain safe, effective, and aligned with pathologists' needs as conditions change.

Harmonizing regulatory frameworks for AI usability

As AI systems become more embedded in pathology, regulatory frameworks must evolve and harmonize to ensure usability and safety without stifling innovation. Currently, any AI tool for diagnostic pathology intended for broad clinical use is likely regulated as a medical device (often an in vitro *diagnostic* software). In the United States and Europe, for example,

AI-based pathology software generally requires clearance or approval by agencies (FDA in the US or a notified body in the EU) before marketing. These approvals demand evidence of analytical validation, clinical performance, and often HFE to show the device can be used by clinicians without causing use errors.

One practical challenge is that many promising pathology AI algorithms are developed in academic or lab settings and may target niche tasks. Such tools may never attract a commercial manufacturer due to high regulatory costs and complexity, despite their potential clinical value. To bridge this gap, experts have pointed to the existing model of laboratory-developed tests (LDTs)—colloquially referred to as "homebrew" diagnostics that laboratories create and validate in-house under CLIA or similar quality standards (Calderaro et al., 2025). It has been argued that pathology departments should be able to develop and deploy "homebrew AI" systems under a quality framework, enabling local validation by pathologists and faster access to AI innovations. However, new rules, such as the proposed FDA LDT regulation, may restrict this pathway. International harmonization of regulatory approaches is thus a key recommendation: pathology AI would benefit from clear, aligned guidelines on what constitutes safe and effective use, how to document usability testing, and how to conduct post-market surveillance.

Global initiatives are underway—for instance, the International Medical Device Regulators Forum (IMDRF) has discussed AI-specific risk management and post-market monitoring, highlighting issues like transparency, data drift, and the need to watch for human factors issues in real-world use. Moving forward, regulators are also grappling with how to handle adaptive AI that learns over time. As of mid-decade, no continuously learning pathology AI has been approved; agencies like the FDA are developing new oversight mechanisms for "adaptive" algorithms (Mahler et al., 2021). The future likely holds a more flexible, risk-based regulatory paradigm—one that still requires rigorous evidence of safety, accuracy, and usability but accommodates the iterative improvement of AI and encourages usability-centric design. By harmonizing standards (across countries and across the AI lifecycle), regulators can ensure that pathologists everywhere benefit from AI tools that are both cutting-edge and user-centered. Toward this end, the FDA finalized its guidance on Predetermined Change Control Plans (PCCP) for AI-enabled device software functions in December 2024, providing a pathway for manufacturers to pre-specify modifications to AI models that may be implemented post-market without a new submission (U.S. Food and Drug Administration, 2024).

Pathologist-centered AI design in future clinical deployment

A consistent theme for future AI in pathology is the primacy of the pathologist's role in the design, deployment, and ongoing use of these technologies. Rather than viewing AI as an autonomous replacement, the field is embracing an "augmented intelligence" model, wherein AI algorithms serve as powerful assistants that enhance the pathologist's diagnostic capabilities. To realize this vision, pathologist-centered design must guide AI development. Studies of AI adoption in pathology have found that trust and familiarity are pivotal: pathologists are more likely to embrace an AI tool if they understand its behavior, have been involved in its evaluation, and see it addressing real pain points in their work

(King et al., 2023). For example, practicing pathologists want to be involved in AI tool design and implementation to ensure those tools genuinely solve pressing problems (which often differ by subspecialty). They also express preferences for AI that integrates smoothly into existing workflows—sometimes operating in the background for efficiency, other times giving the user control, depending on the context (King et al., 2023). Future deployments will therefore focus on customizable integration, allowing AI to adapt to the workflow of each subspecialty and lab. Moreover, pathologists should have a say in when and how AI is introduced into clinical practice. This kind of participatory approach ensures that concerns (like algorithm transparency, error rates, or how an AI handles unusual cases) are addressed *before* full deployment. It also acknowledges that evidence needed to satisfy pathologists goes beyond raw performance metrics—it includes evidence that the AI will improve patient care without disrupting diagnostic processes. In the future, we can expect greater emphasis on training and competencies for pathologists in the AI era. Pathologists will likely need literacy in AI outputs and limitations so they can effectively supervise AI recommendations and recognize when the "assistant" might be wrong. Professional organizations and educators are beginning to outline new competencies (from basic data science understanding to the ethics of AI use) to prepare pathologists for this partnership.

Overall, the next decade of pathology AI will be defined by human-centered innovation. Usability will be continually refined through user feedback loops; regulators will increasingly require (and facilitate) evidence of real-world usability and safety; and pathologists will remain at the heart of the AI design process. By following these recommendations—treating usability as a lifecycle process, aligning regulatory expectations with human factors needs, and keeping pathologists in the loop—the field can ensure that AI tools truly augment pathology practice. This pathologist-centric, usability-focused approach will ultimately accelerate responsible AI adoption, leading to more efficient workflows and improved diagnostic outcomes in pathology.

AI Disclosure

During the preparation of this work, the authors used ChatGPT to proofread the text. After using this tool/service, the authors reviewed and edited the content as needed and take full responsibility for the content of the publication.

References

Alvarado, N., Magee, D., Treanor, D., Ellery, P., Millot, H., Randell, R., 2025. Formative usability testing of artificial intelligence in pathology: the challenge of assessing acceptability. In: Andrikopoulou, E., et al. (Ed.), Intelligent Health Systems – From Technology to Data and Knowledge, IOS Press, United Kingdom, pp. 288–292. http://www.iospress.nl/bookserie/studies-in-health-technology-and-informatics/327. doi:10.3233/SHTI250330.

Asan, O., Choudhury, A., 2021. Research trends in artificial intelligence applications in human factors health care: mapping review. JMIR Hum. Factors. 2292-9495. 8 (2), e28236. doi:10.2196/28236.

Banerji, C.R.S., Shah, A.B., Dabson, B., Chakraborti, T., Hellon, V., Harbron, C., et al, 2025. Clinicians must participate in the development of multimodal AI. eClinicalMedicine. 84. Available from https://www.thelancet.com/journals/eclinm/article/PIIS2589-5370%2825%2900184-1/fulltext.

Berbís, M.A., McClintock, D.S., Bychkov, A., Van der Laak, J., Pantanowitz, L., Lennerz, J.K., Cheng, J.Y., Delahunt, B., Egevad, L., Eloy, C., Farris, A.B., Fraggetta, F., Del Moral, R.G., Hartman, D.J., Herrmann, M.D., Hollemans, E., Iczkowski, K.A., Karsan, A., Kriegsmann, M., Salama, M.E., Sinard, J.H., Tuthill, J.M., Williams, B., Casado-S ánchez, C., Sánchez-Turrión, V., Luna, A., Aneiros-Fernández, J., Shen, J., 2023. Computational pathology in 2030:

a Delphi study forecasting the role of AI in pathology within the next decade. eBioMedicine. 23523964. 88. Available from: http://www.journals.elsevier.com/ebiomedicine/. doi:10.1016/j.ebiom.2022.104427.

Bitkina, O.V., Kim, H.K., Park, J., 2020. Usability and user experience of medical devices: an overview of the current state, analysis methodologies, and future challenges. Int. J. Ind. Ergon. 18728219. 76. Available from: https://www.elsevier.com/locate/ergon. doi:10.1016/j.ergon.2020.102932.

Borsci, S., Macredie, R.D., Martin, J.L., Young, T., 2014. How many testers are needed to assure the usability of medical devices? Expert Rev. Med. Devices. 17452422. 11 (5), 513–525. Available from: http://www.future-drugs.com/loi/erd. doi:10.1586/17434440.2014.940312.

Calderaro, J., Morement, H., Penault-Llorca, F., Gilbert, S., Kather, J.N., 2025. The case for homebrew AI in diagnostic pathology. J. Pathol. 10969896. 266 (4-5), 390–394. Available from: http://onlinelibrary.wiley.com/journal/10.1002/(ISSN)1096-9896. doi:10.1002/path.6438.

Chen, H., Gomez, C., Huang, C.M., Unberath, M., 2022. Explainable medical imaging AI needs human-centered design: guidelines and evidence from a systematic review. Npj Digit. Med. 23986352. 5 (1). Available from: https://www.nature.com/npjdigitalmed/. doi:10.1038/s41746-022-00699-2.

Chong, Y., Bychkov, A., 2025. Commercially available artificial intelligence solutions for gynaecologic cytology screening and their integration into clinical workflow. Cytopathology. 0956-5507. doi:10.1111/cyt.70023.

Chong, Y., Jesús Fernández Aceñero, M., Zaibo, L., Bychkov, A., 2025. Integration of digital cytology in quality assurance programs for cytopathology. Acta Cytol. 0001-5547. 1–22. doi:10.1159/000547486.

Daniels, T., Goldsmith, T., Basu, A., 2021. Overview of a robust human factors engineering process. Biomed. Instrum. Technol. 19435967. 55 (1), 16–20. Available from: https://meridian.allenpress.com/bit/article/55/1/16/454361/Overview-of-a-Robust-Human-Factors-Engineering. doi:10.2345/0899-8205-55.1.16.

Di Nocera, F., Tempestini, G., 2022. Getting rid of the usability/security trade-off: a behavioral approach. J. Cybersecur. Priv. 2624800X. 2 (2), 245–256. Available from: https://www.mdpi.com/journal/jcp. doi:10.3390/jcp2020013.

Drogt, J., Milota, M., Vos, S., Bredenoord, A., Jongsma, K., 2022. Integrating artificial intelligence in pathology: a qualitative interview study of users' experiences and expectations. Mod. Pathol. 15300285. 35 (11), 1540–1550. Available from: https://www.nature.com/modpathol/. doi:10.1038/s41379-022-01123-6.

Engler, M., Johner, C., Krepcke, M., Martinez Torres, M.I., Stangenberg, M., 2024. Usability engineering for medical devices using artificial intelligence and machine learning technology. Available from: https://zenodo.org/records/14203190.

Flood, D., McCaffery, F., Casey, V., McKeever, R., Rust, P., 2015. A roadmap to ISO 14971 implementation. J. Softw. Evol. Process. 20477473. 27 (5), 319–336. doi:10.1002/smr.1711.

Gershkovich, P., 2025. Wearing a fur coat in the summertime: should digital pathology redefine medical imaging? J. Pathol. Inform. 21533539. 18 (100450). doi:10.1016/j.jpi.2025.100450.

Holmström, O., Linder, N., Kaingu, H., Mbuuko, N., Mbete, J., Kinyua, F., Törnquist, S., Muinde, M., Krogerus, L., Lundin, M., Diwan, V., Lundin, J., 2021. Point-of-care digital cytology with artificial intelligence for cervical cancer screening in a resource-limited setting. JAMA Netw. Open. 2574-3805. 4 (3), e211740. doi:10.1001/jamanetworkopen.2021.1740.

IMDRF. 2025. Good Machine Learning Practice for Medical Device Development: Guiding Principles. IMDRF/AIML WG/N88 FINAL:2025. International Medical Device Regulators Forum; (Accessed 27 January 2025).

King, H., Williams, B., Treanor, D., Randell, R., 2023. How, for whom, and in what contexts will artificial intelligence be adopted in pathology? A realist interview study. J. Am. Med. Inform. Assoc. 1067-5027. 30 (3), 529–538. doi:10.1093/jamia/ocac254.

Mahler, M., Auza, C., Albesa, R., Melus, C., Wu, J.A., 2021. Chapter 11—Regulatory aspects of artificial intelligence and machine learning-enabled software as medical devices (SaMD). In: Mahler, M. (Ed.), Precision Medicine and Artificial Intelligence [Internet]. Academic Press, pp. 237–265. Available from: https://www.sciencedirect.com/science/article/pii/B9780128202395000103.

Matthews, G.A., McGenity, C., Bansal, D., Treanor, D., 2024. Public evidence on AI products for digital pathology. medRxiv Available from: https://www.medrxiv.org/. doi:10.1101/2024.02.05.24302334.

McGenity, C., Clarke, E.L., Jennings, C., Matthews, G., Cartlidge, C., Freduah-Agyemang, H., Stocken, D.D., Treanor, D., 2024. Artificial intelligence in digital pathology: a systematic review and meta-analysis of diagnostic test accuracy. Npj Digit. Med. 2398-6352. 7 (1). doi:10.1038/s41746-024-01106-8.

Mirsky, Y., Mahler, T., Shelef, I., Elovici, Y., 2019. CT-GAN: malicious tampering of 3D medical imagery using deep learning. arXiv. Available from http://www.arxiv.org/abs/1901.03597.

Patel, A.U., Williams, C.L., Hart, S.N., Garcia, C.A., Durant, T.J.S., Cornish, T.C., McClintock, D.S., 2023. Cybersecurity and information assurance for the clinical laboratory. J. Appl. Lab. Med. 24757241. 8 (1), 145–161. Available from: https://academic.oup.com/jalm. doi:10.1093/jalm/jfac119.

Pinto Dos Santos, D., Giese, D., Brodehl, S., Chon, S.H., Staab, W., Kleinert, R., et al, 2019. Medical students' attitude towards artificial intelligence: a multicentre survey. Eur. Radiol. 29 (4), 2941640-6.

Pinto, D.G., Bychkov, A., Tsuyama, N., Fukuoka, J., Eloy, C., 2023. Real-world implementation of digital pathology: results From an Intercontinental Survey. Lab. Investig. 00236837. 103 (12), 100261. doi:10.1016/j.labinv.2023.100261.

Puder, A., Henle, J., Sax, E., 2023. Threat assessment and risk analysis (TARA) for interoperable medical devices in the operating room inspired by the automotive industry. Healthcare. 2227-9032. 11 (6), 872. doi:10.3390/healthcare11060872.

Reniewicz, J., Suryaprakash, V., Kowalczyk, J., Blacha, A., Kostello, G., Tan, H., Wang, Y., Reineke, P., Manissero, D., 2024. Artificial intelligence/machine-learning tool for post-market surveillance of in vitro diagnostic assays. New Biotechnol. 18764347. 79, 82–90. Available from: https://www.sciencedirect.com/science/journal/18716784. doi:10.1016/j.nbt.2023.11.005.

Riihiaho, S., 2018. Usability testing. In: Norman, K.L., Kirakowski, J. (Eds.). The Wiley Handbook of Human Computer Interaction. John Wiley & Sons, Ltd, pp. 255–275. Available from: https://onlinelibrary.wiley.com/doi/abs/10.1002/9781118976005.ch14.

Rogers, J., Vedaraju, Y., Hsu, J., Kinskey, J., Long, S.W., Christensen, P., 2025. A comparative usability assessment of computer input devices for navigating digital whole slide images. J. Pathol. Inform. 18, 100449. doi:10.1016/j.jpi.2025.100449.

Russ, A.L., Saleem, J.J., 2018. Ten factors to consider when developing usability scenarios and tasks for health information technology. J. Biomed. Inform. 15320464. 78, 123–133. Available from: http://www.elsevier.com/inca/publications/store/6/2/2/8/5/7/index.htt. doi:10.1016/j.jbi.2018.01.001.

Seh, A.H., Zarour, M., Alenezi, M., Sarkar, A.K., Agrawal, A., Kumar, R., Khan, R.A., 2020. Healthcare data breaches: insights and implications. Healthc. (Switz.). 22279032. 8 (2). Available from: https://www.mdpi.com/2227-9032/8/2/133/pdf. doi:10.3390/healthcare8020133.

Shin, J., Lee, H., 2023. Optimal usability test procedure generation for medical devices. Healthcare. 2227-9032. 11 (3), 296. doi:10.3390/healthcare11030296.

Stowman, C., Cortright, M., Wilburn, B., et al, 2022. Anatomy of a cyberattack: part 3: coordination in crisis, development of an incident command team, and resident education during downtime. Am. J. Clin. Pathol. 157, 814–822.

Tucci, V., Saary, J., Doyle, T.E., 2022. Factors influencing trust in medical artificial intelligence for healthcare professionals: a narrative review. J. Med. Artif. Intell. 50. Available from https://jmai.amegroups.org/article/view/6664.

U.S. Food and Drug Administration. 2024. Marketing Submission Recommendations for a Predetermined Change Control Plan for Artificial Intelligence-Enabled Device Software Functions. Final Guidance.

Vincent, C.J., Blandford, A., 2015. Usability standards meet scenario-based design: challenges and opportunities. J. Biomed. Inform. 15320464. 53, 243–250. Available from: http://www.elsevier.com/inca/publications/store/6/2/2/8/5/7/index.htt. doi:10.1016/j.jbi.2014.11.008.

Zicari, R.V., Ahmed, S., Amann, J., Braun, S.A., Brodersen, J., Bruneault, F., Brusseau, J., Campano, E., Coffee, M., Dengel, A., Düdder, B., Gallucci, A., Gilbert, T.K., Gottfrois, P., Goffi, E., Haase, C.B., Hagendorff, T., Hickman, E., Hildt, E., Holm, S., Kringen, P., Kühne, U., Lucieri, A., Madai, V.I., Moreno-Sánchez, P.A., Medlicott, O., Ozols, M., Schnebel, E., Spezzatti, A., Tithi, J.J., Umbrello, S., Vetter, D., Volland, H., Westerlund, M., Wurth, R., 2021. Co-design of a trustworthy AI system in healthcare: deep learning based skin lesion classifier. Front. Hum. Dyn. 26732726. 3. Available from: https://www.frontiersin.org/journals/human-dynamics/about. doi:10.3389/fhumd.2021.688152.

CHAPTER

16

Regulations and ethical considerations in digital pathology and artificial intelligence

Rama R. Gullapalli[a] *and Anil V. Parwani*[b]

[a]Department of Pathology, University of New Mexico Health Sciences, Albuquerque, NM, United States

[b]Departments of Pathology, The Ohio State University, Columbus, OH, United States

Introduction

Healthcare is inherently data-centric, encompassing various data-generating subdomains such as insurance, pharmacy, administration, healthcare institutions, and different specialties of actual patient clinical practice (Rajkomar et al., 2018b, 2019). Vast amounts of information are generated at each level of healthcare, with the potential to provide unique insights into how medicine is practiced at scale (Rajkomar et al., 2019). Artificial intelligence (AI)-enabled clinical workflows have tremendously improved our ability to collect healthcare data (Chauhan and Gullapalli, 2021; Gullapalli, 2020; Rashidi et al., 2025). However, large-scale data analytics across the healthcare subdomains mentioned above are lagging (Chauhan and Gullapalli, 2021; Gullapalli, 2020; Rashidi et al., 2025; McCaffrey et al., 2025). Computational algorithms based on principles of machine learning (ML) and natural language processing (NLP) are expected to automate big data analytics, identify patterns to improve our understanding of healthcare processes, and enhance the efficiencies of healthcare delivery (Almarie et al., 2025; Beam and Kohane, 2018; Gianfrancesco et al., 2018; Komura and Ishikawa, 2018).

Although data-generating sources within healthcare are vast, clinicians generate significant amounts of unstructured data (e.g., clinical notes during patient encounters) (Rajkomar et al., 2018b; Gianfrancesco et al., 2018). However, especially in developed nations, clinicians rely heavily on radiology and pathology to guide the diagnosis, prognosis, therapeutics, and management of each patient (Abels et al., 2019; Bhargava and Madabhushi, 2016; Louis et al., 2016; Niazi et al., 2019). Radiologists are relatively adept at using technology (e.g., computerized tomography (CT) and magnetic resonance imaging (MRI)), employing it as a key

Advances in Digital Pathology and Artificial Intelligence in Pathology
https://doi.org/10.1016/B978-0-443-40383-5.00018-2

driver of the practice of radiology (Khosravi et al., 2025; Rajpurkar and Topol, 2025). Technological innovation is also playing an increasingly dominant role in the practice of pathology. Both radiology and pathology are image-intensive specialties, making extensive use of image data for patient care via specialist-generated interpretations (McCaffrey et al., 2025; Bhargava and Madabhushi, 2016; Niazi et al., 2019; Khosravi et al., 2025). Though radiology is farther along the path of digitization and management of medical images, pathology is increasingly moving in the same direction. The heavy reliance on images and digitization makes these two healthcare specialties particularly attractive to AI researchers for testing emerging ideas in imaging AI research. Imaging AI algorithms have seen significant research and advances over the past decade (Komura and Ishikawa, 2018; de Laat, 2018; Diakopoulos, 2016; National Academies of Sciences, Engineering, and Medicine, 2025). The confluence of abundant imaging data, ever-increasing cheap and powerful computational capacity, and advancing algorithmic AI research makes radiology and pathology prime targets for disruptive innovation in healthcare AI applications over the next decade (Chauhan and Gullapalli, 2021; Komura and Ishikawa, 2018; Niazi et al., 2019; Cecchini et al., 2025; Parwani, 2019; Parwani and Amin, 2020; Patel et al., 2022).

The second arm of the practice of pathology (in addition to image-intensive anatomical pathology) is the area of clinical laboratory medicine (Gullapalli, 2020; McCaffrey et al., 2025; Sirintrapun, 2025; Zarella et al., 2023). Automation in clinical laboratory medicine has been well underway for many decades, resulting in vastly improved efficiency in the delivery of patient test results (Gullapalli, 2020; McCaffrey et al., 2025; Sirintrapun, 2025; Zarella et al., 2023). In emerging fields such as precision medicine, there is significant interest in the use of genomic and other forms of "omic" data for both diagnosis and prognosis, utilizing information at a molecular level. The new frontier of "omic" technologies is a true "big-data" specialty, with vast amounts of "omic" patient data generated during each encounter. The field of bioinformatics focuses on algorithmic computational methods to manage and interpret such "omic" data in various clinical settings. AI researchers are highly interested in using AI-based methods to understand "omic" data in the context of patient healthcare (Jakka et al., 2025; Khan et al., 2026; Mtoor et al., 2026). Perhaps the ultimate challenge in the use of AI-enabled healthcare is to synthesize both imaging and genomic data from a patient to provide novel insights into clinical outcomes and management (Jakka et al., 2025; Khan et al., 2026; Mtoor et al., 2026). Many such efforts are currently underway.

Although the potential of AI-based algorithms to effectively manage and interpret big data in healthcare is considerable, there are significant downsides to using such a powerful technology without the necessary ethical and moral safeguards (Ananny, 2016; Boddington et al., 2017; Brey, 2012; Ienca et al., 2018; Vayena et al., 2015). There is increasing unease with the unrestricted use of AI in healthcare, especially with regard to ethical issues such as patient privacy, exacerbation of race and gender inequities, and patient safety outcomes. The broader field of AI ethics is focused on the use of AI technologies to ensure development in an ethically and morally appropriate manner to benefit society at-large. Issues surrounding AI technology misuse have common themes across specialties and reflect more specific specialty-centric concerns. Thus the development of AI ethical guidelines *requires* the participation of domain experts (e.g., practicing pathologists) to develop specialty-centric guidelines for the ethical use of AI technologies. Key participants in enabling ethical AI technologies include AI researchers, pathologists, clinicians, institutional administration, professional societies, and, finally, the patients themselves. In addition, some of the key definitions associated with topics of AI and ethics discussed in this paper are listed in Table 16.1 to better inform the reader.

The focus of this chapter is to inform the practicing pathologist about the topics of current concern in AI ethics within the broader field of AI research and ADM, and to educate them on how these AI ethics issues are likely to impact the practice of pathology in the future. In this chapter, we first discuss the issues of AI ethics from the perspective of an AI researcher interested in AI-enabled pathology. We then address issues of AI ethics and research pertinent to a practicing pathologist, including the risks of AI to the practice of pathology. We then focus on issues related to patient privacy and the current state of regulations in the AI space, with a specific focus on pathology. Finally, we discuss AI ethics issues relevant at a professional society or institutional level to guide the safe development and deployment of AI in pathology.

Thinking about ethical AI study designs in pathology

The current section reviews issues in AI pathology research studies, based on principles of ethical AI design.

Thinking about inclusive AI design and biases in pathology AI projects

Pathology is highly data-centric, making use of both clinical and phenotypic (histomorphological) data elements to enable the traditional practice of pathology. Yet, there is an increasing appreciation for the need to place a classical pathology expert diagnosis in a broader context by integrating additional patient data elements as part of the routine diagnostic workflow (e.g., molecular biomarker information at an individual and population level) (Fu et al., 2020; Kather et al., 2020; Moore et al., 2018). There is also a need to include additional data elements, such as lifestyle and socioeconomic data, to improve research categories such as cohort description, methods applied, and patient outcomes within the domains of pathology and precision medicine. Such complex, cross-domain data handling and integration are perfectly suited for the use of AI algorithms. With the increasing amount of scientific information, it becomes important for pathologists to incorporate the latest advances in science into diagnostic paradigms. It is a task that is easier said than done. A key strength of AI and ADM is their ability to enable integrative cross-domain analyses of diverse research datasets on the fly, which are perhaps not easily amenable to the human mind. However, one must be careful while conducting such studies/surveys, as the complexity of these cross-domain datasets may lead to the introduction of biases, which can manifest in two major ways (Gianfrancesco et al., 2018; Parikh et al., 2019; Ahmad et al., 2026).

These include biases inherent to the AI algorithm itself and biases arising within the datasets used for the purpose of training the AI algorithm. The former does not usually have ethical implications; rather, it is necessary to understand the inner workings of an AI algorithm appropriately. For example, the popular *k-means* clustering algorithm works best when the data forms clusters that are roughly spherical and similar in size; however,

TABLE 16.1 A brief description of the key terms used in this review, explaining artificial intelligence (AI), ethics, and regulation in the context of pathology.

Terminology	Explanation
Artificial intelligence	Intelligence is emulated or simulated through the use of technological means. Computational machinery is employed to achieve intellectual autonomy and independence of thought, similar to that observed in humans.
Algorithm	A set of step-wise commands designed to accomplish a specific task, goal, or objective. In AI, algorithms are the programming code that enables the functionality of an objective task and is key to emulating intelligence in an artificial manner.
Bias	Discrimination in favor of or against a set of outcomes in a particular setting. In AI ethics, this often refers to the ability of the AI algorithm to discriminate against individuals, groups, or populations based on the design of the original algorithm.
Big data	"Big data" refers to data produced by an automated and repetitive technological process. Big data may be quantified in terms of the abundance of the data size generated. In pathology, some examples of big data are: (1) a digital pathology whole slide image repository, or (2) databases containing complete blood counts across a population and time.
Data privacy	The moral, legal, and ethical expectations to maintain confidentiality of data collected from either individuals or nonindividual resources are paramount. In pathology, institutions responsible for the collection of patient laboratory data are tasked with the responsibility of ensuring data privacy at both individual and population levels.
Data-shifts	A concept referring to the change in the data distribution between training and real-world datasets in AI algorithm development.
Digital pathology	An emerging paradigm of pathology focuses on the digitization of traditional glass-based slides read by pathologists. Digitized slide data can be stored, viewed, and shared in real-time, leading to enhanced efficiency in the sign-out process.
Ethics	A branch of philosophy studying the concepts of right and wrong in human behavior in a systematized manner.
Machine learning	Computational algorithms that are capable of automated learning processes through iterative feedback of data, without (or with minimal) human intervention.
Total product life cycle	A framework of regulation was developed by the FDA in 2021 for AI or AI-enabled devices. The framework provides oversight from the premarket development stage to the postmarket monitoring stage.
Good machine learning practices	Ten guiding principles for AI/ML development have been developed by organizations such as the FDA and others. It is a key expectation for FDA submissions in the AI/ML space.
HIPAA	HIPAA (Health Insurance Portability and Accountability Act) is a US federal law that establishes national standards to protect sensitive patient health information (PHI).
Predetermined change control plans (PCCPs)	PCCP is the FDA-developed documentation process that describes what modifications will be made to a device and how the modifications will be assessed to ensure the continued safety and effectiveness of the device.

sometimes they may not (Cohen, 2021). This is not an ethical dilemma; it is simply a function of how the algorithm works. Similarly, DL networks typically have tunable variables (known as hyperparameters) integral to the AI algorithm that must be assigned on an ad hoc basis by human beings as part of the AI research study. These are conscious choices on the part of the AI researcher. But in some cases, ingrained and sometimes unconscious biases may creep in, with the potential to result in cascading downstream effects unforeseen by the AI researchers.

From an AI ethics point of view, the key critical issues relating to algorithmic bias are mainly in the context of datasets used for a research study (Gianfrancesco et al., 2018; Parikh et al., 2019; Rajkomar et al., 2018a). Both sample choice and valuation play a role in this regard. For example, if a dataset has category imbalance (i.e., a study that is composed mainly of adult white males due to factors such as sample availability and socioeconomic factors of healthcare access), then the results of the AI algorithm trained on such data may not be accurate when implemented on the population as a whole. Results from such a homogeneous research study might inadvertently disadvantage a minority subpopulation. A second problem is related to under-specification (Table 16.1), which describes a phenomenon in which an AI-training dataset is not provided with all of the necessary parameters. For example, if the genetics of a population were a key factor in the categorization of histology images, not including those details would lead to an incompletely trained AI model. Under-specification can thus lead to faulty correlations in predicting clinical outcomes. Another example: in trying to weigh the importance of various factors in determining the extent of disease, one might consider that a subpopulation that spends less on medical care might be a healthier one. However, it is just as likely, if not more so, that the subpopulation spending less on medical care is from a lower socioeconomic group and simply cannot afford costly care. They are not healthier; they simply do not have access to appropriate care. Issues such as these mandate a deep understanding on the part of the AI researcher seeking to train DL algorithms to improve the practice of pathology. It behooves pathologists to be aware of the importance of such issues that may result in a skewed interpretation while utilizing an AI algorithm in pathology practices. These are issues that are likely to significantly impact the validity of the use of an AI algorithm.

Algorithmic bias and statistical bias, in general, are poorly understood topics. It is perhaps unrealistic to expect a busy practicing pathologist to be well versed in the various nuances of algorithmic and/or statistical bias. The solution to the issue of algorithmic bias may thus fall primarily on the shoulders of regulatory agencies (e.g., the US Food and Drug Administration (FDA), which now identifies Software as a Medical Device (SaMD), available at https://www.fda.gov/medical-devices/digital-health-center-excellence/software-medical-device-samd; last accessed on April 2, 2026) to approve sustainable AI-enabled workflows. However, pathologists must realize the potential for bias to manifest itself over the long-term use of unregulated AI-enabled pathology workflows. AI researchers, as well as vendors, must partner with pathologists to obtain practical perspectives to identify and remedy potential sources of long-term bias in AI pathology algorithms. There are multiple complementary efforts underway in professional pathology organizational committees (e.g., Digital Pathology Association (DPA) and College of American Pathologists (CAP)) tasked with understanding the use of AI within the practice of pathology.

Considering stakeholder concerns: Obtaining consent and raising awareness

In the absence of any defined guardrails and recommendations for the use of AI-based algorithms in clinical practice, issues related to patient consent and awareness have come to the forefront. Although historically there have been rigorous standards in place, monitored by institutional review boards (IRBs) to protect patients with respect to data privacy, there appears to be a stunning lack of regulation and even basic guidelines on how to train an AI algorithm on available pathology datasets. For example, if a researcher trains an AI model on annotated data from a pathology slide archive and launches an independent AI company based on this model, do they own rights to all of the profits generated from it? How should the pathologists who worked on generating the annotation for each patient, based on their expertise, be compensated? This is an issue that has not been dealt with yet. Ethically, one would imagine that at least a part of the profit should be shared toward the continued maintenance of the pathology archive and/or support the department whose pathologists performed the initial annotation to generate a viable AI-based product for the company.

There are other pressing concerns that need a thorough discussion in the context of stakeholder concerns as well. For example, are patients made aware that an AI-based platform may have affected a pathologist's decision? Should the patients be informed about the use of AI-aided decision-making? Are the patients offered a choice to approve or reject the regimen based on their informed consent for the use of an AI-based model? Are the pathologists familiar with, and do they completely understand, the parameters and limitations of an AI-based algorithm? Are AI-based algorithms audited for bias, fairness, transparency, ethics, and risk mitigation, and, if so, how often? Who are the governing bodies overseeing these audits?

In developing and underdeveloped countries, physicians often form informal consultation groups over various social media platforms and apps that frequently have loose privacy settings. These platforms and apps may allow data sharing to varying degrees within the platform/app as well as with third-party vendors. Such data-sharing practices put patient data safety and privacy at high risk. There is often no regulation regarding how patient data is shared within an AI-enabled app that is frequently used without patient consent. How does one address concerns about the de-identification of patient data and patient data privacy in such cases? AI researchers must consider such questions from both pathologist and patient perspectives while designing research studies before AI-enabled algorithms can become mainstream within the discipline of pathology.

Ethical risks of AI usage in pathology and for pathologists

Pathologists in current clinical practice are anxious to understand the scope and impact of AI algorithms. Questions such as "Will AI replace pathologists?" and "How impactful is AI in enabling patient diagnoses?" are commonplace and are often reviewed (with justified concern) by pathologists (Allen, 2019; Jackups, 2017). In this section, we review some of these issues from a practicing pathologist's perspective, with a focus on AI ethics and the important role pathologists have to play in the future of AI algorithm development.

An intensely studied aspect of AI ethics is risk assessment and the evaluation of the dangers associated with the implementation of AI. The popular press and social media are key drivers in fueling the imagination of the public about the future of AI, often in apocalyptic terms. Emerging AI-based technologies such as driverless cars, automated facial recognition, and AI-based deep fakes are indeed a cause for concern regarding the economic and moral outlook of our society as a whole (O'Neil, 2016). Yet, the reality is that AI is here to stay in one form or another, with all of the attendant risks associated with it. A key requirement of the latest iteration of AI, deep learning (DL), is the need for vast amounts of training data for eventual implementation. By training on large amounts of raw data, the algorithmic performance of DL workflows is much better compared to that achieved by previous ADM approaches. In contrast to previous ML algorithms, DL algorithms are able to work on both structured (e.g., laboratory data) and unstructured (e.g., pathologist reports) data to create ADM models capable of attaining high levels of prediction accuracy.

A commonly held notion is that the laboratory-generated data influences 70% of all clinical decisions (Hallworth, 2011). Although the accuracy of the aforementioned claim is contested, it is undeniable that laboratory-generated data (both clinical and anatomical) make up a significant portion of the *quantitative* data associated with a patient's electronic health record (EHR) (Hallworth, 2011). The ready availability of a significant amount of structured and unstructured clinical data in pathology archives and databases is highly attractive to AI researchers (and potential unscrupulous actors) seeking to leverage laboratory-generated data for purposes of benefit (and harm), constituting a potential risk to patient safety. Pathologists, as custodians of laboratory data, will be at the center of heated debates on issues of patient data ownership for purposes of AI research in the future. The profession of pathology must be ready for this battle.

As AI research into pathology accelerates, there is a significant opportunity for practicing pathologists to contribute to such risk assessments to understand the broader implications of AI algorithms within pathology and patient management. Some of these risks (with a specific focus on pathology) are discussed here. We focus on two key issues related to AI risks in pathology: (1) the risk potential of AI to the pathologist workforce and (2) the risk impact of AI on the practice of pathology itself.

Underestimating the risks of AI usage in the practice of pathology

Research into medical applications of AI is proceeding rapidly. Radiology is perhaps at the forefront of many of these AI imaging initiatives. However, other case studies of AI healthcare applications, such as AI-enabled NLP in clinical electronic health record (EHR) text evaluation and AI-based analysis of whole slide images (WSI) in digital pathology, are also moving forward (Hoyt et al., 2016; Miotto et al., 2016; Rajkomar et al., 2018b; Zarella et al., 2019). Economic incentives for low-cost healthcare also drive commercial AI research efforts. Workforce automation is a key economic driver for many industry-driven AI initiatives to reduce overall costs.

Traditionally, pathologists have relied heavily on experience and "Gestalt" in rendering diagnoses. To the extent that pathologists are unwilling to move on from simply rendering a diagnosis based on visual impression, the risk of AI over the next decade to such practitioners is nonnegligible. AI techniques, such as DL are beginning to outperform humans in certain image-based tasks, particularly those that involve a quantitation component. Pathologists must thus expand the scope of their practices and become more integrated within the overall clinical care of the patient. This includes adopting accessory techniques, such as molecular, clinical, and epidemiological data, to provide comprehensive diagnoses for each patient. Future pathology practitioners must develop skills to synthesize information from multiple sources to provide *integrated* prognostic insights and even established high-level therapeutic NCCN guidelines within their reports. Humans have the unique ability to synthesize information across different domains of knowledge with relative ease.

At the pathology practice level, it is also inevitable that the analytical instruments used by pathologists and laboratories will adopt increasing degrees of AI-enabled automation. Some examples of such AI-enabled automation include providing first-pass oversight of diagnostic algorithmic outputs, AI-based quality assurance of laboratory data, and automated assessments in a high-volume clinical laboratory. In the majority of these instances, such developments are driven primarily by the companies responsible for instrument development. This may limit the impact an end-user pathologist or a laboratory has directly on the AI-enabled instrumentation development process. However, as end-user customers, pathologists could significantly influence the adoption of such AI-enabled laboratory instrumentation into clinical practice by adopting due diligence for AI technologies and conducting risk assessments based on principles of ethical AI. Thus an awareness of issues surrounding algorithmic bias and ethical AI should be kept at the forefront while evaluating instruments and technologies in pathology practices. Active development of guidelines for ethical AI and norms of assessment by professional organizations (e.g., the College of American Pathologists) would raise awareness among pathologists, which could help mitigate the risks of adopting AI-enabled technologies into clinical practice in the future. Pathologists and administrators must be aware of the subtle, unanticipated risks posed by AI algorithms in issues related to patient data privacy and the potential for AI-enabled technologies to (unconsciously) deviate from established pathology practice guidelines. The need for a wary and cautious eye on quality and process control by pathologists is unlikely to be automated away anytime soon.

Overestimating the risks of AI usage to the profession of pathology

AI has enabled some truly impressive advances in the automation of defined tasks, particularly through the use of DL-based approaches. The advent of large language models (LLMs) has further accelerated the pace of this automation. True human-like AI (also known as strong AI/AI-hard/AI-complete), capable of emulating human-like awareness and decision-making capabilities, is still not achievable at the current time (Larson, 2021; Mitchell, 2019). The implication of this is that the need for human oversight ("human-in-the-loop" paradigm) is unlikely to become redundant anytime soon, especially in the domain of healthcare. Within the context of pathology, an AI algorithm capable of completely replacing the skills of a highly-trained human pathologist is unlikely to emerge in the next decade. Current DL techniques perform exceedingly well in addressing narrowly defined and well-formed questions in pathology with strict boundaries of performance (Larson, 2021; Mitchell, 2019). These abilities of AI are growing all the time. This is why pathologists must actively endeavor to become "generalists" capable of integrating cross-domain healthcare information content that is of practical value to both the treating physician and the patient. However, pathologists must also be wary of the "hype" and "oversell" commonly associated with many AI research studies, while accurately assessing the claims made by AI researchers. The hype associated with AI has been a well-known issue since the 1960s. AI research has passed through multiple boom-and-bust cycles when the ambitious goals of AI researchers failed to materialize (Larson, 2021; Mitchell, 2019). While it feels as if the recent advances in computer hardware, networking, and data storage capacity may have allowed the field to turn a corner, DL techniques may still prove to be part of one such hype cycle playing itself out now. While AI technologies are no doubt improving in each successive cycle of development, there is still a very high bar that AI-driven technologies need to clear before they are ready for widespread routine application in patient healthcare.

It is instructive to review the study by Frey and Osborne (2017), which assessed the impact of AI technology on workforce displacement across over 700 professions in the United States. Physicians working in healthcare were the 15th least likely profession to be impacted by AI automation, with an assigned low probability score of 0.0042. This finding is not unexpected, as the nature of work performed by physicians is complex, interactive, cross-domain, and multifaceted. Pathologists must thus actively seek skills and expertise to enable cross-domain relevancy in medicine. An oft-repeated quote regarding this issue is worth remembering: "AI will not replace pathologists. However, pathologists who know nothing about AI will be replaced in the future."

The laboratory workflow processes will likely see increased automation with the incorporation of AI algorithms at various steps; however, the pace of change will be, at best, incremental. Instrument vendors in healthcare are, by nature, cautious. Regulatory FDA oversight will also ensure that evolution remains gradual. Labor-intensive and uncompensated steps in laboratory workflow are the areas where AI-enabled automation will be implemented first to improve overall process efficiencies. Equally, one may predict newer job opportunities within AI-enabled pathology to develop, deploy, and maintain these automated AI workflows in the future. However, the specifics of such development are not yet visible.

In summary, the profession of pathology and laboratory medicine must be well informed of the potential risks associated with AI. The only thing predictable about the AI-enabled future of pathology automation is that it will be unpredictable. By adopting a proactive stance toward these technological developments within the AI space, pathologists can be at the forefront of mitigating the risks posed by AI while benefiting from the potential advantages it could confer. Awareness of issues related to AI ethics will ensure that the balanced and informed viewpoint of pathologists is incorporated into the development of AI-enabled technologies in the field of pathology.

Thinking of institutional frameworks to enable ethical AI in pathology

Spending on healthcare worldwide was an estimated ~$9 trillion in 2014 and is projected to grow to ~$24 trillion by the year 2040 (Dieleman et al., 2017). The United States alone spent approximately $3.8 trillion, accounting for nearly 17.7% of its national gross domestic product

(GDP). The United States also had the highest healthcare expenditures per person (~$11,582) in the entire world, which is nearly double compared to the second most expensive country for healthcare (see links at https://www.ama-assn.org/about/research/trends-health-care-spending; last accessed on April 2, 2026). Naturally, there is an increasing push for automation to reduce healthcare costs. The economic incentives of lowering such costs are directly aligned with the potential of AI to help with this process.

Pathologists and laboratory medicine professionals can expect to be squarely in the middle of the upcoming scramble to mine patient healthcare data at scale to enable AI workflows in healthcare. Although the ultimate value of such approaches remains to be determined, it will not impede the desire to acquire patient data for research and commercial purposes. The quantitative (clinical pathology) and (semistructured) anatomical pathology data formats are highly attractive to AI and ML researchers seeking to assess the efficacy of AI algorithms in healthcare. An important emerging question pertains to the ownership of patient data. Ultimately, who "owns" patient data?—Is it the patients themselves? Is it the institution where the data is held? Do pathologists, who generate and curate extensive data residing in institutional databases, have any intellectual property rights over the potential payoffs of collaborative AI algorithm development? Is it even ethical to consider patient data as something that is "ownable," especially with initiatives to mine it using AI?

Conundrums such as these are likely to continue to confront pathology practice over the next decade. Professional organizations such as the College of American Pathologists (CAP) and the American Society for Clinical Pathology (ASCP) have a key role to play in guiding the ethical development of AI in a manner that is appropriate to the practice of pathology. In the next subsections, we describe a basic framework of three key principles governing AI ethics within the specialty of pathology.

Transparency

A key element driving AI and ADM research ethics since the early 1960s is the idea of radical transparency (Diakopoulos, 2016; Botkin, 2018). Algorithmic transparency may be defined as the idea of open availability of all information pertaining to the workings of an AI algorithm. AI algorithmic transparency enables the interactive and relational assessment (moral and ethical) of the way an AI algorithm functions and, eventually, impacts the end-user. Transparency, more generally, is defined as the robust and open availability of all information content related to the function and interactions within a system, both human and nonhuman information content (Diakopoulos, 2016, Diakopoulos (2020)). In healthcare-associated AI research, it is important to implement the idea of radical transparency due to the significant impact of ADM on a patient's healthcare (Botkin, 2018; Vayena et al., 2018). In pathology, an active area of research interest is the use of DL algorithms as agents to aid the anatomical pathologist's workflow. Questions such as "Can DL AI algorithms identify a disease process as efficiently as a human pathologist?" underpin much of the AI research efforts into pathology. Ironically, though, much of the inner workings of DL algorithms remain unknown (O'Neil, 2016). It may come as a surprise to many pathologists that the manner in which AI-DL neural networks work is still a mystery (O'Neil, 2016). When training a DL algorithm on a set of images (anatomical pathology or otherwise) or any other data, the precise imaging data features that a DL algorithm uses to identify and discriminate between different target categories are obscure. Thus, in essence, DL algorithms are a black box in their capabilities at the current time. In light of this, one must ponder when, if ever, we might be confident enough to trust an ADM process with life-and-death decisions that are commonplace in the healthcare domain. Pathologists make such critical calls on a daily basis. Will we ever trust any such call made by an AI algorithm? This is an issue that remains to be answered. What is indisputable, though, is that AI and ADM algorithms show an ever-increasing capability to identify and predict patterns to near-human (or better) accuracies in a variety of fields (Rajkomar et al., 2019). It may well turn out that the performance characteristics of DL algorithms are so well established in the future that we may implicitly trust them without a detailed understanding of their inner workings. However, a transparent development process is the key to achieving general acceptance of AI technologies in the pathology workflow.

Beyond the use of AI and ADM for mere "diagnostic" purposes, there are multiple pathology domains where issues of AI transparency are likely to play an important role in the decade ahead. AI transparency will be critical while seeking to implement algorithmically enabled clinical workflows in a health institution. Although one may rely on the FDA to provide overall regulatory supervision, it is institutions and pathologists who will eventually be responsible and liable for real-world outcomes. Thus it is incumbent on the industry to provide transparency of the ADM process during the development and implementation phases of an AI algorithm. Due to the potential for AI to constantly "learn" as a part of the workflow, developers and vendors must make two-way communication of AI performance a routine part of their normal implementation protocols with end-users. As the AI and ADM protocols are upgraded iteratively, radical transparency must also be maintained to protect patient health. Another scenario where one can envision the need for transparency is in the data used to test and train as a part of the AI development process. AI trained on a local set of data may not necessarily translate at a global implementation level. Transparency of such information will inform the pathologist of potential variability in the performance of the ADM and AI model in the local context and allow them to adjust accordingly. Equally, pathologists must be trained to recognize and deal with such issues. Over the next decade, as our understanding of the use and benefits of healthcare AI expands, our notions and expectations of AI and ADM transparency will also evolve in tandem.

Accountability

A natural corollary of transparency in AI is accountability. Accountability pertains to both human and nonhuman factors and their interactions alike. For additional details on this topic, the reviewer is referred to the article by Kroll (2020). Current AI research initiatives in pathology focus on the eventual use case of AI algorithms as an "assistant" in the normal pathology workflow. This may be attributed to the complexity of a normal pathology workflow process, and it reflects the reluctance of AI researchers and industry to accept full "accountability" in the final decision-making process. Accountability is a shared transactional concept involving multiple entities such as the AI algorithm, the human responsible for the healthcare decision process (e.g., clinicians and/or pathologists), and finally, the institution implementing the AI-enabled workflow process. The eventual goal of a shared accountability process is to assign appropriate "answerability" as a part of normal healthcare delivery. Accountability may be either desired or undesired in the form of a "reward" for a beneficial outcome or "blame" for a nonbeneficial outcome, respectively. However, the key is that there needs to be a formal accountability process in place when considering the implementation of AI-enabled workflows, including pathology. This includes formal documentation of the accountability hierarchy at an institution and oversight detailing who is responsible for what and what sort of outcomes are anticipated due to the implementation and use of algorithmic healthcare AI (Diakopoulos, 2016; Vayena et al., 2018; McCradden et al., 2020; Wieringa, 2020). In addition, periodic review and updating of institutional AI accountability protocols are mandatory to reflect the current state of knowledge of healthcare AI processes, which are constantly evolving. The "black box" nature of neural networks was alluded to in the previous section. In healthcare, where patient safety is paramount, such a black box scenario of an AI algorithm mandates the need for clearly defined hierarchies of accountability to ensure safe patient outcomes over the long term. In seeking to establish the use of AI technologies within the field of pathology, one must establish the different tiers of accountability associated with the use of AI algorithms within an institution. This would include the AI algorithm vendors, the human pathologists using AI technologies, and the institutions adopting AI themselves. Both physicians and institutions must be accountable for the use of AI in an ethical manner and share an equitable burden for the successful long-term use of AI/ADM in patient healthcare (Diakopoulos, 2016; Vayena et al., 2018; McCradden et al., 2020; Wieringa, 2020).

Governance

The third leg of the proposed framework for AI ethics in pathology relates to governance. The potential impact of AI and ADM tools creates enormous financial incentives for research and commercial interests in healthcare. AI has the potential to change the practice of healthcare delivery in the next decade, but it can also pose temptations for unethical entities to take advantage of shortcuts. Establishing and enforcing rules underpinning the governance of AI and policy will be critical to the moral and ethical implementation of AI and ADM in healthcare. Rules guiding governance will thus need to be implemented at multiple scales—national, professional, and institutional (see discussion in the regulatory section).

As domain experts, AI researchers, engineers, and scientists have been at the forefront of assessing the ethical aspects of AI and ADM, as well as the wider impact of technology. A majority of AI researchers and engineers are professionally affiliated with organizations such as the Institute for Electrical and Electronics Engineers (IEEE). IEEE has created a Global Initiative on Ethics of Autonomous and Intelligent Systems under the title Ethically Aligned Design (see https://ethicsinaction.ieee.org/ and https://standards.ieee.org/content/dam/ieee-standards/standards/web/documents/other/ead1e.pdf; last accessed on April 2, 2026). The science and ethics of AI are frequently represented at cutting-edge technical engineering conferences such as the Association for the Advancement of AI (AAAI) and the ACM Conference on Fairness, Accountability, and Transparency (ACM FAccT). However, there is an increasing realization that individuals who are involved in the development (i.e., engineers and AI researchers) of AI technology cannot simultaneously be the sole arbiters of the ethics of these technologies. It will be necessary to incorporate the voices of the end-users of these technologies into the development process. For example, physicians may be well positioned to understand the long-term impact of AI and ADM usage on patient outcomes within an AI-enabled clinical decision support (CDS) system (Gianfrancesco et al., 2018; Miotto et al., 2016). Similarly, a radiologist may be better positioned to understand the weaknesses of an AI algorithm while incorporating clinical information into an AI-enhanced radiological technique (Langlotz et al., 2019). Due to the interdisciplinary and transformative nature of AI and ADM technology, non-AI technical expert voices (e.g., physicians, lawyers, and lawmakers) are required to participate in building a framework of AI ethics to provide domain-specific expertise (Vayena et al., 2015; Vayena and Tasioulas, 2013).

Our colleagues in radiology at the American College of Radiology (ACR) have been at the forefront of launching AI-centric initiatives in the practice of radiology. ACR AI-LAB is an ACR initiative to educate radiologists in the use and implementation of AI tools (Langlotz et al., 2019). This course aims to empower radiologists with basic knowledge to participate directly in the creation, validation, and use of healthcare AI. The College of American Pathologists (CAP) also has similar early-phase initiatives within pathology AI through the creation of various AI and technology committees. The mandate of the CAP AI committee is to create a broad pathology-centric AI strategy. In addition, the committee aims to provide subject-matter expertise to CAP councils and enable the creation of AI laboratory standards in pathology. We propose that AI ethics *must* be a core component of the mandate of this committee moving forward. As more AI and ADM technologies are incorporated into pathology workflows, pathologists must be at the table to guide the development of these technologies in a manner that aligns with the core principles of the profession of pathology. Finally, institutions employing pathologists (i.e., universities and clinical practices) also must be actively engaged in helping to build a framework of AI ethics meaningful in a local context. This would involve initiatives such as the implementation of professional norms and the creation of procedural guidelines for the adoption of ethical AI workflows within an institution. The effectiveness of ethical AI initiatives depends on the awareness of its importance, periodic review and oversight, and reinforcement of the saliency of this issue to employees (clinical and nonclinical healthcare workers) in general.

We have briefly discussed three foundational frameworks to enable and guide ethical AI in the specialty of pathology—transparency, accountability, and governance. These three core principles represent a starting point for the adoption of AI- and ADM-based initiatives in pathology at an institutional level. As the science of AI evolves, pathologists must review and adopt additional measures to enable ethical AI usage in pathology practice in a manner that benefits patients. In addition, as end-users of powerful data-centric technologies like AI and ADM, and as custodians of structured patient data repositories within the field of medicine, pathologists are in a strong position to drive the adoption of ethical healthcare AI in the future. Practicing pathologists must have a high-level awareness of the issues surrounding the regulatory landscape and key concerns related to patient privacy in the context of AI to provide guided inputs as needed at an institutional level. These issues are discussed briefly next.

A brief survey of the privacy concerns and the emerging regulatory landscape surrounding the usage of AI in pathology

In healthcare delivery in the United States, issues pertaining to patient privacy are a key and fundamental concern (and form an overarching guiding principle) (Anderson, 2024; Gilbert et al., 2025). Patient privacy issues introduce risks related to the safeguarding of Protected Healthcare Information (PHI), which has been defined under the HIPAA (Health Insurance Portability and Accountability Act). The 1996 HIPAA Act is expansive in scope and focuses on various aspects of patient privacy, including—setting the national standards to protect PHI, giving individual patients the rights over their healthcare data, standardizing the rules guiding electronic transactions occurring in healthcare, and enabling the enforcement of patient privacy through the privacy and security rule, among many others (Anderson, 2024; Gilbert et al., 2025; Rahimzadeh, 2024; Richards et al., 2021). HIPAA-driven privacy violations are enforced by the HHS Office for Civil Rights (OCR). The use of AI and AI-assisted technologies in the context of healthcare (including pathology) requires some degree of knowledge of the current regulatory landscape (and rules) for the practicing pathologist, especially if one is a part of the decision-making process. Some key principles surrounding the regulatory landscape governing the usage of AI in pathology are summarized as follows:

The key agency responsible for the review and evaluation of AI-assisted tools/devices falls under the purview of the US Food and Drug Administration (FDA). The primary review mechanism for AI-enabled devices remains the 501(k) pathway, which has been traditionally used by the FDA for the evaluation of instrumentation and devices for use in healthcare. However, the FDA has sought to actively implement quicker turnaround times for the review of AI-enabled devices via novel regulatory pathways and mechanisms. The FDA published draft guidance on January 6, 2025, titled "Artificial Intelligence-Enabled Device Software Functions: Lifecycle Management and Marketing Submission Recommendations," which proposes lifecycle considerations and specific recommendations to support marketing submissions for AI-enabled medical devices. As of July 2025, the FDA's public database lists over 1250 AI-enabled medical devices authorized for marketing in the United States, up from 950 devices recorded as of August 2024. The FDA treats clinical AI tools as "Software as a Medical Device" (SaMD). The FDA has enabled a Total Product Life Cycle (TPLC) approach for the review of clinical AI products, which integrates premarket and post-market data about the AI-enabled devices. In August 2024, the FDA provided draft guidance for a policy known as Predetermined Change Control Plans (PCCPs) to ensure a more agile and robust review of devices in the fast-moving clinical AI space. This framework allows manufacturers to pre-specify algorithm update parameters during premarket submission, enabling post-approval modifications within approved safety boundaries. This is of particular importance in a fast-moving field such as clinical AI, where the pace of innovation is rapid (Rahimzadeh, 2024; Aranovich et al., 2025; Mahase, 2025).

In the specific context of the use of AI, digital pathology, and regulatory guidance, practicing pathologists must be aware of multiple issues (Karargyris et al., 2023; Lennerz et al., 2022, 2023). Digital AI tools have the potential to access, use, and disclose protected health information if they are used only for explicitly permitted purposes (e.g., under the supervision of an IRB protocol) within the overarching ambit of HIPAA regulations (Lennerz et al., 2022, 2023; Hill, 2024). The FDA has designated whole slide imaging systems as Class III medical devices, which represent the highest risk category. The regulatory guidance surrounding digital pathology was first provided by the FDA in 2016, when it issued guidance entitled "Technical Performance Assessment of Digital Pathology Whole Slide Imaging Devices." This guidance provides information regarding the nature and scope of the regulatory review necessary for WSI instrument manufacturers. The FDA's Digital Pathology Program conducts regulatory science research to help ensure patients have access to innovative digital pathology devices that are safe and effective. In 2023, the FDA classified the software algorithm device designed to assist users in digital pathology into Class II (special controls), establishing that classifying the device into Class II will provide a reasonable assurance of safety and effectiveness without impeding the rapid pace of innovation occurring in the space of digital pathology.

At the time of this write-up, the final responsibility for the outcomes pertaining to the implementation and use of AI/AI-assisted tools in the United States still rests entirely with the human pathologist at the local clinical laboratory level (Lennerz et al., 2022; Zhang et al., 2025). Pathologists in the United States remain exposed to medical malpractice and corporate negligence claims if they fail to properly evaluate or use AI outputs. As such, the pathologist needs to have a good understanding of the various regulations that are in place at a national level (e.g., the United States versus Europe), state level (i.e., specific laws governing AI at a state level, which may or may not exist), and institutional level (local institutional policies surrounding the use of AI in the clinic). While it may seem overwhelming, a vast majority of the AI/WSI/DP vendors have a good understanding of the regulatory guidelines surrounding the use of their devices, and pathologists need to lean on vendors to gain a high-level understanding of these issues.

It is always a good idea to inquire extensively (and in-depth) regarding the HIPAA compatibility of AI tools and review the extent to which PHI is accessed as part of the clinical workflow. A common-sense approach in the implementation/use of these tools is to adhere to the "minimum necessary" principle—ensuring the tool only accesses the data required for the specific diagnostic task in the lab. Pathologists and medical directors bear the responsibility of ensuring that any third-party AI vendors have the necessary signed Business Associate Agreement, legally binding them to the HIPAA privacy protections that are the national standard for patient privacy protections. Clinical laboratories must adopt a "compliance-by -design" approach, utilizing technical standards like ISO 13485 for quality management and maintaining a Software Bill of Materials (SBOM) to monitor cybersecurity vulnerabilities within the local digital pathology ecosystem, with stringent and constant oversight of the systems. Pathologists must have a good understanding of the performance characteristics of the AI instrumentation/software, including the sensitivity, specificity, and area-under-curve (AUC) data, among many metrics in use (Lennerz et al., 2022, 2023; Hill, 2024). Having a thorough understanding of the nature of the datasets used by the vendors as part of the validation process (to mitigate training biases wherever possible) is an essential element of the evaluation process. It is important for pathologists to verify the quality of datasets used in the development of the AI software, including adherence to the FDA-supported guidelines of "good machine learning practices (GMLP)."

Conclusions

AI research and the ethical implications of AI have become areas of great interest across different scientific fields, including healthcare (Chauhan and Gullapalli, 2021; Gullapalli, 2020; McCaffrey et al., 2025; Cecchini et al., 2025). In this chapter, we have highlighted some of the ethical issues that a researcher needs to consider while conducting AI research in pathology. These include factors such as race, gender, and ethnicity, which play a key role in pathology AI research designs and outcomes. Multiple examples are described, demonstrating that neglecting these factors leads to a downstream exacerbation of existing inequities in healthcare delivery. This is particularly important in pathology, which serves as the "big data" repository of quantitative (and imaging) data measures of patient progress in the practice of medicine and has a central role in personalized patient care. We reviewed potential "risk" scenarios associated with the use of AI in the field of pathology. Even though AI pathology is in its infancy, it behooves the profession to be aware of the potential risks posed by AI workflows to the practice of pathology. By improving awareness of such risks, pathologists can help guide the careful development of these technologies to benefit patients while minimizing potential downsides. We discuss three key foundational principles of AI ethics for professional organizations to adapt to enable the development of AI within the field of pathology—transparency, accountability, and governance. This framework merely represents a starting point for the development of ethical AI in pathology at an institutional and organizational level. As the impact of AI-enabled workflows in pathology continues over the next decade, more elements need to be added to the pathology AI ethics framework in accordance with the specific needs of the field of pathology. Finally, we conducted a brief survey of the emerging regulatory landscape of the use of digital pathology, AI, and privacy issues. Pathologists have a critical role in enabling AI-based workflows in the laboratory and must have a seat at the table to guide the development and implementation of ethical AI and ADM within the practice of pathology. AI- and ADM-based workflows may create incredibly powerful new approaches to the practice of medicine. Pathologists must leverage this once-in-a-generation opportunity to be key drivers of this emerging paradigm shift within the practice of medicine.

References

Abels, E., Pantanowitz, L., Aeffner, F., Zarella, M.D., van der Laak, J., Bui, M.M., Vemuri, V.N.P., Parwani, A.V., Gibbs, J., Agosto-Arroyo, E., Beck, A.H., Kozlowski, C., 2019. Computational pathology definitions, best practices, and recommendations for regulatory guidance: a white paper from the Digital Pathology Association. J. Pathol. 0022-3417. 249 (3), 286–294. doi:10.1002/path.5331.

Ahmad, A., Vallès, Y., Idaghdour, Y., 2026. Bias in AI systems: integrating formal and socio-technical approaches. Front. Big Data. 2624909X. 8. Available from: www.frontiersin.org/journals/big-data#. doi:10.3389/fdata.2025.1686452.

Allen, T.C., 2019. Regulating artificial intelligence for a successful pathology future. Arch. Pathol. Lab. Med. 15432165. 143 (10), 1175–1179. Available from: https://www.archivesofpathology.org/doi/pdf/10.5858/arpa.2019-0229-ED. doi:10.5858/arpa.2019-0229-ED.

Almarie, B., Gonzalez-Gonzalez, L.F., Dos Santos Barbosa, L.A., Lutz, A., Grosse, U., Fregni, F., 2025. Machine learning-enabled medical devices authorized by the US Food and Drug Administration in 2024: regulatory characteristics, predicate lineage, and transparency reporting. Biomedicines. 2227-9059. 13 (12), 3005. doi:10.3390/biomedicines13123005.

Ananny, M., 2016. Toward an ethics of algorithms. Sci. Technol. Human Values. 0162-2439. 41 (1), 93–117. doi:10.1177/0162243915606523.

Anderson, B., 2024. How to bridge innovation and regulation for responsible AI in healthcare. Nat. Med. 1546170X. 30 (5), 1231. Available from: https://www.nature.com/nm/. doi:10.1038/s41591-024-02983-y.

Aranovich, T., Matulionyte, R., Magrabi, F., 2025. Detangling AI transparency in the medical regulation space. J. Law Med. 1320159X. 32 (2), 382–397.

Beam, A.L., Kohane, I.S., 2018. Big data and machine learning in health care. JAMA. 15383598. 319 (13), 1317–1318. Available from: https://jamanetwork.com/journals/jama/articlepdf/2675024/jama_beam_2018_vp_170174.pdf. doi:10.1001/jama.2017.18391.

Bhargava, R., Madabhushi, A., 2016. Emerging themes in image informatics and molecular analysis for digital pathology. Annu. Rev. Biomed. Eng. 15454274. 18, 387–412. Available from: http://arjournals.annualreviews.org/loi/bioeng. doi:10.1146/annurev-bioeng-112415-114722.

Boddington, P., O'Sullivan, B., Wooldridge, M., 2017. Towards a code of ethics for artificial intelligence. In: O'Sullivan, B., Wooldridge, M. (Eds.), Artificial Intelligence: Foundations, Theory, and Algorithms. Springer, Cham, Switzerland, pp. 2365–3051. https://link.springer.com/series/13900.

Botkin, J.R., 2018. Transparency and choice in learning healthcare systems. Learn. Health Syst. 23796146. 2 (1). Available from: http://onlinelibrary.wiley.com/journal/23796146. doi:10.1002/lrh2.10049.

Brey, P.A.E., 2012. Anticipatory ethics for emerging technologies. NanoEthics. 18714765. 6 (1), 1–13. doi:10.1007/s11569-012-0141-7.

Cecchini, M.J., Borowitz, M.J., Glassy, E.F., Gullapalli, R.R., Hart, S.N., Hassell, L.A., Homer, R.J., Jackups, R., McNeal, J.L., Anderson, S.R., 2025. Harnessing the power of generative artificial intelligence in pathology education opportunities, challenges, and future directions. Arch. Pathol. Lab. Med. 15432165. 149 (2), 142–151. Available from: https://meridian.allenpress.com/aplm/article/doi/10.5858/arpa.2024-0187-RA/503226/Harnessing-the-Power-of-Generative-Artificial. doi:10.5858/arpa.2024-0187-RA.

Chauhan, C., Gullapalli, R.R., 2021. Ethics of AI in pathology: current paradigms and emerging issues. Am. J. Pathol. 15252191. 191 (10), 1673–1683. Available from: http://ajp.amjpathol.org/. doi:10.1016/j.ajpath.2021.06.011.

Cohen, S., 2021. Artificial Intelligence and Deep Learning in Pathology. Elsevier, Amsterdam. Chapter 5.

de Laat, P.B., 2018. Algorithmic decision-making based on machine learning from big data: can transparency restore accountability? Philos. Technol. 22105441. 31 (4), 525–541. Available from: http://www.springer.com/philosophy/epistemology+and+philosophy+of+science/journal/13347. doi:10.1007/s13347-017-0293-z.

Diakopoulos, N., 2016. Accountability in algorithmic decision making. Commun. ACM. 0001-0782. 59 (2), 56–62. doi:10.1145/2844110.

Diakopoulos, N., 2020. Transparency. In: Oxford Handbook of Ethics and AI, Oxford University Press, p. 1. https://academic.oup.com/edited-volume/34287.

Dieleman, J.L., Campbell, M., Chapin, A., Eldrenkamp, E., Fan, V.Y., Haakenstad, A., Kates, J., Li, Z., Matyasz, T., Micah, A., Reynolds, A., Sadat, N., Schneider, M.T., Sorensen, R., Abbas, K.M., Abera, S.F., Kiadaliri, A.A., Ahmed, M.B., Alam, K., Alizadeh-Navaei, R., Alkerwi, A., Amini, E., Ammar, W., Antonio, C.A.T., Atey, T.M., Avila-Burgos, L., Awasthi, A., Barac, A., Berheto, T.M., Beyene, A.S., Beyene, T.J., Birungi, C., Bizuayehu, H.M., Breitborde, N.J.K., Cahuana-Hurtado, L., Castro, R.E., Catalá-López, F., Dalal, K., Dandona, L., Dharmaratne, R. D.S.D., Dubey, M., Faro, A., Feigl, A.B., Fischer, F., Fitchett, J.R.A., Foigt, N., Giref, A.Z., Gupta, R., Hamidi, S., Harb, H.L., Hay, S.I., Hendrie, D., Horino, M., Jürisson, M., Jakovljevic, M.B., Javanbakht, M., John, D., Jonas, J.B., Karimi, S.M., Khang, Y.H., Khubchandani, J., Kim, Y.J., Kinge, J.M., Krohn, K.J., Kumar, G.A., Leung, R., Razek, H.M.A.E., Razek, M.M.A.E., Majeed, A., Malekzadeh, R., Malta, D.C., Meretoja, A., Miller, T.R., Mirrakhimov, E.M., Mohammed, S., Molla, G., Nangia, V., Olgiati, S., Owolabi, M.O., Patel, T., Caicedo, A.J.P., Pereira, D.M., Perelman, J., Polinder, S., Rafay, A., Rahimi-Movaghar, V., Rai, R.K., Ram, U., Ranabhat, C.L., Roba, H.S., Savic, M., Sepanlou, S.G., Ao, B.J.T., Tesema, A.G., Thomson, A.J., Tobe-Gai, R., Topor-Madry, R., Undurraga, E.A., Vargas, V., Vasankari, T., Violante, F.S., Wijeratne, T., Xu, G., Yonemoto, N., Younis, M.Z., Yu, C., Zaidi, Z., Zaki, M.E.S., Murray, C.J.L., 2017. Future and potential spending on health 2015-40: development assistance for health, and government, prepaid private, and out-of-pocket health spending in 184 countries. Lancet. 1474547X. 389 (10083), 2005–2030. Available from: http://www.journals.elsevier.com/the-lancet/. doi:10.1016/S0140-6736(17)30873-5.

Frey, C.B., Osborne, M.A., 2017. The future of employment: how susceptible are jobs to computerisation? Technol. Forecast. Soc. Change. 00401625. 114, 254–280. Available from: www.elsevier.com/inca/publications/store/5/0/5/7/4/0/. doi:10.1016/j.techfore.2016.08.019.

Fu, Y., Jung, A.W., Viñas Torne, R., Gonzalez, S., Vöhringer, H., Shmatko, A., Yates, L.R., Jimenez-Linan, M., Moore, L., Gerstung, M., 2020. Pan-cancer computational histopathology reveals mutations, tumor composition and prognosis. Nat. Cancer. 2662-1347. 1 (8), 800–810. doi:10.1038/s43018-020-0085-8.

Gianfrancesco, M.A., Tamang, S., Yazdany, J., Schmajuk, G., 2018. Potential biases in machine learning algorithms using electronic health record data. JAMA Intern. Med. 21686106. 178 (11), 1544–1547. Available from: http://archinte.jamanetwork.com/issues.aspx. doi:10.1001/jamainternmed.2018.3763.

Gilbert, S., Mathias, R., Schönfelder, A., Wekenborg, M., Steinigen-Fuchs, J., Dillenseger, A., Ziemssen, T., 2025. A roadmap for safe, regulation-compliant living labs for AI and digital health development. Sci. Adv. 2375-2548. 11 (20). doi:10.1126/sciadv.adv7719.

Gullapalli, R.R., 2020. Evaluation of commercial next-generation sequencing bioinformatics software solutions. J. Mol. Diagn. 19437811. 22 (2), 147–158. Available from: http://www.sciencedirect.com/science/journal/15251578. doi:10.1016/j.jmoldx.2019.09.007.

Hallworth, M.J., 2011. The '70% claim': what is the evidence base? Ann. Clin. Biochem. 17581001. 48 (6), 487–488. Available from: http://acb.rsmjournals.com/content/48/6/487.full.pdf. doi:10.1258/acb.2011.011177.

Hill, D.L.G., 2024. AI in imaging: the regulatory landscape. Br. J. Radiol. 1748880X. 97 (1155), 483–491. Available from: https://academic.oup.com/bjr/article/97/1155/483/7510846. doi:10.1093/bjr/tqae002.

Hoyt, R.E., Snider, D., Thompson, C., Mantravadi, S., 2016. IBM Watson analytics: automating visualization, descriptive, and predictive statistics. JMIR Public Health Surveill. 2369-2960. 2 (2), e157. doi:10.2196/publichealth.5810.

Ienca, M., Ferretti, A., Hurst, S., Puhan, M., Lovis, C., Vayena, E., Biemba, G., 2018. Considerations for ethics review of big data health research: a scoping review. PLoS One. 1932-6203. 13 (10), e0204937. doi:10.1371/journal.pone.0204937.

Jackups, R., 2017. Deep learning makes its way to the clinical laboratory. Clin. Chem. 15308561. 63 (12), 1790–1791. Available from: http://clinchem.aaccjnls.org/content/clinchem/63/12/1790.full.pdf. doi:10.1373/clinchem.2017.280768.

Jakka, A.L., Chacko, R.M., Vasam, M., Alagarsamy, S., Chandragiri, S.S., Gavini, S.D., Mathew, M.J., 2025. From genomics to clinic: the transformative impact of AI in pharmacogenomics and personalized medicine. Pharmacogenomics. 17448042. 26 (13-14), 573–585. Available from: https://www.tandfonline.com/journals/ipgs20. doi:10.1080/14622416.2025.2591596.

Karargyris, A., Umeton, R., Sheller, M.J., Aristizabal, A., George, J., Wuest, A., Pati, S., Kassem, H., Zenk, M., Baid, U., Moorthy, P.N., Chowdhury, A., Guo, J., Nalawade, S., Rosenthal, J., Kanter, D., Xenochristou, M., Beutel, D.J., Chung, V., Bergquist, T., Eddy, J., Abid, A., Tunstall, L., Sanseviero, O., Dimitriadis, D., Qian, Y., Xinxing, X., Liu, Y., Goh, R.S.M., Bala, S., Bittorf, V., Puchala, S.R., Ricciuti, B., Samineni, S., Sengupta, E., Chaudhari, A., Coleman, C., Desinghu, B., Diamos, G., Dutta, D., Feddema, D., Fursin, G., Huang, X., Kashyap, S., Lane, N., Mallick, I., Mascagni, P., Mehta, V., Moraes, C.F., Natarajan, V., Nikolov, N., Padoy, N., Pekhimenko, G., Vijay Janapa Reddi, G.A.R., Ribalta, P., Singh, A., Thiagarajan, J.J., Albrecht, J., Wolf, T., Miller, G., Huazhu, F., Shah, P., Daguang, X., Yadav, P., Talby, D., Awad, M.M., Howard, J.P., Rosenthal, M., Marchionni, L., Loda, M., Johnson, J.M., Bakas, S., Mattson, P., 2023. Federated benchmarking of medical artificial intelligence with MedPerf. Nat. Mach. Intell. 2522-5839. 5 (7), 799–810. doi:10.1038/s42256-023-00652-2.

Kather, J.N., Heij, L.R., Grabsch, H.I., Loeffler, C., Echle, A., Muti, H.S., Krause, J., Niehues, J.M., Sommer, K.A.J., Bankhead, P., Kooreman, L.F.S., Schulte, J.J., Cipriani, N.A., Buelow, R.D., Boor, P., Ortiz-Brüchle, N., Hanby, A.M., Speirs, V., Kochanny, S., Patnaik, A., Srisuwananukorn, A., Brenner, H., Hoffmeister, M., van den Brandt, P.A., Jäger, D., Trautwein, C., Pearson, A.T., Luedde, T., 2020. Pan-cancer image-based detection of clinically actionable genetic alterations. Nat. Cancer. 26621347. 1 (8), 789–799. Available from: https://www.nature.com/natcancer/. doi:10.1038/s43018-020-0087-6.

Khan, S.N., Danishuddin, Khan, M.W.A., Guarnera, L., Akhtar, S.M.F., 2026. Multi-modal AI in precision medicine: integrating genomics, imaging, and EHR data for clinical insights. Front. Artif. Intell. 26248212. 8. Available from: www.frontiersin.org/journals/artificial-intelligence#. doi:10.3389/frai.2025.1743921.

Khosravi, P., Fuchs, T.J., Ho, D.J., 2025. Artificial intelligence–driven cancer diagnostics: enhancing radiology and pathology through reproducibility, explainability, and multimodality. Cancer Res. 15387445. 85 (13), 2356–2367.

Available from: https://aacrjournals.org/cancerres/article/85/13/2353/763113/Paving-a-Path-to-Clinical-Impact-with-Radiomics. doi:10.1158/0008-5472.CAN-24-3630.

Komura, D., Ishikawa, S., 2018. Machine learning methods for histopathological image analysis. Comput. Struct. Biotechnol. J. 20010370. 16, 34–42. Available from: https://www.csbj.org. doi:10.1016/j.csbj.2018.01.001.

Kroll, J.A., 2020. Accountability in computer systems. In: The Oxford Handbook of Ethics of AI Vol. 1(1), 181–196.

Langlotz, C.P., Allen, B., Erickson, B.J., Kalpathy-Cramer, J., Bigelow, K., Cook, T.S., Flanders, A.E., Lungren, M.P., Mendelson, D.S., Rudie, J.D., Wang, G., Kandarpa, K., 2019. A roadmap for foundational research on artificial intelligence in medical imaging: from the 2018 NIH/RSNA/ACR/The Academy workshop. Radiology. 15271315. 291 (3), 781–791. Available from: https://pubs.rsna.org/doi/pdf/10.1148/radiol.2019190613. doi:10.1148/radiol.2019190613.

Larson, E.J., 2021. The Myth of Artificial Intelligence: Why Computers Can't Think the Way We Do. The Belknap Press of Harvard University Pres, Cambridge, Massachusetts.

Lennerz, J.K., Green, U., Williamson, D.F.K., Mahmood, F., 2022. A unifying force for the realization of medical AI. Npj Digit. Med. 23986352. 5 (1). Available from: https://www.nature.com/npjdigitalmed/. doi:10.1038/s41746-022-00721-7.

Lennerz, J.K., Salgado, R., Kim, G.E., Sirintrapun, S.J., Thierauf, J.C., Singh, A., Indave, I., Bard, A., Weissinger, S.E., Heher, Y.K., De Baca, M.E., Cree, I.A., Bennett, S., Carobene, A., Ozben, T., Ritterhouse, L.L., 2023. Diagnostic quality model (DQM): an integrated framework for the assessment of diagnostic quality when using AI/ML. Clin. Chem. Lab. Med. 14374331. 61 (4), 544–557. Available from: http://www.reference-global.com/toc/cclm/current. doi:10.1515/cclm-2022-1151.

Louis, D.N., Feldman, M., Carter, A.B., Dighe, A.S., Pfeifer, J.D., Bry, L., Almeida, J.S., Saltz, J., Braun, J., Tomaszewski, J.E., Gilbertson, J.R., Sinard, J.H., Gerber, G.K., Galli, S.J., Golden, J.A., Becich, M.J., 2016. Computational pathology: a path ahead. Arch. Pathol. Lab. Med. 15432165. 140 (1), 41–50. Available from: http://www.archivesofpathology.org/doi/pdf/10.5858/arpa.2015-0093-SA. doi:10.5858/arpa.2015-0093-SA.

Mahase, E., 2025. AI in healthcare: what does good evidence and regulation look like? BMJ (Clinical Research Ed.). 17561833. 389, r692. doi:10.1136/bmj.r692.

McCaffrey, P., Jackups, R., Seheult, J., Zaydman, M.A., Balis, U., Thaker, H.M., Rashidi, H., Gullapalli, R.R., 2025. Evaluating use of generative artificial intelligence in clinical pathology practice: opportunities and the way forward. Arch. Pathol. Lab. Med. 1543-2165. 149 (2), 130–141. doi:10.5858/arpa.2024-0208-ra.

McCradden, M.D., Joshi, S., Anderson, J.A., Mazwi, M., Goldenberg, A., Shaul, R.Z., 2020. Patient safety and quality improvement: ethical principles for a regulatory approach to bias in healthcare machine learning. J. Am. Med. Inform. Assoc. 1527974X. 27 (12), 2024–2027. Available from: http://jamia.oxfordjournals.org/content/22/e1. doi:10.1093/jamia/ocaa085.

Miotto, R., Li, L., Kidd, B.A., Dudley, J.T., 2016. Deep patient: an unsupervised representation to predict the future of patients from the electronic health records. Sci. Rep. 2045-2322. 6 (1). doi:10.1038/srep26094.

Mitchell, M., 2019. Artificial Intelligence: A Guide for Thinking Humans, first ed. Farrar, Straus and Giroux, New York.

Moore, D.A., Young, C.A., Morris, H.T., Oien, K.A., Lee, J.L., Jones, J.L., Salto-Tellez, M., 2018. Time for change: a new training programme for morpho-molecular pathologists? J. Clin. Pathol. 14724146. 71 (4), 285–290. Available from: http://jcp.bmj.com/. doi:10.1136/jclinpath-2017-204821.

Mtoor, S., Rashidian, N., Messaoudi, N., Grasso, V., Noel, F., Steindler, M., Jaradat, D., Frigerio, I., Butturini, G., Croner, R., Rawicz-Pruszynski, K., Capelli, G., Spolverato, G., Besselink, M.G., Ishizawa, T., Chouillard, E., Abu-Hilal, M., Kahlert, U., Dagher, I., Gumbs, A.A., 2026. Integrating genomics, radiomics, and pathomics in oncology: a scoping review and a framework for ai-enabled surgomics. Bioengineering. 2306-5354. 13 (1), 117. doi:10.3390/bioengineering13010117.

National Academies of Sciences, Engineering, and Medicine et al., 2025. Disseminating In Silico and Computational Biological Research: Navigating Benefits and Risks: Proceedings of a Workshop. The National Academies Press, Washington, DC.

Niazi, M.K.K., Parwani, A.V., Gurcan, M.N., 2019. Digital pathology and artificial intelligence. Lancet Oncol. 14745488. 20 (5), e253. Available from: http://www.journals.elsevier.com/the-lancet-oncology/. doi:10.1016/S1470-2045(19)30154-8.

O'Neil, C., 2016. Weapons of Math Destruction: How Big Data Increases Inequality and Threatens Democracy, Crown Publishing Group.

Parikh, R.B., Teeple, S., Navathe, A.S., 2019. Addressing bias in artificial intelligence in health care. JAMA. 15383598. 322 (24), 2377–2378. Available from: http://jama.jamanetwork.com/journal.aspx. doi:10.1001/jama.2019.18058.

Parwani, A.V., 2019. Next generation diagnostic pathology: use of digital pathology and artificial intelligence tools to augment a pathological diagnosis. Diagn. Pathol. 17461596. 14 (1). Available from: http://www.diagnosticpathology.org/. doi:10.1186/s13000-019-0921-2.

Parwani, A.V., Amin, M.B., 2020. Convergence of digital pathology and artificial intelligence tools in anatomic pathology practice: current landscape and future directions. Adv. Anat. Pathol. 15334031. 27 (4), 221–226. Available from: http://www.anatomicpathology.com. doi:10.1097/PAP.0000000000000271.

Patel, A.U., Mohanty, S.K., Parwani, A.V., 2022. Applications of digital and computational pathology and artificial intelligence in genitourinary pathology diagnostics. Surg. Pathol. Clin. 18759157. 15 (4), 759–785. Available from: http://www.elsevier.com/wps/find/journaldescription.cws_home/716205/description. doi:10.1016/j.path.2022.08.001.

Rahimzadeh, V., 2024. US regulation of medical artificial intelligence and machine learning (AI/ML) research and development. In: Solaiman, B., Cohen, I.G. (Eds.), Research Handbook on Health, AI and the Law. Edward Elgar Publishing Ltd. https://pubmed.ncbi.nlm.nih.gov/40245231/.

Rajkomar, A., Dean, J., Kohane, I., 2019. Machine learning in medicine. J. Med. 15334406. 380 (14), 1347–1358. Available from: http://www.nejm.org/medical-index. doi:10.1056/NEJMra1814259.

Rajkomar, A., Hardt, M., Howell, M.D., Corrado, G., Chin, M.H., 2018a. Ensuring fairness in machine learning to advance health equity. Ann. Intern. Med. 0003-4819. 169 (12), 866–872. doi:10.7326/m18-1990.

Rajkomar, A., Oren, E., Chen, K., Dai, A.M., Hajaj, N., Hardt, M., Liu, P.J., Liu, X., Marcus, J., Sun, M., Sundberg, P., Yee, H., Zhang, K., Zhang, Y., Flores, G., Duggan, G.E., Irvine, J., Quoc, L., Litsch, K., Mossin, A., Tansuwan, J., Wang, D., Wexler, J., Wilson, J., Ludwig, D., Volchenboum, S.L., Chou, K., Pearson, M., Madabushi, S., Shah, N. H., Butte, A.J., Howell, M.D., Cui, C., Corrado, G.S., Dean, J., 2018b. Scalable and accurate deep learning with electronic health records. Npj Digit. Med. 2398-6352. 1 (1). doi:10.1038/s41746-018-0029-1.

Rajpurkar, P., Topol, E.J., 2025. Beyond assistance: the case for role separation in AI-human radiology workflows. Radiology. 15271315. 316 (1). Available from: https://pubs.rsna.org/toc/radiology/316/1. doi:10.1148/radiol.250477.

Rashidi, H.H., Pantanowitz, J., Chamanzar, A., Fennell, B., Wang, Y., Gullapalli, R.R., Tafti, A., Deebajah, M., Albahra, S., Glassy, E., Hanna, M.G., Pantanowitz, L., 2025. Generative artificial intelligence in pathology and medicine: a deeper dive. Mod. Pathol. 08933952. 38 (4), 100687. doi:10.1016/j.modpat.2024.100687.

Richards, B., Jacobson, S.S., Aquino, Y.S.J., 2021. Regulation of AI in health care: a cautionary tale considering horses and zebras. J. Law Med. 1320159X. 28 (3), 645–654.

Sirintrapun, S.J., 2025. Review and commentary on digital pathology and artificial intelligence in pathology. JCO Clin. Cancer Inform. 24734276. 9. Available from: http://ascopubs.org/journal/cci/. doi:10.1200/CCI-25-00017.

Vayena, E., Blasimme, A., Cohen, I.G., 2018. Machine learning in medicine: addressing ethical challenges. PLoS Med. 1549-1676. 15 (11), e1002689. doi:10.1371/journal.pmed.1002689.

Vayena, E., Salathé, M., Madoff, L.C., Brownstein, J.S., 2015. Ethical challenges of big data in public health. PLoS Comput. Biol. 15537358. 11 (2). Available from: http://www.ploscompbiol.org/article/browseVolume.action?field=volume. doi:10.1371/journal.pcbi.1003904.

Vayena, E., Tasioulas, J., 2013. Adapting standards: ethical oversight of participant-led health research. PLoS Med. 1549-1676. 10 (3), e1001402. doi:10.1371/journal.pmed.1001402.

Wieringa, M. 2020. What to account for when accounting for algorithms: a systematic literature review on algorithmic accountability. In: FAT* 2020—Proceedings of the 2020 Conference on Fairness, Accountability, and Transparency, pp. 1–18. Association for Computing Machinery, Inc Netherlands. doi:10.1145/3351095.3372833. http://dl.acm.org/citation.cfm?id=3351095

Zarella, M.D., Bowman, D., Aeffner, F., Farahani, N., Xthona, A., Absar, S.F., Parwani, A., Bui, M., Hartman, D.J., 2019. A practical guide to whole slide imaging a white paper from the digital pathology association. Arch. Pathol. Lab. Med. 15432165. 143 (2), 222–234. Available from: http://www.archivesofpathology.org/doi/pdf/10.5858/arpa.2018-0343-RA. doi:10.5858/arpa.2018-0343-RA.

Zarella, M.D., McClintock, D.S., Batra, H., Gullapalli, R.R., Valante, M., Tan, V.O., Dayal, S., Oh, K.S., Lara, H., Garcia, C.A., Abels, E., 2023. Artificial intelligence and digital pathology: clinical promise and deployment considerations. J. Med. Imaging. 23294310. 10 (5). Available from: http://medicalimaging.spiedigitallibrary.org/journal.aspx. doi:10.1117/1.JMI.10.5.051802.

Zhang, S., Yujun, L., Liu, W., Chu, Q., Wang, S., Jinyu, L., Chen, Y., 2025. A decade of review in global regulation and research of artificial intelligence medical devices (2015–2025). Front. Med. 2296-858X. 12. doi:10.3389/fmed.2025.1630408.

CHAPTER

17

Conclusions

Zaibo Li and Anil V. Parwani

Department of Pathology, The Ohio State University, Columbus, OH, United States

Digital pathology (DP) and artificial intelligence (AI) have reached a point of maturity at which they are no longer experimental adjuncts but essential components of modern pathology practice. Throughout this book, we have traced the evolution of pathology from a discipline grounded in optical microscopy and physical slides to one increasingly defined by digital images, computational analysis, and integrated data ecosystems. This transition reflects not only technological progress but also a broader response to the growing clinical, operational, and scientific demands placed on pathology in contemporary healthcare systems.

The widespread adoption of DP represents the foundational step in this transformation. By converting glass slides into high-resolution whole slide images (WSI) and embedding them within robust image management systems (IMS) and laboratory information systems (LIS), pathology workflows become more flexible, scalable, and connected. Digital workflows support remote and distributed practice, facilitate rapid access to subspecialty expertise, and enable seamless participation in multidisciplinary care. They also enhance quality assurance (QA) through standardized review processes, audit trails, peer comparison, and image-based proficiency testing. Importantly, DP creates durable, reusable data assets that can support education, research, and continuous improvement long after the immediate diagnostic task has been completed.

Clinical experience to date demonstrates that DP can be safely and effectively deployed for primary diagnosis across many subspecialties, provided that appropriate validation, training, and quality control measures are in place. While challenges remain, particularly in cytology, hematopathology, and others, ongoing advances in scanner technology, image acquisition strategies, and workflow design continue to narrow these gaps. Telepathology, once limited by bandwidth and hardware constraints, has become a routine and reliable modality for frozen section consultation, rapid on-site evaluation, and second-opinion review, expanding access to expert pathology services across geographic and institutional boundaries.

AI builds upon this digital foundation to deliver transformative analytic capabilities. Image analysis algorithms now assist pathologists in tasks ranging from screening and

Advances in Digital Pathology and Artificial Intelligence in Pathology
https://doi.org/10.1016/B978-0-443-40383-5.00015-7

detection to grading, quantification, and prognostic assessment. In doing so, AI has the potential to reduce variability, mitigate fatigue-related errors, and improve efficiency in high-volume diagnostic environments. Beyond these immediate gains, AI enables a shift toward more quantitative and reproducible pathology, where measurements and risk estimates complement traditional descriptive interpretation. The integration of AI into routine pathology practice illustrates how computational tools can be tailored to domain-specific challenges while remaining under the supervision of expert pathologists.

A particularly important theme emerging from DP adoption is the expanding role of histopathologic images as rich, multidimensional data sources. Through imagomics and related approaches, histopathologic images can be linked with genomic, transcriptomic, proteomic, radiologic, and clinical data to generate deeper insights into disease biology and patient outcomes. This convergence positions pathology at the center of precision medicine, not only as a diagnostic endpoint but as a predictive and integrative discipline. As multimodal and foundation models continue to evolve, the boundary between image interpretation, molecular characterization, and clinical decision support will become increasingly fluid.

The rise of generative artificial intelligence (GAI) and large language models (LLMs) further extends this trajectory. These technologies offer new possibilities for report generation, case summarization, decision support, literature synthesis, and workflow automation. When responsibly integrated into DP environments, they have the potential to reduce administrative burdens, enhance communication, and support both trainees and practicing pathologists. However, their adoption also underscores the importance of transparency, explainability, and rigorous governance to ensure that automation enhances, rather than obscures, clinical reasoning.

Equally critical to technological progress are the human, organizational, and ethical dimensions of DP and AI. Successful implementation depends on usability, trust, and alignment with real-world clinical workflows. Pathologists, administrative staff, IT teams, laboratory staff, and trainees must be actively engaged in system design, validation, and continuous refinement. Education and change management are essential to foster digital literacy and to redefine professional roles in an AI-augmented environment. Regulatory oversight, accreditation standards, and institutional governance structures must evolve to address issues of safety, accountability, data privacy, cybersecurity, and algorithmic bias, while still allowing innovation to proceed at a sustainable pace.

Looking ahead, the future of pathology will likely be characterized by hybrid intelligence, in which human expertise and computational systems work in close partnership. Rather than diminishing the role of the pathologist, DP and AI amplify it, enabling pathologists to focus more on complex interpretation, clinical integration, consultation, and leadership in diagnostic medicine. As routine and repetitive tasks become increasingly automated, the cognitive and communicative aspects of pathology will assume even greater importance.

In summary, advances in DP and AI are redefining the practice, scope, and impact of pathology. Their successful integration requires more than the acquisition of new tools; it demands a holistic approach that encompasses technology, workflow, education, ethics, and culture. By embracing these innovations thoughtfully and responsibly, pathology can enhance diagnostic quality, improve patient care, and strengthen its central role in multidisciplinary, data-driven healthcare. This book has provided both a practical guide and a conceptual framework for navigating this transformation, offering insights that will remain relevant as the field continues to evolve in the years and decades ahead.

AI Disclosure

During the preparation of this work, the author(s) used ChatGPT to polish the writing. After using this tool/service, the author(s) reviewed and edited the content as needed and take(s) full responsibility for the content of the publication.

Index

Note: Page numbers followed by "*f*" indicate figures, "*t*" indicate tables.

W

X

Z

Printed and bound by CPI Group (UK) Ltd, Croydon, CR0 4YY
14/09/2026
02216291-0008